Ion Beam Processing of Advanced Electronic Materials

MATERIALS RESEARCH SOCIETY SYMPOSIUM PROCEEDINGS   VOLUME 147

Ion Beam Processing of Advanced Electronic Materials

Symposium held April 25-27, 1989, San Diego, California, U.S.A.

EDITORS:

N. W. Cheung
University of California, Berkeley, California, U.S.A.

A. D. Marwick
IBM Thomas J. Watson Research Center, Yorktown Heights, New York, U.S.A.

J. B. Roberto
Oak Ridge National Laboratory, Oak Ridge, Tennessee, U.S.A.

MRS MATERIALS RESEARCH SOCIETY
Pittsburgh, Pennsylvania

CAMBRIDGE UNIVERSITY PRESS
Cambridge, New York, Melbourne, Madrid, Cape Town,
Singapore, São Paulo, Delhi, Mexico City

Cambridge University Press
32 Avenue of the Americas, New York NY 10013-2473, USA

Published in the United States of America by Cambridge University Press, New York

www.cambridge.org
Information on this title: www.cambridge.org/9781107410442

Materials Research Society
506 Keystone Drive, Warrendale, PA 15086
http://www.mrs.org

First published 1989
First paperback edition 2012

Single article reprints from this publication are available through
University Microfilms Inc., 300 North Zeeb Road, Ann Arbor, MI 48106

CODEN: MRSPDH

ISBN 978-1-107-41044-2 Paperback

Publication of this proceedings was sponsored in part by the U.S. Department of Energy
under contract DE-AC05-84OR21400 with Martin Marietta Energy Systems, Inc. Opinions,
findings, conclusions, or recommendations expressed herein do not necessarily reflect the
views of the U.S. Government.

Contents

*Invited Paper

*Invited Paper

Part V. HIGH-DOSE IMPLANTATION

Part VI. IMPLANTATION IN III-V MATERIALS AND MULTILAYERS

*Invited Paper

*Invited Paper

Part VII.   IMPLANTATION IN ELECTRONIC MATERIALS

*Invited Paper

Preface

The Symposium on Ion Beam Processing of Advanced Electronic Materials was held on April 25–27, 1989, in San Diego, California, U.S.A., as Symposium C of the 1989 Spring Meeting of the Materials Research Society. The purpose of the symposium was to explore the physical and technological basis for future ion beam processing of semiconductor devices. More than 60 papers from 11 countries were presented on fundamental and applied aspects of ion implantation and irradiation of electronic materials. The present symposium follows a similar symposium on this topic held at the 1985 Spring Meeting of the Materials Research Society and published as Volume 45 in the MRS Symposium Proceedings Series.

The symposium included sessions on Shallow Implantation and Solid-Phase Epitaxy, Damage Effects, Focused Ion Beams, MeV Implantation, High-Dose Implantation, Implantation of III-V Compounds and Multilayers, and Implantation in Electronic Materials. Significant progress has been made in the understanding and application of ion implantation at both very low and high energies, in the control and utilization of damage effects in ion beam processing, and in the use of ion beams to fabricate buried layers and submicron structures. The 54 papers included in this proceedings illustrate both the present state of knowledge in ion beam processing of electronic materials and the thrust of current research in the field.

The editors thank invited and contributing authors, session chairmen, reviewers, and MRS staff and program officials for their help in organizing the symposium and publishing this volume. We are particularly indebted to K. C. Brunson and L. W. Hinton of Oak Ridge National Laboratory for their administrative support over the past year.

N. W. Cheung
A. D. Marwick
J. B. Roberto

August 1989

MATERIALS RESEARCH SOCIETY SYMPOSIUM PROCEEDINGS

ISSN 0272 - 9172

MATERIALS RESEARCH SOCIETY SYMPOSIUM PROCEEDINGS

Tungsten and Other Refractory Metals for VLSI Applications, R. S. Blewer, 1986; ISSN 0886-7860; ISBN 0-931837-32-4

Tungsten and Other Refractory Metals for VLSI Applications II, E.K. Broadbent, 1987; ISSN 0886-7860; ISBN 0-931837-66-9

Ternary and Multinary Compounds, S. Deb, A. Zunger, 1987; ISBN 0-931837-57-x

Tungsten and Other Refractory Metals for VLSI Applications III, Victor A. Wells, 1988; ISSN 0886-7860; ISBN 0-931837-84-7

Atomic and Molecular Processing of Electronic and Ceramic Materials: Preparation, Characterization and Properties, Ilhan A. Aksay, Gary L. McVay, Thomas G. Stoebe, 1988; ISBN 0-931837-85-5

Materials Futures: Strategies and Opportunities, R. Byron Pipes, U.S. Organizing Committee, Rune Lagneborg, Swedish Organizing Committee, 1988; ISBN 0-55899-000-3

Tungsten and Other Refractory Metals for VLSI Applications IV, Robert S. Blewer, Carol M. McConica, 1989; ISSN: 0886-7860; ISBN: 0-931837-98-7

PART I

Shallow Implantation and Solid-Phase Epitaxy

POINT DEFECT ENGINEERING APPLIED TO
SHALLOW JUNCTION ULSI PROCESSING

George A. Rozgonyi and J. W. Honeycutt
North Carolina State University
Department of Materials Science and Engineering
Raleigh, NC 27695-7916

Abstract

We describe how a simple qualitative understanding of the interfacial reactions occurring during typical ULSI processes for junction formation, dopant activation, and contact silicidation can be used to eliminate end-of-range interstitial dislocation loops and beneficially impact the diffusion of dopants. Following a brief discussion of the well-documented effects of oxidation and nitridation on extended defects and dopant diffusion, conditions for elimination of implantation-induced defects are specified. Cross-section and plan-view TEM along with angle lapping and chemical etching of implanted and diffused junctions are presented to illustrate the application of point defect engineering to process technology.

Introduction

A number of fundamental materials science and technology problems are associated with the fabrication of low-resistance, shallow source/drain electrical junctions. These junctions, which are crucial to the success of sub-micron ULSI, will require a junction depth of 70nm, as dictated by the scaling criteria for a 0.25µm MOS device. Since silicidation is generally used to reduce the source/drain sheet resistance and increase circuit speed, the 70nm deep junction will be partially consumed during the silicide reaction. It is, therefore, imperative that the dopant distribution, silicon consumption, and silicide metallurgical phase be precisely known. Furthermore, in order to simultaneously produce both ultra-shallow and low leakage junctions, each of the ion implantation, annealing, and silicide formation process steps must be integrated into one intimate and reproducible processing sequence. This process integration must be supported by a tight characterization feedback loop covering structural, chemical, and electrical measurements of each step. In the domain of structural and defect characterization, there is much evidence that interface reactions (metal/Si), thin film growth (oxide, nitride), and free surface/ambient (Ar, N_2, H_2, O_2) effects, along with ion implantation induced damage, will perturb the near-surface point defect concentrations. Consequently, the position and level of electrical activation of dopants, and the size and density of crystal defects can only be properly controlled if the type and concentration of the dominant point defect is known. This paper will make an attempt to place these interactions in perspective and thereby provide a program of point defect engineering for shallow junction/contact technology.

Process-Induced Point Defects

A. Oxidation and Self-Interstitials

It is convenient to discuss process-induced point defects in the context of the thermal oxidation of a silicon surface, which will generally produce an effective injection of silicon self-interstitials. These excess interstitials provide the driving force for measurable growth of oxidation stacking faults (OSF), enhanced diffusion of B and P, and retarded diffusion of Sb (for review see [1-3]). The situation for OSF's is schematically shown in Fig.1, where $C_I{}^i$ is the concentration of self-interstitials injected at the SiO_2/Si interface. The growth or shrinkage of OSF's is determined by the near-surface excess interstitial concentration, γ in Fig.1, which is often complicated in CZ wafers by oxygen precipitation in the bulk[4]. The mechanism of self-interstitial generation at an oxidizing interface is intimately connected with the fact that silicon oxidizes via diffusion of oxygen from the free surface through the SiO_2 layer down to the moving SiO_2/Si interface. It has been speculated that due to interfacial stresses or incomplete reaction, unreacted Si at this interface becomes available to migrate into the substrate, forming the basis for self-interstitial injection and OSF growth[5] as illustrated in the left-hand portion of Fig.1. The same end result of interstitial injection has been arrived at by Tiller, et al.[6]

4

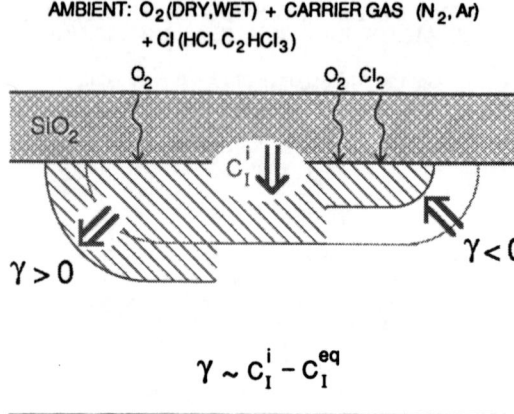

AMBIENT: O_2(DRY,WET) + CARRIER GAS (N_2, Ar)
+ Cl (HCl, C_2HCl_3)

$$\gamma \sim c_I^i - c_I^{eq}$$

Fig.1. Schematic illustration of the effects of near-surface point defect concentrations on growth and shrinkage of oxidation stacking faults. C_I is the interstitial concentration; while superscripts i and eq represent values at the interface and for OSF equilibrium, respectively.

involving interfacial formation of α-cristobalite plus interstitial Si ions, as well as by Tan and Goesele[7] based on SiO_2 viscoelastic flow to accommodate the volume increase associated with formation of new SiO_2 at the interface. However, in this paper we will emphasize the atomistic diffusion models such as that presented in Fig.1, because of parallels with metal silicide/silicon interface reactions to be discussed below.

It was recognized early on that the addition of a chlorine-containing species would lower the growth rate of OSF's and reduce the temperature threshold for the dissolution or retrogrowth of OSF[8-10]. It is believed that upon reaction of Cl with Si at the SiO_2/Si interface, either a vacancy is generated, or simply that the concentration of self-interstitials, C_I, is lowered to the point where the overall point defect balance is closer to the equilibrium state. Thus, OSF shrinkage is enhanced, as illustrated at the right side of Fig.1, and oxidation enhanced diffusion of those dopants which diffuse by an interstitialcy mechanism is reduced.

Table I. Comparison of processing conditions, interdiffusion phenomena, and process-induced point defects produced during oxidation, nitridation, and silicidation.

Reaction	Temp /Time	Diffusion	Scale	Film	Interface
Si Oxidation	900-1100°C-hrs	$O_2\Downarrow$	t ≥1000Å large area	1 phase (SiO_2)	I source V sink
Si Oxidation	≥1100°C - hrs	Si ⇑, $O_2\Downarrow$ (SiO)	t ≥1000Å large area	1 phase (SiO_2)	I sink Vsource
Si Nitridation	≥1000°C - hrs	Si⇑	t ≥1000Å large area	1 phase (Si_3N_4)	I sink V source
Silicidation	≤900°C - secs	Si⇑ or M⇓	t ≤1000Å sub-μ area	multiphase	depends on diffusion(?)

It is interesting to note that for high temperature (T≥1150°C) oxidation, the point defect producing/absorbing properties of a SiO$_2$/Si interface change, i.e. it becomes a vacancy source and/or a sink for self-interstitials[11]. This effect is again attributed to the details of the reaction kinetics, i.e. the solid state diffusion in the growing oxide film. It is believed that at high temperatures, there is a non-negligible flux of either Si[12] or SiO[13] away from the interface and into the oxide, leaving behind vacancies which enhance Sb diffusion[11], and retard P diffusion[12]. Similarly, enhanced OSF shrinkage[14], enhanced Sb diffusion[15,16], and retarded P diffusion[15,16] have been observed for the case of thermal nitridation of Si in NH$_3$, where Si is apparently the dominant diffusing species[17]. Thus one can intuitively predict enhanced/retarded diffusion of dopants and growth/shrinkage of extended defects in Si due to interfacial point defect generation, provided the thin film reaction kinetics are understood. Table I summarizes for typical processing conditions the reaction mechanisms and interface properties for low and high temperature oxidation, thermal nitridation, and silicidation reactions where either metal or Si may be the dominant diffusing species.

B. Ion Implantation Induced Defects

A pervasive type of process-induced defect is that produced at the end-of-range of an amorphizing ion-implantation, following recrystallization via solid phase epitaxy and dopant activation, as shown in the cross-section TEM of Figs. 2a and 2b [18]. These end-of-range dislocation loops are confined to the region of the original amorphous/crystalline interface. They are always of interstitial character, consistent with the intuitive notion of a high concentration of displaced knock-on Si atoms being produced just below the amorphous layer. Options for reducing or even eliminating these interstitial loops have been discussed by Ajmera and Rozgonyi [19] for the particular case of a Ge preamorphized, B-implanted junction. They recommended three process modifications designed primarily to reduce the near-surface concentration of interstitials, namely:

i)   lower the implantation energy of the amorphizing ion,
ii) lower the implant dose, and
iii) control the annealing ambient to produce a non-oxidizing or slightly reducing surface reaction.

These effects are schematically illustrated in Fig.3 where C_I is plotted as a function of depth. A comparison is given for a deep, high dose implant annealed in an oxidizing ambient (· in Fig.3), or a reducing ambient (solid line); versus a shallow, low dose implant (dotted lines). The ambient control, item (iii) above, provides for a minimum injection of interstitials at the free surface if the surface is non-oxidizing; the lower dose recommended in item (ii) lowers the concentration of knock-on interstitial Si atoms; and finally, the lower implantation voltage in item (i) moves the

a   b

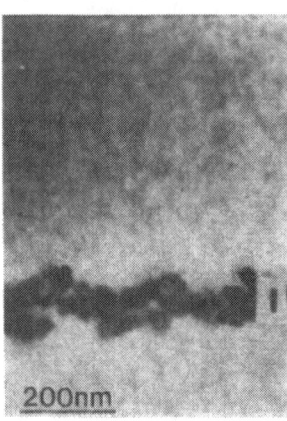

Fig.2. Cross-section TEM micrographs of Si, preamorphized by a triple Si implantation at 77K, see ref.[18]. (a) as-implanted, (b) after RTA at 1150°C/10s.

6

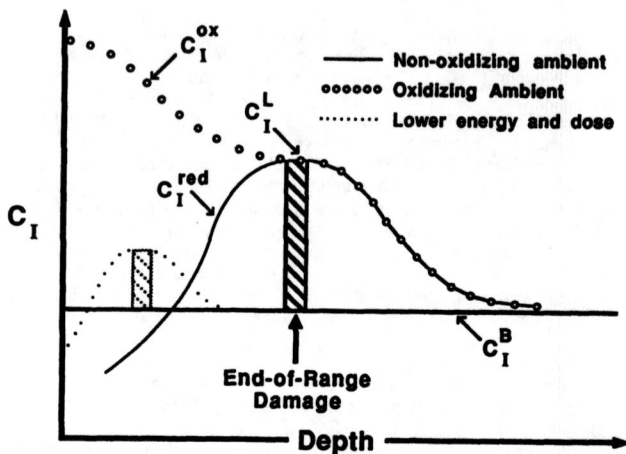

Fig.3. Schematic illustration of silicon self-interstitial concentrations (C_I) during annealing of an ion-implanted Si surface[19]. C_I^B and C_I^L are the interstitial concentrations in the bulk and at the depth of end-of-range dislocation loops, respectively.

a/c interface closer to the free surface, which may act as an effective sink for interstitials. The issue of proximity to a free surface has been addressed by Ganin and Marwick[20] who created shallow and deep bands of end-of-range defects by In implantation at low and high energies, respectively. By etching the surface of the sample with the deeper implant such that the residual damage was at the same depth as that of the shallower implant, they found that the kinetics of end-of-range damage removal is primarily dependent on the concentration of self-interstitials beyond the amorphous/crystalline interface, which is greater for higher energy implants. The above procedures for reducing interstitial concentrations are quite effective in an RTP system at, e.g., 1050°C for 10sec[19], but are generally not able to eliminate the end-of-range defects using more limited thermal budgets. However, by considering processes which offer vacancy injection, as described in the next section, an extension of the interstitial reduction defect engineering options can be offered.

C. Silicidation-Induced Vacancies

Because all the dominant observable process-induced extended defects are of interstitial character, there has been a great deal of interest in processes whose impact in silicon involves vacancy generation. Until recently, either nitridation, or very high temperature thermal treatments have been the means of vacancy injection. Lately, however, it is becoming increasingly evident that silicidation reactions also perturb point defect concentrations (both interstitial and vacancy) in the substrate[21-27]. Since the silicidation reaction occurs directly on the implanted junction region, it offers the most compatible opportunity to control point defect concentrations such that they beneficially influence shallow junction properties (dopant profiles and residual ion-implant damage). The various point defect-generating, point defect-controlled, and dopant-influenced interactions which should be considered for a metal silicide contact to an ion-implanted shallow junction are depicted schematically in Fig.4. The dominant interactions are indicated by solid lines for each process. Namely, the ion implantation and annealing conditions determine the pre-silicidation junction depth and the level of residual damage. Likewise, the metallization and subsequent silicidation anneal result in a silicide film with a predictable sheet resistance and silicon consumption. The dashed lines in Fig.4 depict the interactions which are less obvious and have been shown to vary with, e.g., the particular dopant present in the substrate, or the details of the implant process. Most importantly, the silicidation reaction is a likely candidate for defect engineering using the beneficial effects of vacancy generation to counteract the effects of implantation damage annealing processes dominated by interstitials.

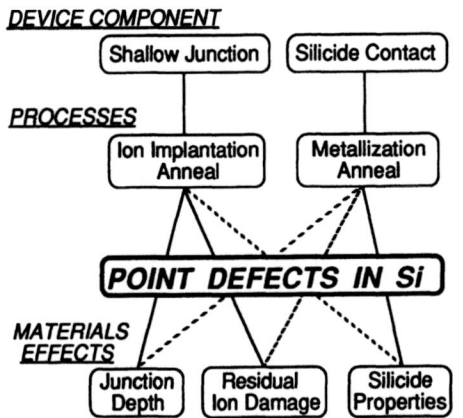

DEVICE COMPONENT

PROCESSES

MATERIALS EFFECTS

Fig.4. Illustration of the point defect controlled or dopant-influenced interactions occuring during silicided shallow junction processing.

As in the case of oxidation and nitridation, interfacial point defect generation during silicidation is believed to be directly determined by the details of the interdiffusion between Si substrate atoms and the other reacting species, i.e. the metal film (see Table I). For example, observations of pile-up (segregation) of implanted As in Si during growth of a near-noble metal silicide layer (PtSi, Pd$_2$Si) have been explained[21,22] by the predominant diffusion of metal atoms to the silicide/silicon interface (analogous to oxygen motion during oxidation). It was speculated that the near-noble metal atoms diffuse into interstitial sites in the silicon lattice during the reaction, weakening the silicon-silicon covalent bonds, and thus leading to dissociation of silicon atoms from their lattice sites and production of vacancies and/or self-interstitials[21]. Additional evidence of point defect generation during silicidation was provided by Ohdomari, et al.[22], who saw shrinkage of implantation-induced extrinsic dislocation loops during Pd$_2$Si formation, presumably due to recombination of the silicidation-induced vacancies with "interstitial" atoms contained in the loops. Although the phenomenon of dopant segregation under the silicide is considered a useful by-product in silicidized shallow junction processes, there are other thermal stability, resistivity, and silicon consumption issues which favor other silicides under which dopant pile-up does not occur. Due to their lower resistivites, as well as their superior thermal stabilities, TiSi$_2$ and CoSi$_2$ are the prime candidates for silicide contacts to shallow junctions. In contrast to the dopant pile-up observed for near-noble metal silicides, dopant incorporation into the silicide and subsequent loss to the ambient can be a serious problem, especially for TiSi$_2$ formation on boron-implanted junctions[26]. A summary of the resistivities, calculated silicon consumption ratios, and diffusing species for Ti, Co, and Pt silicides is given in Table II[29].

Table II. Comparison of silicon consumption ratios, resistivities, and diffusing species of Ti, Co, and Pt silicides.

Metal	Silicide	Formation Temp. (°C)	$\frac{t_{Si}}{t_{metal}}$	$\frac{t_{Si}}{t_{silicide}}$	Diffusing Species	Resistivity ($\mu\Omega$-cm)
Pt	PtSi	500-600	1.32	.67	Pt [35]	28-35
Ti	TiSi$_2$(C54)	650-850	2.24	.90	Si [36]	13-16
Co	CoSi$_2$	600-800	3.63	1.04	Co [37]	15-20

Recently, the observations of Ohdomari, et al.[22] of a reduction in ion implantation induced damage during Pd_2Si reactions have been extended to $TiSi_2$ by Maex, et al.[23] and Wen, et al.[24,25]. In refs. [24,25] cross-sectional TEM revealed complete annihilation of Ge implantation-induced end-of-range damage due to $TiSi_2$ formation, as shown in Fig. 5. Complete damage removal was observed only when the silicide sheet resistance, which is a measure of both the thickness and the resistivity of the film, was about 3Ω/sq or less. Fig.5b shows the annihilation of end-of-range damage during a silicidation anneal of a 30nm Ti layer at 650°C for 10sec plus 850°C for 10sec. A conceptual model was proposed to explain this effect, as shown in Fig.6, based on the Ti-Si reaction occurring by the predominant diffusion of silicon atoms through the growing silicide. As silicon atoms break away from the substrate, a vacant lattice site is left behind. These vacancies then diffuse to the damaged region and recombine with interstitial atoms contained in dislocation loops. The dislocations thus shrink by climb processes until they are eventually completely dissolved.

An alternate "marker" experiment for the qualitative identification of point defects involves the comparison of diffusion profiles or junction depths as a function of process conditions. A particularly useful structure for comparing the relative effects of silicidized and passive or masked areas is shown in Fig. 7. This figure documents the influence of a vacancy flux as observed by Jiang and Honeycutt[30], who delineated a retarded B diffusion profile under a $TiSi_2$ interface. This optical micrograph of an angle-lapped, polished, and preferentially etched surface contains a blanket layer of B, implanted into a Ge-preamorphized substrate at 10keV, 1×10^{15}/cm², with patterned $TiSi_2$ regions initially ~60nm thick. Note that the p⁺ implanted surface is not etched during 10 sec in a Secco[31] solution. This allows delineation of a concentration (diffusion depth) difference of ~20nm when the silicide region is compared with that under the masking oxide. These procedures were previously used by Hu[27], who observed enhanced diffusion of buried layers of both Sb and B during a long time (10hr) 1050°C anneal of a $TaSi_{1.8}$ layer, which was known to absorb Si to form the stoichiometric $TaSi_2$ phase.

Although the model of vacancy injection due to silicon diffusion away from the substrate interface is heuristically appealing, its application may not be universal to all reactions where

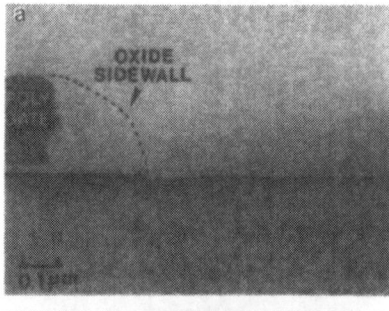

Fig.5. Cross-sectional TEM micrographs of Ge preamorphized, B implanted junction regions of an MOS device[24,25]. The Ge implantation end-of-range damage present in the form of extrinsic dislocation loops in (a) is annihilated during two-step annealing of a 30nm Ti film at 650°C for 10 sec plus 850°C for 10 sec to form $TiSi_2$ (b).

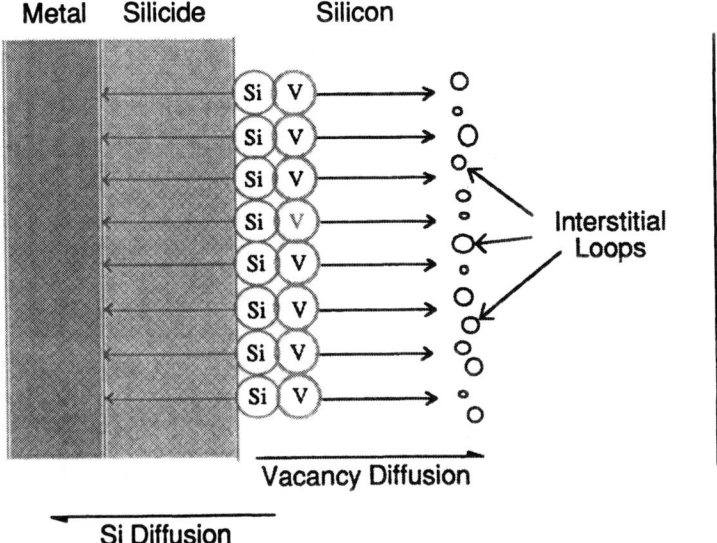

Fig.6. Schematic illustration describing the model for vacancy injection and ion implantation damage removal during TiSi$_2$ formation[24,25].

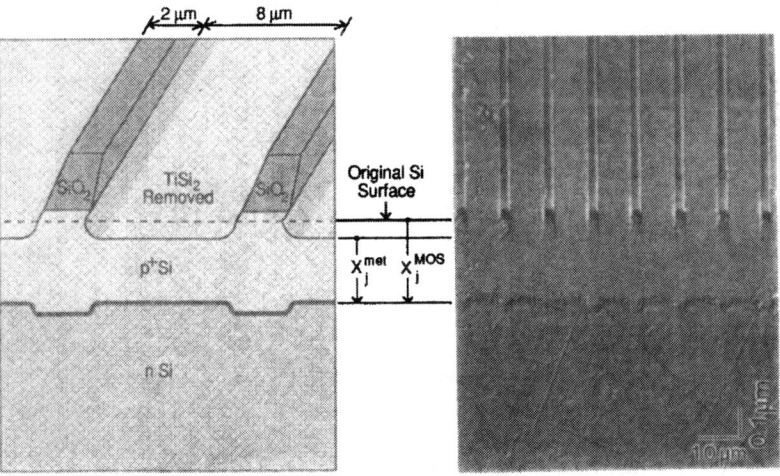

Fig.7. Nomarski optical micrograph and schematic diagram of a bevel polished and preferentially etched sample illustrating retarded B diffusion under a TiSi$_2$ layer formed at 650°C for 120sec.

silicon is the dominant diffusing species. Fahey and Dutton[26] have pointed out that oxidation of a deposited WSi_2 layer, where silicon atoms diffuse through the silicide to form SiO_2 at the surface does not produce an enhanced diffusion of buried layers under the silicide. Also, recent results by the authors of the present paper have shown that ion implantation damage removal may not occur in processes very similar to those of Wen, et al.[24] where complete damage removal was expected from considerations of the silicide thickness and annealing times and temperatures. Figure 8 shows a cross-section TEM micrograph of a sample after two-step Ti-silicidation, for a process identical to that of Wen, et al.[24] with the exception of a lower Ge energy and dose (60keV, $3x10^{14}/cm^2$, compared to Wen's 85keV, $1x10^{15}/cm^2$). The thickness of the silicide in Fig.8 is ~60nm, which according to the previous results, should have totally annihilated the end-of-range damage, especially since the residual damage is less severe and closer to the surface than for the deeper Ge implant of Fig.5. We note, however, that the silicide sheet resistance of this sample was about $4\Omega/sq$, greater than the maximum of $3\Omega/sq$ necessary for damage removal as reported by Wen et al.[24] for a silicide thickness of ~60nm. Multiplying the silicide thickness by the sheet resistance yields an electrical resistivity of ~$24\mu\Omega$-cm, considerably higher than the expected value of ~$15\mu\Omega$-cm (see Table II). X-ray and electron diffraction were used to verify that the low resistivity C54 form of $TiSi_2$ was the only phase present. The reason for the unusually high resistivity of this silicide film has not been determined, but wafer cleanliness (native oxide) prior to metallization and/or RTA ambient purity (oxygen), resulting in an impure silicide is suspected. The higher resistivity is indicative of an interfacial reaction for which an idealized Si diffusion model would not hold, or that an increase in interfacial vacancy recombination had occurred. Any mechanism which raises the resistivity may also act to reduce the level of vacancy injection. As discussed by Fahey et al.[3,26], a silicon/thin film interface may not only act as a point defect generation source, but also as a sink for the interfacially generated point defects. In addition to the effects of diffusing species, stress, etc., the efficiency of interfacial recombination processes should depend on the atomic arrangement at the interface, which obviously varies between deposited and thermally grown films, and between the various silicide phases. Thus one might speculate that the interface between Si and an epitaxial or preferentially oriented $CoSi_2$ film, the lattice parameters of which differ by only 1.2%, may be a less efficient point defect source/sink than e.g., a randomly oriented, polycrystalline $TiSi_2$/Si interface. Such a comparison between various silicide reactions as point defect generation sources/recombination sinks is in progress at NCSU.

Fig.8 Cross-section TEM micrograph showing persistence of residual end-of-range damage in a Ge-preamorphized (60keV, $3x10^{14}/cm^2$), B-implanted (10keV, $1x10^{15}/cm^2$) sample, following two-step annealing of a 30nm Ti layer, process identical to that of Wen, et al.[24] except for lower Ge^+ energy and dose.

Even if one accepts the idea that vacancy injection occurs during Ti silicidation, the defect removal mechanism is still unclear. For example, consider the two-step silicidation process of Wen, et al.[24] which resulted in complete annihilation of implantation damage. The thickness of the silicide (and thus the quantity of vacancies injected) is determined by the metal deposition thickness and its reaction with Si during the first 650°C/10sec anneal (resulting in a high resistivity silicide possibly containing the C49 $TiSi_2$ phase). Unreacted metal and reaction by-products (TiO_xN_y) are then removed with a selective etch. Thus, if damage removal or enhanced diffusion depends only on the Ti-Si chemical reaction itself, it should occur prior to the final 850°C/10sec anneal, which is necessary to achieve low resistivity. In fact, this is indicated in the diffusion front of Fig.7. The vacancies created during the low temperature silicide reaction may also be driven towards the end-of-range loops during the high temperature resistivity anneal. However, there will also be an opportunity for the injected vacancies to recombine at the surface or in the bulk before reaching the depth of the end-of-range damage. Current knowledge of interstitial and vacancy recombination kinetics in Si is still in a dynamic state of development[3], and more precise experimental data are needed to assist in formulation of accurate models[34]. Lur, et al.[32,33] have examined a wide variety of silicides (both metal and Si diffusers) and concluded that damage removal can occur after the silicidation reaction is complete. They attribute the defect annihilation not simply to vacancy formation during the reaction by the diffusion of Si substrate atoms, but also to stress relief after the silicide is formed. The various models of point defect generation during oxidation, combined with the results of Ahn, et al.[35], which showed enhanced diffusion of Sb and enhanced OSF shrinkage during annealing of a deposited nitride layer, where no chemical reaction occurred, also suggest that factors other than the reaction itself, such as stress, may play an important role in point defect generation kinetics. In consideration of all of the above results, both the growth of thin films and their mere presence will impact substrate point defect concentrations. It is obvious that further investigations of both the effects of silicide formation on dopant diffusion and extended defects, as well as the point defect generation mechanisms are necessary.

Conclusion

In conclusion, it appears that the determination of the true driving force for defect and dopant profile control during silicidation will require a high level of process purity and control of silicide interface reactions. We see these processing requirements evolving to a point where they parallel those which are currently used for SiO_2 gate dielectrics. Because of the large number of silicides currently under study and the variety of chemical, physical, and metallurgical phases possible, the resolution of this problem will continue to require an extensive array of structural, chemical, and electrical diagnostic tools to be applied to each system. Accepting these difficulties, it is still possible to apply a defect engineering approach to process integration of shallow junctions and their low resistance contacts using the concepts outlined in this review.

Acknowledgements

Portions of this work were funded by the Microelectronics Center of North Carolina, the Semiconductor Research Corporation, Northern Telecom Electronics, and the NCSU NSF Engineering Research Center for Advanced Electronic Materials Processing (CDR 8721505). The authors wish to thank C. Osburn, G. McGuire, R. Fair, P. Smith, and S. Chevacharoenkul of MCNC, C. Carter of CREE Research, and Z. Xiao and H. Jiang of NCSU for useful discussions and assistance. J.W. Honeycutt is supported by a SRC Graduate Research Fellowship.

12

References

1. S.M. Hu, *MRS Symp. Proc.*, **2**, 333 (1981).
2. D.A. Antoniadis, *J. Electrochem. Soc.*, **129**, 1093 (1982).
3. P.M. Fahey, P.B. Griffin, and J.D. Plummer, *Rev. Mod. Phys.*, **61**, 289 (1989).
4. S.M. Hu, *J. Appl Phys.*, **51**, 3666 (1980).
5. S.M. Hu, *J. Appl. Phys.*, **45**, 1567 (1974).
6. W.A. Tiller, *J. Electrochem. Soc.*, **128**, 689 (1981).
7. T.Y. Tan and U. Goesele, *Appl. Phys. Lett.*, **39**, 86 (1981).
8. H. Shiraki, *Jpn. J. Appl. Phys.*, **14**, 747 (1975).
9. H. Shiraki, *Semiconductor Silicon 1977*, H.R. Huff and E. Sirtl, eds. (Electrochemical Society, Pennington, NJ, 1977), p.546.
10. C. Claeys, E.L. Laes, G.J. Declerck, and R.J. van Overstraeten, *Semiconductor Silicon 1977*, H.R. Huff and E. Sirtl, eds. (Electrochemical Society, Pennington, NJ, 1977), p.773.
11. T.Y. Tan and B.J. Ginsberg, *Appl. Phys. Lett.*, **42**, 448 (1983).
12. R. Francis and P.S. Dobson, *J. Appl. Phys.*, **50**, 280 (1979).
13. T.Y. Tan and U. Gosele, *Appl. Phys. Lett.*, **40**, 616 (1982).
14. Y. Hayafuji, K. Kajiwara, and S. Usui, *J. Appl. Phys.*, **53**, 8639 (1982).
15. S. Mizuo and H. Higuchi, *J. Electrochem. Soc.*, **130**, 1942 (1983).
16. P. Fahey, G. Barbuscia, M. Moslehi, and R.W. Dutton, *Appl. Phys. Lett.*,**46**, 784 (1985).
17. Y. Hayafuji and K. Kajiwara, *J. Electrochem. Soc.*, **129**, 2102 (1982).
18. C. Carter, W. Maszara, D.K. Sadana, G.A. Rozgonyi, J. Liu, and J. Wortman, *Appl. Phys. Lett.*, **44**, 459 (1984).
19. A.C. Ajmera and G.A. Rozgonyi, *Appl. Phys. Lett.*, **49**, 19 (1986).
20. E. Ganin and A. Marwick, These Proceedings.
21. M. Wittmer and K.N. Tu, *Phys. Rev. B*, **29**, 2010 (1984).
22. I. Ohdomari, K. Konuma, M. Takano, T. Chikyow, H. Dawarada, J. Nakanishi, and T. Ueno, *MRS Symp. Proc.*, **54**, 63 (1985).
23. K. Maex, R. De Keersmaecker, C. Claeys, J. Vanhellemont, P.F.A. Alkemade, *Semiconductor Silicon 1986*, H.R. Huff, T. Abe, and B. Kolbesen, eds. (Electrochemical Society, Pennington, NJ, 1986), 346.
24. D.S. Wen, P.L. Smith, C.M. Osburn, and G.A. Rozgonyi, *Appl. Phys. Lett.*, **51**, 1182 (1987).
25. D.S. Wen, P.L. Smith, C.M. Osburn, and G.A. Rozgonyi, *J. Electrochem. Soc.*, **136**, 466 (1989).
26. P. Fahey and R.W. Dutton, *Appl. Phys. Lett.*, **52**, 1092 (1988).
27. S.M. Hu, *Appl. Phys. Lett.*, **51**, 308 (1987).
28. C.M. Osburn, T. Brat, D. Sharma, D. Griffis, S. Corcoran, S. Lin, W.-K. Chu, and N.Parikh, *J. Electrochem. Soc.*, **135**, 1490 (1988).
29. M.A. Nicolet and S.S. Lau, "Formation and Characterization of Transition-Metal Silicides," in *VLSI Electronics: Microstructure Science*, **6**, Norman G. Einspruch, ed. (Academic Press, 1983), p.329.
30. H. Jiang and J. W. Honeycutt, unpublished result.
31. F. Secco d'Aragona, *J. Electrochem. Soc.*, **119**, 948 (1972).
32. R.B. Fair and R. Subrahmanyan, proceedings of the May, 1989 meeting of the Electrochemical Society, *ULSI Science and Technology / 1989*, to be published.
33. W. Lur, J.Y. Cheng, C.H. Chu, M.H. Wang, T.C. Lee, Y.J. Wann, W.Y. Chao, and L.J. Chen, *Nucl. Instrum. and Meth.*, **B39**, 297 (1989).
34. W. Lur, J.Y. Cheng, and L.J. Chen, These Proceedings.
35. S.T. Ahn, H.W. Kennel, J.D. Plummer, and W.A. Tiller, *J. Appl. Phys.*, **64**, 4914 (1988).
36. R. Pretorius, A.P. Botha, and J.C. Lombard, *Thin Solid Films*, **79**, 61 (1981).
37. S.P. Murarka and D.B. Fraser, *J. Appl. Phys.*, **51**, 342 (1980).
38. F.M. D'Heurle and C.S. Petersson, *Thin Solid Films*, **128**, 283 (1985).

IS THE END-OF-RANGE LOOPS KINETICS AFFECTED BY SURFACE PROXIMITY OR ION BEAM RECOILS DISTRIBUTION?

E. Ganin and A. Marwick
IBM Research Division, T.J. Watson Research Center
Yorktown Heights, NY 10598

Abstract

We studied formation and annihilation of dislocation loops formed beyond the amorphous/crystalline interface after indium and boron dual implantation and subsequent annealing in the 800-1100°C temperature range. The residual damage for low (40 keV) and high (200 keV) energy In implants were compared. The depth of the amorphous region in the sample implanted with the higher energy ions was reduced by using anodic oxidation and etching, to equate it with that of the sample implanted by lower energy ions. This enabled the study of the effect of surface proximity on residual disorder upon annealing. The damage was strongly dependent on the energy of In ions. No end-of-range damage was observed for the low energy implant. High energy implantation resulted in end-of-range dislocation loops, stable below 1050°C. The loops kinetics was neither affected by their proximity to the surface, nor by In precipitation. Monte-Carlo full cascade simulation has been used to estimate the depth distribution of interstitials and vacancies produced by In implant.

Introduction

Ion implantation of silicon (Si) by heavy ions is known to cause formation of a continuous amorphous layer, provided the implant dose is exceeding the amorphization threshold. Below the amorphization threshold, formation of "buried" amorphous layers, not extending to the surface, is usually observed. The amorphous layer regrows epitaxially during subsequent annealing, leaving dislocation loops beneath the original amorphous/crystalline (a/c) interface. The loops are considered to nucleate from the excess interstitials created by the implantation [1]. This particular type of residual radiation disorder, called end-of-range damage, was extensively studied in the recent years [2]. The "amorphization technique" is widely used in shallow junctions formation to prevent channeling of light ions like boron [3,4]. Therefore, it is important to control the residual damage, induced by amorphization. Complete removal of end-of-range damage is desirable since the presence of dislocation loops in the depletion region of a junction detriments the device performance. However, the end-of-range loops are harmless, when remote from the depletion region [4]. Complete elimination of dislocation loops has been reported to take place at sufficiently high thermal budgets [3-6]. However, the mechanism of defect annihilation is still not well understood. The conditions, under which complete annihilation occurs, seem to be dependent on a type of ion species used for the amorphization, their dose and energy. It has been reported repeatedly that for thin amorphous layers, when the end-of-range loops are close to the surface, the annihilation occurs at temperatures lower than in the case of thick layers [5,6]. Thus, it has been suggested [6], that the proximity of end-of-range loops to the free surface plays a dominant role in defect annihilation. However, in those reports, the significance of ion beam energy has not been taken into consideration. Using different energies may have implications on the residual damage, others than variations in the amorphous layer depth. It is well known that, the higher the ion beam energy, the thicker the amorphous layer formed during implantation. However, the question remains, whether the energy of the implanted beam affects the residual damage distribution beyond its unavoidable effect on the amorphous layer depth. Is indeed the facilitated damage annihilation, observed for thin amorphous layers, caused by surface proximity, or the implant beam energy is affecting the end-of-range loops kinetics? This paper is an attempt to answer that question.

Experimental

Indium was implanted into n-type (100) Si substrates with energies of 40 keV and 200 keV and doses ranging from 5×10^{13} /cm^2 to 3×10^{14} /cm^2. Boron (B) was subsequently implanted into amorphized Si at 5 and 17 keV respectively, with 1×10^{15} / cm^2 dose. Plan-view and cross-sectional TEM (XTEM) was applied to study the end-of-range loops formation and annihilation at different rapid thermal annealing (RTA) cycles. RBS channeling (1 MeV, He$^+$ beam) was used to determine the thickness of the amorphous layer. Anodic oxidation was performed by using Si wafer as anode placed into electrolyte solution. Self-limiting oxidation of Si at room temperature occurs, when the current flows from Pt cathode to the Si anode. The oxide, grown on Si, was measured by ellipsometer and removed by HF etch. SIMS analysis was applied for concentration-depth profiling of In dopant. RTA annealing was performed in a Heatpulse-410 apparatus, using Ar ambient in a temperature range 850-1100°C with the increments of 25°C.

Results

Fig. 1 demonstrates residual structure produced by In ions implanted into Si at 40 keV (Fig.1a) and 200 keV (Figs.1b, 1c) at doses 5×10^{13} /cm^2 and 3×10^{14} /cm^2 and annealed at 950°C for 10sec. No end-of-range dislocation loops are present in the case of low energy implant (Fig.1a), while the same dose of the 200 keV implant creates a wide defected region confined between ~46 nm and 140 nm distance from the surface. (Fig.1b). Increasing the 200 keV implant dose to 3×10^{14} /cm^2 results in a narrow band of end-of-range loops beneath the a/c interface at ~160 nm depth. Apparently, the low dose 200 keV implantation was not sufficient to produce continuous amorphous layer extending to the surface and buried amorphous layer was produced. To make a fair comparison between the residual structure produced by the low and the high energy implants, one has to start with similar initial structure, which implies having continuously amorphous layers. Inevitably, the dose required for amorphization is higher in the 200 keV case. That is why in the following experiments we concentrated mainly on two samples: i) 40 keV/ 5×10^{13} /cm^2 (Sample 1) and ii) 200 keV/ 3×10^{14} /cm^2 (Sample 2).

Random and aligned RBS spectra of as-implanted samples showed that the amorphous layer thickness was ~56 and 160 nm, corresponding to the cases presented in Figs. 1a and 1c prior to annealing. B implantation was needed to obtain p$^+$ junctions for the electrical characterization.

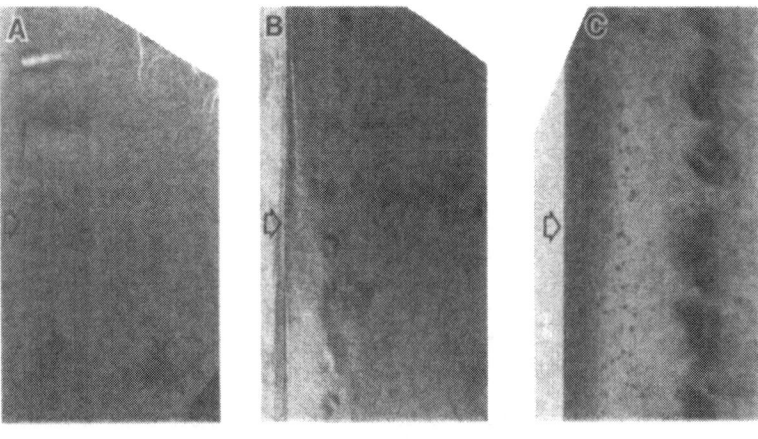

100 nm

Fig. 1 XTEM micrographs of (100) Si implanted and RTA annealed at 950°C for 10 sec
: a)In is implanted at 40 keV/5 × 10^{13}/cm^2; B is implanted at 5 keV/1 × 10^{15}/cm^2;
b)In is implanted at 200 keV/5 × 10^{13}/cm^2; B is implanted at 17 keV/1 × 10^{15}/cm^2;
c)In is implanted at 200 keV/3 × 10^{14}/cm^2; B is implanted at 17 keV/1 × 10^{15}/cm^2.
Free surface is indicated by arrows.

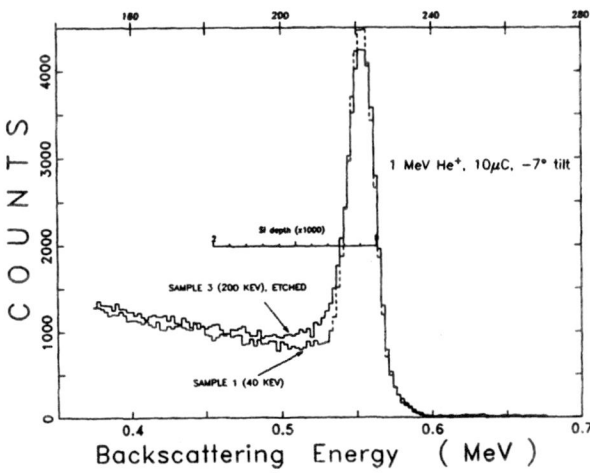

Fig. 2 Aligned RBS spectra of implanted samples: Sample 1 implanted with In at 40
keV/5 × 10^{13}/cm^2 and B at 5 keV/1 × 10^{15}/cm^2; Sample 3 implanted with In at 200
keV/3 × 10^{14}/cm^2 and B at 17 keV/1 × 10^{15}/cm^2, exposed to anodic oxidation and
etch.

However, since the damage produced by B ions implanted into amorphous Si is negligibly small in comparison with the one produced by In, for the experiment described here, the presence of B is not significant.

The maximum concentration of In, as shown by SIMS, was 1.8×10^{19} /cm^3 and 3×10^{19} /cm^3 for samples 1 and 2, respectively. These values are above the solubility limit of In in Si [7], and precipitation of In occurs upon annealing. Indium precipitates (5-10 nm), which are readily visible in sample 2 (Fig. 1c), have a certain crystallographic relationship with Si lattice, as reported previously [8]. The precipitation was not pronounced in sample 1. However, at a certain annealing conditions, In clusters were detected with high resolution XTEM.

The end-of-range dislocation loops beneath the amorphous layer in sample 2 were stable in a temperature interval of 800°C-1050°C. Beyond 1050°C the loops begin to disappear and by 1075°C complete removal of end-of-range damage was achieved. On the contrary, no loops were observed in sample 1 at any of the 10 sec annealing cycles in the same temperature interval.

To decouple the effect of the amorphous layer thickness, i.e. the end-of-range loops proximity to the surface, from the energy of the implanted beam, part of the thicker amorphous layer was removed. This was done by anodically oxidizing sample 2 and then removing the oxide. The process was repeated 15 times until the a/c interface was brought to the same distance from the surface as in the sample 1. This sample, obtained by etching away the top layer of sample 2, will be referred as sample 3. Again, the 10 sec RTA annealing cycles were repeated for sample 3 and compared with the original sample 2. RBS channeling results for samples 1 and 3, presented in Fig. 2., show equally thick amorphous layers in both samples.

Plan view TEM micrographs, corresponding to samples 2 and 3 after 10 sec anneal at 950°C, are presented in Figs. 3a and 3b. The end-of-range loops structure of both samples is identical, despite the different proximity of the surface.

To verify that precipitates do not affect the annealing kinetics of end-of-range loops, one sample was anodically oxidized to remove only ~50 nm from the surface, leaving the precipitates inside the sample. The subsequent annealing of that sample resulted in the end-of-range structure similar to the cases of samples 2 and 3.

Discussion

The experiment described above demonstrated that in spite of the close proximity of the a/c interface to the free surface in sample 3, the end-of-range loops kinetics remained mainly unchanged, as compared with sample 2. Therefore, the difference between samples 1 and 2 in defect generation upon annealing, should be attributed to the energy and dose of the implanted In ions and not to the proximity of the surface.

Monte-Carlo full cascade simulation with TRIM [9] for 10000 ions have been used to estimate the depth distribution of the vacancies and interstitials produced by 40 and 200 keV In ion implantation. The radiation damage distribution is shown in Fig. 4, as the interstitial concentration, normalized to one implanted ion.

One can estimate the amount of the damage by simple integration of the recoils beneath the amorphous layer, assuming that only the ion recoils penetrating beyond the a/c interface contribute to the residual damage. This integral is illustrated by a cross-hatched area for each curve.

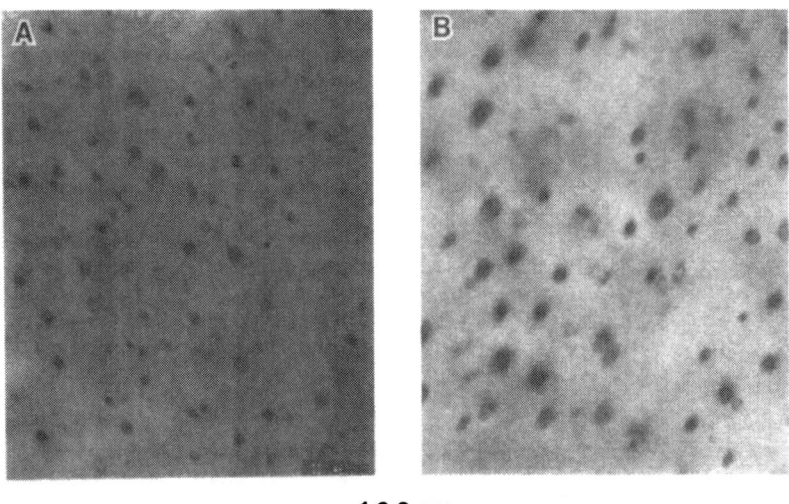

1 0 0 nm

Fig. 3 Plan view TEM micrographs of implanted sample after 950°C/10 sec anneal: a) Sample 2 implanted with In at 200 keV/3 × 10¹⁴/cm² and B at 17 keV/1 × 10¹⁵/cm²; b) Sample 3 implanted as sample 2, subsequently subjected to anodic oxidation and etch.

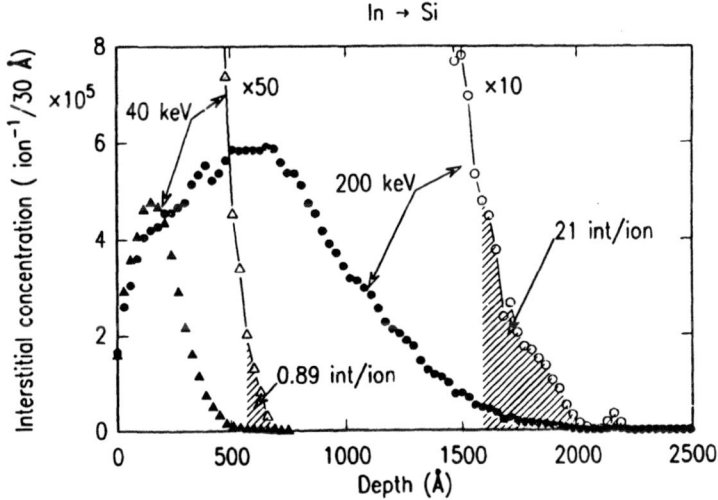

Fig. 4 Damage distribution in Si implanted by In at 40 and 200 keV.

The number of interstitials produced by 200 keV In ions is more than 20 times the one created by 40 keV beam. Taking into account the dose and the depth below the amorphized region in which the interstitials are distributed, as defined from Fig.4 , one can derive the total number of interstitials as 2.25×10^{19} /cm^3 and 1.3×10^{21} /cm^3 in samples 1 and 2 respectively. One should keep in mind that the end-of-range dislocation loops are formed from excess interstitials, and those are only a small fraction of the total number of interstitials created by radiation [1]. The excess interstitials are those left after recombination with vacancies, including dopant ions. However, we make a straightforward assumption, that the number of excess interstitials at the end of the annealing cycle is roughly proportional to the initial number of interstitials created by radiation.

This experiment demonstrated that the effect of surface proximity on residual radiation damage is negligibly small if any, as compared with the effect of recoils distribution, which strongly depends on energy of the primary beam. However, we can not rule out the possibility of the surface serving as a sink for the excess interstitials, if passivation of the surface by a native oxide is prevented. Generally, this is not the case, since formation of a native oxide is usually unavoidable.

Summary

Surface proximity was not found to affect the residual damage distribution for the case of In implanted Si. The role of ion beam energy in residual damage formation is significant, far beyond its effect on amorphous layer thickness. Lower energy beams are generating lower concentration of excess interstitials, as compared with higher energy beams.

Acknowledgements

It is a pleasure to acknowledge valuable discussions with G.A. Sai-Halasz, A. Michel, T. Sedgwick and D. Sadana.

REFERENCES

1   J. Thornton, P.L.F. Hemment, and I.H. Wilson, Nucl. Instrum. Methods B, 19/20, p.307 (1987)
2. K.S. Jones, S. Prussin, and E.R. Weber, Appl. Phys. A 45, p.1 (1988)
3. M.C. Ozturk, J.J. Wortman, C.M. Osburn, A. Ajmera, G.A. Rozgonyi, E. Frey, W.K. Chu, and C. Lee, IEEE Trans. Electron Devices, ED-35, p.659 (1988)
4. E. Ganin, B. Davari, D. Harame, G. Scilla and G.A. Sai-Halasz in Processing and Characterization of Materials, Using Ion Beams, edited by L.E. Rhen, J.E. Greene, F.A. Smidt (Mater.Res.Soc.Proc. vol.128, Pittsburgh, PA 1989)
5. A.C. Ajmera and G.A. Rozgonyi, Appl. Phys. Lett. vol.49, p.1269 (1986)
6. E.R. Myers, Proc. of the 46-th Ann. Meeting of EMSA edited by G.W Bailey, p.906 (1988)
7. F.A. Trumbore, Bell. Syst. Tech. J. Vol.39, p.205 (1960)
8. E. Ganin and W. Krakow, in High Resolution Microscopy of Materials, edited by W. Krakow, F.A. Ponce, D.J. Smith (Mater.Res.Soc.Proc. vol.139, Pittsburgh, PA 1989)
9. J.P. Biersack and L.G. Haggmark, Nucl. Intrum. Methods, vol.174, p.257 (1980)

SHALLOW JUNCTION FORMATION IN As-IMPLANTED Si BY LOW-TEMPERATURE RAPID THERMAL ANNEALING

M. K. EL-GHOR, S. J. PENNYCOOK, and R. A. ZUHR
Solid State Division, Oak Ridge National Laboratory, Oak Ridge, TN 37831-6057

ABSTRACT

Shallow junctions were formed in single-crystal Si(100) by implantation of As at energies between 2 and 17.5 keV followed by conventional furnace annealing or by rapid thermal annealing (RTA). Cross-sectional transmission electron microscopy (XTEM) showed that defect-free shallow junctions could be formed at temperatures as low as 700°C by RTA, with about 60% dopant activation. From a comparison of short-time and long-time annealing, it is proposed that surface image forces are responsible for the efficient removal of end-of-range (EOR) dislocation loops.

INTRODUCTION

Recent developments in semiconductor technology require the formation of shallower device structures [1]. Ion implantation has been the method adopted to produce fairly accurate junction positions with the required dopant concentration. However, post-implantation annealing is necessary in order to recrystallize the amorphous layers and electrically activate the dopant, during which structural defects nucleate in the form of dislocation loops. These are the main source of junction leakage and basically have two origins. The first is due to the EOR damage which occurs at the original position of the amorphous/crystalline interface. The second may occur at the mean projected range due to the release of interstitials trapped within the regrown epitaxial layer [2,3]. The nature of the EOR damage depends on ion implantation conditions, especially temperature and the mass of the ions. At high implant temperatures and large ion masses, dense damage will be induced. This damage will form loops upon annealing, which will coalesce by climb and glide mechanisms until a planar dislocation network is formed. The only way to remove this stable network is through the bulk diffusion of intrinsic point defects, which leads to dislocation climb. RTA was developed to maximize bulk Si diffusion for the removal of extended defects while minimizing the smearing of the dopant profile; a short-time high-temperature anneal favors the higher activation energy process (self diffusion) over lower activation energy processes (dopant diffusion). In this paper we show that it is possible to efficiently remove extended defects through surface interaction, avoiding entirely the need for significant bulk diffusion.

Arsenic is used as a common dopant in the formation of metal-oxide-semiconductor devices. This has triggered much research to study the damage produced during implantation of As^+ as well as defects remaining after annealing processes [2–14]. In the case of B^+ [15] and BF^+ [16–18] implantation, the EOR damage has been reported to anneal out at temperatures as low as 800°C [18], the proposed mechanism for defect removal being diffusion of point defects [15]. Pennycook et al. [2,3] have studied in detail the evolution of the dislocation network at the projected range in the case of Sb^+-implanted Si and have shown that the loops can be removed by a different mechanism, not bulk diffusion of Si, but glide of loops to the free surface under the action of image forces. In the present work, we use the same mechanism to eliminate both projected-range and EOR damage in the case of shallow As+ implants.

EXPERIMENTAL

P-type Si wafers of (100) orientation with resistivities of 6–8 Ω-cm were implanted at energies between 2 and 17.5 keV and doses of 3×10^{14} to 1×10^{16} As+/cm^2. Implantation was carried out at room temperature at a tilt angle of 7° off beam incidence with an average current density of ~2 μA/cm^2. Beam energies between 2 and 8 keV were obtained using a deceleration lens system used for ion beam deposition [19] while the 17.5 keV was produced by implantation of molecular As$_2$ at 35 keV (source extraction voltage). Post-implantation annealing was carried out in a conventional, quartz tube furnace while RTA was performed using a graphite strip heater, in both cases under a dry N$_2$ ambient. Annealing temperatures and times were 600 to 900°C and 40 s to 30 min, respectively. Structural characterization was done using XTEM, while Van der Pauw measurements provided information about the electrical activation of dopants.

RESULTS

Figure 1 compares the microstructure resulting from shallow (17.5 keV) As+ implantation of Si at a high dose of ~9×10^{15} cm^{-2} for furnace annealing for 30 min or RTA for 40 s at temperatures of 800 and 900°C. The furnace anneal at 800°C shown in Fig. 1(a) reveals the EOR damage in the form of dislocation loops at about 400 Å below the surface, while on increasing the temperature to 900°C [Fig. 1(b)], the loops disappeared. Similar results were obtained for RTA. An 800°C anneal for 40 s [Fig. 1(c)], showed the presence of the band of loops which again disappeared after the short time anneal at 900°C, as seen in Fig. 1(d). The dark lines in Fig. 1(d) are extinction fringes from dynamical imaging conditions so as to maximize the sensitivity to extended defects.

The EOR damage was moved closer to the surface by implanting at a lower energy of 8 keV as shown in Fig. 2. Again, a comparison between furnace annealing and RTA was carried out, but now a lower temperature was required for removal of the loops. They are present after annealing at 700°C [Fig. 2(a) and (c)], however, they are absent after annealing at 800°C [Fig. 2(b) and (d)]. Again we see defect removal independent of annealing time. The temperature required is lower than for the 17.5 keV implant, which correlates directly with the EOR damage lying closer to the surface, 235 Å compared to 400 Å for the 17.5 keV implant. Thus, the annealing temperature depends sensitively on the depth of the damage from the surface. This depth is also dependent on dose, for example by reducing the dose of the 17.5 keV implant a factor of 30 to 3×10^{14} cm^{-2}, the EOR damage was moved from 400 to 300 Å below the surface. In this case, the damage could be completely removed by 800°C annealing, whereas at the higher dose (Fig. 1) 900°C was required.

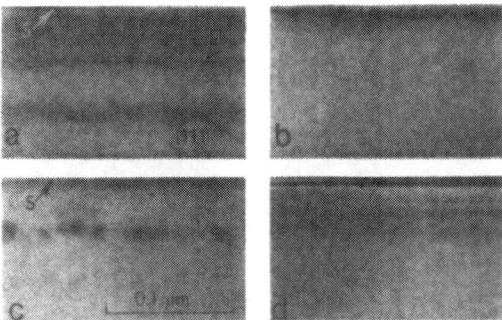

Fig. 1. As+ (17.5 keV, 8.8×10¹⁵ cm⁻²) implanted into single-crystal Si(100) after furnace annealing at (a) 600°C/30 min and 800°C/30 min, (b) 600°C/30 min and 900°C/30 min, and after RTA at (c) 800°C/40 s and (d) 900°C/40 s.

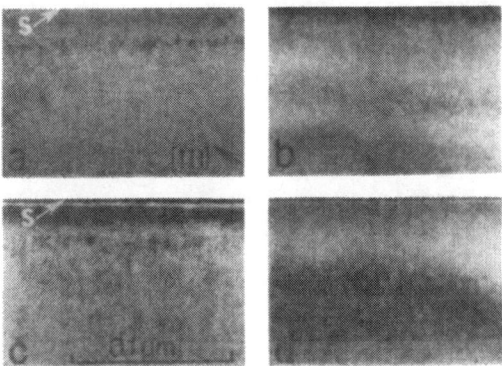

Fig. 2. As+ (8 keV, 4×10¹⁵ cm⁻²) implanted into single-crystal Si(100) after furnace annealing at (a) 600°C/30 min and 700°C/30 min, (b) 600°C/30 min and 800°C/30 min, and after RTA at (c) 700°C/60 s, and (d) 800°C/40 s.

Figure 3 shows an extremely shallow implant using an energy of 2 keV and a dose of 3×10¹⁴ cm⁻². In this case, loops visible after annealing at 600°C for 30 min at a depth of 70 Å below the surface have disappeared completely after annealing at a temperature as low as 700°C for only 40 s, as shown in Fig. 3(b). In fact, the incomplete band of loops visible after the 600°C anneal indicates that much of the damage has already escaped at this temperature.

Sheet resistivity measurements on the samples implanted at 17.5 keV are reported in Table 1. For the high-dose case, a resistivity of 95 $\Omega/\square$ was measured, which is considered to be a good value for such a high concentration sample. Due to the limited depth resolution of the Rutherford backscattering (RBS) technique, the peak As concentration was determined from RBS measurements of the total implanted dose, the depth distributions being calculated by Monte Carlo simulation using the TRIM code [20] (Fig. 4). For the low-dose sample, the peak concentration was calculated to be ~2.0×10²⁰ cm⁻³, which is just below the maximum As+

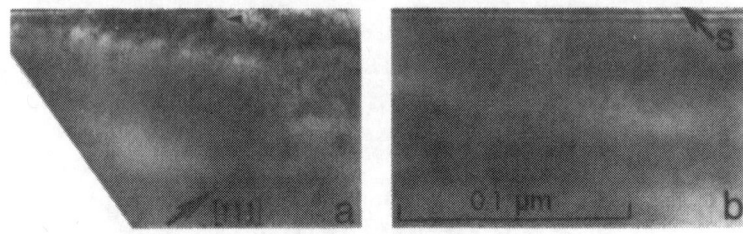

Fig. 3. As+ (2 keV, 3×10^{14} cm^{-2}) implanted into single-crystal Si(100) after furnace annealing at (a) 600°C/30 min showing a band of discrete dislocation loops (arrowed) and after RTA at (b) 700°C/40 s.

Table 1. Sheet resistance measurements of As+ implanted into Si at an energy of 17.5 keV after RTA at 900°C/40 s.

	Dose (cm^{-2})	
$R_s(\Omega/\square)$	3×10^{14}	8.8×10^{15}
Exp.	430	95
Theor.	267	–

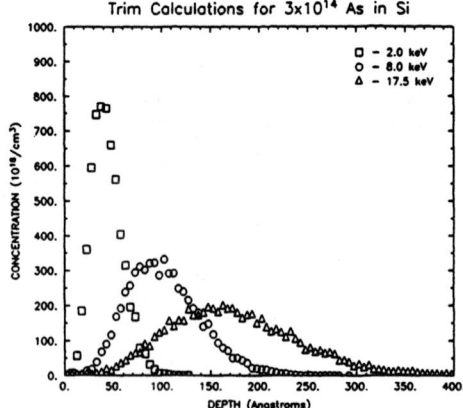

Fig. 4. TRIM calculations for the distribution of 3×10^{14} As+ in Si at energies of 2, 8, and 17.5 keV.

concentration (2.3×10^{20} cm^{-3}) which is electrically active at 900°C [21]. Thus, we would expect it to be possible to fully activate the low concentration sample at 900°C giving a sheet resistivity of 260 $\Omega/\square$. However, our measured value of 430 $\Omega/\square$ indicates only about 60% activation of the implanted As, possibly due to enhanced As clustering because of interaction with oxygen very close to the surface [10–12]. Similar results have been reported by other workers [9].

DISCUSSION

With implantation at conventional energies, the EOR damage removal takes place through the annihilation of extended defects via the incorporation of intrinsic point defects. The activation energy for defect removal is therefore the same as for Si bulk diffusion, approximately 5 eV [3]. This allows RTA to be used to advantage for forming shallow junctions. The high-temperature, short-time anneal favors a high activation energy process (defect removal) over lower activation energy processes, in particular, dopant diffusion leading to profile broadening. Recently it has been shown that defect removal can be achieved at temperatures lower than expected from these considerations when the damage band is close to the sample surface [16–18], and it has been proposed that this is due to an enhanced point defect concentration in the vicinity of the surface [13]. It is the purpose of this paper to point out that there is a second possible mechanism for the removal of defects close to a surface which avoids entirely the need for significant bulk diffusion. This is the direct glide of dislocation loops to the surface under the action of image forces. This mechanism has been observed to remove isolated dislocation loops at the projected range in Sb-implanted samples [2,3].

There are two conditions which must be satisfied for image forces to lead to defect removal. First, the damage must be in the form of isolated dislocation loops and not in the form of an extended dislocation network. Second, these loops must lie within roughly twice their diameter from the sample surface [22–23]. A detailed study of the evolution of the EOR damage was reported by Mader and Michel [24], who found that the damage indeed coalesced into an array of dislocation loops of Burgers vector a/2<110>, all Burgers vectors being inclined to the surface, as shown in Fig. 5. Only with further annealing did these loops undergo further reactions leading to the formation of a stable planar dislocation network. If, at the stage when the loops are still isolated, they are sufficiently close to the surface, we would expect them to be capable of direct glide to the free surface. The key point concerning this mechanism is that glide, once initiated, is a runaway process since the image forces increase as the loop moves closer to the surface. Therefore, we expect the insensitivity to annealing time which we see experimentally. The shallower the original amorphous layer, the smaller the critical loop size required for the image force mechanism to work and, therefore, the lower the temperature required to coalesce the damage to this critical size, as observed experimentally.

If we extrapolate the energy lower than the 2-keV minimum reported here, we come to the region of beam-assisted deposition. We already show evidence with the 2-keV implant of significant loop removal at only 600°C, which is a typical substrate temperature used for these deposition schemes. It seems very likely that the image force mechanism plays an important role in eliminating damage during low-energy beam-assisted deposition.

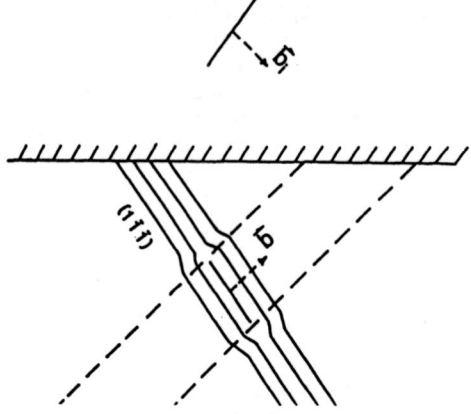

Fig. 5. Schematic of an interstitial loop lying on a (1$\bar{1}\bar{1}$) plane with Burgers vector b= a/2[$\bar{1}$01] indicating attraction of image force. Loop glides to surface along the cylinder shown by dashed lines.

CONCLUSIONS

We have shown that defect-free shallow junctions can be formed with good electrical characteristics by As$^+$ implantation into Si followed by low-temperature RTA. For a 2-keV implant, complete defect removal was seen after annealing at only 700°C for 40 s. We propose that the mechanism for defect removal is the glide of isolated dislocation loops to the free surface under the action of image forces. This mechanism becomes more efficient the shallower the implant.

ACKNOWLEDGMENTS

The authors are grateful to O. W. Holland for helpful discussions and C. W. Boggs, J. T. Luck, and J. L. Moore for technical assistance. M. K. El-Ghor acknowledges the financial support of a fellowship award provided by Oak Ridge Associated Universities. This research was sponsored by the Division of Materials Sciences, U.S. Department of Energy under contract DE-AC05-84OR21400 with Martin Marietta Energy Systems, Inc.

REFERENCES

1. J. M. Shannon, *Nucl. Instrum. and Methods* **182**, 545 (1981).
2. S. J. Pennycook, R. J. Culbertson, and J. Narayan, *J. Mater. Res.* **1**, 476 (1986).
3. S. J. Pennycook and R. J. Culbertson, *Mat. Res. Soc. Symp. Proc.* **52**, 37 (1986).
4. T. O. Sedgwick, *J. Electrochem. Soc.* **130**, 484 (1983).
5. T. E. Seidel, D. J. Lischner, C. S. Pai, R. V. Knowell, D. M. Maher, and D. C. Jacobson, *Nucl. Instrum. and Methods Phys. Res. Sect. B* **7/8**, 251 (1985).

6. M. Delfino, D. K. Sadana, A. E. Morgan, and P. K. Chu, *J. Electrochem. Soc.* **133**, 1900 (1986).
7. R. T. Hodgson, J. E. E. Baglin, A. E. Michel, S. Mader, and J. C. Gelpey, *Laser-Solid Interactions and Transient Thermal Processing of Materials*, ed. by J. Narayan, W. L. Brown, and R. A. Lemons (North-Holland, New York, (1983), p. 355.
8. J. Narayan, O. W. Holland, R. E. Eby, J. J. Wortman, V. Ozguz, and G. A. Rozgonyi, *Appl. Phys. Lett.* **43**, 957 (1983).
9. A. Kamgar, W. Fichtner, T. T. Sheng, and D. C. Jacobson, *Appl. Phys. Lett.* **45**, 754 (1984).
10. N. R. Wu, D. K. Sadana, and J. Washburn, *Appl. Phys. Lett.* **44**, 782 (1984).
11. S. N. Kumar, G. Chaussemy, B. Canut, and A. Laugier, *Appl. Phys. Lett.* **53**, 2167 (1988).
12. M. Tamura, M. Horiuchi, I. Ito, and T. Abe, *Appl. Phys. Lett.* **52**, 1210 (1988).
13. G. A. Rozgonyi, these proceedings.
14. E. Ganin, B. Davari, D. Harame, G. Scilla, and G. A. Sai-Halasz, *Appl. Phys. Lett.* **54**, 2127 (1989).
15. A. C. Ajmera and G. A. Rozgonyi, *Appl. Phys. Lett.* **49**, 1269 (1986).
16. M. C. Ozturk and J. J. Wortman, *Appl. Phys. Lett.* **52**, 281 (1988).
17. E. Myers, M. C. Ozturk, J. J. Wortman, and J. J. Hren, *Appl. Phys. Lett.* **53**, 228 (1988).
18. E. Myers, S. N. Hong, G. A. Ruggles, J. J. Wortman, and J. J. Hren (unpublished).
19. R. A. Zuhr, S. J. Pennycook, T. S. Noggle, N. Herbots, T. E. Haynes, and B. R. Appleton, *Nucl. Instrum. and Methods Phys. Res. Sect. B* **37/38**, 16 (1989).
20. J. P. Biersack, *Nucl. Instrum. and Methods* **174**, 257 (1980).
21. A. Lietoila, J. F. Gibbons, and T. W. Sigmon, *Appl. Phys. Lett.* **36**, 765 (1980).
22. P. P. Groves and D. J. Bacon, *Philos. Mag.* **22**, 83 (1970).
23. J. Narayan and J. Washburn, *J. Appl. Phys.* **43**, 4862 (1972).
24. S. Mader and A. E. Michel, *Phys. Status Solidi a* **33**, 793 (1976).
25. R. O. Schwenker, E. S. Pam, and R. F. Lever, *J. Appl. Phys.* **42**, 3195 (1971).

DAMAGE REMOVAL OF LOW ENERGY ION IMPLANTED BF₂ LAYERS IN SILICON

E. Myers* and J.J. Hren, Department of Material Science and Engineering
S-N Hong and G.A. Ruggles,Department of Electrical
and Computer Engineering
North Carolina State University, Raleigh NC 27695
* Currently with National Semiconductor, Santa Clara CA

ABSTRACT

Recent results indicate that thermal budgets associated with ion implantation induced end of range damage removal is affected by the presence of a free surface. Low energy BF_2 implants (6 keV) were done into both single crystal and Ge preamorphized silicon substrates. Rapid thermal processing was used to study the residual end of range defect structure in the temperature range from 700 to 1000°C. 6 keV, 5E14 cm^{-2} BF_2 implantation resulted in formation of continuous amorphous layers approximately 10 nm deep with a mean B penetration of approximately 7 nm. Conventional TEM analysis found the structures to be completely free of any spanning "hairpin" dislocations or stacking faults associated with the BF_2 implant for all the annealing temperatures. For anneals between 700°C and 900°C end of range damage formation resulted, but the size of the dislocation loops remained small. Annealing at 1000°C, 10 seconds showed no evidence of residual end of range damage. Location of the end of range damage region close to the free surface was found to decrease the thermal budget required for the removal of ion implantation induced radiation damage.

INTRODUCTION

Ultra large scale integration (ULSI) CMOS technology requires junction depths on the order of 75 nm.[1] Formation of n+-p junctions can be achieved by low energy As ion implantation, however no scheme has been developed for shallow B doped p+-n junctions. Boron is a light element which has a larger projected range than As and diffuses rapidly in silicon. The combination of these factors, lead to deep B profiles even at moderate implant energies.

During the VLSI era various schemes were developed so that B remained the p-type dopant of choice. Processing techniques included implantation of BF_2 and preamorphization prior to the dopant implant. While these schemes resulted in shallower doping profiles, they were not without detrimental side effects.

BF_2 implantation has been used for years to decrease the effective energy of the B implant. Only $^{11}/_{49}$ ths of the beam energy is transferred to the B ion. Thus a 6 keV BF_2 implant results in a B doping profile typical of an 1.3 keV implant. Implantation of BF_2 introduces a substantial F concentration into the substrate. Fluorine itself is a small ion and readily migrates during low temperatures anneals.[2] Fluorine segregates to the end of range damage and readily out diffuses from a free surface. Fluorine precipitates on the end of range dislocation loops and has been used as a marker during SIMS profiling.[3] Simultaneously, F has been shown to segregate to the top most surface of the substrate during regrowth. High F concentration effects the stacking fault energy of silicon resulting in formation of numerous stacking faults. Preamorphization of the substrate by Si-self implantation prior to the BF_2 implant has been reported.[2,3] Implantation of BF_2 into Si-self amorphized substrates resulted in a high density of extended defects. These defects include ion implantation induced damage and spanning dislocations. The density of spanning dislocations were found to increase when followed by a BF_2 dopant implant.[4,5]

Preamorphization using Ge implantation prior to the dopant implant has also been reported.[6,7] Ge is a heavier ion and is reported to be isoelectronic in Si. Advantages of Ge preamorphization include creation of continuous amorphous layers at room temperature implants using a much lower dose and a single energy implant. Regrowth of amorphous layers created by Ge implantation were shown to regrow much cleaner than Si-self amorphized layers. Few spanning dislocations were observed and complete removal of the residual end of range damage was shown at lower temperatures.[6]

EXPERIMENTAL

Samples used in this study were blanket 6 keV, 5E14 cm^{-2} BF_2 implants into unpatterned n-type silicon wafers with resistivities of 0.2-0.3 Ω-cm. Annealing was done in an AG associates Heatpulse 210-T tungsten lamp rapid thermal annealer, under flowing argon. The anneal cycle utilized a 5 sec. ramp up, 10 sec. anneal cycle. The samples were place face down on a silicon wafer during the anneal. Temperature was monitored using a thermal couple embedded in a silicon chip placed near the sample.

Transmission electron microscopy was done using a 200 keV Hitachi H-800 scanning transmission electron microscope (STEM) operating in the transmission mode. Rutherford backscattering spectrometry was provided by Dr. N Parikh from the University of North Carolina at Chapel Hill.

RESULTS AND DISCUSSION

TEM imaging (not shown) found the as-implanted amorphous layer created by the 6 keV, 5E14 cm^{-2} BF_2 implantation to be 10 nm deep. Mean B penetration was found to be approximately 7 nm using the TRIM monte carlo simulation program.[8] The amorphous-crystalline interface was very sharp, indicating the implant will regrow without formation of spanning hairpin type defects.[9]

6 keV BF_2 Implantation into crystalline Si

Figure 1a is a bright field micrograph following the 700°C, 10 sec. anneal. A continuous band of defects, with a mean depth approximately of 10 to 15 nm are present. The defects are located just beyond the original amorphous-crystalline interface. The defect band is comprised of isolated clusters on the order of 5 nm that have not coalesced into extended end of range dislocation loops. Dark field imaging (not shown) did not provide additional information on the type of defect

Figure 1b and 1c are bright field micrographs of the microstructure following the 900°C 10 sec RTA. The dark objects in figure 1b is gold that was deposited on the substrate surface prior to TEM sample preparation. The band of defects located at or near the original amorphous-crystalline interface have coalesced into discrete dislocation loops The loops are located approximately 10 nm from the surface.. Figure 1c is a plan view micrograph showing the distribution and density of the end of range dislocation loops. The diameter of the loops are on the order of 15 nm and are separate from neighboring loops. They have not interacted and formed a dislocation network seen in higher energy implants.[10]

Figure 1d is a plan view micrograph of the highest annealing temperature analyzed, 1000°C for 10 seconds. No extended end of range dislocation loops were observed. The entire field of the TEM foil which was transparent to the electron beam did not show evidence of any ion implantation induced end of range dislocation loops.

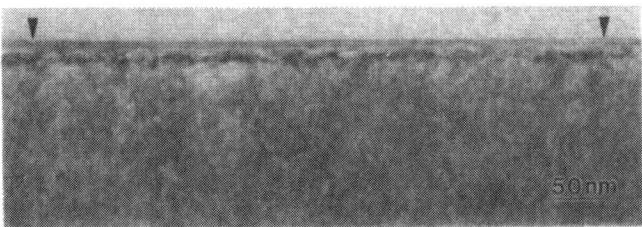

Figure 1a   Cross-section TEM micrograph of a 6 keV BF_2, 5E14 cm^{-2} implant into crystalline Si following a 10 sec, 700°C rapid thermal anneal. The end of range damage has not formed individual, distinguishable dislocation loops.

Figure 1b   TEM micrograph showing the same 6 keV BF_2 implant following a 900°C, 10 sec. RTA, showing end of range damage dislocation loops.

Figure 1c   Plan view TEM micrograph of the same sample as figure 2, showing the distribution of the end of range damage. Note how the dislocation loops are discrete and have not interacted to form a dislocation network.

Figure 1d   Plan view TEM micrograph of a 6 keV BF$_2$, 5E14 cm^{-2} implant
             annealed at 1000°C for 10 sec. No evidence of any
             extended defects are present.

27 keV Ge preamorphized and 6 keV BF$_2$ implanted Si

Figures 2a and 2b are of the 27 keV, 3E14 cm^{-2} Ge preamorphization done
prior to the common 6 keV , 5E14 cm^{-2} BF$_2$ implant. Rutherford backscattering
spectrometry found the amorphous layer created by the Ge implant to be 38 nm deep
and continuous. The Ge energy is sufficient to completely contain the as-implanted
B profile within the amorphous layer.[11]

Figure 2a is a bright field micrograph following an anneal at 700°C for 10
seconds. A wide band of indistinguishable defects about 25 nm wide, beginning at
approximately 43 nm from the surface extend down to approximately 70 nm. The
defect band is comprised of isolated clusters on the order of 5 nm that have not
coalesced into extended end of range dislocation loops. The 6 keV BF$_2$ implant
which can not be seen, appears transparent to the TEM when coupled with the Ge
implant.

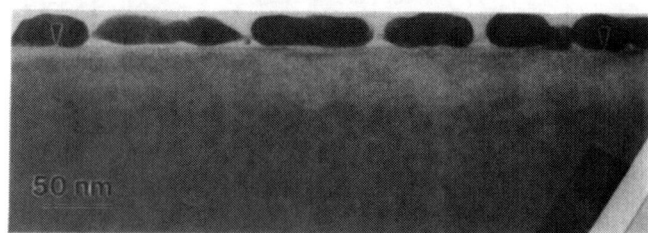

Figure 2a   6 keV BF$_2$, 5E14 cm^{-2} implant into Ge 27 keV, 3E14 cm^{-2}
             preamorphized Si following a 700°C, 10 sec. rapid thermal
             anneal. The end of range damage zone is wide with very
             small (<5 nm) point defect clusters.

Figure 2b shows that the defect band, seen following the 700°C anneal, coalesced into isolated dislocation loops during the 900°C, 10 sec anneal. The loops are approximately 15 nm in diameter and are located 50 nm below the surface, just beyond the amorphous-crystalline interface found by RBS.

Plan view imaging found no evidence of extended defects following the 1000°C, 10 sec anneal. Regrowth of the amorphous layer occurred without retention of any end of range damage.

Figure 2b   Same sample as figure 5, but the anneal temperature was increased to 900°C, for 10 sec. Small dislocation loops are retained near the original amorphous-crystalline interface.

Effect of the free surface

Figure 3 shows a proposed model relating the effect of the free surface and the temperature dependent point defect concentration gradient to the enhanced removal of end of range damage. Free surfaces act as a "perfect" sink for point defects, pinning the point defect concentration. As a result a point defect gradient exists from the surface extending into the substrate. The depth at which the intrinsic point defect concentration is reached depends on the temperature of the substrate. At room temperature the intrinsic point defect concentration is less, resulting in a shallower transition zone. As the temperature is elevated to the various annealing temperatures the transition zone increases in width to compensate for the increase in the intrinsic point defect concentration. If the transition zone expands to include the end of range damage an additional driving force is added to the dissolution kinetics. Qualitatively this explains why a decrease in the minimum energy required to annihilate the end of range damage has been observed as the implant energy decreases.

CONCLUSION

Low energy ion implantation damage can be annealed out of the substrate at significantly lower temperatures than end of range damage located deeper in the substrate. The actual mechanism for the enhanced defect removal is not yet quantified. Possible mechanisms include the different stress state felt by the dislocations when they are in the proximity of a free surface. The free surface acts to exsert a pulling force on the dislocation loop resulting in more energetic climb mechanism. Another possible mechanism is the effect the surface has on point defects, as described above. What ever the mechanism(s) responsible for the enhanced dissolution, as the implant energies become lower there is a greater tendency for removal of the end of range damage to occur at lower temperatures.

While no clear winning method or scheme for achieving 75 nm p+-n junctions has been identified, the appeal of ion implantation remains high. Precise dopant concentration control is unachievable by any other processing technology.

32

Therefore, continued investigations into low energy damage creation, removal and diffusion are required.

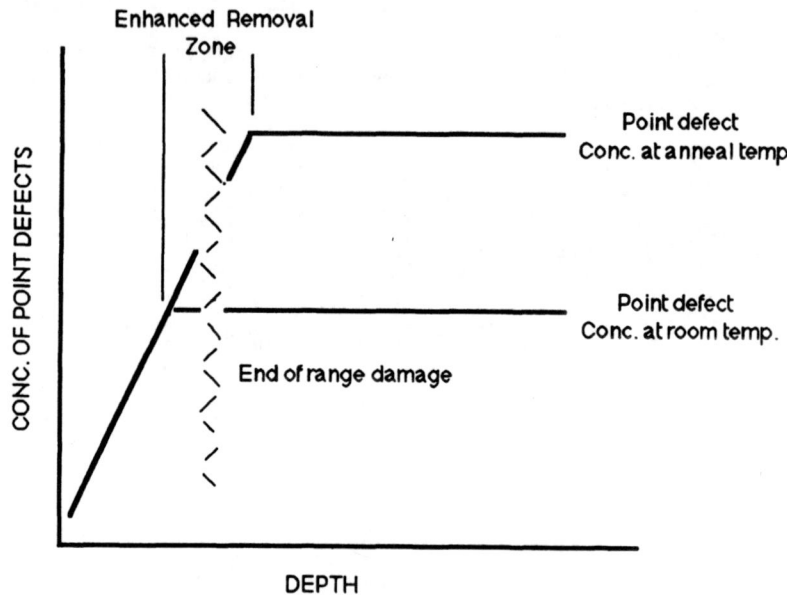

Figure 3   Schematic representation showing the effect of the free surface and point defect concentration gradient on the enhanced removal of ion implanted end of range dislocation loops.

REFERENCES

1) G. Baccarani, M.R. Wordeman and R.H. Dennard, IEEE Trans. Electron Devices **ED-31**, 452(1985).
2) Y. Kim, H.Z. Massoud and R.B. Fair, Appl. Phys. Lett. **53**, 2197(1988).
3) R.G. Wilson, J. Appl. Phys. **54**, 6869(1983).
4) C. Carter, W.P. Maszara, D.K. Sadana, G.A. Rozgonyi, J. Liu and J.J. Wortman, Appl. Phys. Lett. **44**, 459(1984).
5) D.M. Maher, R.V. Knoell, M.B. Ellington and D.C. Jacobson, in Proc of Mat. Res. Soc., **52**, 93(1986).
6) D.K. Sadana, E. Myers, J. Liu, T. Finsted and G.A. Rozgonyi, in Proc of Mat. Res. Soc., **23**, 303(1984).
7) T. Seidel. R. Knoell, G. Poli and B. Schwartz, J. Appl. Phys, **58**, 683(1985).
8) J.F. Ziegler, J.P. Biersack and U. Littmark, eds., The Stopping and Range of Ions in Solids, Perigamon Press, New York, 1985.
9) W. Maszara, D.K. Sadana, G.A. Rozgonyi, T. Sands, J. Washburn and J.J. Wortman, in Proc of Mat. Res. Soc.,**35**, 277(1984).
10) J. Narayan and O.W. Holland, Electrochem. Soc., **131**, 2651(1984).
11) S-N Hong, G.A. Ruggles, J.J. Paulos and J.J. Wortman, submitted to IEEE-ED.

REMOVAL OF END-OF-RANGE ION IMPLANTATION DEFECTS IN SILICON BY NEAR NOBLE AND REFRACTORY SILICIDE FORMATION

W. LUR, J.Y. CHENG and L.J. CHEN
Department of Materials Science and Engineering, National Tsing Hua University, Hsinchu, Taiwan, Republic of China

ABSTRACT

Complete removal of end-of-range (EOR) defects in ion implanted silicon has been achieved by the formation and growth of near noble silicides (CoSi$_2$ and NiSi$_2$) and refractory silicides (MoSi$_2$ and WSi$_2$). Continued generation of vacancies during the silicide growth was found to be essential to reduce EOR defects. The results are consistent with the suggestion of the presence of a vacancy diffusion barrier near the EOR defects.

INTRODUCTION

Ion implantation has been found to be compatible with the various processing steps involved in the silicon device planar technology. For many applications the doses of the implanted ions are frequently high enough to render the surface layer of silicon amorphous. A post-implantation thermal annealing is essential to activate the dopants and regrow the amorphous layer. A zone of interstitial loops was often formed near the original amorphous/crystalline (a/c) interface due to residual end-of-range (EOR) damage from the ions.[1] The removal of EOR defects in ion implanted silicon is a long standing problem. The EOR damage is expected to be detrimental to the junction properties if it is contained within the space charge region of the junction. As a result, the issue is of particular relevance to the present and future shallow junction devices. Metal silicides have been widely used in microelectronics devices. In applications as contacts to source and drain regions and as electrodes at the gate level, silicides are often formed on ion implanted silicon. It has been recognized for some time that residual defects formed in ion implanted silicon could be reduced by silicide formation. Recently, it has been demonstrated that the EOR ion implantation damage in a shallow p$^+$ junction can be eliminated using titanium silicide.[2] It was suggested that silicon atoms diffuse into the silicide layer thereby injecting vacancies into the substrate during the Ti silicidation to annihilate EOR defects.[2] In many silicide-forming systems, such as most near noble metals on silicon, the dominant moving species were found to be metal atoms. If the EOR defects were eliminated primarily by the vacancies left behind by silicon diffusion, it would severely restrict the effectiveness of near noble silicides to remove the defects. It is therefore to be of interest to investigate the effects of silicide formation, in which metal atoms are dominant diffusing species, on the removal of EOR defects. Metal atoms were found to be the dominant diffusing species during NiSi$_2$ and CoSi$_2$ formation. In this paper, we report effective removal of EOR defects in ion implanted silicon by CoSi$_2$ and NiSi$_2$ formation.[5,6] For comparison, data on the removal of EOR defects by the formation of two refractory silicides, MoSi$_2$ and WSi$_2$, are also included. Mechanisms for the annililation of EOR defects by silicide formation are discussed. We note that previous studies showed that complete annihilation of EOR defects by silicide formation depends on a number of factors, which include distance between silicide/Si interface and location of the original a/c interface and/or EOR defects, annealing temperature and time, defect complexity, grain sizes of silicide, and proximity to the sample surface.[5][6]

EXPERIMENTAL PROCEDURES

3-8 Ω-cm, 3 inches in diameter, n type Si wafers were used in this study. Following a standard cleaning procedure, wafers were implanted with 25 keV B^+, 110 keV BF_2^+, 65 keV P^+, 65 keV Si^+ or 150 keV As^+ to a dose of 5 x 10^{15} /cm^2 at room temperature without substrate cooling unless otherwise specified. The ion energies were chosen so that their projected ion ranges are approximately equal. The wafers were oriented 7° off the incident beam direction to minimize the channeling effect. Some of the samples were oxygen plasma cleaned to remove the hydrocarbon contaminants at the surface which were induced during ion implantation. Some of the samples were annealed at 600-1000 °C in N_2 ambient for 1/2 h to regrow the amorphous layers and activate the dopants prior to metal depositions. Thin Ni, Co, Mo, and W films, 30 nm in thickness unless otherwise mentioned, were then electron gun deposited onto the wafers in a vacuum of better than $1x10^{-6}$ Torr. The samples were then annealed isothermally in an oil-free vacuum furnace at 600-1000 °C in a vacuum of better than $2x10^{-6}$ Torr. The annealing time at each temperature was 1 h unless indicated otherwise. Both cross-sectional and plan-view thin foils were prepared for TEM examination.

RESULTS AND DISCUSSION

For 110 keV BF_2^+, 65 keV P^+, 65 keV Si^+ and 150 keV As^+ implantation, the a/c interfaces were found by TEM to locate at distances about 140, 130, 140 and 200 nm from the silicon surface, respectively, in as implanted samples. For 25 keV B^+ implanted samples, a defect band centered at a distance about 150 nm from the surface was observed. It is to be noted that for 30-nm-thick (Co, Ni) and (Mo, W) films on silicon, about 110- and 80-nm-thick Si are consumed to form ($CoSi_2$, $NiSi_2$) and ($MoSi_2$, WSi_2), respectively.

1. $CoSi_2$ formation on (001)Si

A.   800 °C annealing
The density of EOR defects was considerably reduced in B^+, P^+, and As^+ implanted samples. A high density of defects, on the other hand, remained in BF_2^+ implanted samples. As the annealing time was increased to 4 h, a further decrease in density of EOR defects was found in B^+, P^+, and As^+ implanted samples but not in BF_2^+ implanted samples. Examples are shown in Fig. 1.

B.   900 °C annealing
Total elimination of EOR defects was achieved in B^+ and P^+ implanted silicon. In As^+ implanted samples, a low density of dislocation loops was still evident as seen in Fig. 2. The EOR defects in BF_2^+ implanted silicon were found to be most difficult to remove. In samples annealed for 5 min, all EOR defects were removed in P^+ implanted samples, whereas a low density of dislocation loops remained in B^+ and As^+ implanted samples. Examples are shown in Figs. 1 to 3.

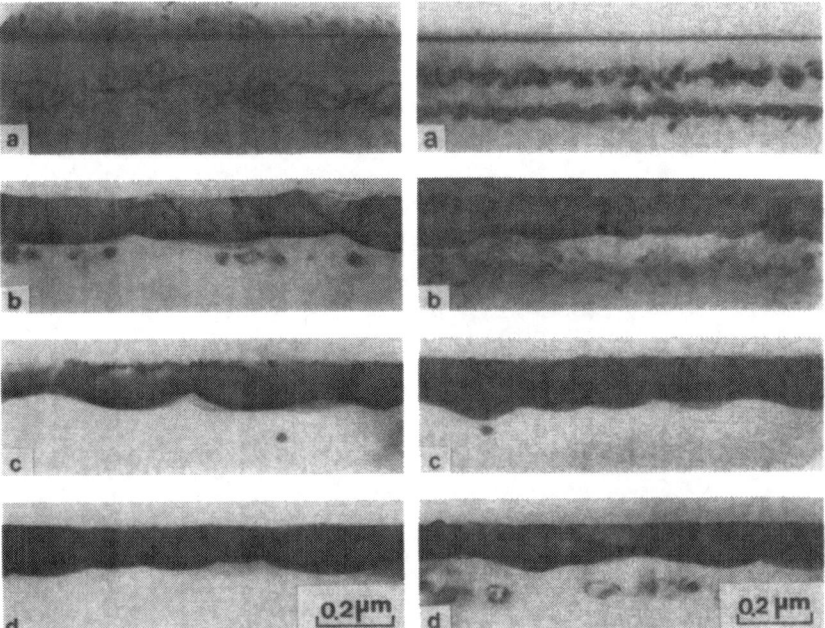

Fig. 1 Cross-sectional TEM micrograph (XTEM), bright field (BF), P⁺, (001)Si, (a) without metal overlayer, 800 °C, 1 h, (b) Co, 800 °C, 1 h, (c) Co, 800 °C, 4 h, (d) Co, 900 °C, 5 min.

Fig. 2 XTEM, BF, As⁺, (001)Si, (a) without metal overlayer, 800 °C, 1 h, (b) Co, 800 °C, 1 h, (c) Co, 800 °C, 4 h, (d) Co, 900 °C, 5 min.

2. NiSi₂ formation on (111)Si

The density of EOR defects was considerably reduced in P⁺ and As⁺ implanted samples after 800 °C annealing. Examples are shown in Fig. 4. A high density of EOR defects was still present in BF₂⁺ implanted samples annealed at 900 °C.

3. MoSi₂ and WSi₂ formation on ion implanted silicon

MoSi₂ and WSi₂ are isomorphic to each other with almost identical lattice parameters. They also behaved similarly in the removal of EOR defects.

A high density of EOR defects were present in all samples annealed at 800 °C up to 4 h. EOR defects were completely annihilated in Si⁺ implanted samples after 900 °C annealing. A high density of EOR defects remained in P⁺ and As⁺ implanted samples annealed at 900 °C for 5 min. As the annealing time was increased to 15 min, all EOR defects were removed. For B⁺ and BF₂⁺ implanted samples, no significant reduction of EOR defects were found after 900 °C annealing. However, they were completely removed after 1000 °C annealing. Examples are shown in Fig. 5 and 6.

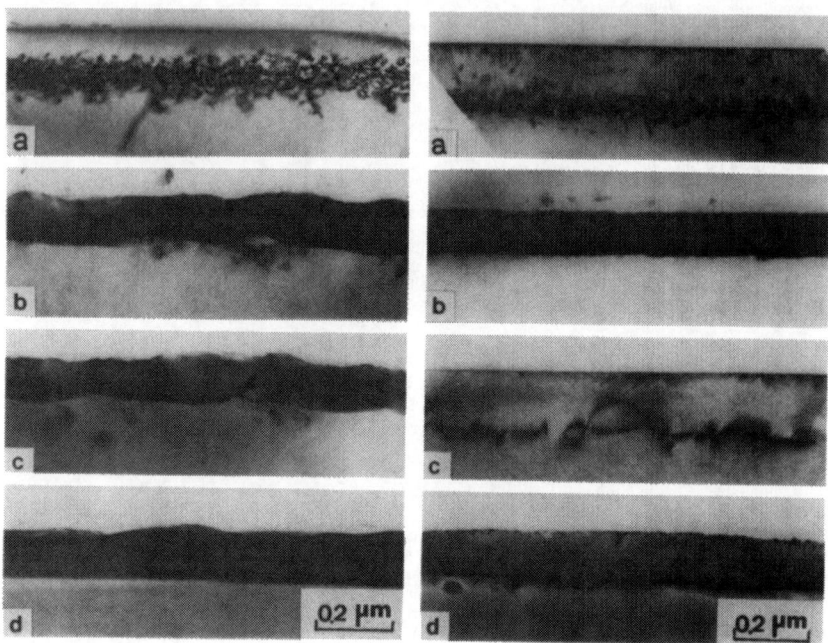

Fig. 3 XTEM, BF, B⁺, (001)Si, (a) without metal overlayer, 800 °C, 1 h, (b) Co, 800 °C, 1 h, (c) Co, 900 °C, 5 min, (d) Co, 900 °C, 1 h.

Fig. 4 XTEM, BF, (111)Si, 800 °C, 1 h, (a) 80 keV, As⁺, (b) Ni, 80 keV, As⁺, (c) 65 keV, P⁺, (d) Ni, 65 keV, P⁺.

Two discrete layers of defects were observed in As⁺ implanted (001)Si annealed at 800 °C. The upper layer of defects was no longer present following $MoSi_2$ formation at 800 °C. EOR defects were completely removed by $MoSi_2$ formation at 900 °C.

The results showed that total elimination of EOR defects is achievable for both near noble and refractory silicides formed on silicon under appropriate annealing conditions in samples implanted by elemental ions. It has been proposed that the reduction of EOR defects by silicide formation is due to injection of vacancies originated either from volume shrinkage or silicon diffusion during silicide formation.[2][6] Metal atoms were found to be the dominant diffusion species during both $CoSi_2$ and $NiSi_2$ formation, which diminishes the relative weight of the mechanism of injecting vacancies left behind by Si diffusion during formation to remove EOR defects.

The migration energy of vacancies in defect-free silicon was generally estimated to be about 0.3 ev. The diffusion distance is of the order of μm in samples annealed for 1 h at a temperature as low as 700 °C. Since the interstitial defects located at a distance less than 0.1 μm from the silicide/Si interfaces were by and large intact in 800 °C annealed samples, it is apparent that the generation of a high density of vacancies during silicide formation alone is not sufficient to reduce EOR defects. The direction of vacancy flow and presence of a vacancy diffusion barrier due to the strain field of EOR defects are the possible complicating factors. It seems that continued generation of vacancies during silicide growth,

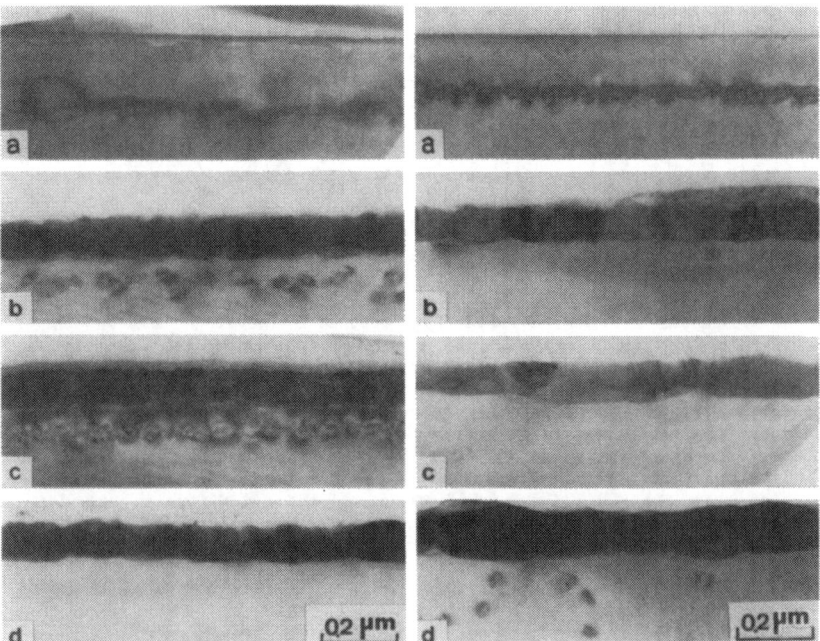

Fig. 5 XTEM, BF, As⁺, (001)Si, (a) without metal overlayer, 900 °C, 1 h, (b) Mo, 800 °C, 1 h, (c) Mo, 900 °C, 5 min, (d) Mo, 900 °C, 15 min.

Fig. 6 XTEM, BF, 65 keV, 5 X 10¹⁵ /cm², Si⁺, (001)Si, (a) without metal overlayer, 900 °C, 1 h, (b) Mo, 900 °C, 1 h, (c) W, 900 °C, 1 h, and (d) Mo, 110 keV, BF₂⁺, 900 °C, 1 h.

accompanied by stress relief at the silicide/Si interface which directs the direction of vacancy flow, was essential to reduce EOR defects. The compressive stress exerted on the silicon layer immediately beneath the silicide layer would favor the in-diffusion of vacancies formed during the formation and growth of $CoSi_2$. More effective removal of EOR defects by annealing at 900 °C as well as by prolonged annealing at 800 °C is consistent with the suggestion of the presence of a vacancy diffusion barrier near the EOR defects. The nature of the vacancy diffusion barrier is not clear at this time. However, a previous study showed that fluorine from BF_2, boron and gold were gettered at EOR defects exhibiting a double peak in concentration, which presumably mimics the strain filed of the EOR defects.[7] A minimum in strain may exist within the EOR defects if the strain fields of defects overlap and reduce their own strain. It is worthwhile to note that difference in reduction of defects in P⁺ and As⁺ implanted silicon is likely related to the location of the original a/c interface from the surface being deeper than that of P⁺ implanted samples.

Dopants were known to redistribute during silicide formation for metal films on ion implanted silicon.[8,9] The redistribution of dopants depends on doping species, silicide phase, solubility and diffusivity of doping atoms. It was found that some dopants prefer to stay in silicide layers as well as at interface and redistribute at low temperatures.[9] It was also shown that P and As atoms have high diffusivities compared with B atoms which appear to be rather immobile.[10] The P and As atoms were observed to redistribute homogeneously in most silicide layers. However, for B^+ and BF_2^+ implanted samples, it has been found that the fluorine redistribution profiles are strongly influenced by the magnitude and distribution of damage that remains after annealing and therefore fluorine profile has peaks near the surface and silicide/silicon interface.[5,11] On the other hand, boron atoms were found to shift deeper into the silicon substrate after annealing. The annihilation of EOR defects in ion implanted samples by silicide formation is likely related to the dopant depletion in the silicon substrate.

SUMMARY AND CONCLUSIONS

In summary, we demonstrated that complete removal of EOR defects is achievable in silicon implanted by elemental dopants by the formation and growth of near noble and refractory silicides. It is suggested that continued generation of vacancies during silicide growth, accompanied by stress relief at the silicide/Si interface which directs the direction of vacancy flow, is essential to reduce EOR defects. More effective removal of EOR defects by annealing at 900 °C as well as by prolonged annealing at 800 °C is consistent with the suggestion of the presence of a vacancy diffusion barrier near the EOR defects. Effects of dopant out-diffusion and dopant redistribution during silicide formation on the removal of EOR defects are discussed.

ACKNOWLEDGMENT

The research was supported in part by the Republic of China National Science Council.

REFERENCES
1.   J.M. Poate and J.S. Williams, in Ion Implantation and Beam Processing, eds. J.S. Williams and J.M. Poate (Academic, Sydney, 1984) p. 13.
2.   D.S. Wen, P.L. Smith, C.M. Osburn and G.A. Rozgonyi, Appl. Phys. Lett. 51, 1182 (1987).
3.   M.A. Nicolet and S.S. Lau, in Materials and Process Characterization, eds. N.G. Einspruch and G.B. Larrabee (Academic, New York, 1983) p. 329.
4.   F.M. d'Heurle and P. Gas, J. Mater. Res. 1, 205 (1986).
5.   W. Lur and L.J. Chen, J. Appl. Phys. 64, 3505 (1988).
6.   W. Lur, J.Y. Cheng, C.H. Chu, M.H. Wang, T.C. Lee, Y.J. Wann, W.Y. Chao, and L.J. Chen, Nucl. Instrum. Methods B39, 297 (1989).
7.   T.E. Seidel, D.J. Lischner, C.S. Pai, R.V. Knoell, D.M. Maher, and D.C. Jacobson, Nuclear Instrum. Methods B7/8, 251 (1985).
8.   I. Ohdomari, K. Konuma, M. Takano, T. Chikyow, H. Kawarada, J. Nakanishi, and T. Ueno, Mater. Res. Soc. Symp. Proc. 54, 63 (1986).
9.   S.P. Murarka, and D.S. Williams, J. Vac. Sci. Technol. B 5, 1674 (1987).
10.   P. Gas, V. Deline, F.M. d'Heurle, A. Michel, and G. Scilla, J. Appl. Phys. 60, 1634 (1986).
11.   R.G. Wilson, J. Appl. Phys. 54, 6879 (1983).

ION INDUCED CRYSTALLIZATION AND AMORPHIZATION
OF SILICON

KENNETH A. JACKSON
AT&T Bell Laboratories, Murray Hill, NJ 07974

ABSTRACT

Extensive experimental investigations which have been reported on the ion-induced motion of the interface between the crystalline and amorphous phases of silicon are shown to fit a model based on a single defect. The model accounts for the temperature and flux dependence of the interface motion. The defects annihilate each other by binary recombination, and have a motion energy of 1.2 eV. The defect is believed to be the dangling bond in the amorphous phase.

INTRODUCTION

A thin (1000 Å) amorphous layer can be created on the surface of a silicon crystal by bombardment with silicon ions. Subsequent irradiation with high energy ions which pass through the amorphous layer can result in motion of the amorphous-crystal interface at temperatures as low as 150°C. Detailed experimental measurements [1-5] have shown that crystallization occurs at higher temperatures and lower ion fluxes, whereas amorphization proceeds at lower temperatures and higher fluxes.

A model which accounts for these results is presented below, based on the following picture. Each ion disturbs the atoms near its path as it passes through the sample. Most of the atoms return to their original positions after the ion passes, but some of the atoms that were in the crystal at the interface relax into non-crystalline positions, so that each ion results in some growth of the amorphous phase. Each ion also leaves behind defects which provide atomic mobility, permitting some regrowth of the thermodynamically stable crystal phase into the metastable amorphous phase. The net rate of motion of the interface, R, is then given by

$$R = (a<N>\Lambda)/\tau_j - V_\alpha \emptyset \qquad (1)$$

where the first term on the RHS is the rate of crystallization and the second is the rate of amorphization. $<N>$ is the average defect density at the interface, "a" is the lattice parameter, Λ is the volume of crystal created by a defect each time it jumps and τ_j is the time between defect jumps. V_α is the volume of amorphous material created at the interface by one ion. It will be assumed that the defects created by an ion are uniformly distributed across the cross-section of a cylinder of diameter l_0 about its path. The defect density created at the interface inside the cylinder by an ion is then:

$$N_0 = N_1/l_0^2 \qquad (2)$$

where N_1 is the number of defects created by an ion per unit path length [3].

The average time between the arrival of defects in the area l_0^2 is given by

$$\tau_0 = 1/(l_0^2 \varnothing) \tag{3}$$

where $\varnothing$ is the incident flux of ions/cm^2 sec.

DEFECT CONCENTRATION

We will assume that the defects created by the ion beam annihilate each other in pairs, at a rate given by:

$$dN/dt = -N^2 \sigma^2 a/\tau_j \tag{4}$$

where σ^2 is the capture cross section for one defect by another. τ_j, the defect jump period, has the form:

$$1/\tau_j = \nu_0 \exp(-E/kT) \tag{5}$$

After the first ion arrives, the defect concentration is given by

$$N = N_0/(1 + \gamma t/\tau_0) \tag{6}$$

where

$$\gamma = N_0 \sigma^2 a \tau_0/\tau_j \tag{7}$$

is the ratio of the defect decay rate to the ion arrival rate. There are two limiting cases: for large γ, the defects produced by one ion substantially disappear before the next ion arrives; for small γ, the next ion arrives before there has been much defect annihilation, so the defect population builds. It will reach a steady state concentration, however, since the annihilation rate increases as the defect density increases. After the ith ion arrives, the defect concentration is given by

$$N_i = N_0^i/[1 + N_0^i \gamma(t - i\tau_0)/N_0 \tau_0] \tag{8}$$

where N_0^i is the defect concentration immediately after the arrival of the ith ion.

The steady state value of N_0^i can be determined using Eq. 8 and the average steady state defect density can then be calculated by integrating Eq. 8:

$$\langle N_{ss} \rangle = (N_0/\gamma)\ln\{1 + (\gamma/2)[1 + (1 + 4/\gamma)^{1/2}]\} \tag{9}$$

The net crystallization rate can now be obtained using the average steady state defect concentration given by Eq. 9 in Eq. 1.

$$R = (\Lambda l_0^2 \varnothing/\sigma^2) \ln\{1 + (\gamma/2)[1 + (1 + 4/\gamma)^{1/2}]\} - V_\alpha \varnothing \tag{10}$$

For large γ, the logarithmic term can be approximated as $\ln\gamma$, and the growth rate is given by

$$R = \Lambda l_0^2 \varnothing \ln(\gamma)/\sigma^2 - V_\alpha \varnothing \tag{11}$$

In Eqs. 10 and 11, both the crystallization rate and the amorphization rate are linear with $\emptyset$ and independent of temperature, except for the logarithm term. γ is temperature dependent through τ_j as given by Eq. 5.

COMPARISON WITH EXPERIMENT

For a given flux $\emptyset$, Eq. 11 has the form:

$$R = R_0 - b/T \qquad (12)$$

which is the functional form of the experimental data [3] shown in Fig. 1.

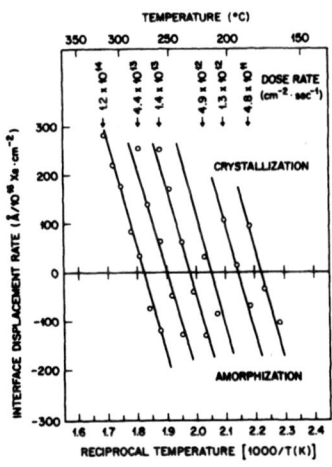

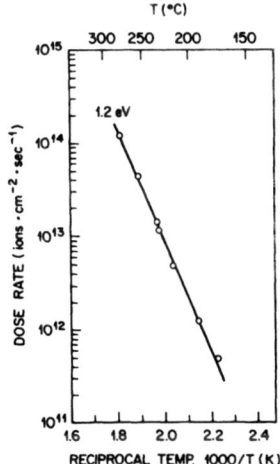

Figure 1. Interface displacement rate for various temperatures and ion fluxes for 1.5 MeV Xenon ions. The interface displacement has been normalized to a dose of 10^{16} ions/cm². The data are from Ref. 3, and the lines were calculated from Eq. 10.

Figure 2. Dose rate dependence of the temperature for zero growth rate for 1.5 MeV Xe ions [3], corresponding to the zero crossings in Fig. 1.

For zero growth rate, corresponding to the zero crossings in Fig. 1, the linear dependencies on flux in Eqs. 10 or 11 cancel out and the only flux and temperature dependencies are from the logarithm term. From Eq. 10, $R = 0$ for:

$$1 + (\gamma/2)[1 + (1 + 4/\gamma)^{1/2}] = \exp[V_\alpha\sigma^2/\Lambda l_0^2] = \beta \qquad (13)$$

Using Eqs. 2, 3, 5 and 7, Eq. 13 becomes:

$$\emptyset = [N_1\sigma^2 a v_0 \exp(-E/kT)/l_0^4][\beta/(1-\beta)^2] \qquad (14)$$

which has the form:

$$\emptyset = \emptyset_0 \exp(-E/kT) \qquad (15)$$

which fits the experimental data [3] shown in Fig. 2. The slope of the Arrhenius plot in Fig. 2 gives $E = 1.2$ eV, the activation energy for the defect motion.

BEAM PULSING EXPERIMENTS

Figure 3 illustrates beam pulsing experiments reported by Linnros et al. [6]. The ion beam was turned on for 1/3 of the period and off for 2/3 of the period. The pulsing frequency was changed, keeping this duty cycle constant. The sample was irradiated with a total of 10^{16} ions/cm^2. At low frequencies, the interface displacement was the same as it would have been if the beam had been left on for a total dose of 10^{16} ions/cm^2. However, at higher frequencies, the interface displacement corresponds to a dose of 10^{16} ions/cm^2 at 1/3 the flux.

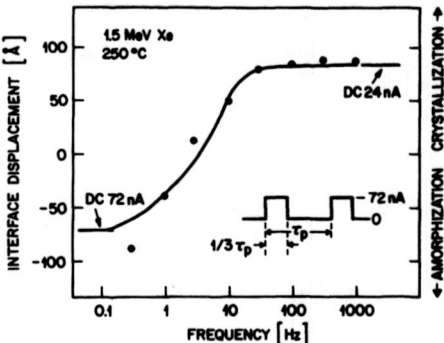

Figure 3. Frequency dependence of the interface displacement at 250°C for a pulsed beam with a peak current of 72 nA, corresponding to a peak flux of 5×10^{13} 1.5 meV Xe ions/cm^2 sec. The beam was pulsed with a constant duty cycle of 1/3. The interface displacements are normalized to a fluence of 10^{16} ions/cm^2. The data are from Ref. 6, and the interface displacement was calculated from Eq. 1 using the time dependence of the defect concentration.

These observations can be understood as follows. At low pulsing frequencies, the average defect density, while the beam is on, is essentially the same as it would be if the beam were left on continuously. And so interface displacement is the same for 10^{16} ions/cm^2 as if the beam had been left on. At high frequencies however, the beam is turned off and on many times before the next ion is due to arrive, and so the time between ion arrivals is increased by a factor of 3, corresponding to a factor of three decrease in flux.

The solid line in Fig. 3 is the result of computer calculations, based on Eq. 8, of the average defect concentration as a function of frequency. The calculations fit the data for $\tau_0 = 1/45$ sec. For a flux of 5×10^{13} ions/cm^2 sec, this gives a value of $l_0 = 95$Å from Eq. 3. Using $N_1 = 1.3\times10^8$ defects/cm [7] in Eq. 2 gives the defect density created by each ion at the interface: $N_0 = 1.4\times10^{20}$ defects/cm^3, or about 0.3%.

Only two of the four remaining parameters v_0, σ^2, Λ and V_α can be determined from the data in Figs. 1 and 2. The curves in Figs. 1, 2 and 3 were calculated using $\gamma = 6$, $\Lambda = 8\times10^{-25}$cm^3, $V_\alpha = 5\times10^{-22}$cm^3, $\sigma^2 = 2.93\times10^{-15}$ cm^2 and $v_0 = 4.26\times10^{15}$ sec^{-1}. All of these are reasonable values.

THERMAL CRYSTALLIZATION

At temperatures above about 400°C, the crystalline phase grows into the amorphous phase at a rate which has an Arrhenius temperature dependence, with an activation energy of 2.7 eV [8]. Assuming that the same defect is responsible for thermal annealing and for ion beam induced interface motion, Eq. 4 can be modified to take into account the thermal generation of defects [9]. The data for thermal epitaxial regrowth can then be fitted, using the crystallization rate in Eq. 3. This gives a formation energy for the defect of 1.5 eV. Taking into account both thermally generated and ion generated defects [9] then provides a very satisfactory fit to the experimental measurements [4] in the transition region from ion beam induced to thermal crystallization, as illustrated in Fig. 4.

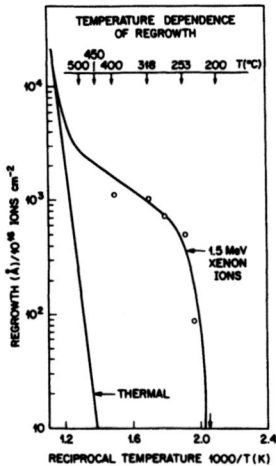

Figure 4. Temperature dependence for thermal regrowth and for regrowth with a flux of 6×10^{12} Xenon ions/cm². The vertical axis is the interface displacement during the time required for a dose of 10^{16} ions/cm². The data are from Ref. 4. The arrow indicates the temperature of zero growth for ion beam induced growth. The short linear section in the transition region should not be interpreted as giving an activation energy.

CONCLUSION

The model presented here accounts extremely well for the experimental data, and therefore makes a compelling case for the role of a single defect in controlling ion induced crystallization. There are two primary candidates for the defect. One is a combination of vacancies and interstitials in the crystalline phase. The known properties of these defects make them unlikely candidates for the defect identified here.

The other possibility is that the defect is a dangling bond in the amorphous phase. A dangling bond can only be eliminated by another dangling bond, and, in the interface region its motion can rearrange the amorphous structure into the crystalline structure. It is believed that the dangling bond in the amorphous phase is the defect which is responsible for ion beam induced epitaxial crystallization, and that these experiments provide important new information about the dangling bond defect.

ACKNOWLEDGMENT

The author wished to thank W. L. Brown, D. M. Maher, G. H. Gilmer and L. C. Kimerling for valuable discussions.

44

REFERENCES

1. J. Linnros, Ph.D. Thesis, Chalmers University of Technology, Goteborg, Sweden (1985).

2. J. S. Williams, R. G. Elliman, W. L. Brown and T. E. Seidel, Mat. Res. Soc. Symp. Proc. **37**, 127 (1985).

3. W. L. Brown, R. G. Elliman, R. V. Knoell, A. Leiberich, J. Linnros, D. M. Maher, and J. S. Williams, Microscopy of Semiconductor Materials, Inst. of Phys. Conf. Series ed. by A. G. Cullis, 61 (1987).

4. R. G. Elliman, J. S. Williams, W. L. Brown, A. Leiberich, D. M. Maher, and R. V. Knoell, Nucl. Inst. & Methods, **B19/20**, 435 (1987).

5. A. Leiberich, D. M. Maher, R. V. Knoell and W. L. Brown, Nucl. Inst. & Methods, **B19/20**, 457 (1987).

6. J. Linnros, W. L. Brown, and R. G. Elliman, Mat. Res. Soc. Symp. Proc., **100** (1988) 369.

7. J. P. Biersack and L. J. Haggmark, Nucl. Inst. & Methods **174**, 257 (1980).

8. Csepregi, W. K. Chu, H. Muller, J. W. Mayer and T. W. Sigmon, Radiat. Eff. **28**, 277 (1976).

9. K. A. Jackson, J. Mater. Res. **3** (1988) 1218.

DIFFUSION, SEGREGATION, AND RECRYSTALLIZATION
IN HIGH-DOSE ION-IMPLANTED Si

S. J. PENNYCOOK
Solid State Division, Oak Ridge National Laboratory, Oak Ridge, TN 37831-6024

ABSTRACT

Using the new technique of Z-contrast scanning transmission electron microscopy (STEM), we have been able to study the segregation of Sb at an advancing SPE growth interface and the resulting interface breakdown. The first direct information is obtained on Sb diffusion in the amorphous phase, which is many orders of magnitude enhanced over tracer crystalline values. This controls both the dopant incorporation and the stability of the resulting supersaturated alloy. These results are compared to the behavior of the low melting point substitutional diffusers and the interstitial diffusers.

INTRODUCTION

It is well known that solid-phase-epitaxial (SPE) growth can be used to incorporate dopants in Si at concentrations well above their equilibrium solubility with minimum dopant redistribution [1-4], (at least up to some limiting concentration [5]). It is also becoming clear from the push towards shallow junctions, that the removal of extended defects can be accomplished at significantly lower temperatures due to the close proximity of the surface [6,7]. We have reported complete removal of extended defects at temperatures as low as 700°C [8]. It therefore becomes increasingly important to understand the limitations of the low-temperature epitaxial growth process for the incorporation of high dopant concentrations, and also the stability of the resulting highly doped layer.

A number of effects have been observed during low-temperature epitaxial growth of highly doped silicon. For the low melting point dopants such as In, interfacial segregation and a push-out effect has been observed [5], and also a competing amorphous-to-crystalline (a/c) transformation [9]. For the high melting point dopants such as Sb, the limiting concentration is determined by interface breakdown leading to the formation of extended defects [10]. Provided dopant concentrations are below these limiting values, SPE growth can proceed to the surface, resulting in substitutional dopant concentrations which are well above the solubility limit. Unfortunately, these alloys show greatly enhanced dopant deactivation due to diffusion coefficients which are greatly enhanced over tracer values [11]. In this paper we show that all these effects have a common origin; they result from a diffusion coefficient in amorphous silicon which is greatly enhanced over tracer crystalline values. Understanding and measuring diffusion in amorphous silicon now becomes the key to predicting the breakdown of epitaxial growth and the stability of highly doped shallow junctions.

Our measurements of diffusion coefficients in amorphous silicon have been made possible by a new electron microscopy imaging technique, Z-contrast STEM. A small probe scans the sample and the intensity scattered to high angles is detected by an annular detector, which is placed around the incident beam and used to form an image. The cross section for scattering into the detector shows almost the Z^2 dependence expected for Rutherford scattering, giving the images strong chemical sensitivity [12]. For dopants such as In and Sb in Si, the concentrations at which SPE growth breaks down can be directly imaged for the first time and correlated with the normal TEM images of the structure [13]. The Z-contrast images can be taken with the incident beam in either a channeling or a random orientation. For random incidence the image is sensitive to composition but not to structure so that amorphous and crystalline phases are indistinguishable. Such an image is therefore ideal for studying segregation at an a/c interface during SPE growth. It can also detect precipitation of these dopants in crystalline, amorphous, or polycrystalline silicon. Under channeling conditions, with a beam size below the crystal lattice spacing, the lattice can be directly resolved while preserving the chemical sensitivity of the technique [14]. This represents a fundamentally new method of high-resolution electron microscopy which has several advantages over conventional techniques [15].

RECRYSTALLIZATION OF Sb-IMPLANTED Si

Using the Z-contrast STEM technique, we have been able to detect for the first time the segregation of Sb at an advancing SPE growth front. Figures 1-3 illustrate how the concentration ahead of the a/c interface gradually builds up until interface breakdown is induced, and the remainder of the growth leads to highly twinned single crystal. No sign of precipitation is seen in the Z-contrast image of the segregated Sb, although with further annealing both precipitation and further Sb redistribution occur [16]. The initial segregation of atomically dispersed Sb seen here is controlled by the diffusion coefficient of Sb in the amorphous phase ahead of the interface D'_a and by the segregation coefficient k'. Since we see a definite segregation peak ahead of the growing interface, this immediately implies that Sb diffusion in amorphous Si is much faster than the tracer value for crystalline Si; although the segregation may not have reached its equilibrium level, and the SPE growth velocity v is dependent on Sb concentration, we can directly estimate values of k', v, and D'_a from the micrographs in Figs. 1 and 2 and the corresponding line traces shown in Fig. 4. If C_0 is the Sb concentration in the amorphous phase, the peak concentration ahead of the interface is $\sim C_0/k'$, and the width of the segregated peak x $\sim D'_a/v$. An average value of k' ~ 0.7 is obtained, much greater than the equilibrium value of 0.023 [17] and, interestingly, the same as found in the

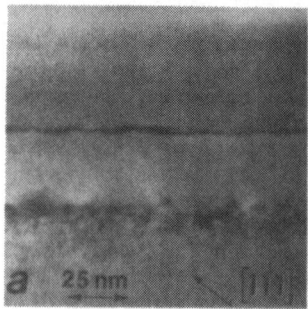

Fig. 1. Sb (80 keV, 1.5×10^{-16} cm^{-2}) implanted Si(100) annealed at 575°C for 10 min; (a) conventional TEM image showing end-of-range damage and a/c interface, (b) Z-contrast STEM image showing segregated Sb building up at the a/c interface.

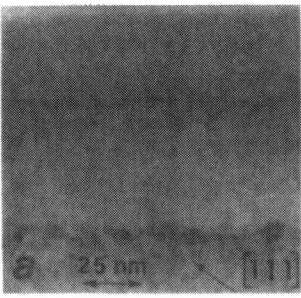 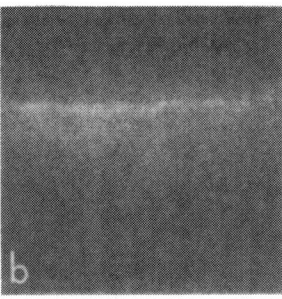

Fig. 2. Same sample as Fig. 1 annealed at 575°C for 15 min, showing (a) onset of interfacial breakdown via nucleation of twins, (b) a higher concentration of segregated Sb.

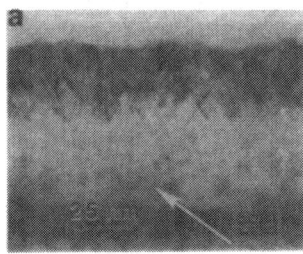 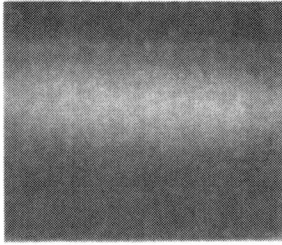

Fig. 3. Same sample after annealing at 575°C for 20 min, showing twins propagating toward the surface and segregated Sb reducing again.

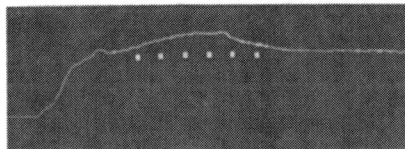 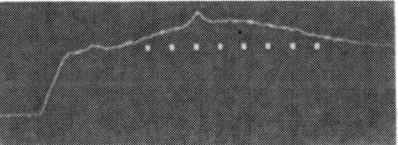

Fig. 4. Z-contrast intensity traces after (a) 10 min, (b) 15 min annealing (obtained digitally from Figs. 2 and 3). The Gaussian form of the implanted Sb is seen above the dotted line and corresponds to a maximum as-implanted Sb concentration of ~6 at.% [12]. The small segregation peak grows as it moves through the peak of the Sb profile.

case of laser annealing from the melt [18]. This arises since the value of D/v is also the same in the two cases, ~3 nm for the solid phase recrystallization here, and ~3.7 nm in the liquid phase recrystallization work.

The incorporation of dopants at metastable concentrations is clearly controlled by the same underlying physics for both liquid and solid phase recrystallization, as originally suggested by Campisano [4]. However, the limiting concentration for growth free of extended defects is determined by the mechanism of interface breakdown, which is different for solid and liquid phase recrystallization. In the case of Sb seen in Fig. 3, it is quite clearly the strain-induced nucleation of twins. Strain was the mechanism suggested by Williams and Elliman [5], although this is not the case for the low melting point dopants.

The value of D'_a we extract from these profiles is ~2.5×10^{-15} cm^2 s^{-1}, a factor of 3×10^8 over the tracer crystalline value at 575°C [19]. This enhancement contrasts markedly to the behavior of the fast diffusers Cu, Ag, and Au [20], where diffusion coefficients in amorphous and crystalline Si were found to be quite similar. This point will be discussed later. The enhanced Sb diffusion is very reminiscent of the behavior of Sb immediately following SPE growth, where a transient enhanced diffusion has been observed. The Sb concentration in the recrystallized Si greatly exceeds equilibrium solubility, so that diffusion will lead to precipitation and diffusion coefficients can be measured from precipitation kinetics [11]. Figure 5 summarizes the results [21]. A concentration-dependent, enhanced diffusion is observed characterized by an activation energy of only 1.8 eV, less than half the tracer value of 4.02 eV [19]. The value of D'_a obtained from the segregation experiments is only a factor of 30 above the enhanced diffusion expected for the same concentration of Sb immediately following recrystallization.

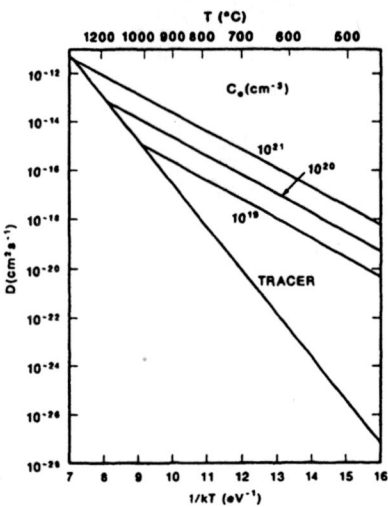

Fig. 5. Arhennius plot of enhanced Sb diffusion coefficients observed following recrystallization for three dopant concentrations C_0 compared to tracer values.

RECRYSTALLIZATION OF In-IMPLANTED Si

The case of low melting point dopants is complicated by the fact that if precipitation occurs the liquid droplets are exceedingly mobile and can migrate at velocities exceeding SPE growth velocities. Thus, precipitation in the amorphous phase ahead of the interface could lead to dopant redistribution as the advancing SPE growth front pushes the mobile droplets ahead of it, as was first observed by RBS [5]. In this case segregation is controlled by the droplet mobility and not by the diffusion coefficient of dispersed In atoms in the amorphous Si. This indicates the importance of Z-contrast observations to determine the state of the In. An additional complication is the competing a/c transformation, which is greatly enhanced by a high In concentration. Nygren et al. [9] demonstrated that this transformation can occur independently of SPE growth and they proposed a mechanism based on migrating liquid droplets. This was completely confirmed by Z-contrast observations of In precipitates in a largely untransformed amorphous Si matrix as shown in Fig. 6 [22]. This allowed an estimate to be made of the diffusion coefficient of In in amorphous Si from the size distribution of the precipitates, giving $D'_a \sim 2.7 \times 10^{-15}$ cm^2 s^{-1}. This figure, which must be a lower limit, represents an enhancement of 10^7 over tracer crystalline values. As in the Sb case, if we drop the concentration so that growth proceeds completely through the In dopant profile, we can estimate the diffusion coefficient in the crystalline phase immediately after recrystallization from the size distribution of the precipitates seen in the crystal. From Fig. 7 we obtain a value of $D'_c \sim 3.2 \times 10^{-15}$ cm^2 s^{-1}, a similar order of magnitude to that found for the amorphous phase. Although the concentrations are different and in neither case has it been possible to follow the kinetics in detail, the similarity of these values is a strong indication that, as in the Sb case, the enhanced diffusion is not strongly dependent on the phase of the Si matrix.

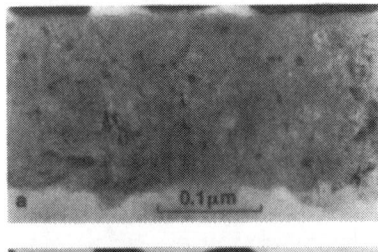

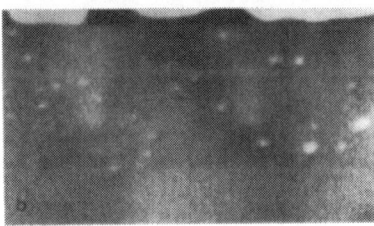

Fig. 6. In (125 keV, 2×10^{15} cm^{-2}) implanted into preamorphized Si(100) and annealed at 600°C for 30 min. (a) Conventional TEM image showing largely amorphous matrix with some crystallites of Si and In precipitates, (b) Z-contrast STEM image showing In precipitates alone.

Fig. 7. Conventional TEM image of In precipitates after recrystallizing In (125 keV, 3×10^{14} cm^{-2}) implanted Si(100) at 600°C for 30 min.

We have also used a low concentration sample to look for In segregation at the advancing SPE growth front. In the absence of the competing a/c transformation we would expect segregation effects similar to those observed in the Sb case. However, Fig. 8 appears to show that this is not the case, the segregated In being in the form of precipitates which simply increase in size as the interface advances. Since it is known that liquid In drops can efficiently transform the amorphous phase to polycrystalline Si, this observation is very surprising. It possibly indicates that, at the growth temperature, the segregated In concentration is below the solubility limit of the amorphous phase and precipitation occurs during subsequent sample cooling.

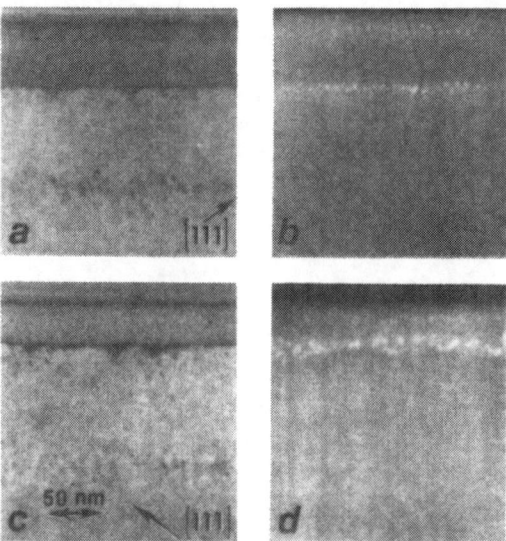

Fig. 8. Conventional (a,c) and Z-contrast (b,d) STEM images of In (125 keV, 1×10^{15} cm^{-2}) implanted Si(100) after annealing at (a,b) 575°C for 10 min, and (c,d) 575°C for 15 min. The band of precipitates close to the sample surface is due to surface sputtering during sample preparation

DISCUSSION

From the results presented here it is clear that although the processes competing with SPE growth are different for the high and low melting point dopants, in both cases diffusion coefficients before and after recrystallization are similar, and greatly enhanced over the tracer values for crystalline Si. Previous work on the fast diffusers Cu, Ag, and Au also indicated similar diffusion coefficients in the amorphous and crystalline phases, but no enhancement over tracer values [20]. These observations can be simply reconciled as follows. If we assume that diffusion mechanisms are essentially similar in the amorphous and crystalline phases, but that the defect density in the amorphous phase is very much higher, then all defect mediated diffusion will be enhanced in the amorphous phase. This would result in the observed enhanced diffusion of the substitutional dopants seen here (which diffuse through interaction with vacancies and Si interstitials) and also the observations of enhanced interdiffusion of amorphous Si/Ge multilayers [23]. The rapid diffusion of Cu, Ag, and Au in crystalline Si is due to true interstitial diffusion, and this component would not be strongly affected by the presence of excess vacancies or Si interstitials.

The transient-enhanced diffusion of the substitutional dopants after recrystallization is then due to the metastable incorporation of these high defect densities during recrystallization, in exactly the same way that the dopants themselves are incorporated at metastable concentrations. (Thus we might expect the defect concentration to pile up ahead of the a/c interface, causing the diffusion coefficient D'_a inferred from the segregation data to be higher than would be inferred from a precipitation or profile broadening experiment.) The concentration dependent diffusion coefficient seen following recrystallization (Fig. 5), indicates that the concentration of trapped defects scales linearly with dopant concentration. This suggests that the defect density in amorphous Si may also be dependent on dopant concentration, which could be directly investigated through further segregation experiments.

The apparent similarity of diffusion mechanisms in the amorphous and crystalline phases seems quite reasonable since only the long-range order is absent in the amorphous phase, and locally the tetrahedral bonding is maintained. Similarly, a much higher equilibrium (or possibly metastable) defect density also seems entirely reasonable, and metastable incorporation of these defects during recrystallization would be expected. The resolution and sensitivity of Z-contrast STEM can provide much insight into impurity incorporation during SPE growth, giving fundamental information on diffusion coefficients and solubilities in the amorphous phase. These properties control not only the recrystallization but the stability of the resulting supersaturated alloy, and are therefore critical for the formation of shallow junctions.

ACKNOWLEDGMENTS

It is a pleasure to thank O. W. Holland and J. C. McCallum for discussions, R. J. Culbertson for ion implantation, and C. W. Boggs, J. T. Luck, and T. C. Estes for technical assistance. This research was sponsored by the Division of Materials Sciences, U.S. Department of Energy under contract DE-AC05-84OR21400 with Martin Marietta Energy Systems, Inc.

REFERENCES

1. P. Blood, W. L. Brown, and G. L. Miller, *J. Appl. Phys.* **50**, 173 (1979).
2. J. L. Regolini, T. W. Sigmon, and J. F. Gibbons, *Appl. Phys. Lett.* **35**, 114 (1979).
3. A. Lietoila, J. F. Gibbons, T. J. Magee, J. Peng, and J. D. Hong, *Appl. Phys. Lett.* **35**, 532 (1979).
4. S. U. Campisano, G. Foti, P. Baeri, M. G. Grimaldi, and E. Rimini, *Appl. Phys. Lett.* **37**, 719 (1980).
5. J. S. Williams and R. G. Elliman, *Appl. Phys. Lett.* **40**, 266 (1982).
6. E. Myers, M. C. Ozturk, J. J. Wortman, and J. J. Hren, *Appl. Phys. Lett.* **53**, 228 (1988).
7. G. A. Rozgonyi, these proceedings.
8. M. ElGhor, S. J. Pennycook, and R. A. Zuhr, these proceedings.
9. E. Nygren, J. S. Williams, A. Pogany, R. G. Elliman, G. L. Olson, and J. C. McCallum, *Mat. Res. Soc. Symp. Proc.* **74**, 307 (1987).
10. J. Narayan, O. W. Holland, and B. R. Appleton, *J. Vac. Sci. Technol. B* **1**, 871 (1983).
11. S. J. Pennycook, R. J. Culbertson, and J. Narayan, *J. Mater. Res.* **1**, 476 (1986).
12. S. J. Pennycook, S. D. Berger, and R. J. Culbertson, *J. Microscopy* **144**, 229 (1986).
13. S. J. Pennycook and J. Narayan, *Appl. Phys. Lett.* **45**, 385 (1984).

14. S. J. Pennycook and L. A. Boatner, *Nature* **336**, 565 (1988).
15. S. J. Pennycook, D. E. Jesson, and M. F. Chisholm, *Proc. 6th Oxford Conference on Microscopy of Semiconducting Materials* (in press).
16. S. J. Pennycook, *Mat. Res. Soc. Symp. Proc.* **52**, 37 (1986).
17. F. Trumbore, *Bell Syst. Tech. J.* **39**, 205 (1960).
18. C. W. White, S. R. Wilson, B. R. Appleton, and F. W. Young, Jr., *J. Appl. Phys.* **51**, 738 (1980).
19. R. B. Fair, *Impurity Doping Processes in Silicon*, ed. by F. F. Y. Wang, North-Holland, New York, 1981, p. 315.
20. J. M. Poate, J. Linnros, F. Priolo, D. C. Jacobson, J. L. Batstone, and M. O. Thompson, *Phys. Rev. Lett.* **60**, 1322 (1988).
21. S. J. Pennycook and R. J. Culbertson, *Advanced Processing of Semiconductor Devices, SPIE Proc. No. 797*, ed. by Sayan D. Mukherjee, SPIE, Bellingham, Washington, 1987, p. 69.
22. S. J. Pennycook, R. J. Culbertson, and S. D. Berger, *Mat. Res. Soc. Symp. Proc.* **100**, 411 (1988).
23. S. M. Prokes and F. Spaepen, *Appl. Phys. Lett.* **47**, 234 (1985).

LATTICE STRAIN FROM HOLES IN HEAVILY DOPED Si:Ga.

K. L. KAVANAGH*, G. S. CARGILL III**, R. F. BOEHME** AND J. C. P. CHANG*
*Department of Electrical and Computer Engineering, University of California at San Diego, La Jolla, CA. 92093
**IBM Research Division, T. J. Watson Research Center, Yorktown Heights, New York 10598.

ABSTRACT

Heavily doped Si:Ga has been prepared by liquid phase epitaxy (LPE) and by ion-implantation with rapid thermal annealing (RTA) or laser annealing (LA). Peak substitutional Ga concentrations obtained by each technique were 1.5, 2.5 and 2.9 $\times 10^{20} \mathrm{cm}^{-3}$, respectively. Substitutional fractions (>90%) were similar in the three types of samples, and the conductivity scaled with the total Ga concentration. A lattice expansion per substitutional Ga atom in Si of $+0.9 \pm 0.1 \times 10^{-24} \mathrm{cm}^3$/atom was measured by double crystal x-ray diffraction. The average nearest neighbor Si-Ga bond length measured with extended x-ray absorption fine structure (EXAFS) was 0.237 ± 0.004 nm, indistinguishable, to within experimental error, from the intrinsic Si-Si bond length, 0.235 nm. Combining these two results the lattice strain per hole in the Si valence band was calculated, $+0.4 \pm 0.8 \times 10^{-24} \mathrm{cm}^3$. This result complements the lattice contraction per electron in the Si conduction band $(-1.8 \pm 0.4 \times 10^{-24} \mathrm{cm}^3)$ already reported for Si:As [G. S. Cargill III, J. Angilello and K. L. Kavanagh, Phys. Rev. Letters **61**, 1748 (1988)].

INTRODUCTION

The effect of substitutional impurities and other point defects on lattice strain in semiconductors remains an area of active research. Lattice parameter variations and point defect cluster dynamics have been studied using x-ray diffraction and transmission electron microscopy [1-3]. However, very few direct measurements of the the local atomic structure at isolated point defects have been reported [4,5].

Using extended x-ray absorption fine structure (EXAFS) we have suceeded in measuring the nearest neighbor Si-Ga bond lengths in heavily doped Si:Ga. Combining this result with measurements of the overall lattice expansion per Ga atom, the lattice strain per hole in the valence band have been determined. These are the first measurements of this kind for an acceptor impurity in Si.

EXPERIMENTAL PROCEDURE AND RESULTS

Si:Ga samples were prepared by liquid phase epitaxy (LPE) and by ion implantation with rapid thermal annealing (RTA) or laser annealing (LA). The LPE samples were obtained from M. Cardona (Max Planck Institut, Stuttgart). They were grown on (111) Si wafers at a temperature of 450°C from Ga saturated Si melts to thicknesses of 0.6 and 1.8 microns [6]. Measurements with Rutherford backscattering spectrometry (RBS) and ion channeling showed that the Ga was > 95% substitutional and uniformly distributed at concentrations of 1 and $1.5 \times 10^{20} \mathrm{cm}^{-3}$.

The Si wafers used for the ion implanted samples were (100) Czochralski grown, boron doped with a resistivity of 10–20 Ω·cm. Gallium was implanted at room temperature in a direction 7° from the surface normal at an energy of 100 keV. Ion doses of 3 and 6×10^{15}/cm^2 were obtained as measured by RBS. Rapid thermal annealing was carried out on the 3×10^{15}cm^{-3} implanted samples using an arc lamp system. The optimal processing parameters, 640°C for 2 sec, were chosen based on minimum sheet resisitivity. The final Ga concentration profile is shown in Fig. 1 (a). The substitutional fraction was greater than 95% with a peak substitutional Ga concentration of 2.5×10^{20}cm^{-3}. This result for the maximum Ga solubility after rapid thermal annealing is comparable to reports in the literature [7].

The Si implanted to a dose of 6×10^{15}cm^{-3} was laser annealed. This was carried out using a pulsed, frequency-doubled Nd-doped yttrium-aluminum-garnet (Nd:Yag) laser, wavelength 532 nm, pulse length 5 ns, using a gaussian energy profile, spatially filtered and focussed to a beam diameter of 1.5 mm. The sample was scanned under the laser in a hexagonal pattern such that adjacent rows overlapped by 30%. The unusually small pulse length greatly reduced the Si melt duration (30 ns) and hence the fraction of Ga which segregated to the surface during the liquid phase regrowth (40%). By comparison, the same type of sample annealed with a 40 ns pulse length ruby laser (melt duration 140 ns) or a 50 ns Nd:Yag laser both resulted in complete segregation of the Ga. For the 5 ns laser annealed sample the surface layer (40 nm) containing segregated Ga was removed by ion sputtering. This was done to avoid interference between EXAFS signals from metallic Ga and Si:Ga. The final Ga concentration profile of the laser annealed Si:Ga layers is shown in Fig. 1 (b). The layer thickness was approximately 150 nm with a total Ga concentration of 3.2×10^{15}/cm^2, approximately the same as that of the RTA sample. The Ga substitutional fraction was greater than 90% with a peak substitutional concentration of $2.9\pm0.3\times10^{20}$cm^{-3} near the surface. This value is somewhat less than 4.5×10^{20}cm^{-3}, the highest substitutional Ga concentration reported in Si obtained by laser annealing (ruby laser, 15 ns pulse width, room temperature sample.) [8].

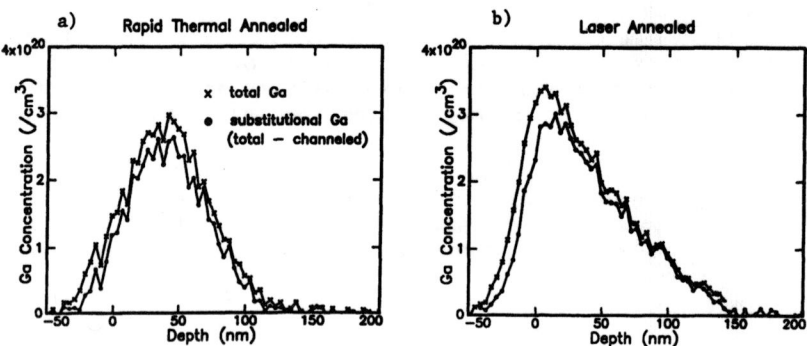

Fig.1. Total and substitutional-Ga concentration profiles of Si:Ga samples prepared by ion-implantation with (a) rapid thermal or (b) laser annealing.

A comparison of the resistivity of the LPE and ion-implanted samples as measured by the four point probe technique showed them to be very similar (12-16 Ω·cm). Transmission electron microscopy (TEM) detected no precipitates, misfit dislocations or other extended defects in the LPE or LA material. End of range dislocation loops were present in the RTA sample but again no extended defects were detected in the heavily Ga-doped regions nearer the surface. Therefore, the Si:Ga layers were assumed to be coherent with the substrate.

Double crystal x-ray rocking curves were recorded from CuK_α (400) or (333) symmetric reflections, depending on the sample substrate orientation. A Si reference crystal with the same orientation was used as the first crystal monochromater. Results from the LPE (333) and ion implanted (004) samples are shown in Fig. 2. In each case smaller peaks or a marked asymmetry on the larger substrate peak is observed towards smaller angles, indicating an expansion in the lattice in the Si:Ga region. The intensity of the Si:Ga peak is greatest for the LPE samples, as expected from the greater layer thickness. The strain in the thicker LPE material can be calculated directly from the peak separation. However, rocking curve simulations must be used to determine lattice strain in the thinner samples. Simulated rocking curves based on a dynamical scattering algorithm [9] have been overlaid with the data (solid lines). The strain profiles associated with each simulation in the case of the ion-implanted samples are shown in the inserts of Fig. 2. Uniform strain profiles, $\Delta d/d$ equal to 1.60 and $1.35 \pm 0.05 \times 10^{-4}$, were used to fit the data of the LPE samples. The results are relatively good. Likewise, a relatively good fit to the LA ion-implanted rocking curve data was obtained with a strain profile corresponding to the shape of the Ga concentration profile (Fig. 1). However, the strain profile is not unique. Comparable fits could be obtained with significantly different strain profiles, examples of which are shown in Fig. 3. In each case, the peak strain magnitude and depth position cannot be uniquely determined without further experiments. We have therefore assigned a large error to the strain associated with the peak in the Ga concentration for the ion-implanted and laser annealed sample. In the case of the RTA sample, expansion associated with the end of range loops dominated the strain profile. A reliable value for the strain per Ga atom could not be obtained. However, the peak strain associated with the end of range loops was about 7×10^{-4}.

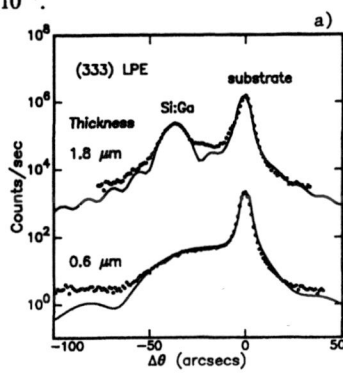

Fig. 2. Double crystal x-ray rocking curves of Si:Ga samples prepared by (a) liquid phase epitaxy (LPE) and ion implantation with (b) rapid thermal or (c) laser annealing. The solid lines are simulated spectra. In (a) uniform strain profiles of 1.60 and $1.35 \pm 0.05 \times 10^{-4}$, 1.8 and 0.6 micron thick layers, respectively, were used in the simulations. The inserts in (b) and (c) show the strain profiles used in the simulation of the ion implanted data.

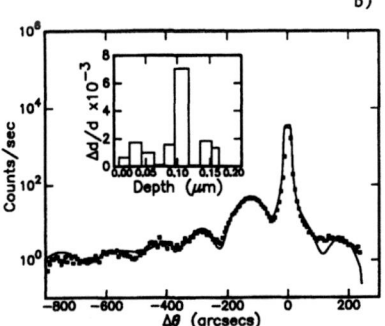

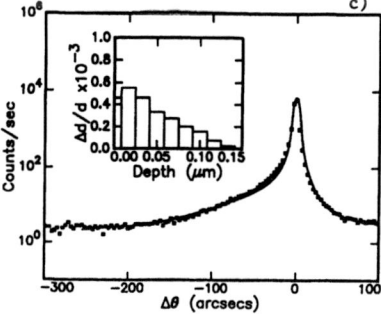

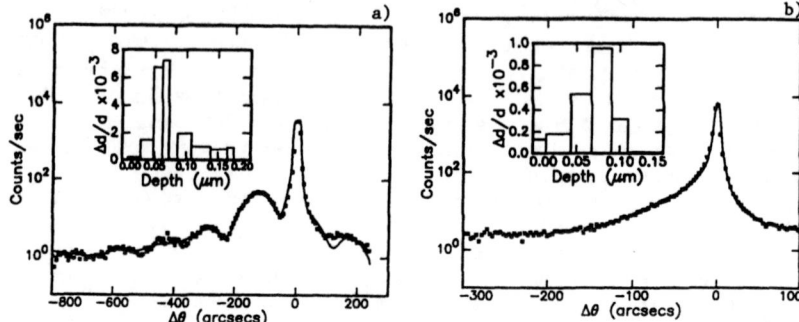

Fig. 3. Double crystal rocking curves of Si:Ga samples prepared by ion implantation. The data is the same as in Fig.2 (b) and (c) but the the strain profiles used in the simulated spectra (solid lines) are significantly different.

The lattice expansions, $\Delta d/d$, at the peak Ga concentration for each sample are listed in Table 1. $\Delta a/a$ can be obtained from $\Delta d/d$ with an appropriate correction for Poisson's expansion (0.56 for (100) and 0.70 for (111) substrates) [10]. $\Delta a/a$ and the coefficient of lattice expansion, $\beta_{tot} = (\Delta a/a)/N_{Ga}$ for each sample are also listed in Table 1. The results for β_{tot} from the LPE and ion-implanted samples agree to within experimental error with an average value of $0.9 \pm 0.1 \times 10^{-24} \, cm^3$.

Total yield EXAFS measurements were made at the Cornell synchrotron facility (CHESS). The background subtracted and normalized EXAFS data taken from the 1.8 micron LPE sample and the ion implanted RTA and LA samples are shown in Fig. 4 as a function of wavenumber. Two shell fits to this data have been overlaid. The average value calculated for the Si-Ga nearest neighbor distance, 0.237 ± 0.004 nm, is indistinguishable, to within experimental error, from the Si-Si bond distance, 0.235 nm, but is notably smaller than Si-As bonds in heavily doped Si:As, 0.241 ± 0.002 nm [5]. The result is also smaller than 0.243 or 0.240 nm published values of the Ga and Si tetrahedral covalent radii predicted on the basis of covalent-solid lattice parameters [11] and on quantum mechanical calculations [12], respectively. However, it agrees with more recent results from total energy calculations by Pandey [13].

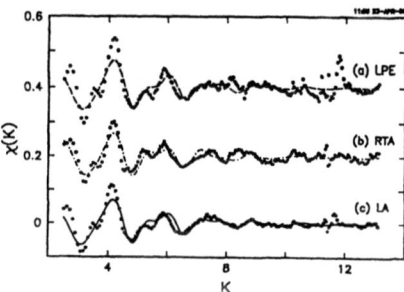

Fig. 4. Extended x-ray absorption fine structure (EXAFS) data, background subtracted and normalized, plotted as a function of wavenumber for three types of Si:Ga samples: (a) liquid phase epitaxy, (b) rapid thermal annealed and (c) laser annealed. The lines are two shell fits to the data.

Sample	$\Delta d/d$	$\Delta a/a$	N_{Ga}	β_{tot}
	$\times 10^{-4}$	$\times 10^{-4}$	$\times 10^{20}$ cm^{-3}	$\times 10^{-24}$ cm^3
LPE (1.8 μm)	(333) 1.60 ± 0.05	1.15 ± 0.04	1.5 ± 0.2	0.84 ± 0.1
LPE (0.6 μm)	(333) 1.35 ± 0.05	0.94 ± 0.04	1.0 ± 0.1	0.99 ± 0.1
LA	(004) 5 ± 4	2.8 ± 2	2.9 ± 0.3	1.0 ± 0.7
RTA	(004) 70 ± 5 (loop dominated)	39 ± 2	2.5 ± 0.2	

Table 1. Strain and concentration data for Si:Ga samples prepared by liquid phase epitaxy (LPE) and ion implantation with laser annealing (LA) and rapid thermal annealing (RTA). Tabulated are: the difference in layer spacing measured by from double crystal x-ray diffraction, $\Delta d/d$, the subsitutional Ga concentration, N_{Ga}, from Rutherford backscattering, the unconstrained lattice parameter difference, $\Delta a/a$, calculated from $\Delta d/d$ and the lattice expansion per Ga, $\beta_{tot} = (\Delta a/a)/N_{Ga}$.

DISCUSSION

We assume, following Yokota [14], that the net fractional change in the Si lattice constant per substitutional Ga atom, β_{tot}, originates from two independent components (i) β_{siz}, created by the local strain associated with different atomic radii, and (ii) β_h, resulting from the change in the band structure and hence atomic bond lengths created by additional holes. As in ref. 5, to estimate β_{siz} we use the EXAFS result for the Si-Ga bond length in Si:Ga, d_{SiGa}, to predict the "natural" bond length of a hypothetical SiGa zincblende structure [15]. Thus:

$$d_{SiGa}^{nat} = 4/3\left[d_{SiGa} - 1/4\ d_{SiSi}\right] = 0.2377 \pm 0.005 \text{ nm}$$

and

$$a_{SiGa}^{nat} = (4/3^{\frac{1}{2}})\ d_{SiGa}^{nat} = 0.5489 \pm 0.01 \text{ nm}$$

Then assuming Vegard's Law,

$$\beta_{size} = \frac{a_{Si} - a_{SiGa}}{a_{Si}\ N_{Ga}} = \frac{0.5482 - 0.5489 \pm 0.01}{0.5482\ (2.5 \times 10^{22} \text{cm}^{-3})} = 0.5 \pm 0.8 \times 10^{-24} \text{cm}^3$$

The lattice strain created by holes in the valence band, β_h, can then be calculated.

$$\beta_h = \beta_{tot} - \beta_{size} = +0.4 \pm 0.9 \times 10^{-24} \text{cm}^3.$$

The experimental error associated with the result for β_h is such that the sign of the strain is inconclusive. And although it is clear that Ga causes an overall expansion in the Si lattice the proportion of this expansion due to a size effect versus the holes is inconclusive. However, this is the first direct measurement of the Si-Ga bond length and of β_h and it provides significant limits on the magnitude of β_h. It is hoped that future refinements in the experimental procedures will reduce the experimental error.

CONCLUSIONS

Lattice expansion in heavily doped Si:Ga has been studied by EXAFS and double crystal x-ray diffraction. The Si:Ga was prepared by liquid phase epitaxy (LPE) and ion implantation with rapid thermal (RTA) or laser annealing (LA). Peak subsitutional Ga concentrations obtained by each technique were 1.5, 2.5, $2.9 \times 10^{20} \mathrm{cm}^{-3}$, respectively. Lattice expansion was observed in all three types of samples with an average expansion per substitutional Ga atom of $0.9 \pm 0.1 \times 10^{-24} \mathrm{cm}^3$. The average nearest neighbor Si-Ga bond length was 0.237 ± 0.004 nm indistinguishable, to within experimental error, from the intrinsic Si-Si bond length, 0.235 nm. Combining these two results the lattice strain per hole in the valence band is $+0.4 \pm 0.9 \times 10^{-24} \mathrm{cm}^3$.

ACKNOWLEDGEMENTS

We acknowledge the assistance of M.O.Thompson (Cornell) with the laser annealing, M. Cardona (Max-Planck Institut, Stuttgart) for supplying the LPE material, R. Singer (MIT) for the rapid thermal annealing and useful discussions with J. Angilello, S. R. Herd, L. Hobbs, R. Hodgson, K. C. Pandey, J. Tersoff, A. Segmuller and C. G. Van de Walle. KK is grateful for an IBM postdoctoral fellowship.

REFERENCES

1. R. J. Culbertson and S.J. Pennycook, Nucl. Inst. Meth. **B13**, 490 (1986).
2. P. Becker and M. Scheffler, Acta. Crystall. A **40**, C-341 (1984)
3. B.C. Larson and J.F. Barhorst, J. Appl. Phys. **51**, 3181 (1980).
4. B. Pajot and A. M. Stoneham, J. Phys. C, **20** 5241 (1987).
5. G. S. Cargill III, J. Angilello, and K. L. Kavanagh, Phys. Rev. Letts. **61**, 1748 (1988).
6. W. H. Apple, Ph.D. thesis, Univ. Stuttgart, 1985, unpublished.
7. H. B. Harrison, Y. H. Li, G. A. Sai-Halasz and S. Iyer, Mat. Res. Soc. Proc. **71**, 223 (1986).
8. C. W. White, S. R. Wilson, B. R. Appleton, F. W. Young, J. Appl. Phys. **51**, 738 (1980).
9. S. Bensoussan, C. Malgrange and M. Sauvage-Simkin, J. Appl. Crystallogr. **20**, 222 (1987).
10. J. Hornstra and W. J. Bartels, J. Cryst. Growth, **44**, 513 (1978).
11. L. Pauling, <u>The Nature of the Chemical Bond,</u> Cornell University Press, 1960, page 246.
12. J. A. Van Vechten and J. C. Phillips, Phys. Rev. B, **2**, 2160 (1970).
13. K. C. Pandey, private communication.
14. I. Yokota, J. Phys. Soc. Japan. **19**, 1487 (1964).
15. C. K. Shih, W. E. Spicer and W. A. Harrison, Phys. Rev. B, **31**, 1139 (1985).

PART II

Damage Effects

REDUCTION OF EXCESS SELF-INTERSTITIALS IN SILICON BY GERMANIUM AND SILICON IMPLANTATION-INDUCED DAMAGE

PAUL FAHEY
IBM Research Division, Watson Research Center
Box 218
Yorktown Heights, NY 10598

ABSTRACT

We have investigated a phenomena first reported by Pfiester and Griffin, that the presence of implanted Ge in Si can substantially reduce excess self-interstitial concentrations [J. R. Pfiester and P. B. Griffin, Appl. Phys. Lett., 52, 471 (1988)]. By studying the effects of Ge implantation on P diffusion, we are able to deduce that residual implantation damage can act as an efficient sink for self-interstitials. This effect can also be produced by Si self-implantation, demonstrating that there is nothing unique about the chemical indentity of Ge in reducing self-interstitial concentrations. Our experiments provide solid evidence that there is no unexpectedly strong interaction of Ge with self-interstitials, a situation that would undermine the validity of previous Ge diffusion experiments aimed at studying Si self-diffusion. Our experimental results show that the effect of Ge implantation on P diffusion is a complicated function of implantation conditions. Diffusion is affected by the order of the P and Ge implants as well as by changes in implant energies and doses.

INTRODUCTION

Ge is expected to be a relatively inert element in Si (when present in dilute concentrations) for a number of reasons: Si and Ge are perfectly miscible; the tetrahedral covalent bonding radius of Ge is close to that of Si (0.122 nm vs. 0.118 nm, respectively [1]); and being in the same column of the periodic table as Si, Ge introduces no net charge to the Si lattice. Because of these considerations, it was first proposed by McVay and DuCharme [2] that Ge might be used as a tracer to study Si self-diffusion. The study of Ge diffusion in Si for this purpose is still an active area of research (see [3] for the most recent update on these studies). These experiments all support the proposal of McVay and DuCharme that Ge is a relatively inactive element in Si. In contrast, recent work by Pfiester et al. [4,5] shows implantation of Ge into Si can greatly reduce the high-concentration diffusion of P and reduce or eliminate the oxidation-enhanced diffusion of B. Both of these phenomena are caused by the presence of excess Si self-interstitials in the bulk. In recognition of this, Pfiester and Griffin [4] stated that their work calls into question the validity of the most basic assumption of Ge diffusion studies: Ge and Si interact with native point defects in essentially the same manner.

The work of Pfiester et al. is obviously of technological as well as academic interest. Excess interstitials may be generated in many ways during process steps used in making integrated circuits. Elimination of enhanced dopant diffusion caused by excess interstitials is highly desirable.

The aim of this work is to identify the mechanism responsible for the effects of Ge implantation on dopant diffusion. We focus mainly on the effects of Ge implantation on P diffusion, rather than oxidation-enhanced diffusion. Studies have shown that the oxidation process can be modified when high concentrations of Ge are present at the oxidizing surface [6-8]. The question of Ge's affecting point-defect generation at the surface by modifying the oxidation process is a separate issue from the one concerning interactions of point defects with Ge atoms. Recent work by LeGoues et al. [9] has demonstrated that oxidation of a chemically deposited layer of $Si_{50}Ge_{50}$ suppresses oxidation-enhanced diffusion of B, presumably by changing point-defect surface reactions. The experiments of Pfiester et al, [4,5] are not complete enough to differentiate between surface and bulk effects. For example, Pfiester and Griffin reported that annealing a Ge-implanted sample for 4 hr at 1000°C prior to B implantation did not change the reduction in oxidation-enhanced diffusion; however, no TEM was performed and it cannot be decided whether this means the damage from the Ge implant was not annealed or Ge at the surface altered point-defect generation.

By studying the effects of Ge implants on P diffusion we are able to separate out bulk from surface effects. Convincing evidence is presented that shows residual damage from Ge implantation causes suppression of excess interstitials and that this effect can also be brought about by creation of damage with Si implantation.

EXPERIMENTAL DETAILS

Wafers used in this study are Czochralski-grown ⟨100⟩ substrates of 10-20 Ω cm. All implantations were performed at room temperature with no screen oxides present. Amorphized depths were measured using RBS analysis. Dopant profiles were determined by either spreading resistance or SIMS analysis. The junction depths for P profiles analyzed by spreading resistance were defined as those locations where the carrier concentration was measured to be 10^{16} cm^{-3}. Spreading resistance measurements give P penetration depths significantly shallower than those found by SIMS; however, the qualitative behavior of diffusion deduced from the two profiling techniques is consistent for all cases where the same sample was measured by both techniques. Unless otherwise noted, all annealed samples were given the following thermal treatment in Ar: 800-900°C ramp, 10 min. ; 900°C, 20 min. ; 900-800°C ramp, 15 min.

SELF-INTERSTITIAL GENERATION DURING P DIFFUSION

In this section, we explain why we believe silicon self-interstitials are generated during P diffusion and the mechanisms by which they are generated. It is well known that when high concentrations of P are present in the bulk (typically, concentrations exceeding 5×10^{19} cm^{-3}), profiles measured after diffusion show a characteristic shape consisting of a low-concentration, fast-diffusing tail connected to the high concentration region by a "kink" in the profile (see for example Fig. 2). Lower concentrations of P exhibit simple Gaussian diffusion behavior. For high-concentration conditions, there is good evidence that P generates interstitials as part of its diffusion mechanism. Fahey *et al.* [10] have shown this with the following simple experiment. Buried marker layers of lightly doped P and Sb are formed in the silicon substrate by implantation and epitaxy (peak concentrations $< 5 \times 10^{18}\,\mathrm{cm}^{-3}$). Above these layers, high concentrations of P are in-diffused using a POCl$_3$ source (surface concentration $\sim 4 \times 10^{20}\,\mathrm{cm}^{-3}$). Then, some areas of the samples are etched to remove the heavily P-doped layers. This leaves the buried marker layers in some places on the sample with heavily P-doped layers above them and in other regions with only undoped Si above them. After annealing in Ar, it is observed that the diffusion of the buried Sb marker layers is retarded underneath the heavily P-doped layers compared to the regions where the heavy P doping was etched away. (In the latter regions, only normal thermal diffusion of the Sb layer is observed.) Under the same conditions, the buried marker layers of P show the opposite behavior of enhanced diffusion below the heavily P-doped layers. This same qualitative behavior of diffusion enhancement and retardation occurs for the case in which self-interstitials are generated by thermal oxidation. This indicates that as P diffuses from the heavily doped layer, silicon self-interstitials are being generated. The experiment of Fahey *et al.* showed that self-interstitials are generated by high concentration P diffusion when there is no oxidation of the surface, no in-diffusion, and no precipitation (surface concentrations are just below solid solubility limits [11,12]). It is likely that P diffusing in the interstitial state P$_i$, generates Si interstitials I, in the process of returning to the substitutional state P$_s$:

$$\mathrm{P}_i \rightleftharpoons \mathrm{P}_s + I . \tag{1}$$

Equation 1 implies that by reducing the number of Si interstitials, the diffusion of P can be slowed by reducing the number of mobile P atoms in the P$_i$ state. Such a correlation has been made by Finetti *et al.* [13]. They showed that P diffusion can be reduced during POCl$_3$ depositions by diffusing through a polysilicon layer into the underlying Si substrate. Furthermore, they correlated the extent of the characteristic tail in the P concentration profiles with the amount of excess interstitials present by monitoring diffusion of B buried marker layers: the smaller the P tail, the less enhanced diffusion of the B layer. We interpret this to mean that the smaller the P tail, the less excess interstitials are present. Changes in P diffusion, as revealed by profile analysis, are used in this study as a measure of the effectiveness of Ge implantation in reducing excess self-interstitial concentrations.

In these experiments, P is introduced into the silicon substrate by ion implantation rather than chemical deposition. It is likely that excess interstitials are also created by the implantation process. Numerous studies have shown that enhanced diffusion resulting from excess point defects produced by ion implantation is a transient phenomenon, typically lasting only a few seconds. The kick-out reaction of Eq. 1 on the other hand lasts for the entire time of diffusion in our experiments. The distinction between the two sources of point defects is made in our discussion comparing the effectiveness of Ge implants in reducing P diffusion to their effectiveness in reducing transient-enhanced diffusion of B.

DEPENDENCE OF P DIFFUSION ON Ge IMPLANT CONDITIONS

A matrix of experiments was designed to illustrate the dependence of P diffusion on Ge and P implantation conditions. The results of these experiments show that P diffusion is affected by the order of the implant sequence, i.e, P first followed by Ge or vice versa, as well as by variations in the implantation parameters of P and Ge for a given amorphization condition.

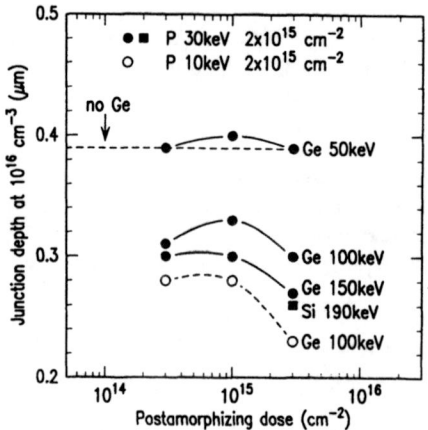

Figure 1: Effects of postamorphization on P diffusion.

Postamorphizing implants

By postamorphizing implants we mean that P is implanted first, followed by Ge implantation. The effects of postamorphizing Ge implants on P diffusion are summarized in Fig. 1. Samples were implanted with P at energies of 10 keV or 30 keV to a dose of 2×10^{15} cm^{-2}. Following this, some samples were implanted with either Si or Ge. After implantation, samples received a 900°C ramped anneal. Spreading resistance measurements were made on samples and the junction depths determined as described in the section on experimental details. No difference in junction depths is observed between samples implanted with only P at energies of 10 keV or 30 keV; the horizontal dashed line indicates that the final junction depth after anneal is 0.39μm for samples which did not receive Si or Ge implants. We first discuss the results for the 30 keV P implants. Within experimental error (junction depths are reproducible within $\pm 0.015 \mu$m), 50 keV Ge implants have no effect on P diffusion for Ge implant doses between 3×10^{14} cm^{-2} and 3×10^{15} cm^{-2}. Increasing the Ge implant energy to 100 keV and 150 keV for the same doses as the 50 keV Ge implants results in significant reduction of P diffusion. Note also the important result that the largest reduction in P diffusion occurs by implanting Si, rather than Ge. (These Si implant conditions do not amorphize the substrate; see the section on implant damage for further discussion.) Figure 2 shows results from SIMS analyses of the as-implanted 30 keV P implant and the profiles after annealing for the Ge conditions specified in the figure. For the case where there is no Ge implanted, the P profile coincides with the profile shown in Fig. 2 for a 50 keV Ge implant condition.

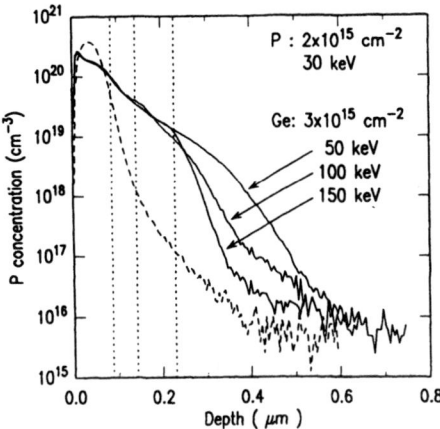

Figure 2: P profiles after anneal for postamorphizing conditions. Vertical dotted lines show the progressively deeper amorphized depths for 50, 100 and 150 keV Ge implants, respectively (see Fig. 4).

Examining the results in Fig. 1 for the 10 keV P implants, we find the junctions of the 10 keV P implants are surpisingly shallower than the junction depths of 30 keV P implants under identical conditions of 100 keV Ge implants. As mentioned before, there is no difference in final junctions between 10 keV and 30 keV P implants when there is no Ge present. When Ge is implanted into the silicon in addition to P, the final junction depths after anneal are evidently very sensitive to the P implant conditions.

Preamorphizing implants

Figure 3 summarizes the results for conditions where the substrate is amorphized by Ge implantation first, followed by P implantation. Only the P implant condition of 10 keV, $2 \times 10^{15} \text{cm}^{-2}$ was studied. All samples received the same anneal as the postamorphized samples described in the preceding section. Qualitatively, Fig. 3 shows that for a given Ge implant energy, increasing the implanted Ge dose reduces P diffusion. Correspondingly, for a given Ge implant dose, increasing the Ge implant energy reduces P diffusion. Not shown in Fig. 3 is the junction depth after anneal for a sample first implanted with Si at 190 keV to a dose of $3 \times 10^{15} \text{cm}^{-2}$ followed by a P implant at 30 keV to a dose of $2 \times 10^{15} \text{cm}^{-2}$. These implant conditions are identical to those for the Si implant shown in Fig. 1 except for the order of the P and Si implantations. Within experimental error, the junctions as measured by spreading resistance are the same between the two cases; however, SIMS profiling shows the higher concentration regions of the P tail to be slightly deeper (~ 20nm) for the case in which Si implantation preceded P implantation.

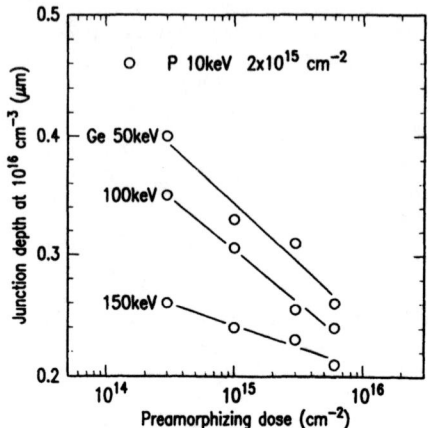

Figure 3: Effects of preamorphization on P diffusion.

Comparison between pre- and postamorphizing conditions

Important differences exist between Figs. 1 and 3. For example, based on information from Fig. 1 alone, one might conclude that 50 keV Ge implants are not effective in reducing P diffusion; however, Fig. 3 shows P diffusion can be reduced dramatically by 50 keV Ge implants when the Ge implant dose is raised above 3×10^{14} cm^{-2}. Comparing results in Figs. 1 and 3 for the equivalent P implant conditions of 10 keV, 2×10^{15} cm^{-2}, we find that for the same implanted Ge dose, postamorphization consistently produces shallower junctions than preamorphization. This is an unexpected result. The as-implanted P profiles are shallower in the case of preamorphizations because of the reduction of channeling. If the effects of Ge implants on P diffusion are assumed to be the same regardless of the order of P and Ge implantations, then the preamorphized samples would be expected to give either the same or shallower junction depths after annealing than the postamorphized samples; our results are contrary to the expectations of this assumption. Pfiester and Griffin [4] reported in their study that changing the order of B and Ge implantation did not affect results for reduction of oxidation-enhanced diffusion. Our experiments show that the order of dopant and Ge implantation can be very important. It may be significant that the B implants used by Pfiester and Griffin were nonamorphizing (100 keV, 1×10^{13} cm^{-2}) while our P implants amorphize the substrate.

We have also performed TEM examinations and made RBS measurements to investigate differences in the crystalline state that result from pre- and postamorphization conditions. Discussion of TEM results is deferred to the section on implant damage. Figure 4 shows the amorphized depths for samples that received 30 keV P implants followed by Ge implants. (P implanted to a dose of 2×10^{15} cm^{-2} at energies of 10 and 30 keV produces amorphized depths of 41 and 70 nm respectively.) Also shown are the data of Ozturk et al. [14] from measurements made on samples implanted with Ge only. The data for 50 keV Ge implants agree very well between the two studies. With

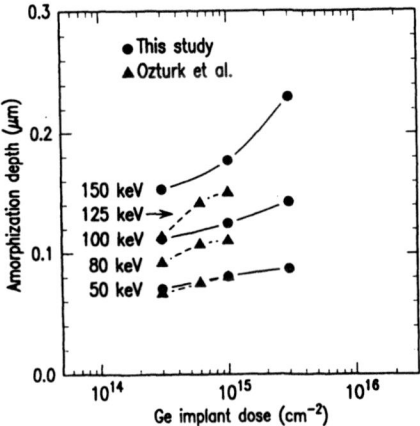

Figure 4: Amorphized depths. Samples were postamorphized with Ge, following P implantation at 30 keV to a dose of $2 \times 10^{15}\,\mathrm{cm^{-2}}$. Also shown are the data of Ozturk *et. al* [14] for samples implanted with Ge only.

the exception of the data point for a 100 keV, $3 \times 10^{14}\,\mathrm{cm^{-2}}$ Ge implant, the amorphized depths measured by Ozturk *et al.* lie between the curves generated in this study, as expected; however the amount of increase in amorphization depths with increased Ge implant dose seems anomalous, especially for the 150 keV data. Preliminary results for samples preamorphized with Ge, or amorphized with Ge only, show amorphization depths shallower than those shown in Fig. 4. The latter conditions also result in a relatively weak dependence of amorphization depth on implant dose. This difference between pre- and postamorphization conditions is still under investigation.

EFFECTS OF Ge IMPLANTATION ON TRANSIENT-ENHANCED DIFFUSION OF B IMPLANTS

The effectiveness of Ge implantation in reducing transient-enhanced diffusion of implanted B was also investigated. This part of our study is more limited in scope than the examination of P diffusion.

Fig. 5 shows the effect of Ge preamorphization on a high concentration B implant. The implant conditions are those specified in the figure. The same 900°C ramped anneal used in the P diffusion study was performed on these samples. The as-implanted profiles for cases with and without Ge implants show that the Ge preamorphization is quite effective in eliminating B channeling. After annealing, the preamorphized sample shows a slightly shallower junction than the sample that received no Ge implantation. Simulations with SUPREM III show that the shallower junction results only because the initial B profile is shallower in the preamorphized sample, not because of any reduction in transient-enhanced diffusion. We have also examined the effect of preamorphizing with 50 keV Ge implanted to doses of $3 \times 10^{14}\,\mathrm{cm^{-2}}$ or $1 \times 10^{15}\,\mathrm{cm^{-2}}$. The $1 \times 10^{15}\,\mathrm{cm^{-2}}$ dose results in a slight increase in enhanced diffusion. We have also found that first recrystallizing the substrate with a 30 min. anneal at 600°C in Ar does not eliminate transient-enhanced diffusion. Furthermore, there is no difference in final junction depths when samples are annealed in an O_2 or Ar ambience.

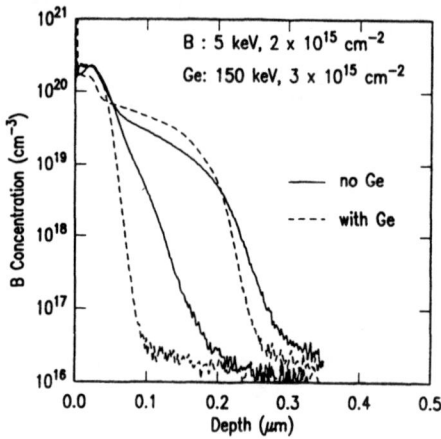

Figure 5: Effect of Ge preamorphization on diffusion of a B implant.

DAMAGE AS THE UNDERLYING CAUSE OF DIFFUSION EFFECTS

The basic premise of this work is that residual damage resulting from Ge implantation is the cause of the reduction in high-concentration P diffusion that has been reported by Pfiester et al. [4,5]. Before summarizing the evidence from the present study in favor of this view, it is first necessary to discuss what the nature of the damage might be and by what mechanism does it lead to reduced diffusion.

The generation of excess interstitials is responsible for the rapid diffusion of high concentration P and also for oxidation-enhanced diffusion. There are two ways in which the concentration of excess self-interstitials can be suppressed by residual implantation damage: (1) recombination of interstitials with vacancy-type defects, and (2) absorption of interstitials at interstitial-type defects. Ion implantation is expected to produce both interstitial-type and vacancy-type defects. From studies of samples amorphized by Si implantation (utilizing the technique of triple-crystal x-ray diffraction), Servidori et al. [15] found strong evidence that interstitial dislocation loops below the original amorphous/crystalline interface are extremely efficient sinks for excess interstitials. By first forming the loops by self-amorphization and annealing, then implanting P into the single-crystal Si, they were able to reduce the transient-enhanced diffusion of P and to show that this reduction is due to the absorption of interstitials at the dislocation loops. Much less information about residual vacancy-type defects and their effects on diffusion have been reported; in part this may be due to the fact that these defects do not image readily in TEM. In a separate work from the one cited previously, Servidori et al. [16] showed that there is a vacancy-rich layer near the surface in samples amorphized with Si and subsequently recrystallized. They also showed dopant diffusion to be affected by the presence of this layer. More recently, Uedono et al. [17] detected vacancy-type defects from B implantation using a variable-energy positron beam technique, but no relation to dopant diffusion was made.

Evidence from the P diffusion study

The most important inference that can be drawn from the results in Fig. 1 is that Ge itself is not responsible for the reduction in P diffusion: large doses of Ge can be present in the same implanted layer as P with no effect, and Si can reduce P diffusion in the absence of Ge. Another important point worth making is that implanting Si at 190 keV to a dose of $3 \times 10^{15} \mathrm{cm}^{-2}$ does *not* amorphize the silicon. RBS analysis shows that the 190 keV Si implantation does not change the initial amorphous layer thickness of 70 nm that results from the 30 keV P implants. In addition, high channeling yields from the RBS spectrum indicate that the substrate below the amorphized layer is highly dislocated. These observations support the contention that residual damage from the Si and Ge implants is responsible for reductions in P diffusion. The differences between pre- and postamorphization conditions also support the idea that it is implantation damage, rather than interactions of Ge with interstitials or P atoms, that is responsible for reduced diffusion. The location of the Ge with respect to the P is the same in Fig. 3 as it is in Fig. 1 for the 10 keV P implants (except for the difference in the fraction of the total dose channeled in the postamorphized case), yet the diffusion results are different. It is likely differences in residual damage between pre- and postamorphizations are responsible for the different diffusion behavior. We discuss preliminary TEM results comparing the two amorphization conditions in the following subsection.

It is not clear from the dependence of P diffusion on Ge implant conditions whether the defects responsible are interstitial or vacancy in nature. The trends in these experiments show that higher energies and larger doses lead to more pronounced reduction in P diffusion. It is known that higher energies and larger doses lead to more interstitial-type defects at the original amorphous/crystalline interface [14,18]. On the other hand, it is expected that the amount of vacancy-type defects will increase in the same way. The much reduced diffusion of the 10 keV P implants compared to 30 keV implants in Fig. 1 is an interesting effect. The work of Servidori et al. [16] showed that the effects of implant damage on dopant diffusion depend on the location of the dopant with respect to vacancy-rich or interstitial-rich damage zones. It could be the case that the 10 keV P implants in Fig. 1 lie within a vacancy-rich layer near the surface while the 30 keV P implants lie below it; the projected range for the 10 keV implants is ~ 16 nm compared to ~ 36 nm for the 30 keV P implants. Similarly small difference in the location of dopants led to differences in diffusion behavior in the experiment of Servidori et al.

Evidence from TEM studies

TEM examinations show two kinds of interstitial-type defects in both pre- and postamorphized samples. The biggest difference between the two amorphization conditions is the proportion of each type of defect in the two cases. In postamorphized samples, most of the defects are complete (prismatic) dislocation loops. These loops grow during anneal and are able to react with one another. Cross sections show these defects grow towards the surface and can intersect it, leaving half loops visible. In the preamorphized samples, most of the defects are stacking fault (Frank) loops. Cross sections show these loops to be confined to layers of a few hundred nm at the original amorphous/crystalline interface. Comparison with SIMS profiles shows a tendency of P to pile up at either side of this band of defects. When comparing different implant conditions (i.e., energy and dose) for a given amorphization method, there appear to be large differences in the density and extent of the defects; however, the images from TEM do not have a simple correlation with the diffusion effects observed. This work is still in progress.

Explanation of B diffusion results

Our data demonstrate a distinct difference in the ability of Ge preamorphization to reduce transient-enhanced diffusion of B compared to high-concentration P diffusion. Little effectiveness in reducing B junctions was found for the Ge preamorphization conditions of 150 keV, 3×10^{15} cm^{-2} (Fig. 5). These same Ge implant conditions appear to be quite effective in reducing diffusion of P implanted at energies of 10 or 30 keV to a dose of 2×10^{15} cm^{-2}. (Profiles and junction depths for 10 keV implants are shown in Figs. 2 and 3.)   The important question is why transient-enhanced diffusion of B may be reduced less than high concentration P diffusion, even though both phenomena are believed to be caused by the presence of excess self-interstitials. All studies indicate that annealing of B implants releases a large amount of interstitials in the first few seconds of diffusion. On the other hand, for P diffusion, interstitials are generated during the entire time of diffusion. It may simply be the case that interstitials from the B implants are generated before the internal sinks from Ge implants have time to form. In addition, it should be recognized that for a given efficiency of interstitial sinks in the bulk, a large enough concentration of excess self-interstitials will overwhelm these sinks. In view of the short duration of excess interstitials, the large enhancements of B diffusion are evidence of very large excess interstitial concentrations. Implantation-induced defects may not be able to suppress these large concentrations.

CONCLUSION

The experimental results presented in this paper provide sound evidence that residual damage from Ge implants can substantially reduce excess self-interstitial concentrations. The fact that these reductions can also be brought about by Si implantation further confirms the picture of damage acting as a sink for excess interstitials and demonstrates there is nothing unique about the chemical identity of Ge in this regard. These findings should go a long way to allaying concerns that previous studies using Ge as a tracer for Si self-diffusion are invalid because of some unexpectedly strong interaction between Ge and native point defects. Our results also show that some of the conclusions drawn from the small number of experiments of Pfiester et al. [4,5], are not true in general.

For example, that the order of dopant and Ge implantation does not affect dopant diffusion. Other conclusions in their studies may not be generally valid because the effects of modification of surface point-defect generation by Ge vs. gettering of point defects by bulk damage were not unamibiguously separated. In our study, which focused on the influence of bulk damage on excess interstitials, we have demonstrated that effects on P diffusion by Ge amorphization depend on the order of Ge and P implants as well as the implant parameters (i.e., energy and dose) for a given amorphization condition. Our experiments demonstrate that implant-induced damage alone, without any modification of point-defect surface reactions by Ge, can greatly reduce excess interstitial concentrations. Creation of the damage responsible for these effects is a complicated function of the implant conditions.

ACKNOWLEDGMENT

I would like to thank the following people for contributing to this study: M. Dimeo and S. Pollack for implantations, D. Sieloff and G. Scilla for SIMS measurements, G. Coleman for RBS measurements, S. Mader and C. Stanis for TEM examinations.

References

[1] L. Pauling in The Nature of the Chemical Bond, Third Edition (Cornell University, Ithaca, New York, 1960), p. 247.

[2] G. L. McVay and A. R. DuCharme, J. Appl. Phys, 44, 1409 (1973).

[3] P. Fahey, S. S. Iyer, and G. J. Scilla, Appl. Phys. Lett., 54, 843 (1989).

[4] J. R. Pfiester and P. B. Griffin, Appl. Phys. Lett., 52, 471 (1988).

[5] J. R. Pfiester, M. E. Law, and R. W. Dutton, IEEE Electron Dev. Lett., 9, 343 (1988).

[6] O. W. Holland, C. W. White, and D. Fathy, Appl. Phys. Lett., 51, 520 (1987).

[7] D. Fathy, O. W. Holland, and C. W. White, Appl. Phys. Lett., 51, 1337 (1987).

[8] F. K. LeGoues, R. Rosenberg, and B. S. Meyerson, Appl. Phys. Lett., 54, 644 (1989).

[9] F. K. LeGoues, R. Rosenberg, and B. S. Meyerson, Appl. Phys. Lett., 54, 751 (1989).

[10] P. Fahey, R. W. Dutton, and S. M. Hu, Appl. Phys. Lett., 44, 777 (1984).

[11] G. Masetti, D. Nobili, and S. Solmi, in Semiconductor Silicon 1977, edited by H. R. Huff and E. Sirtl (Electrochem. Soc., Princeton, 1977), p. 648.

[12] D. Nobili, A. Armigliato, M. Finetti, and S. Solmi, J. Appl. Phys, 53, 1484 (1982).

[13] M. Finetti, G. Masetti, P. Negrini, and S. Solmi, IEE Proc., 127, 37 (1980).

[14] M. C. Ozturk, J. J. Wortman, C. M. Osburn, A. Ajmera, G. A. Rozgonyi, E. Frey, W. K. Chu, and C. Lee, IEEE Trans. Electron. Dev., 35, 660 (1988).

[15] M. Servidori, S. Solmi, P. Zaumseil, U. Winter, and M. Anderle, J. Appl. Phys. 65, 98 (1989).

[16] M. Servidori, R. Angelucci, F. Cembali, P. Negrini. S. Solmi, P. Zaumseil, and U. Winter, J. Appl. Phys., 61, 1834 (1987).

[17] A. Uedono, S. Tanigawa, J. Sugiura, and M. Ogasawara, Appl. Phys. Lett., 53, 25 (1988).

[18] A. C. Ajmera and G. A. Rozgonyi, Appl. Phys. Lett., 49, 1269 (1986).

THE ION IMPLANTED ARSENIC TAIL IN SILICON

S.E. BECK*, R.J. JACCODINE*, AND C. CLARK**
* Sherman Fairchild Center #161, Lehigh University, Bethlehem, PA 18015
** AT&T Bell Laboratories, 555 Union Blvd., Allentown, PA 18103

ABSTRACT

Rapid thermal annealed tail regions of shallow junction arsenic implants into silicon have been investigated. Tail profiles have been produced by an anodic oxidation and stripping technique after implantation to fluences of 10^{14} to 10^{16} cm^{-2} and by implanting through a layer of silicon dioxide. Electrical activation and diffusion have been achieved by rapid thermal annealing in the temperature range of 800 to 1100 °C. Electrically active defects remain after annealing. Spreading resistance and deep level transient spectroscopy results are presented. The diffusion of the arsenic tail is discussed and compared with currently accepted models.

INTRODUCTION

Arsenic has been used extensively as an n-type dopant in silicon. Many empirical models and mechanisms have been proposed for arsenic diffusion in silicon including clustering [1-6], precipitation [7], concentration enhanced diffusion [4,5], and diffusion by ionized point defects [8]. As the lateral dimension of VLSI device structures shrink, it becomes necessary to reduce the junction depth of the device as well. Therefore, a better understanding of the physical nature and annealing characteristics of the implanted tail region becomes essential. We define the implant tail as the region beyond the projected range plus the straggle.

A previous study [9] examined the furnace annealing characteristics of high dose arsenic implants and the associated tail region. The diffusion profile for the tail was found to be significantly different from the total. Concentration dependent diffusion was proposed as the mechanism for the enhanced profile motion. Results from anneals above 1050 °C supported a declustering mechanism and diffusion assisted by a doubly charged vacancy as proposed by Seidel, et al. [10].

The purpose of the present work is to further investigate the implanted arsenic tail in silicon after rapid thermal annealing. Electrically active concentration profiles of arsenic in silicon from bare silicon and through oxide implants are examined. Currently accepted models for arsenic diffusion in silicon are compared with these results.

EXPERIMENTAL DETAILS

Arsenic implanted tails were prepared by two different methods. Both methods were designed to eliminate the majority of the implant damage to a depth of the projected range plus the straggle. Doses of 10^{14}, 10^{15}, and 10^{16} cm^{-2} were implanted into bare <100> p-type silicon. The 10^{14} and 10^{15} cm^{-2} implants were performed at 50 keV while the 10^{16} implant was performed at 100 keV. The high concentration portion of the profile for the above samples was removed by first anodizing the sample and then stripping in 10% hydrofluoric acid. After successive strips the tail region of the implantation was all that remained.

Samples were also produced by implanting through a surface oxide. An 80 nm thermal oxide was grown on <100> n-type silicon samples. Arsenic was implanted into these samples at fluences of 10^{15} and 5 X 10^{15} cm^{-2} at 140 keV. The oxide thickness and implant energy were selected so that the peak of the implant distribution would be within the oxide layer and the silicon dioxide - silicon interface would be at the projected range plus the straggle. This was done to keep most of the implant damage within the oxide. After implantation the thermal oxide was removed by a 10% hydrofluoric acid etch.

Mat. Res. Soc. Symp. Proc. Vol. 147. ©1989 Materials Research Society

All samples were coated with a layer of spin-on-glass to prevent the evaporation of arsenic from the surface. Anneals were performed in a Heatpulse 210-M rapid thermal annealer for 10 to 300 seconds in the temperature range of 800 to 1100 °C. A nitrogen ambient was used for all anneals.

RESULTS AND DISCUSSION

Implants into bare silicon

The anneal temperature plays a significant role in the diffusion and annealing characteristics of the ion implanted samples. Spreading resistance measurements provided the electrically active concentration profiles of the implanted and annealed samples. At 800 °C the tails diffused deeper into the original substrate than the total profiles. Fig. 1a shows typical profiles for the 10^{16} cm^{-2} implants after a 10 second anneal at 800 °C. For anneals at and above 950 °C the total profile diffuses deeper into the silicon than the tail. This behavior is shown in Fig. 1b for the 10^{16} cm^{-2} implants after a 10 second anneal at 1080 °C. The behavior shown in Fig. 1 was also typical of the high dose furnace annealed samples [9] and of the 10^{15} cm^{-2} samples after rapid thermal annealing. These data suggest that before the total profile can diffuse the implant damage must anneal, whereas the tail diffuses in a manner consistent with concentration dependent diffusion.

The temperature effect was also studied by deep level capacitance transient spectroscopy (DLTS). Mesa structure diodes with aluminum contacts were fabricated on the 10^{15} and 10^{16} cm^{-2} samples. Fig. 2 shows the spectra for the 10^{15} cm^{-2} total profiles after 800 and 1080 °C anneals for 10 seconds. A third defect appears in the spectrum after the 1080 °C anneal. The tail profile created from the 10^{15} cm^{-2} implant exhibited the same defect characteristics as the total profile.

Table I lists the energy levels and defect concentrations derived from the spectra for the 10^{15} cm^{-2} tail and total profile samples. Hautojarvi et al. [14] report a large number of vacancies throughout the arsenic distribution and beyond after implantation. As can be seen, the defects that appear at 1080 °C are arsenic and vacancy related. The defects present after the 800 °C anneal are not identifiable, but are probably related to vacancies since there are many vacancies present after implantation. These results are consistent with a declustering mechanism with vacancies being released above 1050 °C as suggested by Seidel et al. [10].

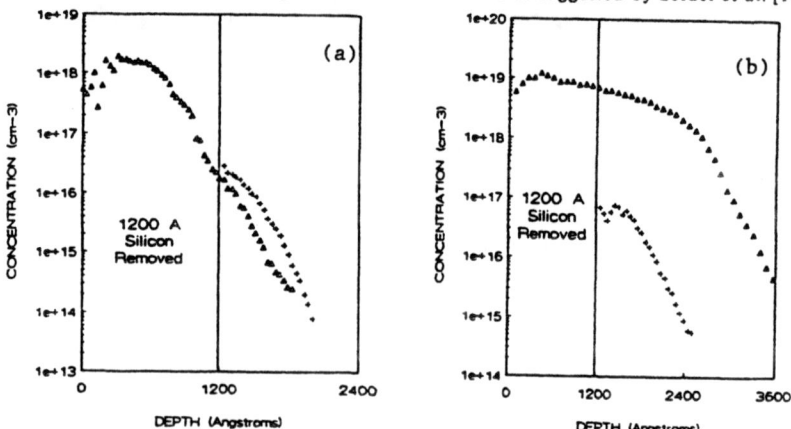

Fig. 1: Typical tail (+) and total (△) profiles for the 10^{16} cm^{-2} implants. These profiles are for 10 second anneals at (a) 800 °C and (b) 1080 °C. The tail precedes the total profile in (a) and not in (b).

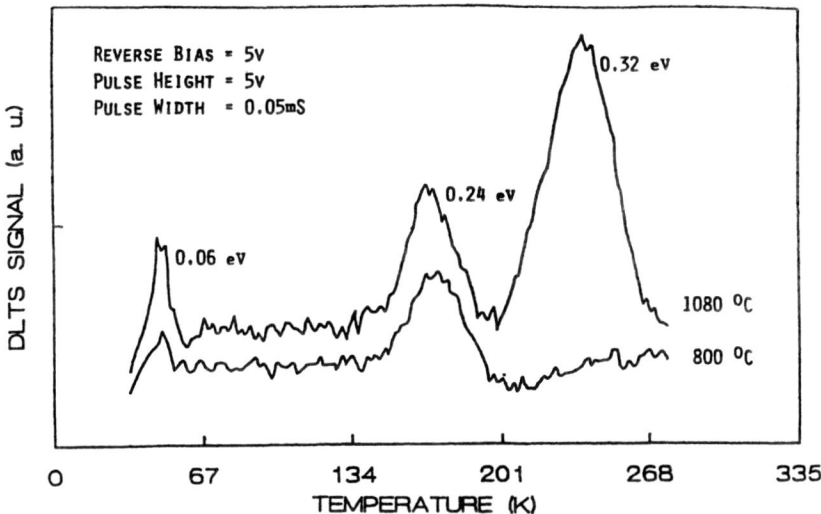

Fig. 2: Typical DLTS spectra for 800 and 1080 °C, 10 second annealed samples. Results from the 10^{15} cm^{-2} total profile are shown. A third defect appears after the 1080 °C anneal.

TABLE I. Results from DLTS study of rapid thermal annealed 10^{15} cm^{-2} implanted samples. Study conditions are defined in Fig. 2.			
SAMPLE	**$E_c - E_t$**	**DEFECT CONCENTRATION**	**COMMENTS**
800/1E15 TAIL	0.06 eV	1.4E12 cm-3	Unknown
	0.28 eV	1.1E12 cm-3	Unknown
TOTAL	0.04 eV	5.2E11 cm-3	Unknown
	0.29 eV	1.0E12 cm-3	Unknown
1080/1E15 TAIL	0.08 eV	1.7E12 cm-3	Unknown (Harris [11])
	0.32 eV	1.1E12 cm-3	As related (Troxell [12])
	0.35 eV	1.8E12 cm-3	Vacancy complex (Song et al. [13])
TOTAL	0.06 eV	1.0E12 cm-3	Unknown
	0.24 eV	1.2E12 cm-3	Doubly Charged V-V(Troxell [12])
	0.32 eV	2.3E12 cm-3	As related (Troxell [12])

The 10^{16} cm^{-2} tail and total profile samples were studied in the same manner. As above, a peak appeared at the high temperature end of the spectra after the 1080 °C anneal. These defects are arsenic or vacancy related.

Electrical activation is dependent upon anneal temperature, anneal time, and implant dose. For the 10^{15} cm^{-2} and above implants electrical activation and diffusion increases with annealing temperature, as shown in Fig. 3. A study of electrical activation and diffusion during 950 °C anneals for times between 10 and 300 seconds was conducted. Results from the 300 second anneal of the total profiles for all doses are presented in Fig. 4. This figure shows that the electrical activation is both dependent upon anneal time and implant dose. When the implant dose exceeds the reported solid solubility limit, as in the case of the 10^{16} cm^{-2} implant [15], diffusion and electrical activation are concentration dependent with the electrically active profile being driven into the substrate by the high peak concentration.

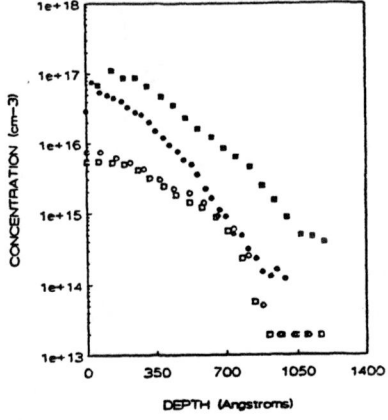

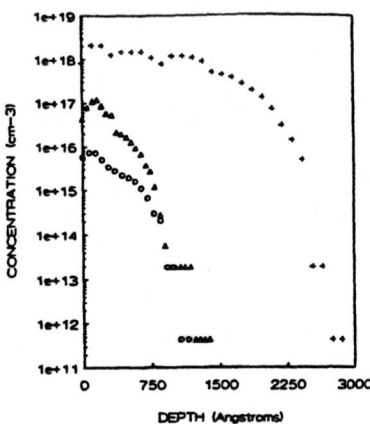

Fig. 3: Spreading resistance profiles for the 10^{14} cm^{-2} (open) and the 10^{15} cm^{-2} (filled) samples annealed for 10 seconds at 800 $^{\circ}$C (circles) and 1080 $^{\circ}$C (boxes). Diffusion is not observed for the 10^{14} cm^{-2} samples.

Fig. 4: Spreading resistance profiles for samples annealed for 300 seconds at 950 $^{\circ}$C. (o) 10^{14} cm^{-2}, (△) 10^{15} cm^{-2}, and (+) 10^{16} cm^{-2}. The 10^{16} cm^{-2} implant exceeds the solid solubility limit for arsenic in silicon.

Very little diffusion was observed for the 10^{14} cm^{-2} implant. There are two possible reasons for the small amount of electrical activation and diffusion. First, this low dose implant does not exceed the reported solid solubility limit [15] for the temperature range studied and therefore may diffuse with a diffusivity that is independent of concentration. Alternatively, since the dose is below that needed to form a continuous amorphous layer, it is possible that solid phase epitaxial growth does not occur. Solid phase expitaxy could account for the difference in profiles above the 10^{15} cm^{-2} concentration.

Through oxide implants

Spreading resistance measurements were also performed on silicon samples implanted through an 80 nm silicon dioxide layer and rapid thermal annealed. Results for 10 second anneals are shown in Fig. 5. For the 10^{15} cm^{-2} samples a decrease in the electrical activation is observed for anneals above 800 $^{\circ}$C. In the case of the 5 X 10^{15} cm^{-2} implant, the electrical activation is about the same for the 800 and 950 $^{\circ}$C anneals and increases at 1080 $^{\circ}$C.

Recoil implanted oxygen is known to exist in the silicon substrate after implantation [16]. Oxygen concentrations greater than 10^{20} cm^{-3} have been calculated in the substrate near the interface, decaying exponentially with depth. The data in Fig. 5 suggest that this recoil implanted oxygen plays a role in the annealing characteristics for the above implants. It has been suggested [17] that the arsenic complexes with the recoiled oxygen and becomes electrically inactive.

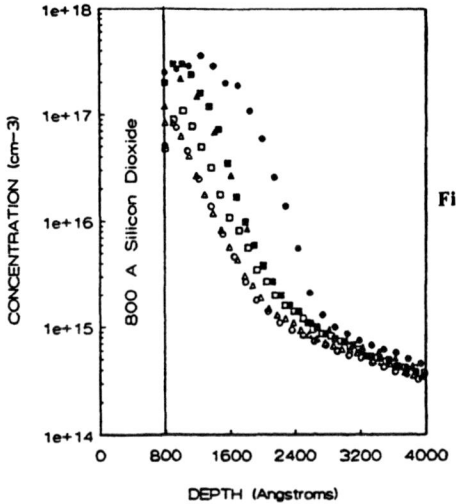

Fig. 5: Spreading resistance profiles for the through oxide implanted and annealed samples: 10^{15} cm^{-2} (open) and 5×10^{15} cm^{-2} (filled). Samples were annealed at 800 (□), 950 (▲), and 1080 $^{\circ}$C (o) for 10 seconds. The 5×10^{15} cm^{-2} implant exceeds the solid solubility limit for arsenic in silicon.

The results for the two different implantation doses differ. Implantating at 5×10^{15} cm^{-2} produces a layer above the reported solid solubility limit [15]. Therefore excess arsenic drives the diffusion, while clustering and solid phase epitaxy also occur. For the low dose case solid solubility is not exceeded and the recoil implanted oxygen plays a greater role.

CONCLUSIONS

A complete model of ion implanted arsenic diffusion in silicon must incorporate many effects. These effects include damage annealing, clustering, point defect interactions, and concentration dependence. Our results suggest that clustering or precipitation in the damaged surface region slow the movement of the total profile up to 900 $^{\circ}$C. The diffusion of the tail and 950 $^{\circ}$C and above total profile suggest concentration dependent diffusion. Our results also support a declustering mechanism for anneals above 1080 $^{\circ}$C. When arsenic is implanted through an oxide the recoil implanted oxygen tends to complex with the arsenic, effecting its diffusion and electrical activation.

REFERENCES

1. S.M. Hu, in Atomic Diffusion in Semiconductors, edited by D. Shaw (Plenum, London, 1973), Chap. 5.
2. R.O. Schwenker, E.S. Pan, and R.F. Lever, J. Appl. Phys. 42, 3195 (1971).
3. R.B. Fair and G.R. Weber, J. Appl. Phys. 44, 273 (1973).
4. R.B. Fair and J.C.C. Tsai, J. Electrochem. Soc. 122, 1689 (1975).
5. M.Y. Tsai, F.F. Morehead, J.E.E. Baglin, and A.E. Michel, J., Appl. Phys. 51, 3230 (1980).
6. E. Guerrero, H. Potzl, R. Tielert, M. Grasserbauer, and G. Stingeder, J. Electrochem. Soc. 129, 1826 (1982).
7. D. Nobili, A. Carabelas, G. Celotti, and S. Solmi, J. Electrochem. Soc. 130, 922 (1983).
8. R. Angelucci, G. Celotti, D. Nobili, and S. Solmi, J. Electrochem. Soc. 132, 2726 (1985).
9. S.E. Beck, R.J. Jaccodine, A.J. Filo, and R. Irwin in Defects in Electronic Materials, edited by M. Stavola, S.J. Pearton, and G. Davies (Mater. Res. Soc. Proc. 104, Pittsburgh, PA 1988) pp. 219-222.

78

10. T.E. Seidel, C.S. Pai, D.J. Lischner, D.M. Maher, R.V. Knoell, J.S. Williams, B.R. Penumalli, and D.C. Jacobson in Energy-Beam Solid Interactions and Transient Thermal Processing, edited by D.K. Biegelsen et al. (Mater. Res. Soc. Proc. 35, Pittsburgh, PA 1985)
11. R.D. Harris, PhD thesis, Lehigh University, 1981.
12. J.R. Troxell, PhD thesis, Lehigh University, 1979.
13. L.W. Song, B.W. Benson, and G.D. Watkins, Phys. Rev. B 33, 1452 (1986).
14. P. Hautojarvi, P. Huttunsen, J. Makinen, E. Punkka, and A. Vehanen in Defects in Electronic Materials, edited by M. Stavola, S.J. Pearton, and G. Davies (Mater. Res. Soc. Proc. 104, Pittsburgh, PA 1988) pp 104-110.
15. A. Lietoila, J.F. Gibbons, and T.W. Sigmon, Appl. Phys. Lett. 36, 765 (1980).
16. L.A. Christel, J.F. Gibbons, and S. Mylroie, Nucl. Instrum. Methods 182/183, 187 (1981).
17. S. Alexandrova and D.R. Young, J. Appl. Phys. 54, 174 (1983).

EFFECT OF RECOIL IMPLANTATION OF OXYGEN
ON BORON ENHANCED DIFFUSION IN SILICON

D.Fan and R.J.Jaccodine
Sherman Fairchild Center for Solid State Studies, Lehigh University,
Bethlehem, PA 18015

In device fabrication, dopants are frequently implanted into silicon through silicon dioxide masks. A consequence of this technique is the co-implantation of recoiled oxygen into the substrate. This study investigates the effect of recoiled oxygen on the widely observed transient enhanced boron diffusion. Comparison of the spreading resistance profiles of annealed through-oxide and directly implanted samples reveals that transient enhanced diffusion of boron can be suppressed by the former process. Continued annealing of the through-oxide implanted silicon recovers the enhanced diffusion of boron. This behavior is believed to be due to precipitation of recoiled oxygen. The mechanisms leading to the above observations are discussed and transmission electron microscopy support presented.

INTRODUCTION

Annealing boron-implanted silicon results in transient enhanced boron diffusion in the tail region. This diffusion tail creates difficulties in shallow junction fabrication for scaled-down devices. Progress has been made toward understanding the mechanism leading to the anomalous diffusion effect. Though still controversial, it is generally agreed that boron diffusion is enhanced by the interaction with excess silicon interstitials released from the implantation damage during annealing[1,2].

The study of transient enhanced diffusion of boron has been limited to the case of direct implantation of boron. Standard process steps however frequently involve implantation of dopants into silicon through a surface layer of oxide. This inevitably results in a large amount of knock-on oxygen incorporated in the silicon substrate along with the expected implantation-induced damage[3]. These oxygen atoms could greatly effect boron enhanced diffusion. On one hand, precipitation of the supersaturated oxygen would emit extra silicon interstitials and further enhance boron diffusion. On the other hand, interaction between silicon interstitials and oxygen, as was suggested for the trapping of carbon interstitials at interstitial oxygen sites[4], could suppress the enhanced diffusion of boron. In this paper, the possible effects will be investigated by comparing the annealing characteristics between silicon directly and through-oxide implanted with boron. It will be shown that oxygen does play an import role in effecting the diffusion of ion-implanted boron.

EXPERIMENTAL

N-type <100> oriented, 20 ohm-cm, FZ silicon was used in this study. Half of the wafers were thermally oxidized to grow 870A surface oxides. All the wafers were then implanted with boron at 25keV to $3 \times 10^{15} cm^{-2}$. The implant creates a dopant distribution which peaks at a depth of about 870A in both the through-oxide implanted and directly implanted silicon. For comparison purposes, the oxidized wafers were stripped and the non-oxidized wafers etched down, by anodically oxidizing and oxide stripping, to a depth of 870A. All the processed wafers were capped with spin-on-glass before annealing.

The anneals were carried out using both furnace and rapid thermal annealers at 800 to 1000°C for various times. Diffusion profiles were determined from spreading resistance measurements. The estimation of the as-implanted tail profile was also obtained by measuring the spreading resistance profile after 800°C annealing of a 1500A surface etched silicon(non-oxidized). The nearly damage-free tail region of the implant would result in little movement of the tail, as will be seen in Fig,1. Excellent agreement between secondary ion mass spectrometry and spreading resistance profiles has been demonstrated previously[5]. Defect structures of selected specimens after annealing of the implanted region were studied by plan-view transmission electron microscopy (TEM).

RESULT AND DISCUSSION

Fig.1 shows the spreading resistance profiles for the 1500A surface etched silicon after 800°C anneals for various times. Profiles were plotted down to the junctions. As was similarly demonstrated in a previous paper[5], the 1500A surface etching removes the majority of the implantation damage, resulting in negligible boron diffusion during annealing. The nearly invariant profiles/junctions therefore correspond to the initial profile of the implant.

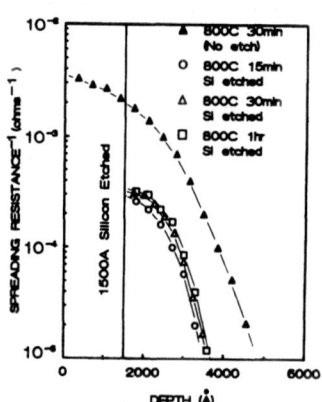

Fig.1. Spreading resistance profiles of a 1500A etched boron-implanted sample annealed at 800°C for various times. Results for an annealed, unetched sample are shown for comparison. Transient-enhanced diffusion is eliminated by etching away the implan-tation-damaged region.

Fig.2a shows the spreading resistance profiles for the 870A surface etched silicon after 800, 900, and 1000°C anneals. The annealing times were chosen approximately corresponding to the enhanced diffusion transients reported in the literatures[6,7]. Apparently, 870A surface etching does not remove all the implantation damage, permitting enhanced diffusion in the remaining silicon substrate during subsequent annealing. The arrow indicates the initial junction obtained from Fig.1. The corresponding diffusion profiles for the annealed through-oxide implanted silicon are shown in Fig.2b. It is clearly seen that the

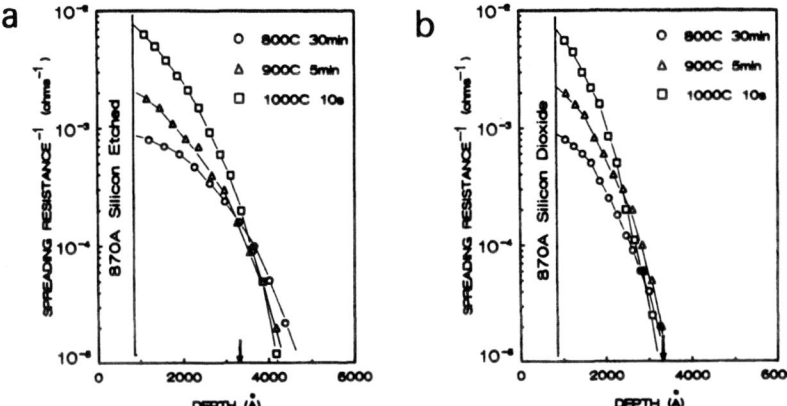

Fig.2. Spreading resistance profiles of samples a)directly implanted with boron and etched to 870A, and b)implanted with boron through 870A oxide, after annealing at 800, 900, and 1000°C for times approximately corresponding to the boron enhanced diffusion transient. Implantation-damage enhanced diffusion is clearly suppressed in the through-oxide implanted silicon.

widely observed transient enhanced diffusion of ion-implanted boron is markedly reduced in the case of implantation through oxide. The result also indicates that not only annealing of implantation damage but also possible oxygen precipitation have been ineffective in enhancing boron diffusion.

Fig.3 compares the plan-view TEM micrographs of the directly and through-oxide implanted silicon samples rapid thermal annealed at 1000°C for 10 seconds. The direct implanted samples, Fig.3a, shows scattered elongated dislocations/dipoles which have been previously associated with the annealing of rod-like defects during and after transient-enhanced diffusion of boron[8]. Fig.3b, however, shows that annealing of through-oxide implanted silicon result in a high

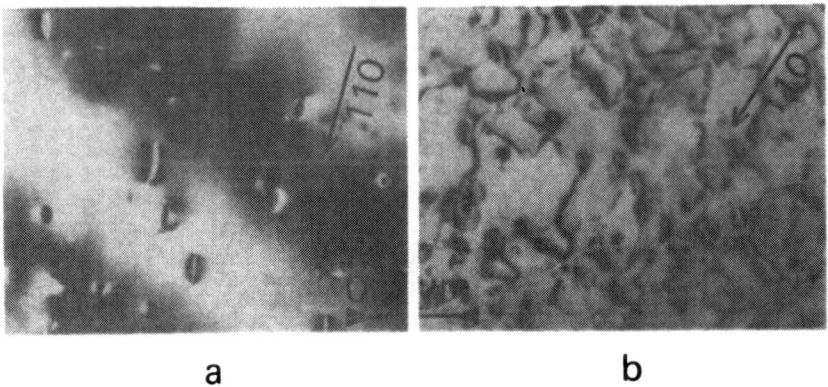

a   b

Fig.3. Plan-view transmission electron micrographs of samples a)directly implanted and etched, and b)through-oxide implanted, after rapid thermal annealing at 1000°C for 10 seconds. Dislocation density is much higher in the through-oxide implanted silicon.

density of irregularly shaped dislocation loops and stacking faults. These defects have been identified as "interstitial" type. We believe the highly dense dislocations are a result of rapid condensation of excess silicon interstitials released during annealing of implantation damage. This condensation of silicon intertstitials is probably initiated due to the relaxation of localized stress, induced by either oxygen or oxygen clusters. By correlating the TEM observation with the spreading resistance diffusion profiles, it is proposed that the suppression of boron enhanced diffusion in the through-oxide implanted silicon is due to the formation of the high density dislocation loops, limiting the interactions between boron and silicon interstitials. Additional studies currently underway will clarify the early stage mechanisms leading to the formation of the highly dense dislocation loops.

The fact that precipitation of oxygen, which normally would enhance boron diffusion by emitting silicon interstitials, seems ineffective during annealing of the through-oxide implanted silicon is of interest. In fact, the following results show that an enhanced diffusion of boron, which we attribute to oxygen precipitation, does occur after an incubation period. Fig.4 compares the 1000°C isothermal annealing profiles for directly and through-oxide implanted silicon. In Fig.4a, the implantation damage-enhanced diffusion of boron occurs within the initial 10 seconds rapid thermal anneal or shorter, after which the diffusion is normally slow. In contrast, in the through-oxide implanted silicon, as seen in Fig.4b, the initial damage-enhanced diffusion is suppressed for a certain period of time after which, pronounced diffusion of boron is again observed. The effective first order diffusivity

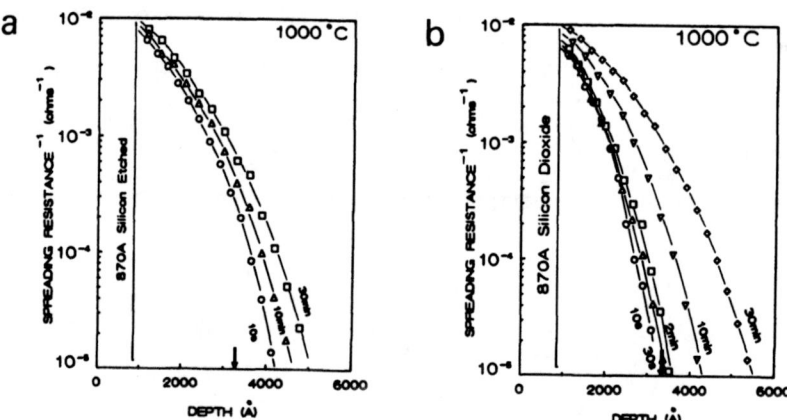

Fig.4. Spreading resistance profiles after 1000°C isothermal anneals of samples a)directly implanted and etched, and b)through-oxide implanted. The diffusion is normally slow in the directly implanted silicon after annealing beyond enhanced diffusion transient (approximated by 10 seconds). Note that in the through-oxide implanted silicon enhanced diffusion of boron is initially suppressed.

for the through-oxide implant, annealed after the incubation period, is ten times greater than that of boron directly implanted, annealed after the enhanced diffusion transient (approximated by 10 seconds). This is shown in Fig.5. The time scale is much greater for the enhanced diffusion afterward in the through-oxide implanted

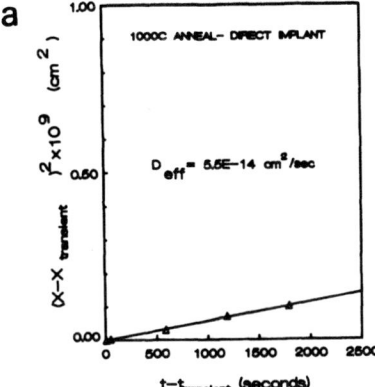 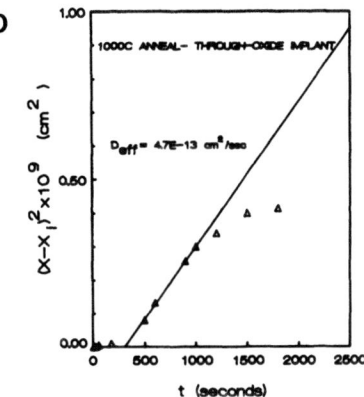

Fig.5. Diffusivities of a)directly implanted and b)through-oxide implanted boron obtained from Fig.4 using first order approximation. The abrupt increase in diffusivity for boron in the through-oxide implanted samples, after 4-5 minutes anneal at 1000°C, is about ten times greater than the normal diffusivity obtained in the directly implanted silicon after annealing beyond the enhanced diffusion transient.

silicon than for the implantation damage-enhanced diffusion in the directly implanted silicon. The diffusion after incubation period is therefore unlikely a result of re-emission of the initially trapped implantation damage-released silicon interstitials. We attribute this enhanced diffusion to the emission of silicon interstitials from oxygen precipitation.

It has been reported that during annealing of oxygen-containing silicon at elevated temperatures, an incubation period for oxygen precipitation nucleation exists[9]. This observation has been explained by the exigent volume effect, in which the nucleation of precipitates is retarded in a homogenized silicon until small-sized dislocations form by accumulation of silicon interstitials[10]. The model is not applicable to the present case since high-density dislocation loops, which certainly provide heterogeneous nucleation sites for oxygen precipitation, already exist shortly after the annealing begins. We believe that, in the through-oxide implanted silicon, the ineffectiveness of oxygen precipitation in enhancing boron diffusion is mainly caused by the trapping effects of the high density dislocations. For instance, silicon interstitials emitted from oxygen precipitates, that are nucleated at the dislocations, can be trapped by the dislocations. A more plausible explanation is the trapping of oxygen atoms at the dislocation sites where the nucleation of oxygen precipitates is even suppressed. The onset of the oxygen precipitation-enhanced diffusion would then imply saturation of trapped species. Detailed mechanism can not be provided at the present time; however, our most recent study[11] indicates that the process leading to the onset of the oxygen precipitation-enhanced diffusion of boron possesses an activation energy of about 2.4eV, implying that diffusion of oxygen is involved. This seems to support the above postulations for enhanced diffusion incubation, where migration of recoiled oxygen to the dislocation is a necessity.

84

SUMMARY

We have shown that implantation damage-enhanced boron diffusion can be suppressed when implanting boron through an oxide. Oxygen precipitation--enhanced boron diffusion, having much higher diffusivity, exhibits a long incubation period. The incubation for the oxygen precipitation-enhanced diffusion in the through-oxide implanted silicon is attributed to dislocation trapping effects. The onset of the precipitation-enhanced diffusion probably indicates saturation of dislocation trapping capabilities in a complex manner.

REFERENCES

1. J.Huang,D.Fan,and R.J.Jaccodine, J.Appl.Phys.63,5521(1988).
2. T.O.Sedgwick,A.E.Michel,V.R.Deline,and S.A.Cohen, J.Appl.Phys.63, 1452(1988).
3. L.A.Christel,J.F.Gibbons, and S.Mylroie, Nucl.Inst.and Meth.182/183, 187(1981).
4. M.T.Asom,J.L.Benton,R.Sauer,and L.C.Kimerling, Appl.Phys.Lett.51, 256(1987).
5. D.Fan,J.Huang,and R.J.Jaccodine, Appl.Phys.Lett.50,1745(1987).
6. A.E.Michel,W.Rausch,A Ronsheim,and R.H.Kastl, Appl.Phys.Lett.50, 416(1986).
7. R.Angelucci,P.Negrini,and S.Solmi, Appl.Phys.Lett.49,1468(1986).
8. J.Huang, Ph.D. Dissertation, Lehigh Un iversity, 1986.
9. N.Inuoe,K.Wada,and J.Osaka, Semiconductor Silicon 1981, Eds. H.R.Huff, R.J.Kriegler, and Y.Takeishi(Electrochem. Soc., Pennington, NJ, 1981), p.282.
10. T.Y.Tan, Materials Research Society Symposium Proceedings, Vol.59, p269(1986).
11. D.Fan and R.J.Jaccodine, to be published.

TEMPERATURE DEPENDENCE OF DAMAGE IN BORON-IMPLANTED
SILICON

G. OTTAVIANI", F. NAVA", R. TONINI", S. FRABBONI", G.F. CEROFOLINI"" AND P.
CANTONI"""
" Physics Department, via Campi 213/A, 41100 Modena, Italy,
"" Enichem, via Medici del Vascello 26, 20138 Milano, Italy
"""INFN, via Irnerio 46, 40126 Bologna, Italy.

ABSTRACT

We have performed a systematic investigation of boron implantation at 30
keV into <100> n-type silicon in the 77 -300 K temperature range and mostly at
9×10^{15} cm^{-2} fluence. The analyses have been performed with ion channeling and
cross sectional transmission electron microscopy both in as-implanted samples
and in samples annealed in vacuum furnace at 500 °C and 850 °C for 30 min. We
confirm the impossibility of amorphization at room temperature and the
presence of residual damage mainly located at the boron projected range. On
the contrary, a continuous amorphous layer can be obtained for implants at 77
K and 193 K; the thickness of the implanted layer is increased by lowering the
temperature, at the same time the amorphous-crystalline interface becomes
sharper. Sheet resistance measurements performed after isochronal annealing
shows an apparent reverse annealing of the dopant only in the sample implanted
at 273 K. The striking differences between light and heavy ions observed at
room temperature implantation disappears at 77 K and full recovery with no
residual damage of the amorphous layer is observed.

INTRODUCTION

Boron is the only used p-type dopant in silicon device technology and it
is generally introduced in the crystal by ion implantation. In current Very
Large Scale Integration devices, where, less than 5000 A shallow junctions are
needed, the implanting energies are below 50 keV and for practical reasons the
sample is held at room temperature. Experimental investigations show that the
amorphization threshold [1] lies, at room temperature, around 10^{16} cm^{-2}, the
maximum concentration being well above the boron solid solubility limit [2].
Moreover in this energy range significant channeling tails [3] are present.
The electrical activation of the boron atoms, together with the recovery
of the defects occurs at temperatures as high as 800 °C, several hundred degrees higher than the temperature needed to reach the same results after
heavier atom (arsenic or phosphorus) implantation.
Reverse annealing [4-6] of the electrically active boron atoms has been
observed in silicon samples implanted at room temperature. Various mechanisms
have been proposed to explain this phenomenon; some of them assume the formation of boron-silicon compounds, others of boron-defect (vacancy or interstitial silicon) complexes.
The presence of an amorphous layer seems to be mandatory in order to
achieve a full electrical activation with a low temperature process. The amorphous layer can be obtained by using a pre-amorphization technique either with
heavy atoms as Si, Ge, Sn etc. or by using molecular ions as for instance
BF_2, BF, BF_3 etc. A possibility to obtain amorphous layer directly with light
ions is to use low temperature implantation.
The purpose of this paper is to characterize silicon layers implanted
with 30 keV boron ions at 77 K and to compare defect recovery and electrical
activation behaviour with those of samples implanted at 193 K and 273 K.

EXPERIMENTAL

Czochralski-grown, <100> oriented, n-type, 10-15 ohm cm resistivity silicon crystals were used for the experiments. The ^{11}B was implanted at 30 keV with a fluence of 9×10^{15} cm^{-2}. All the implants were performed with a current density of 10^{-7} A/cm^2; and the targets were tilted 8 degrees off the normal incidence. Samples implanted at liquid nitrogen, dry ice and ice temperatures are referred to as 77 K, 193 K and 273 K, respectively, though a modest heating (not higher than 5 °C) can occur due to the combined effect of poor thermal coupling and of the impinging power. After implantation the samples were furnace annealed for 30 min in the 400 - 1000 °C temperature range.

The electrical activity of the annealed samples were measured by four point probe resistivity measurements. 1 MeV ^{4}He$^+$ backscattering in channeling conditions [7], with the detector at 130 degrees out of normal, was used to follow the amorphization-crystallization phenomena. The corresponding microstructure as a function of the depth were examined using cross section transmission electron microscopy (TEM) with a Philips EM400T microscope operating at 100 kV. ^{11}B concentration was checked with ^{11}B(p,^{4}He)^{8}Be nuclear reaction [7].

RESULTS

Fig. 1 shows the sheet resistance values measured on samples implanted with 9×10^{15} cm^{-2} fluence and after 30 min isochronal annealing at various temperatures. While the 77 K implanted sample reaches the lowest sheet resistance value after 500 °C and remains almost constant, the 273 K sample has a peak around 600 °C. It is worth noting that the 77 K samples reach a sheet resistance lower than the others.

Fig. 2 shows the backscattering spectra in <100> channeling condition obtained from samples implanted at three different temperatures at 9×10^{15} cm^{-2} fluence. An amorphous layer having a thickness of 1700 A at 77 K and 1400 A at 193 K is clearly identified. At the contrary the implantation at 273 K shows only a peak localized at the boron projected range. The back edge of the amorphous peak of the sample implanted at 77 K, is much sharper than the back edge obtained after 193 K implant.

Furnace annealing at 500 °C for 30 min produces a complete recovery of the damage in the 77 K implanted sample and has almost no effect in the 273 K sample. The sample implanted at 193 K shows, in the region of the original amorphous peak back edge a well defined peak due to residual disorder.

Treatment at higher temperature, 850 °C 30 min, produces defect coalescence and increases , in the samples implanted at 193 K and 273 K, the dechanneling efficiency. The high temperature annealing has no effect on the 77 K implanted sample.

TEM observations on cross-sectioned samples of the three as-implanted specimens and the respective low- and high-temperature furnace treatments were performed. In Figs. 3a,b,c, three micrographs taken on the as-implanted samples are reported.

The 273 K as-implanted specimen (Fig. 3a) exhibits the presence of a 1000 A thick damaged region centred around a depth of about 1000 A, whereas the 193 K and the 77 K as-implanted specimens show amorphous layers which extend from the surface to a depth of 1250 A and 1500 A respectively (Figs. 3b and c).

According to ion channeling measurements, the roughness of the amorphous-crystalline interface is quite different in the two amorphised samples. The 193 K implanted specimen shows a rough and wavy interface followed by a 450 A thick tail of clusters of point defects which, according to their depth position, can be identified as clusters of interstitial type (Fig. 3b).

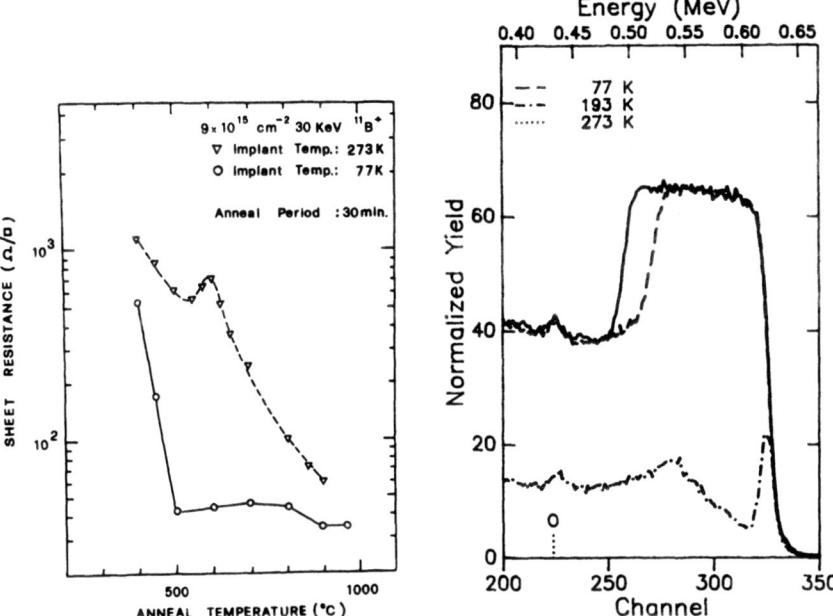

Fig. 1 Sheet resistance measurements performed on samples implanted at 77 K and 273 K. Isochronal heath treatment for 30 min were performed.

Fig. 2. Backscattering spectra obtained in channeling conditions in samples implanted at 77 K, 193 K and 273 K.

The same band of defects, below a very smooth interface, is detected in the 77K implanted sample, but in this specimen their number density and extension in depth (about 250 A) is smaller than in the previous case (Fig. 3c).

Furnace annealing at 500 °C for 30min does not produce an observable variation of the defective band detected on the 273 K implanted specimen (Fig. 4a).

The same thermal treatment performed on the 193 K implanted sample, accomplishes a complete recrystallization of the amorphous layer, but some dislocations in the regrown crystallized film are detected. These extended defects are created when the advancing and recrystallizing amorphous-crystalline growth front intersects a small crystallite embedded in the amorphous region. The dimensions of the clusters of point defects beyond the regrown layer is apparently increased by this annealing (Fig. 4b). These clusters are also detected in the low temperature annealed sample implanted at 77 K, but in this case the regrown crystallized layer is defect free (Fig. 4c).

In Figs. 5a,b,c, we report the results of the TEM analysis performed on the three high temperature annealed samples. The effect of this thermal treatment on the 273 K implanted specimen is to condense the defects present in the high stressed layer detected in the low temperature annealed sample; large dislocation loops at the deeper side of this band are visible (Fig. 5a). In the 77 K implanted specimen this high temperature annealing is able to induce a complete recovery of the displacement damage (Fig. 5c), whereas the 193 K implanted sample still shows extended defects in and beyond the regrown crystalline film (Fig. 5b) .

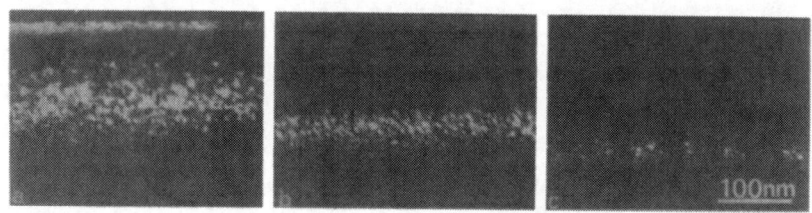

Fig. 3: [110] cross-section weak beam dark field images (g,5g), g=[111] of the three as-implanted specimens; a), b) and c) refer to samples implanted at 273 K, 193 K and 77 K respectively. The arrow marks the surfaces of the three specimen.

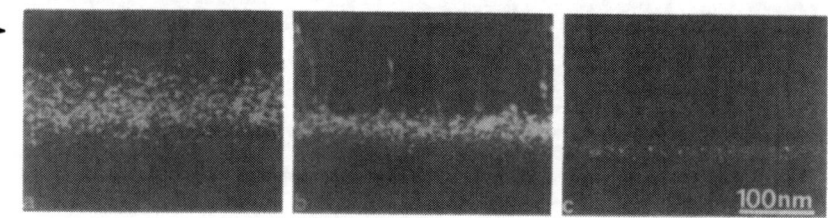

Fig. 4: [110] cross-section weak beam dark field images (g,3g) g=[220] of the same specimens shown in Fig. 3 and annealed at 500 °C for 30 min. The arrow marks the surfaces of the three specimen.

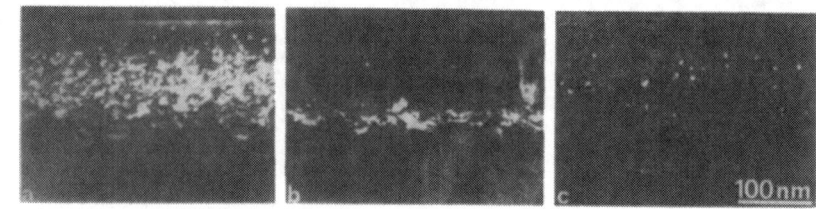

Fig. 5: [110] cross-section weak beam dark field images (g,5g) g=[111] of the same specimens shown in Fig. 3 and annealed at 850 °C for 30 min. The arrow marks the surfaces of the three specimens.

In these amorphized samples starting at a distance of 600 A from the surface, a band of small coherent precipitates of a SiB_x phase is also detected. The precipitation phenomenon is qualitatively explained by the high boron concentration present in the regrown layer which, at this temperature, is well above the value of the solid solubility of this element in silicon.

A few diodes have been fabricated using n-type <100> floating zone material with a 3000 ohm cm resistivity. The implantation was carried out at 77 K at $2x10^{15}$ cm^{-2} fluence. Furnace annealing was performed at 550 °C for 30 min. The back contact was made by an aluminum evaporated film. In reverse bias condition, the current-voltage characteristics present a sharp increase around 90 V where the total depletion of the wafer occurs. At 80 V the reverse current density is around 100 - 200 nA/cm^2, almost the same as observed on diodes obtained with arsenic implantation.

Similar diodes, made with 273 K implantation, present reverse I-V characteristics three order of magnitude higher, even after 800 °C annealing.

DISCUSSION AND CONCLUSIONS

Most of the experiments have been performed using high dose ion implantation to elucidate the role of the temperature on the residual damage. The backscattering spectra as well as the TEM cross sections indicate that an amorphous layer can be obtained only by implanting at low temperatures and moreover the quantity of the defects at the amorphous-crystalline interface strongly depends upon the temperature. Implantation at 193 K produces an amorphous layer but a very poor crystalline-amorphous interface which strongly affect the quality of the regrown layer.

The amorphization phenomena cannot be explained with the nuclear interactions alone, and the implanting temperature plays a fundamental role. It allows diffusion and recombination of the defects (8) and also a more effective quenching of the thermal spike (9).

The quality of the amorphous layer strongly affects the electrical properties of the layer. The two curves in Fig. 1 show a completely different behaviour for the two samples implanted at 273 and 77 K. The fact that the 77 K sample completely recovers after 500 °C 30 min treatment indicates that the "reverse annealing" of boron is a result of the interactions of the dopant and the defects created during ion implantation and not an intrinsic property of the B-Si system. The 77 K implanted sample shows also a slight increase in sheet resistance at temperatures around 600 - 700 °C, which is probably due to the small concentration of the defects present. This point however needs to be clarified. Further annealing, above 800 °C, produces a decrease in the resistance and it is probably due to diffusion of boron in single crystal silicon. At low temperatures, when boron redistribution does not occour, the value of the sheet resistance is consistent with an active dopant concentration of the order of $3x10^{20}$ cm^{-3}, a concentration well above the solid solubility limit. Since the atomic boron concentration is around 10^{21} cm^{-3}, at high temperatures precipitates are present, most likely due to SiB_x complexes. Similar experiments [10] performed at room temperature by implanting boron in silicon preamorphized samples show an increase in sheet resistance after annealing at temperatures higher than 800 °C. The behaviour observed in our samples is quite different, but similar to laser-annealed samples. The high value for the solid solubility is due to the non-equilibrium processes occurring during the amorphous-crystalline transformation. The low temperatures involved, of the order of 500 °C, do not allow the whole distribution and precipitation of boron. Similar results have been obtained [10] by room temperature boron implantation in preamorphized silicon.

In conclusion we have shown that low temperature implantation can produce amorphous layers at doses lower than the ones necessary at higher temperatures and this affects the crystalline quality of the regrown layer.

Moreover the amorphous crystalline interface is sharper at lower temperatures. Full recovery is obtained in the samples implanted at 77K after 500 °C temperature annealing; much higher temperatures are needed after 273 K temperature implantation. At 273 K a band of defects is always present.

References

1. F.F. Morehead, Jr. and B.L. Crowder, Radiat. Eff._6_, 27 (1970).

2. A. Armigliato, D. Nobili, P. Ostoja, M. Servidori and S. Solmi, in Semiconductor Silicon 1977 (The Electrochem. Soc. Inc.,Princeton, NJ 1977) p.638.

3. A.E. Michel, R.H. Kastl, S.R. Mader, B.J. Masters and J.A. Gardner, Appl. Phys. Lett., _44_, 404 (1984).

4. D.E. Davies, Appl. Phys. Lett., _14_, 227 (1969).

5. R.R. Hart and O.J. Marsh, Appl. Phys. Lett., _15_, 206 (1969).

6. J. Huang, D. Fan and R.J. Jaccodine, J. Appl. Phys., _63_, 5521 (1988).

7. W.K. Chu, J.W. Mayer and M.A. Nicolet, Backscattering Spectrometry, (Academic Press, New York, 1978) .

8. J.S. Williams, R.G. Elliman, W.L. Brown and T.E. Seidel, Phys. Rew. Lett., _55_, 1482 (1985).

9. G.F. Cerofolini, L. Meda and C. Volpones, J. Appl. Phys., _63_, 4911 (1988).

10. E. Landi, S. Guimaraes and S. Solmi, Appl. Phys.,_A44_, 135 (1987).

Plasma Immersion Ion Implantation for Impurity Gettering in Silicon

H. Wong, X. Y. Qian, D. Carl,
N. W. Cheung and M. A. Lieberman,
Department of Electrical Engineering and Computer Sciences,
University of California, Berkeley, CA 94720;

I. G. Brown and K.M. Yu,
Lawrence Berkeley Laboratory, Berkeley, CA 94720.

Abstract

We have utilized plasma immersion ion implantation (PIII) to demonstrate effective gettering of metallic impurities in silicon wafers. Metallic impurities such as Ni, Cu or Au were intentionally diffused into Si as marker impurities. The Ar or Ne atoms were ionized in an electron cyclotron resonance (ECR) plasma chamber. The ions were accelerated by a negative voltage applied to the wafer and implanted into the wafer. The as-implanted saturation dose can be as high as $5 \times 10^{16} \mathrm{cm}^{-2}$. After an annealing step at 1000°C for 1 hour in a N_2 ambient, the retained doses and the amount of gettered impurities were measured with Rutherford backscattering spectrometry (RBS). With a retained Ar dose in $10^{15} \mathrm{cm}^{-2}$ range after annealing, the gettered Ni, Cu and Au were $3.0 \times 10^{14} \mathrm{cm}^{-2}$, $3.0 \times 10^{14} \mathrm{cm}^{-2}$ and $4.4 \times 10^{13} \mathrm{cm}^{-2}$ respectively.

INTRODUCTION

A new ion implantation technique for surface modification, the *plasma immersion ion implantation,* (PIII) has recently been demonstrated [1-2]. In this technique, the substrate is exposed to a plasma containing the ions to be implanted. A high negative voltage applied to the substrate holder accelerates the ions towards the substrate. Using an electron cyclotron resonance (ECR) plasma, PIII systems are capable of providing extremely high dose rate for implantation. Since the ion energy is controlled by the applied voltage, very low energy ion implantation or deposition are possible as well [3-4]. In this paper, we demonstrate that PIII can be used to getter metallic impurities in silicon. It is well-known that implantation of various ions, such as Ar and Ne, can form gettering centers in silicon [5]. This backside gettering technique is costly with conventional ion implanters because of the high dose needed. However, the simplicity and high dose rate capability of PIII makes this gettering step an attractive process.

PIII SYSTEM

The schematic of our experimental PIII system is shown in Figure 1, which consists of an electron cyclotron resonance (ECR) plasma source, an implantation chamber, a high voltage power supply, charge and voltage measurement circuits, and diagnostic apparatus.

A. ECR plasma chamber

The plasma chamber was pumped to a base pressure of about 10^{-7} Torr. Working gas was then fed into the chamber through a gas pressure controller. A constant pressure at 10^{-3} Torr range was maintained during operation. The plasma was excited by 2.45 GHz microwave at the ECR condition. A magnetic field at about 800 Gauss for the resonance was generated by two water cooled coils surrounding the vessel. The output of a microwave generator (not shown in the figure) was controlled by adjusting the magnetron current and guided into the ECR chamber through a quartz window. Two bolometers were used to monitor the forward and reflected microwave power from the generator. A Langmuir probe for measuring ion density, an optical emission spectrometer for detecting excited gas species and a standing wave power meter for checking tuning condition were connected to the system. With a microwave power of 700 watts and an Ar gas pressure of 10^{-3} Torr, the ion density inside the ECR chamber was $1-5\times10^{12}$cm^{-3}. In the implantation chamber, an ion density of 10^{11}cm^{-2} was detected.

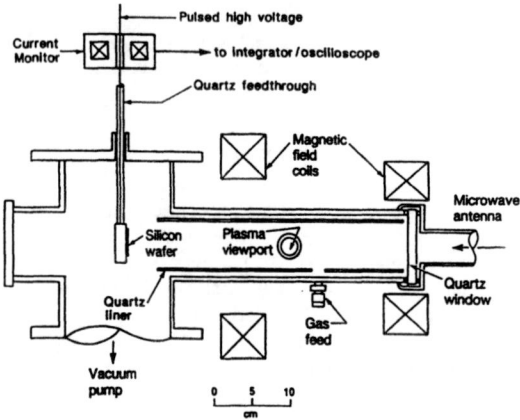

Figure 1.   Schematic of the PIII set-up in this study.

B. Pulsed voltage ion implantation

A pulse generator delivers negative voltage pulses to the wafer holder, which accelerates the positive ions towards the wafer. The negative voltage pulses were created by discharging a storage capacitor through a spark gap. The firing of the spark gap was controlled by a triggering signal. The gap was flushed by dry N_2 for stable performance during operation. Since the voltage on the wafer holder decreases as the charge is gradually neutralized by the implanted ions, the ion energy is not mono-energetic. The average potential that the ions experience is about one half of the peak potential. In addition, charge transfer effect in the plasma during acceleration will further reduce the effective implantation energy [6]. A current monitor placed between the high voltage pulse generator and the implantation chamber senses the current flow through the wafer holder. Its output is fed to an integrator to obtain the charge as a function of time. Shown in Figure 2(a) is the equivalent circuit of the charge measurement circuitry. Figure 2(b) is a typical charge output obtained with a storage capacitance of 500 pF charged to -38 kV at an Ar gas pressure of 10^{-3} Torr. The corresponding wafer voltage shown in Figure 2(c) was measured by an oscilloscope through a 1-to-1000 reduction resistor. The duration of each implant pulse is about 1 µsec. The measured total charge per pulse is about 17µC, which is close to the total charge stored in the capacitor (19µC). The accumulated charge divided by wafer holder area and electron charge will give an estimated dose. Due to secondary electron emission and other losses discussed later in this paper, the actual implanted dose can be much less then the estimated dose. Nevertheless, the estimated dose can be treated as an upper-limit dose, which is useful for dose dependence studies. With pulse mode operation, the time-average ion current is proportional to the pulse rate. The maximum pulse rate of our experimental set-up is 30Hz.

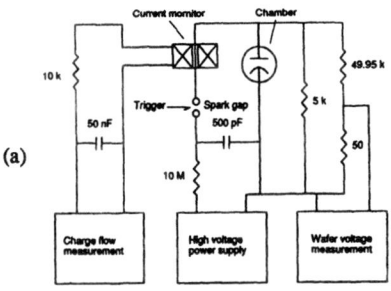

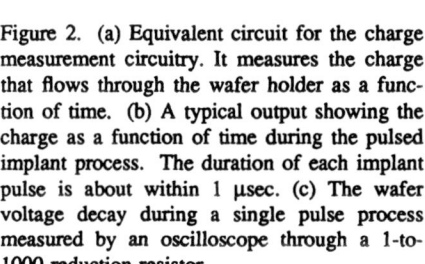

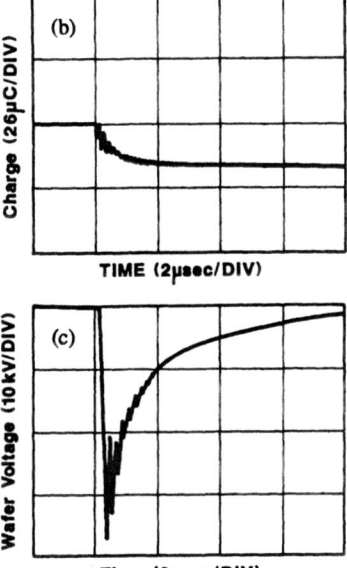

Figure 2. (a) Equivalent circuit for the charge measurement circuitry. It measures the charge that flows through the wafer holder as a function of time. (b) A typical output showing the charge as a function of time during the pulsed implant process. The duration of each implant pulse is about within 1 μsec. (c) The wafer voltage decay during a single pulse process measured by an oscilloscope through a 1-to-1000 reduction resistor.

C. DC voltage ion implantation

Instead of using pulsed voltage, an alternative approach for PIII is to apply a negative DC potential directly to the wafer holder. In this configuration, the charge rate was measured by a current meter. The dose was simply controlled by the implantation time. In this way, extremely high implantation dose can be achieved in a time period of seconds. To prevent excessive wafer heating by high current obtained in DC operation mode, the ion flux has to be reduced by lowering the microwave power to $\approx$ 70 Watts.

GETTERING RESULTS

For the implant gettering experiments, p-type CZ silicon wafers with (100) orientation and resistivity of 15 ohm-cm were used. Ar or Ne was used for the gettering implant, while Ni, Cu or Au were chosen as the marker impurities. The metallic impurity was first evaporated on the backsides of the wafers and then diffused into the wafers at 1000°C for 1 hour. After surface cleaning, the surface of the wafers were subjected to ion implantation with PIII at a negative voltage of 20-40 kV, gas pressure of about 10^{-3} Torr, and a microwave power of 700 watts. The gettering thermal annealing was performed at 1000°C for 1 hour in a N_2 ambient. The implanted dose and the amount of gettered metallic atoms were measured with Rutherford backscattering spectrometry (RBS).

A. Pulsed PIII gettering

Two RBS spectra of Au doped wafers with and without the Ar implantation are shown in Figures 3(a) and 3(b). The gettering effect of Au at the silicon surface due to PIII is evident. No gettered Au could be measured by RBS without the Ar implant, while a layer of Au atoms with areal density of 4.4×10^{13} cm^{-2} was found after an Ar implant with a retained dose of 1.1×10^{15} cm^{-2} after annealing. The Ar to gettered Au ratio was about 25.

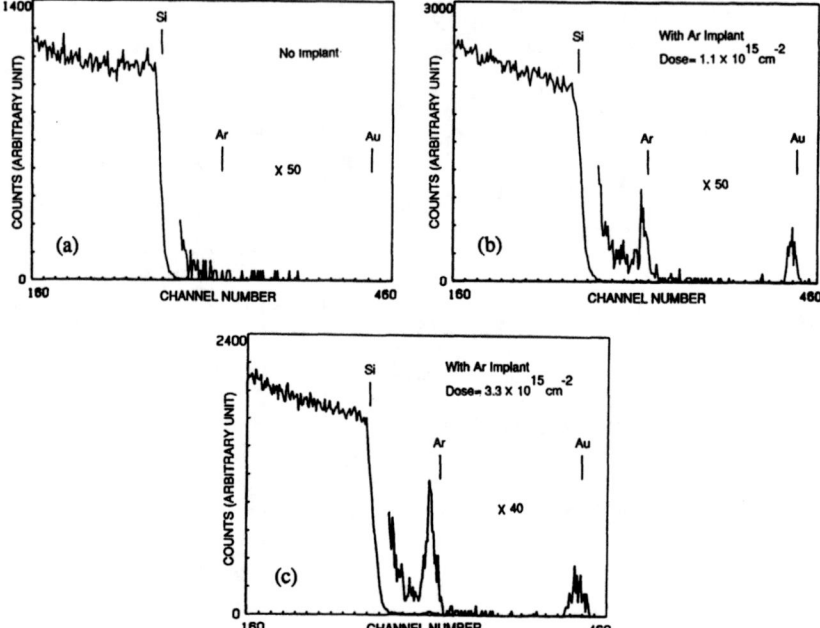

Figure 3.   RBS spectra of Au-doped Si wafers. (a) without implant;   (b) with Ar implant at a dose of $1.1 \times 10^{15} cm^{-2}$ and peak voltage of -38 kV. The amount of gettered Au is $4.4 \times 10^{13} cm^{-2}$. The positions of surface peaks for Ar and Au are marked in the figure. The portion of the spectra containing Ar and Au peaks are magnified 50 times. (c) with Ar implant in DC configuration at a dose of $3.3 \times 10^{15} cm^{-2}$. The applied DC voltage was -40 kV. The amount of gettered Au is $6.2 \times 10^{13} cm^{-2}$. The positions of surface peaks for Ar and Au are marked in the figure. The portion of the spectra containing Ar and Au peaks are magnified 40 times.

In Table 1, we summarize the amount of gettered impurities by PIII of Ar and Ne. No marker metallic impurities were detected by RBS in the non-implanted samples. More efficient gettering was found for Ni and Cu impurities, probably due to the larger diffusivities of these metals in silicon. The Ne doses cannot be measured by RBS due to its light mass. However, the gettering of Au is still obvious by comparing samples with and without implantation. From the RBS spectra, the Ar profile was found to peak at about 200 Å below the silicon surface. It should be mentioned that some control silicon wafers without metallic diffusion were also implanted with Ar PIII and analyzed with RBS. No measurable marker impurities were detected in these control wafers after identical annealing conditions.

B. DC PIII Gettering

Impurity gettering was also achieved with PIII in the DC operation mode. Figure 3(c) shows the RBS spectrum of an Ar implanted sample with the wafer holder kept at -40 kV DC during implantation. After annealing at 1000°C for 1 hour, the retained Ar dose was $3.3 \times 10^{15} cm^{-2}$ and the areal density of gettered Au was $6.2 \times 10^{13} cm^{-2}$.

TABLE I. Amount of gettered impurities with PIII of Ar and Ne. All samples were annealed at 1000°C for 1 hour in N_2 ambient after PIII.

Retained implant dose (cm^{-2})	Amount of gettered impurity (cm^{-2})
Ar 1.1×10^{15} (pulse)	Au 4.4×10^{13}
Ar 1.2×10^{15} (pulse)	Ni 3.0×10^{14}
Ar 2.4×10^{15} (pulse)	Cu 3.0×10^{14}
Ne 4.7×10^{16} * (pulse)	Au 4.5×10^{13}
Ar 3.3×10^{15} (DC)	Au 6.2×10^{13}
No implant	Gettering not detected by RBS

*Estimated from the total number of discharge pulses and the amount of charge per pulse. The retained dose is much less as discussed in text.

DISCUSSIONS

It was found that the as-implanted Ar is only a fraction of the measured charge that flows through the wafer holder. For example, with measured charge of 8×10^{15} unit charge per cm^2 for pulsed mode operation at -20kV, the as-implanted Ar dose is only 1.2×10^{15}cm^{-2}. Similar results have been reported by Tendys, et al [2]. This discrepancy can be due to that secondary electron emission from the negatively biased substrate and the implanted Ar ions may have multiple ionization states. In addition, surface sputtering and Ar out-diffusion will remove part of the implanted Ar from the substrate. We also found that the as-implanted Ar dose shows saturation as the amount of total charge increases. By increasing the wafer holder voltage from -20 kV to -38 kV, an increase of Ar saturation dose was observed. Evidently the deeper implantation of higher energy ions suppressed the out-diffusion effect and reduced the percentage of shallow implants which was removed by sputtering.

We noticed that for most gettering samples the retained doses after thermal annealing at 1000°C for 1 hour is in the low 10^{15} range in spite of very different as-implanted doses. More than 50 percent of the Ar can be retained for samples with 10^{15} as-implanted doses. However, the percentage of retained dose is getting lower when the dose gets higher. A pair of RBS spectra before and after annealing is shown in Figure 4. The as-implanted Ar dose of this sample was 4.5×10^{16}cm^{-2}. After annealing, the retained dose was only 3.3×10^{15}cm^{-2}. By comparing the positions of the Ar peaks, it is clear that the retained Ar profile is in the deeper wing of the as-implanted profile. This suggests the substantial Ar loss is an out-diffusion mechanism during the gettering annealing step. It's interesting to note that the gettered Au profile coincides with as-implanted Ar profile, both peak at the depth of 280Å. This observation suggests that the metallic impurities were trapped by implantation defects instead of the implant species.

With our present experimental set-up, 10^3 pulses will be needed for a dose of 10^{15}cm^{-2} over a 50cm^2 wafer holder area. It will takes several minutes for a 30 Hz pulse generator. Since the pulse duration is only about 1 μsec, the pulse rate can be raised by many orders for high dose rate requirements. In the case of DC, our preliminary data shows that an Ar dose of 4.2×10^{16}cm^{-2} was obtained in 10 seconds. This dose rate is extremely high comparing with conventional ion implanters.

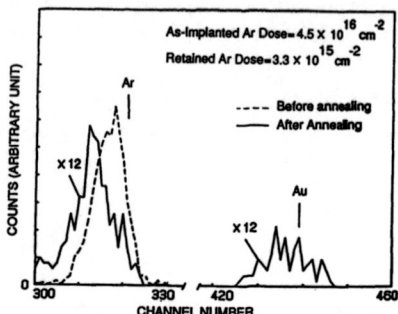

Figure 4.   RBS spectra of an Au-doped Si wafer before and after annealing. The Ar was implanted in DC mode at -40 kV for 1 minute. The thermal annealing was performed at 1000°C for 1 hour in N_2 ambient. The after-annealing Ar and Au yields are magnified 12 times.

CONCLUSIONS

We have demonstrated plasma immersion ion implantation of Ar and Ne in both pulsed and DC modes with an ECR plasma source. The as-implanted saturation dose can be as high as $5 \times 10^{16} cm^{-2}$. This novel implantation technique was successfully applied to getter Au, Ni and Cu in silicon. Considerations for charge collection, surface sputtering, out-diffusion, and the effect of extremely high dose rate implantation are also discussed.

Reference

[1]   J. R. Conrad, J. L. Radtke, R. A. Dodd, F. J. Worzala, and N. C. Tran, *J. Appl. Phys.* 62, 4591, (1987).

[2]   J. Tendys, I. J. Donnelly, M. J. Kenny and J. T. A. Pollock, *Appl. Phys. Lett.* 53 (22), 2143, (1988).

[3]   B. Mizuno, I. Nakayama, N. Aoi, M. Kubota, and T. Komeda, *Appl. Phys. Lett.* 53 (21), 2059, (1988).

[4]   Toshiro Ono, Chiharu Takahashi and Seitaro Matsuo, *J. Appl. Phys.* 23, L534, (1984).

[5]   T. E. Seidel, in "Material Issues in Silicon Integrated Circuit Processing", M. Wittmer, J. Stimmell, and M. Strathman, eds. (Materials Research Society, Pittsburgh, 1986), 3.

Impurity Gettering by Implanted Carbon in Silicon

H. Wong and N.W. Cheung

Dept. of EECS, University of California, Berkeley, CA 94720

K.M. Yu

Lawrence Berkeley Laboratory and University of California, Berkeley, CA 94720

P.K Chu

Charles Evans & Associates, Redwood City, CA 94063

J. Liu

IBM Thomas J. Watson Research Center, Yorktown Heights, NY 10598

ABSTRACT

We have observed strong gold gettering by implanted carbon in silicon. It was found that the gettering agents in carbon implanted layers are point defects associated with singular carbon atoms. The positions of the gettered Au atoms were found to be distorted substitutional sites. A point-defect gettering model is proposed to explain our findings.

INTRODUCTION

Impurity gettering is a technique which removes the metallic impurities from the active device region in integrated circuits (IC) fabrication [1-4]. The yield in IC manufacturing can be strongly affected by the effectiveness of the gettering technique employed. Most of the gettering schemes utilize extended defects as gettering sites which can facilitate the precipitation of metallic impurities. These are the cases in intrinsic gettering with oxygen precipitates and backside gettering with implantation damage [5,6]. Recently, it has been demonstrated that implanted carbon in silicon can form strong gettering centers [7,8]. The gettering behavior of carbon was found to be fundamentally different from extended defects related gettering. In this work, we have investigated the gettering mechanism of Au by implanted carbon in silicon.

EXPERIMENTAL PROCEDURES

N-type (100) Czochralski (CZ) silicon wafers with resistivity of 3-5 ohm-cm were used in this experiment. To study the gettering effects, Au was first doped into the wafers as marker impurity. This was achieved by first evaporating 100Å of Au on the backside of the Si wafers and then diffusing it into the wafers at 900°C for 1 hour in a N_2 ambient. Both 100keV C ions and 2.4 - 2.8MeV C ions were used in this study. The C doses ranged from $5 \times 10^{14} cm^{-2}$ to $2 \times 10^{16} cm^{-2}$. The implanted wafers were subsequently annealed in inert ambient for times from 1 to 12 hours. The annealing temperature was 900 – 1000°C.

A preliminary study was made to compare the gettering effectiveness of the implantation of various ions and their residual damages. C, O, N, or Ar ions of 100keV energy were used and the implant doses were $10^{16} cm^{-2}$ for all samples. To investigate the role of oxygen in carbon related gettering, a set of samples were implanted with various doses of carbon together with $10^{16} cm^{-2}$ oxygen. The energies of C and O ions were both 100keV in this dual implant experiment. Since the projected ranges of 100keV C and O ions in silicon are 0.22 μm and 0.24 μm respectively, the C and O profiles in the dual implants overlapped one another.

We have used secondary-ion mass spectrometry (SIMS), Rutherford backscattering spectrometry (RBS), and cross-section transmission electron microscopy (XTEM) to study the carbon implanted layers and their gettering effects. In the SIMS analysis, a Cs^+ beam was used for the profiling. The concentrations of C, O and Au were simultaneously monitored as a function of depth. In some cases, the Si signal was also monitored to check matrix effects on the uniformity of sputtering rate. For all the samples examined by SIMS, no change of the Si matrix sputtering rate was found in the implanted layer due to the implanted species. RBS was used to measure the amounts of the gettered Au and the residual damages for samples implanted with 100keV ions. RBS channeling analyses were also carried out along the <100>, <110>, and <111> axes to probe the lattice location of the gettered Au atoms.

RESULTS

a. SIMS measurements of the gettering effect

The gettering of Au by implanted carbon in silicon is clearly demonstrated in the SIMS profiles shown in Figures 1(a) and (b). Figure 1(a) shows the profiles of C, Au and O measured from a Au-doped silicon sample implanted with 2.8MeV C for a dose of 10^{16} cm^{-2}. The annealing condition was 930°C for 1 hour. The amount of gettered Au by the C implanted layer is about 2×10^{13} cm^{-2}. Figure 1(b) was measured from a sample implanted with 2.4MeV carbon at a lower dose of 5×10^{14}cm^{-2}. Again, the gettering of Au is evident. For both samples, the profiles of gettered Au coincide with that of the implanted carbon, and the concentrations of the gettered Au and implanted C are approximately proportional to one another, which holds for carbon concentration ranging from 10^{17}cm^{-3} to 5×10^{20}cm^{-3}. This observation strongly suggests that the gettering is related to the presence of implanted carbon atoms. If Au gettering is through residual defects, the gettering sites should occur mostly in a defect band which peaks around the tail part of the implant profile [12,13].

Another interesting observation is that a large amount of oxygen atoms were also trapped in the carbon implanted layer after thermal annealing. The amount of trapped oxygen was much more than the amount of gettered Au. The oxygen were diffused into the carbon implanted region from the CZ Si substrates which had a background oxygen concentration in the high 10^{17}cm^{-3}

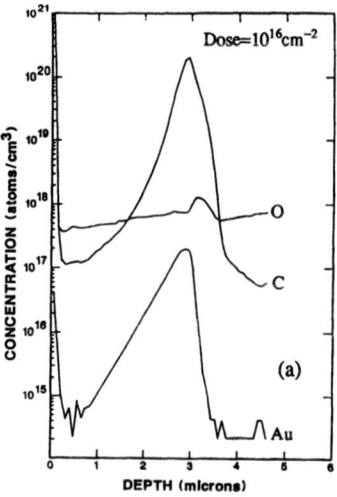
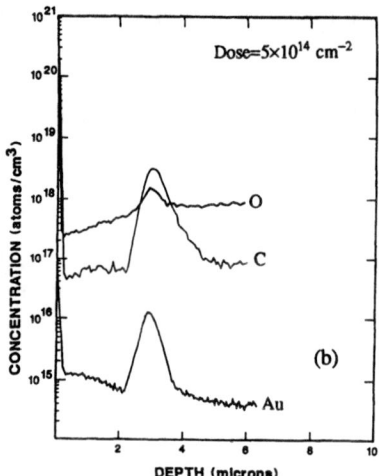

Figure 1. SIMS profiles of C, Au, and O measured on carbon implanted samples after thermal annealing. Au was doped into the samples as marker impurity. (a) 2.8MeV carbon implant at a dose of 10^{16}cm^{-2}. Annealed at 930°C for 1 hour. (b) 2.4MeV carbon implant at a dose of 5×10^{14}cm^{-2}. Annealed at 1000°C for 1 hour.

range. It is well known that carbon can facilitate oxygen precipitation in Si crystals [14-17]. This observation is a direct evidence showing the synergism of carbon and oxygen in intrinsic gettering for CZ Si substrates.

b.   XTEM study of C implanted layers

The samples studied with XTEM were implanted with 2.4MeV C and annealed at 1000°C for 1 hour. For C doses of about 10^{15}cm^{-2}, no features were found throughout the carbon implanted depths. No extended defects could be observed even for doses as high as 10^{16}cm^{-2}. Figure 2 is a bright field XTEM micrograph showing a high dose carbon implanted layer. The implantation was made at 2.4MeV for a dose of 2×10^{16} cm^{-2}. The peak C concentration in this case was more than 5×10^{20}cm^{-3} (1 at%). No extended defects, such as dislocations or precipitates, were found. Only a shady band with a thickness of about 0.5μm could be seen near the projected range of the implant profile. The diffraction patterns from this layer showed no features different from single crystalline silicon. This shady band was probably due to the difference in atomic density between the carbon-rich layer and the silicon matrix. It is well-known that for implant dose higher than 10^{14}cm^{-2}, even light ions such as boron can create an band of dislocations loops after high temperature annealing [12,13]. The anomalous behavior of implanted carbon is under investigation. One tentative explanation is that the implanted carbon atoms sink the excess Si self-interstitials that are responsible for the dislocation growth [14-17], and thereby quench the formation of dislocations.

The absence of any extended defects in the C implanted samples shows that carbon induced gettering is not through dislocations or precipitate interfaces, but through point defects or small C-Si complexes.

Figure 2. XTEM micrograph (bright field) showing a carbon implanted layer after a 1 hour annealing at 1000°C. The implant energy is 2.4MeV and the dose is 2×10^{16} cm^{-2}. About two microns of silicon above the implanted layer was sputtered off during sample preparation.

c.   RBS study of damage and gettering

We have used RBS/channeling to measure the gettering effect of various implanted ion species in Si, and compared the gettering effects with the residual damages. A 2.14MeV ^{4}He beam was used in this study. The samples were implanted to a dose of 10^{16}cm^{-2} at 100keV for C,

Table I. Comparison of the gettering effects of various ion species implanted at 100keV. The annealing temperature is 1000°C for all the samples. The amounts of the gettered Au are measured with RBS.

Ion	Amount of Gettered Au(10^{13} cm^{-2})
Argon	7 (±10%)
Nitrogen	0.4 (±90%)
Oxygen	0.23 (±50%)
Carbon	5 (±10%)
Carbon (FZ)*	13 (±10%)

*   in Float-zone wafer

N, O, or Ar ions, and subsequently annealed at 1000°C for 1 hour. RBS was then used to measure the amount of gettered Au in the implanted layers. Table I summarizes the amounts of gettered Au by the implantation of various implant species. It is clear that the amount of gettered Au by Ar and C implantations are more than one order of magnitude higher than N and O implants. In the case of carbon implant, the amount of gettered Au is about 5×10^{13} (±10%) cm^{-2} in CZ wafers and 1.3×10^{14}cm^{-2} (±10%) in FZ wafers.

The residual damages after the implantation and thermal annealing were studied with the RBS/channeling. The channeling yield from the implanted depth can serve as a measure of the amount of disorder in the crystal. Figure 3 shows the RBS spectra under channeling condition along <100> direction for samples whose gettering effects are listed in Table I. Comparing Table I and Figure 3, we notice that although the crystalline damage due to C implantation is lower than those of N and O, C is more than ten times more effective in gettering Au. On the other hand, Ar implantation gives both strong gettering effect and crystalline damage. This finding indicates that the gettering effect produced by ion implantation is dependent more on ion species than on implantation damage alone.

Figure 3. Backscattering spectra along <100> axial channel for Ar, O, N, and C ion implanted samples, measured with 2.14MeV ^{4}He. The energies and doses for all the ions are 100keV and 10^{16} cm^{-2} respectively.

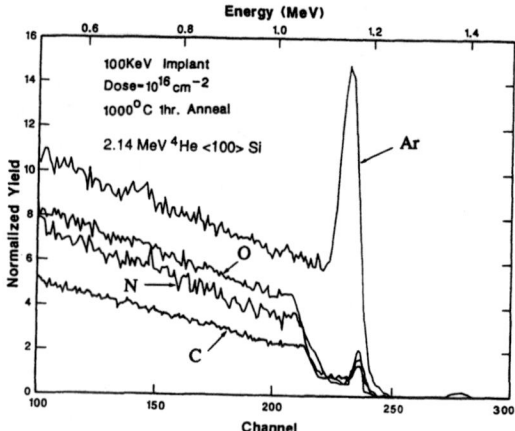

d. Lattice location study of gettered Au atoms

The triangulation method with RBS/channeling has been widely used to determine the lattice location of foreign atoms in crystals [18,19]. We have conducted angular scan measurements along <100>, <110>, and <111> axial channels to probe the location of the gettered Au atoms in carbon implanted Si. The samples used in this study were implanted with 100keV carbon at a dose of $10^{16}cm^{-2}$ and annealed at 1000°C for 12 hours. A 1.5MeV ^{4}He beam was used for this study. Shown in Figure 4 is a channeling spectrum along <100> axis. The amount of Au trapped in the C implanted layer was about $1.3 \times 10^{14}cm^{-2}$, as determined from the random spectrum. Also marked in Figure 4 is the silicon region which corresponds to the depth where Au was gettered.

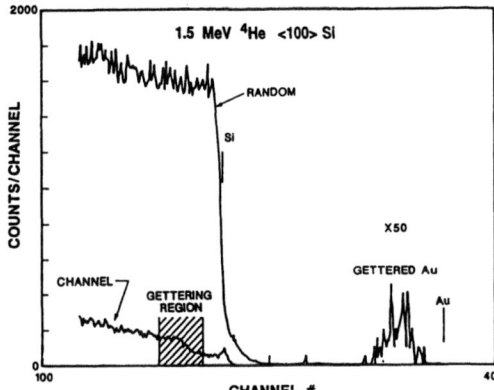

Figure 4. Backscattering spectra along <100> axial channel and random direction with 1.5MeV ^{4}He beam. Only Au signal under channeling condition is shown. The sample was implanted with 100keV carbon at a dose of $10^{16}cm^{-2}$, and subsequently annealed at 1000°C for 12 hours.

Shown in Figure 5(a), (b) and (c) are the angular yield profiles around the three major axes for the Au signal and Si substrate signal. The yields were normalized to the yield at random orientation. Decreases in backscattering yield are observed along all three axial channeling directions. Both angular scans around <100> and <110> directions exhibit yield enhancement off the channel axes due to ion-flux peaking, while <111> direction shows no flux peaking effect. The strongest enhancement occurs at about 0.4° off the <100> channel axis (Figure 5(a)). The peak yield value is nearly 50% above the random yield. The strong flux peaking effect displayed in the angular profiles around <100> and <110> axes indicates that large portion of the gettered Au are preferentially located in the Si lattice. The shape of the angular yield profiles is consistent with the atoms occupying distorted substitutional sites [20,21]. The channeling half-angle, $\psi_{1/2}$, of the Au <100> angular profile is approximately 0.1°. Using Lindhard's continuum potential approximation [22], the displacement from the lattice string is estimated to be 0.6Å. More detailed studies are underway to determine the fraction of Au residing in such positions.

e. Dose Dependence of Au gettering

Figure 6 shows the amount of gettered Au as a function of the carbon implant dose. Both SIMS data for 2.4MeV carbon implanted samples and RBS data for 100keV carbon implanted samples are included. The samples were annealed at 1000°C for in a N_2 ambient. The anneal time was 1 hour, except for the group in dash line which were annealed for 12 hours. Figure 6 clearly shows that the amount of gettered Au is approximately linear with dose in the dose range from $5 \times 10^{14} cm^{-2}$ to $2 \times 10^{16}cm^{-2}$. For the group with annealing time of 12 hours, the C:Au ratio is about 100:1. This linear dose dependence is in accordance with SIMS profiling results, which show the profiles of the gettered Au follow those of the implanted C. Such linear relation suggests

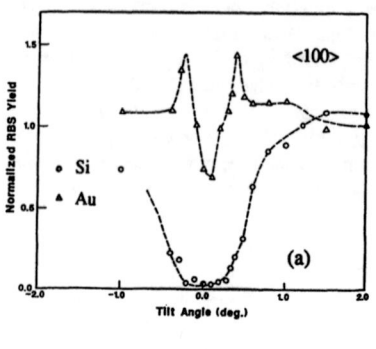

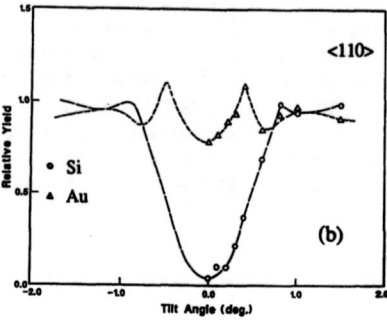

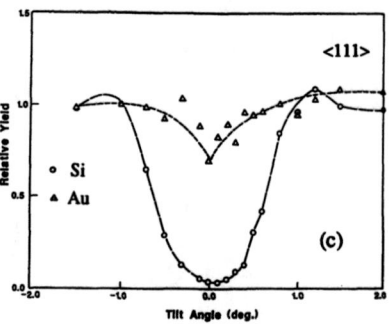

Figure 5. Angular yield profiles around (a) <100>, (b) <110>, and (c) <111> axial channels for Au signal (diamond) and Si substrate signal (circle). The dash lines are drawn only to guide the eyes. A 1.5MeV ^{4}He beam was used.

Figure 6. The amount of gettered Au as a function of carbon implant doses. The samples were annealed at 1000°C in a N_2 ambient. Both RBS data from 100keV implanted sample and SIMS data from MeV implanted samples are included.

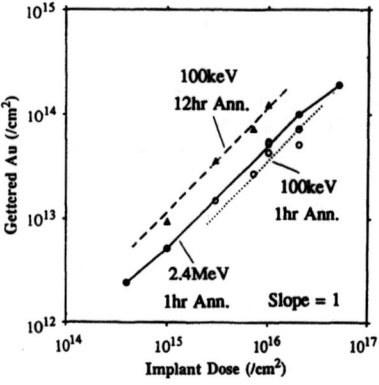

that the gettering agents are related to singular carbon atoms. If a gettering agent is created by two carbon atoms, the amount of gettering should be proportional to $(dose)^2$, since the chance for two carbon atoms to meet is proportional to $(concentration)^2$.

f. Effect of oxygen in carbon induced gettering

As shown by SIMS analysis, a large amount of oxygen atoms were also trapped in carbon implanted layers after annealing. It is therefore important to investigate the role of oxygen in carbon related gettering. Figure 7 shows the amount of gettered Au measured in samples which were implanted with various doses of carbon together with 10^{16} cm^{-2} of oxygen. RBS was used to detect the gettered Au atoms. The best linear fit to the gettering data from single carbon implants is also shown for comparison. As shown in Figure 7, the implanted oxygen atoms, with a peak concentration as high as 8×10^{20} cm^{-3}, actually decrease the gettering effect of the carbon implant by 30%. This $(C + O)$ dual implant experiment demonstrates that that it is the carbon atoms that account for the strong gettering effect of implanted carbon, rather than the oxygen atoms, or oxygen precipitates.

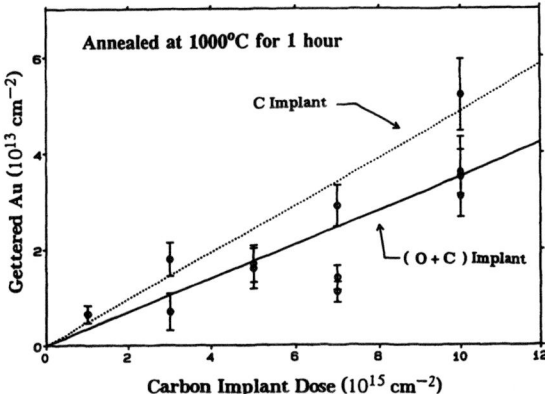

Figure 7. The amount of gettered Au in samples with various carbon implant doses and an additional oxygen dose of 10^{16} cm^{-2}. Both carbon and oxygen implantations were done at 100KeV. The samples are annealed at 1000°C for 1 hour. The solid line is the best linear fit to the dual-implantation data, while the dotted line is the best linear fit to the single carbon implantation results shown in Figure 6.

POINT DEFECT COMPLEX MODEL

The gettering results of carbon presented above can be summarized as follows: 1) The impurity gettering is through point defects or small complexes; 2) The gettered Au atoms sit at heavily distorted substitutional sites; 3) The concentration of gettering sites is linearly proportional to the concentration of carbon atoms; 4) C atoms in silicon also trap oxygen atoms, but the presence of oxygen reduces the gettering efficiency of carbon.

The point defect nature of carbon induced gettering rule out all models involving precipitates, i.e. oxygen or carbon precipitates, as the gettering agent. In fact, it has been demonstrated in an earlier study that the gettering effect of oxygen precipitation layer is much too weak to explain the gettering effect of carbon [7,8].

A point defect gettering model has been proposed earlier [8], in which carbon getters by decreasing silicon self-interstitial concentration. It is known that carbon in silicon can shrink the lattice constant and thereby creates free volumes for carbon/self-interstitial agglomerates (Si_xC) [14-17]. On the other hand, Au in silicon is mostly substitutional but diffuses as an interstitial [23,24]. A decrease in Si self-interstitial concentration can enhance Au concentration through the kick-out mechanism [25]. The linear dependence of gettered Au on carbon implant dose is a natural consequence of this model.

This model assumes that the gettered Au atoms reside on substitutional sites. However, our lattice location experiments show that the gettered Au atoms are away from the lattice sites by as much as 0.6Å. This model also predicts that the Au concentration should be inversely proportional to the concentration of Si self-interstitials. Since Si self-interstitials have much higher diffusivity than carbon atoms in silicon [26], the profiles of the gettered Au should be much broader than the carbon profiles, which again contradicts our experimental findings.

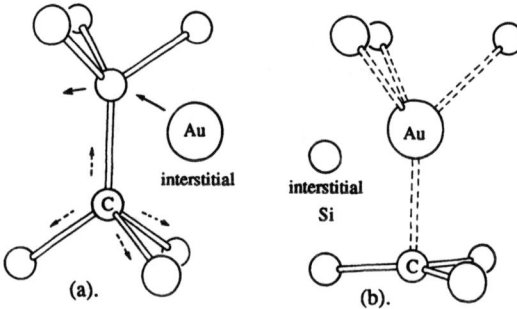

Figure 8. A point-defect model in which substitutional C is the latent gettering agent. The unlabeled atoms are silicon atoms. (a). An interstitial Au atom near a substitutional C before been gettered. (b). A substitutional Si atom next to C is replaced by the interstitial Au. The C atom collapses into the (111) plane perpendicular to the C-Au direction to release the strain. The gettered Au atom is then at a distorted substitutional site.

To explain all the findings on the gettering behavior of carbon, we propose the following model. In this model, both C atoms and the trapped Au atoms sit in adjacent distorted substitutional sites, as illustrated in Figures 8(a) and (b). It is known that carbon in silicon occupy mostly substitutional sites. However, the C-Si bond distance is only 82% of the bond distance of Si-Si bond (as in β–SiC). The lattice is then under strain (Figure 8(a)). When an interstitial Au atom is next to the substitutional C, a Si atom bonded to C is replaced by the Au atom, releasing a Si self-interstitial (Figure 8(b)). Because the interaction between Au and C is very weak, the C atom subsequently collapses into the (111) plane where the other three Si atoms reside. This process releases the strain in lattice and leaves a large distorted substitutional site for the Au atom. The three Si bonds next to the Au may also collapse into (111) plane, giving even more room for the Au atom.

This simple model agrees with all the experimental results shown in this work. Its point-defect nature explains the XTEM results and the channeling results on damage study. The lattice position of the gettered Au is in accordance with the channeling lattice location study. This model proposes that the substitutional carbon atoms in silicon are the latent gettering agents. One supporting evidence is that higher Au concentration was found in carbon-rich regions after Au diffusion in float-zone silicon [27]. Since carbon also trap oxygen, the presence of oxygen will reduce the amount of carbon atoms available for gettering metallic impurities.

CONCLUSIONS

We have studied various aspects of impurity gettering by carbon in silicon. Using SIMS, XTEM, and RBS/channeling, it is concluded that the gettering effect of carbon in silicon is not through extended defects, but through point defects related to singular carbon atoms. The positions of the gettered Au atoms were found to be heavily distorted substitutional sites. A point-defect model is proposed to explain our findings.

REFERENCES

[1] T. E. Seidel, *Proc. Mater. Res. Soc.* **71**, 3 (1986).

[2] A. Ourmazd, *Proc. Mater. Res. Soc.* **59**, 331 (1986).

[3] C. W. Pearce, L. E. Katz and T. E. Seidel, in "Semiconductor Silicon 1981", H. R. Huff, R. J. Kriegler, and Y. Takeishi, eds. (Electrochem. Soc. Pennington, 1981), 705.

[4] H. R. Huff and F. Shimura, *Solid State Tech.* **28** (3), 103 (1985).

[5] D.K. Sadana, *Proc. Mater. Res. Soc.* **36**, 245 (1985).

[6] T. E. Seidel, R. L. Meek, and A. G. Cullis, *J. Appl. Phys.* **46** (2), 600 (1975).

[7] H. Wong, N. W. Cheung, P.K. Chu, *Appl. Phys. Lett.*, **52** (11), 889 (1988).

[8] H. Wong, N. W. Cheung, P. K. Chu, J. Liu and J. W. Mayer, *Appl. Phys. Lett.* **52** (12), 1023 (1988).

[9] J. Lerouille, *Phys. Stat. Sol.* **67** (A), 177 (1981).

[10] J. Kishino, M. Kanamori, N. Yoshihiro, M. Tajima, and T. Tizulea, *J. Appl. Phys.* **50**, 8240 (1979).

[11] U. Gosele, *Proc. Mater. Res. Soc.* **59**, 419 (1986).

[12] D. K. Sadana and J. Washburn, and C. W. Magee, *J. Appl. Phys.* **54** (6), 3479 (1983).

[13] K. S. Jones, Ph.D Dissertation, University of California at Berkeley, (1987).

[14] H. Foll, U. Gosele, and B. O. Kolbesen, *J. Cryst. Growth*, **40**, 90 (1977).

[15] H. Foll, U. Gosele, and B. O. Kolbesen, in "Semiconductor Silicon 1977", H. R. Huff and E. Sirtl, eds., (Electrochem. Soc. Pennington , 1977), 565.

[16] H. Foll, U. Gosele, and B. O. Kolbesen, *J. Cryst. Growth*, **52**, 907 (1981).

[17] J.A. Baker and T.N. Tucker, *J. Appl. Phys.* **39** (9), 4365 (1968).

[18] B. Domeij, G. Fladda, and N.G.E. Johnson, *Radiat. Effects*, **6**, 155 (1970).

[19] J. U. Andersen, O. Andreasen, J.A. Davies, and E. Uggerhoj, *Radiat. Effects*, **7**, 25 (1971).

[20] S.T. Picraux, W.L. Brown, and W.M. Gibson, *Phys. Rev.*, **6**, 1382 (1972).

[21] S.T. Picraux, in "New Uses of Ion Accelerators", J.F. Ziegler, ed. Ch. 4, (Plenum Press New York, 1975).

[22] J. Lindhard, *Dansk. Vid. Selsk., Mat. Fys. Medd.*, **34**, 14 (1965).

[23] W. Frank, U. Gosele, H. Mehrer, and A. Seeger, in "Diffusion in Crystalline Solids," G. E. Murch and A. S. Norwich, eds. (Academic Press, New York, 1984), 64.

[24] W.M. Bullis, *Sol. State Electr.* **9**, 143 (1966).

[25] U.Gosele, W. Frank, A. Seeger, *Appl. Phys.* **23**, 361 (1980).

[26] T.Y. Tan, U. Gosele, *Appl. Phys. A*, **37**, 1 (1985).

[27] M. J. Hill and P. M. Van Iseghem, in "Semiconductor Silicon 1977", H. W. Huff and E. Sirtl, eds. (Electrochem. Soc. Princeton, 1977), 715.

CRYSTAL STABILITY AND MICROSTRUCTURAL EVOLUTION IN POLYCRYSTALLINE SI FILMS DURING ION IRRADIATION

Harry A. Atwater[a] and Walter L. Brown[b]
[a]California Institute of Technology, Pasadena, CA 91125
[b]A.T.&T. Bell Laboratories, Murray Hill, N.J. 07974.

ABSTRACT

Amorphous Si is nucleated heterogeneously at grain boundaries during irradiation of polycrystalline Si by 1.5 MeV Xe$^+$ ions for temperatures of 150-225 °C. Following formation at grain boundaries, the amorphous Si layer grows at a rate comparable to the growth rate of a pre-existing amorphous-crystal interface, resulting in a decrease in average grain size and a marked change in the grain size distribution. The heterogeneous nucleation kinetics of amorphous Si are strongly dependent on grain boundary structure. A simple atomistic model for amorphous phase formation, which suggests that the nucleation kinetics are dependent on the point defect mobilities and grain boundary structure, is related to the experimental results.

INTRODUCTION

Ion beams have been employed to enhance or modify the kinetics of many solid phase processes, such as impurity diffusion, crystallization, impurity segregation, and grain growth. Hence understanding the stability of crystals with respect to phase transformations during non-equilibrium processing such as steady state ion irradiation is of considerable interest. Although many studies have been undertaken to investigate the question of phase stability under irradiation, little progress has been made in relating the structure of defects, such as surfaces and grain boundaries to phase stability.

Grain boundary structure can have a significant impact on kinetics and phase stability in polycrystalline materials. Anomalously high grain boundary mobilities are seen for boundaries with a high density of coincidence sites[1]. Grain boundary structure can also significantly affect phase stability, as evidenced by studies of <011> tilt boundaries in Bi at the melting temperature which suggested that a quasiliquid layer is present for boundary misorientations of greater than 15°[2]. Molecular dynamics simulations indicate that grain boundary disordering may occur below the melting temperature, and that boundary melting is orientation-dependent[3].

High energy ion irradiation of elemental semiconductors provides an excellent experimental system for study of the relation of grain boundary and surface structure to phase stability, since the target is compositionally homogeneous, and the nuclear energy loss is constant with depth in the range where observations of interface motion can be conveniently made. The critical ion flux and temperature conditions for reversal of an existing planar amorphous-crystal interface in Si, from crystallization to amorphization, have been specified[4]. In polycrystalline Si and Ge irradiated at higher temperatures, grain growth occurs, and the grain boundary migration rate is weakly temperature-dependent and proportional to the density of atomic displacements at grain boundaries[5]. Here we report on microstructural evolution and the relation between phase stability and the structure of grain boundaries in polycrystalline Si during ion irradiation, in the regime of ion flux and temperature where interface reversal occurs at a pre-existing amorphous-crystal interface.

Mat. Res. Soc. Symp. Proc. Vol. 147. ©1989 Materials Research Society

Bright Field   Dark Field Amorphous Phase

———————— 100 nm

Figure 1: During irradiation of a columnar polycrystalline Si film with 1.5 MeV Xe+ and a flux of 3.5 x 10^{12}/cm²-sec at 178 °C, amorphization is initiated at high-angle grain boundaries, as seen in the bright field TEM micrograph in (a). In (b), the dark field image was formed with diffracted intensity from the first amorphous ring in Si.

EXPERIMENTAL

Si films 1000 Å-thick were deposited by chemical vapor deposition onto thermally oxidized Si wafers. Columnar polycrystalline films with average grain diameter of approximately 800 nm were produced by annealing at 1000 °C for 5 hours. These samples were irradiated with 1.5 MeV Xe+ with ion fluxes in the range 3.2 x 10^{12} - 4.0 x 10^{12}/cm²-sec. One series of irradiations was conducted in which the substrate temperature varied from 150-261 °C, and ion flux and dose were held constant at 3.5 ± 0.3 x 10^{12}/cm²-sec and 3 x 10^{15}/cm², respectively. A series of isothermal irradiations was performed at T = 178.5 ± 0.5 °C, and constant flux of 3.5 ± 0.5 x 10^{12}/cm²-sec. For this series, the ion dose was varied in the range 1 x 10^{14} - 3 x 10^{16}/cm². After irradiation, samples were examined using transmission electron microscopy.

RESULTS AND DISCUSSION

For the above irradiation conditions, nucleation of amorphous Si occurs heterogeneously at high angle grain boundaries in polycrystalline Si for temperatures less than 225 °C. Figure 1 clearly indicates growth of amorphous Si at grain boundaries for irradiation with 1.5 MeV Xe+ and a flux of 3.5 x 10^{12}/cm²-sec at 178 °C. Figure 2 illustrates the variation of the film microstructure and grain size distribution with Xe+ ion dose. Note that after irradiation with 3 x 10^{15}/cm² Xe+, the average grain size actually increases with respect to its value for 1.7 x 10^{15}/cm² since small grains disappear with greater frequency in this dose interval(i.e, this is the interval in which the initially average-sized grain disappears).

The variation of amorphous fraction in the film, χ, with ion dose at T = 178 °C is shown in Fig. 3(a). When replotted as in Fig. 3(b), a constant nucleation rate is expected to yield a linear dependence of $log[log1/(1 - \chi)]$ with $log[Dose]$ in the absence of any restriction in the availability of nucleation sites. The deviation from linearity is a manifestation of the depletion of available grain boundary sites for nucleation of amorphous Si[6].

The stability of polycrystalline Si with respect to amorphization was investigated in the range of temperatures 150-261 °C. Results of these irradiations are summarized in Table I. Previously, we have measured in this range the displacement rate of a pre-existing, planar amorphous-crystal interface of (001) orientation (for both amorphization and crystallization). Figure 4 is a comparison of the measured amorphous fraction with the amorphous

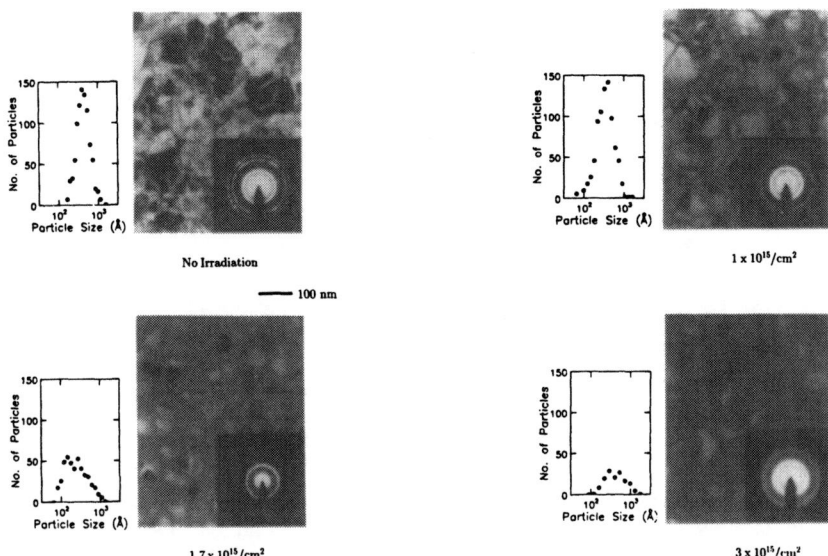

Figure 2: Microstructure of an initially polycrystalline 1000 Å thick Si film after 1.5 MeV Xe⁺ ion irradiation at 178 °C for various doses is shown. Also shown are diffraction patterns and the corresponding grain size distributions.

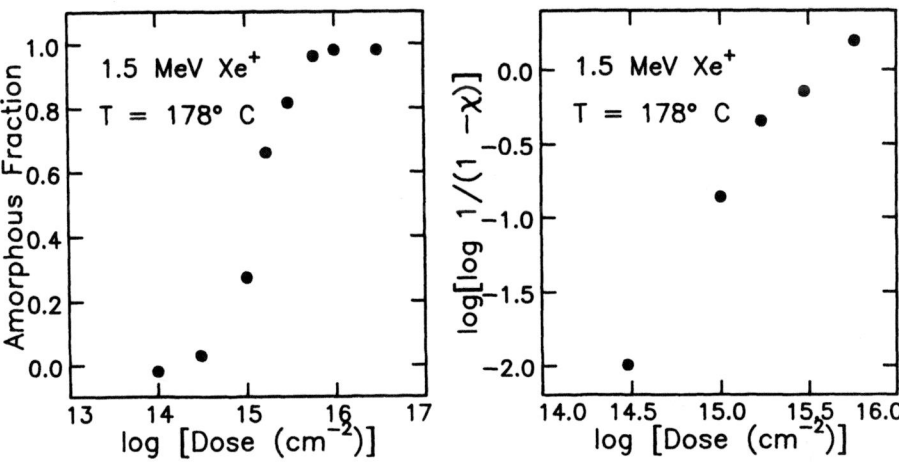

Figure 3: Variation of amorphous fraction with Xe⁺ ion dose in a 1000 Å thick Si film after 1.5 MeV Xe⁺ irradiation at 178 °C and a flux of 3.5 x 10^{12}/cm²-sec.

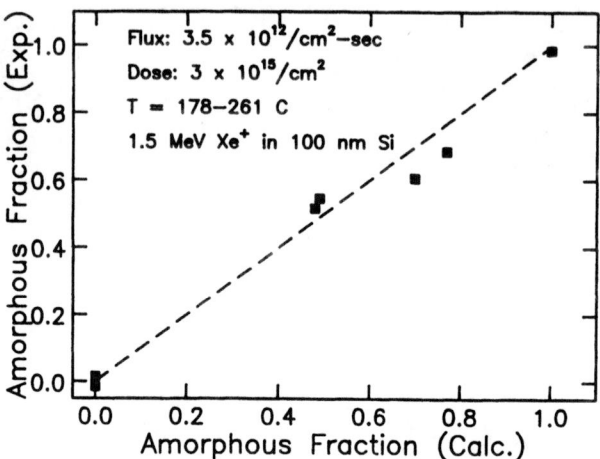

Figure 4: Comparision of the measured amorphous fraction with the amorphous fraction calculated assuming the measured initial grain radius, and an amorphous layer growth rate equal to that of a pre-existing amorphous-crystal interface under the same irradiation conditions. A constant dose of 3 x 10^{15}/cm^2 Xe$^+$ was employed for temperatures in the range 178-261 °C.

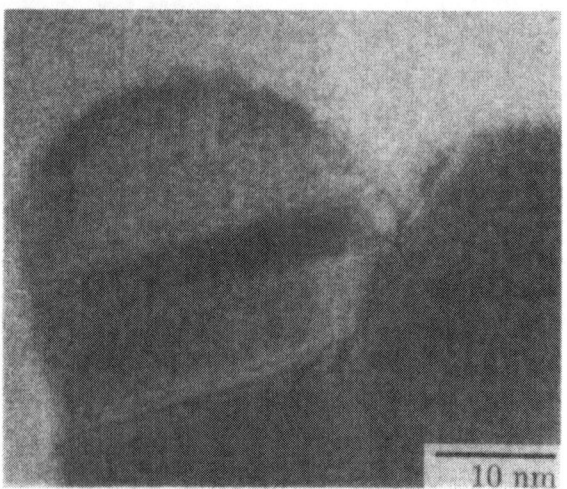

Figure 5: High-resolution electron micrograph of a $\Sigma = 27$ symmetric tilt boundary and a $\Sigma = 3$ twin boundary about the <110> zone axis. The 1000 Å thick Si film was irradiated with 1.5 MeV Xe$^+$ ions at 178 °C and a flux of 3.5 x 10^{12}/cm^2-sec. The micrograph indicates that the amorphous phase nucleated at the high angle boundary at the top, and has begun nucleation at the $\Sigma = 27$ symmetric tilt boundary. No amorphous phase was detected at $\Sigma = 3$ twin boundaries for these irradiation conditions.

fraction calculated assuming the measured initial grain radius, and an amorphous layer growth rate equal to that of a pre-existing amorphous-crystal interface under the same irradiation conditions, for a constant dose of $3 \times 10^{15}/\mathrm{cm}^2$. This correlation indicates that following nucleation, the growth rate at a non-planar amorphous-crystal interface in polycrystalline silicon is essentially identical to the growth rate at a pre-existing planar interface.

Table I

Morphology of polycrystalline Si films irradiated with 1.5 MeV Xe$^+$.

Temperature	Dose	Flux	Morphology
(°C)	cm^{-2}	cm^{-2}-sec^{-1}	
150	3.4×10^{15}	$3.5 \pm 0.3 \times 10^{12}$	Completely Amorphous
177	3.0×10^{15}	$3.5 \pm 0.3 \times 10^{12}$	Crystalline Islands in Amorphous Matrix
179	3.0×10^{15}	$3.5 \pm 0.3 \times 10^{12}$	Crystalline Islands in Amorphous Matrix
201	3.0×10^{15}	$3.5 \pm 0.3 \times 10^{12}$	Crystalline Islands in Amorphous Matrix
208	3.0×10^{15}	$3.5 \pm 0.3 \times 10^{12}$	Amorphization at Boundaries
224	3.0×10^{15}	$3.5 \pm 0.3 \times 10^{12}$	Amorphization at Boundaries
229	3.0×10^{15}	$3.5 \pm 0.3 \times 10^{12}$	Completely Polycrystalline
261	3.0×10^{15}	$3.5 \pm 0.3 \times 10^{12}$	Completely Polycrystalline
178	1.0×10^{14}	$3.8 \pm 0.3 \times 10^{12}$	Completely Polycrystalline
178	3.0×10^{14}	$3.8 \pm 0.3 \times 10^{12}$	Completely Polycrystalline
178	1.0×10^{15}	$3.8 \pm 0.3 \times 10^{12}$	Amorphization at Boundaries
178	1.7×10^{15}	$3.8 \pm 0.3 \times 10^{12}$	Crystalline Islands in Amorphous Matrix
178	3.0×10^{15}	$3.8 \pm 0.3 \times 10^{12}$	Crystalline Islands in Amorphous Matrix
178	5.8×10^{15}	$3.8 \pm 0.3 \times 10^{12}$	Crystalline Islands in Amorphous Matrix
178	1.0×10^{16}	$3.8 \pm 0.3 \times 10^{12}$	Completely Amorphous
178	3.0×10^{16}	$3.8 \pm 0.3 \times 10^{12}$	Completely Amorphous

CRYSTAL STABILITY-BOUNDARY STRUCTURE RELATION IN SI DURING IRRADIATION

The majority of grain boundaries in our polycrystalline Si films are high angle grain boundaries, whose boundary structure and orientation relationship are difficult to determine. These boundaries dominate the microstructural evolution of the film during ion irradiation. However, we were able to determine the structure and orientation relationships for certain special grain boundaries, and to determine qualitatively the relation between boundary structure and amorphization kinetics.

The nucleation rate of amorphous Si at $\Sigma = 27$ symmetric tilt boundaries along the $<110>$ zone axis is much lower than the rate at adjacent high angle boundaries. Moreover, $\Sigma = 3$ symmetric tilt (twin) boundaries along $<011>$ appear to be stable against amorphization. These observations, illustrated in Fig. 5, are a qualitative indication of the relationship between grain boundary structure and stability under irradiation.

We speculate that the grain boundary structure, and the relative mobilities of vacancies and self-interstitials play a key role in determining the amorphization kinetics. The above experiments clearly indicate that amorphous phase formation is nucleated heterogeneously at grain boundaries prior to homogeneous nucleation in the crystal. In single crystals under conditions of uniform irradiation, growth of the amorphous phase is initiated at crystal surfaces, rather than in the bulk[4]. Grain boundary dislocations can act as sinks for vacancies and self-intersitial atoms during steady state ion irradiation, and the grain boundary sink strength should be, in principle, boundary structure-dependent. Models for symmetric tilt boundaries about [110] using Keating potential [7] and total energy calculations [8] indicate that grain boundary energy is orientation-dependent. The $[110]\Sigma = 27$ tilt boundary can be resolved as a array of edge dislocations, while the [110] $\Sigma = 3$ tilt boundary contains no edge dislocations, and thus is not expected to have a sink strength for point defects. Experimental evidence suggests that the silicon self-interstitial is very mobile even at cryogenic temperatures[9], while the vacancy motion, although weakly activated, is considerably slower[10]. Hence for vacancies and interstitials generated by collision cascades in the vicinity of a grain boundary, the boundary will act as a more effective sink for the more mobile interstitial, leading to an excess vacancy concentration within a characteristic vacancy diffusion distance of the grain boundary. In the context of this picture, amorphous Si is nucleated when the vacancy concentration exceeds a critical value. Although it cannot be shown to be unique, this atomistic view of point defect mobility and boundary structure-dependent sink efficiency accounts for the experimentally observed heterogeneous amorphization kinetics.

In summary, we have shown that the amorphous phase is nucleated heterogeneously at structural defects, such as grain boundaries, in Si during MeV ion irradiation. The kinetics of amorphization of polycrystalline Si are grain boundary structure dependent, but once the amorphous phase is nucleated, growth occurs in manner similar to growth at planar amorphous crystal interfaces. The observed kinetics suggest that the relative mobilities of point defects and their interactions with sinks can lead to an excess vacancy concentration in the vicinity of a grain boundary which mediates the nucleation of the amorphous phase.

References

[1] J.W. Rutter and K.T. Aust, Acta Met. **13** 181 (1965).

[2] M.E. Glicksman and C.L. Vold, Surf. Sci., **31**, 50 (1972).

[3] J.Q. Broughton and G.H. Gilmer, Phys. Rev. Lett., **56**, 2692 (1986); and references cited therein.

[4] A. Leiberich, D.M. Maher, R.V. Knoell and W.L. Brown, Proceedings of the 5th International Conference on the Ion Beam Modification of Materials, Catania, (1986); R.G. Elliman, J. Linnros, and W.L. Brown, Mat. Res. Soc. Symp. **100**, 363 (1988).

[5] H.A. Atwater, C.V. Thompson and H.I. Smith, Phys. Rev. Lett., **60**, 112 (1988); H.A. Atwater, C.V. Thompson and H.I. Smith, Mat. Res. Soc. Symp. **100**, 345 (1988).

[6] *Theory of Transformations in Metals and Alloys*, J.W. Christian, Pergamon, New York, (1975), pp. 526-534.

[7] J.T. Wetzel, A.A. Levi, and D.A. Smith, in *Grain Boundary Structure and Related Phenomena*, edited by Y. Ishida, Japan Institute of Metals International Symposium, Vol. 4, (Japan Institute of Metals, Miyagi, 1986, pp. 1061-1067.

[8] M. Kohyama, R. Yamamoto, and M. Doyama, in *Grain Boundary Structure and Related Phenomena*, edited by Y. Ishida, Japan Institute of Metals International Symposium, Vol. 4, (Japan Institute of Metals, Miyagi, 1986, pp. 99-106.

[9] G.D. Watkins, in *Radiation Damage in Semiconductors*, Dunod, Paris, 1965, p. 97.

[10] J.A. Van Vechten, Phys. Rev. B, **10**, 1482 (1974).

AMORPHIZATION AND ANNEALING OF
6H SiC IMPLANTED WITH N-TYPE, P-TYPE
OR ISOVALENT DOPANTS

J. A. SPITZNAGEL*, SUSAN WOOD*, W J. CHOYKE**, R. P. DEVATY** AND
J. RUAN**
*Westinghouse Science & Technology Center, 1310 Beulah Rd., Pittsburgh, PA 15235
**University of Pittsburgh, Department of Physics, Pittsburgh, PA 15260

ABSTRACT

Ions of boron, phosphorous, titanium and neon were implanted into (0001) oriented 6H SiC crystals at 300 K. Implantation energies and fluences were chosen to produce equal (calculated) displacements per atom at similar peak damage depths and a randomized (metaminct or amorphous) zone extending to the front surface. RBS/channeling was used to test the amorphization criteria. Dopant effects on regrowth kinetics and microhardness have been determined by isochronal annealing.

INTRODUCTION

SiC is a wide bandgap semiconductor with many potential applications in high temperature electronic devices. Dopant incorporation during growth by sublimation or CVD methods or by subsequent diffusion is difficult and ion implantation is receiving increasing attention. Previous studies have not provided consistent data for evaluation of as-implanted damage profiles, amorphization conditions or regrowth kinetics for different dopants [1-11]. The purpose of this study was 1) to further test the usefulfuness of a relatively simple amorphization criteria by extension of earlier studies on N and Al [12,13], and 2) to determine whether the electronic nature of the dopant affects regrowth of the amorphous layer and recovery of properties such as near-surface microhardness.

EXPERIMENTAL PROCEDURES

Silicon carbide <0001> crystals were selected from the D149 and D143 sublimation growth run samples, and determined by transmission Laue patterns to be mixed polytype but mostly 6H SiC. These samples are completely transparent, but slightly yellow in color. Those utilized for RBS/channeling measurements were approximately 1 cm x 0.5 cm and 1 mm thick. They are compensated, slightly n-type, with an impurity concentration of ~5 x 10^{17}/cm^3. Crystal faces were lapped flat and parallel to the (0001) plane with a 0.1 μm diamond finish. An oxidation and acid stripping process was then used to remove any residual damaged layer left after the final diamond polish.

Reference samples for channeling were provided by placing a tantalum mask over ~1/3 of the specimen surface during implantation. Implants were made at 300 K. Current densities were typically 0.5 μA/cm^2.

The calculation of deposited damage energy, $S_D(x)$, and amorphization criteria for SiC have been discussed previously [12]. The calculated $S_D(x)$ curves, Figs. 1 and 2 of reference [12], were used to determine first, the fluences of B, P, Ti and Ne required to produce equal numbers of displaced atoms in the SiC crystals at a peak damage depth, x_m, of 0.1 μm. Based on previous results [12], a critical deposited damage energy of 2 x 10^{21}keV/cm^3 was then used to estimate the fluences required to produce an amorphous zone extending from x_m to the front surface using Equation 1. The resultant implantation conditions are listed in Table I.

$$\text{Critical Fluence} = \frac{2 \times 10^{21} \text{ keV/cm}^3}{S_D(x) \times 10^7} \qquad (1)$$

TABLE I.

Implantation Conditions for B, P, Ti and Ne Ions Calculated to Produce
an Amorphous Zone Extending from the Peak Damage Depth at
0.1 Microns to the Front Surface of a SiC Crystal at 300 K

Ion	Kinetic Energy, keV	Fluence (ions/cm^2) from Equation (1)
B$^+$	50	3×10^{15}
Ne$^+$	135	2×10^{15}
P$^+$	150	9×10^{14}
Ti$^+$	250	5×10^{14}

Post-implantation isochronal (1800 s) annealing was performed at temperatures of 573-1233 K in vacuum at a pressure of 2.7×10^{-4} Pa. All crystals were analyzed by RBS/channeling and microhardness measurements after each anneal stage.

A 1.5 MeV He beam from a 2 MV Van de Graaff was used for the RBS measurements. Beam currents were normally 15 nA and the spot size was 0.8 mm, with a beam divergence of <0.03 degrees. RBS measurements with made at 168°. Each RBS spectrum obtained from an implanted or implanted and annealed crystal was directly compared with a spectrum from the corresponding non-implanted portion of the crystal, and normalized with respect to the randomly aligned backscattering yield.

Hardness indentations parallel to a [1210] facet were obtained on a Tukon hardness tester calibrated for loads ≳ 10 g. A Knoop indenter was used with loads of 10-300 g. Actual Knoop Hardness Numbers (KHN) were calculated from micrographs of the indents taken on a Reichert optical microscope at 1000x.

RESULTS AND DISCUSSION

RBS/channeling spectra of implanted and annealed crystals are shown in Fig. 1. The neon fluence, chosen primarily to study effects on microhardness, is a factor of 3 higher than the critical value given in Table I. This fluence was clearly sufficient to produce a randomized or metaminct distribution of Si atoms from R_p to the front surface of the crystal. Similar atomic displacements have also occurred in the carbon sublattice. The direct backscattering carbon peak, however, cannot be readily unfolded from the composite dechanneling spectrum at low channel numbers because of the relative scattering efficiencies and signal to noise ratio. The displaced Si atoms are stable and very little lattice recovery occurs during successive 0.5 h anneals at increasing temperature up to 1173 K.

RBS/channeling measurements of displacements in the Si sublattice after implantation of P, B, and Ti to the conditions shown in Table I are also summarized in Fig. 1. While the phosphorus implant produced amorphicity over most of the expected depth range, the boron and titanium implants resulted in much narrower direct backscattering peaks. Also, the experimental x_m values show some variability. These differences are probably related, in part, to the efficiency of energy transfer for the three species. The efficiency of these ions for displacing either Si or C atoms is different because of significant mass differences with respect to both elements in the host lattice.

Table II gives the calculated maximum fraction of kinetic energy transferred in "hard sphere" collisions of ions with Si and C atoms. The calculations suggest the likelihood of different amounts of displacement damage on the two sublattices. It is not possible, however, to quantify the relative sublattice displacements without a detailed calculation of the energy transfers through multiple collisions of the "primary knock-on atoms" or PKA's and their progeny [14]. This has not been performed in the current investigation. Qualitatively,

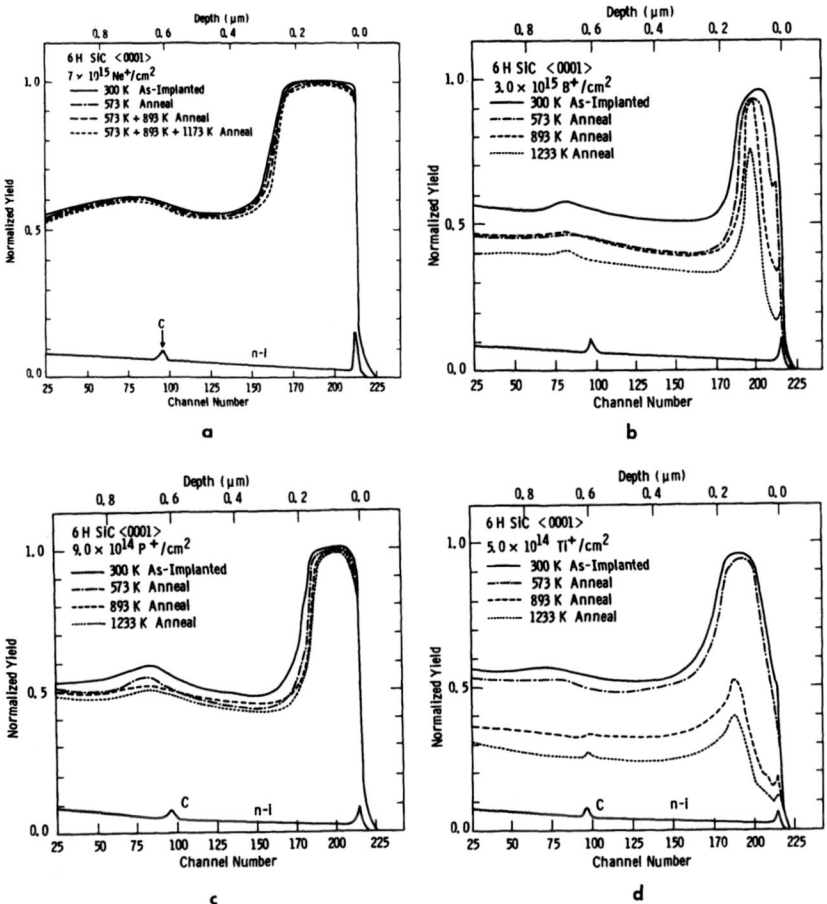

Figure 1. RBS/Channeling spectra from unimplanted (n-i) and implanted
and annealed (0001) 6H SiC a) neon b) boron c) phosphorous
d) titanium

B and N ions preferentially displace carbon atoms, while Al, P, and Ti ions should preferentially displace silicon atoms. Neon ions are expected to displace Si and C atoms approximately equally. Examination of the collective RBS/channeling direct backscattering peaks and TEM results (not shown) from both these data and the Al and N implantation studies reported previously indicates that the amorphization criterion holds for Al^+, N^+, P^+,(and Ne^+), and but slightly underestimates the fluences required for B^+ and Ti^+. Table II. For the latter two ions, the mass of the implanted ion is either smaller than that for a carbon atom or larger than that for a silicon atom.

TABLE II
Calculated Maximum Fraction of Kinetic Energy Transferred in "Hard-Sphere"
Collisions of Ion of Mass m_1 with Silicon and Carbon Atoms

Ion	M_1	Maximum Fraction[a] of Kinetic Energy Transferred		Observed Extent of Amorphicity To x = 0
		T_{Si}	T_c	
B	10.8	0.80	0.99	No
N	14	0.89	0.99	Yes
Ne	20.2	0.97	0.94	Yes
Al	26.98	0.99	0.85	Yes
P	30.97	0.99	0.80	Yes
Ti	47.90	0.93	0.64	No

a) Calculated from $T = (4M_1M_2)/(M_1+M_2)^2$.

Annealing/recovery behaviors are also documented in Fig. 1. It is interesting to note that, while some recovery was observed after the 573 K anneal for all three specimens, it accelerated at the higher temperatures for the B and Ti implants, with the latter showing the greatest reduction in damage. Clearly, the recovery behavior of the B and Ti implanted SiC specimens is not the same. This suggests that either the rearrangement of bonds to permit crystal regrowth is dependent upon the electronic nature of the dopant, or that the as-deposited damage structure was sufficiently different for the two ion species. In all three specimens, recovery (i.e., regrowth of single crystal SiC) occurs from both sides of the amorphous zone. However, in the P-implant, which had a broad a-SiC region extending almost to x = 0, minimal regrowth is observed from the front surface.

As described elsewhere [12], integration over the area under the Si direct backscattering peak gives the effective number of displaced silicon atoms projected per unit area of the implanted surface. The fractional damage recovery with annealing can also be described in terms of the change in area under the peak and provides an easy means for comparing effects for different ions and fluences. Fig. 2a shows the results for the P^+, B^+, and Ti^+ implanted and annealed SiC specimens. Both the extent of recovery after the isochronal anneals and the rate of recovery are different for the three crystals. Comparison with Fig. 2b, however, indicates that some of the differences may be explicable entirely on the basis of different initial damage states prior to annealing [12]. The annealing behavior of the Ti^+ implanted sample is very similar to the Al^+ doped crystal which was also implanted to produce a buried amorphous layer near x_m, but insufficiently heavily for the amorphous region to reach the surface. Similarly, the annealing behavior of the P^+ implanted sample closely resembles the data for the Al^+ implant of highest fluence. For the B^+ implanted sample, the situation is less clear. The annealing behavior is consistent with an initial damage profile between those shown for an Al^+ implanted crystal with a amorphous region extending to the surface and a more heavily doped and damaged region. However, this does not agree with the RBS/Channeling curve for the as-implanted boron doped crystal. The possibility of chemical/electronic effects of the dopant on regrowth of the SiC crystal for the boron cannot be entirely ruled out.

Knoop microhardness measurements for the B^+, P^+, Ti^+ and Ne implanted and annealed SiC crystals are presented in Fig. 3. It was not possible to obtain a valid hardness number for the boron doped crystal in the as-implanted condition because of extensive lateral fracturing due to the very brittle nature of the surface. The hardness data indicate a definite dependence of the near-surface mechanical properties on the dopant. Phosphorus and titanium (and neon) implantation produced softening. Boron implantation hardened and embrittled the surface. The boron effect is similar to the hardening observed in the low fluence implantation of aluminum reported earlier [13] but the effect is much larger for boron. The hardening is probably attributable to a combination of solid solution, precipitation and solute-defect strengthening mechanisms in the SiC crystals. Annealing at 1233 K results in complete recovery to the unimplanted hardness level.

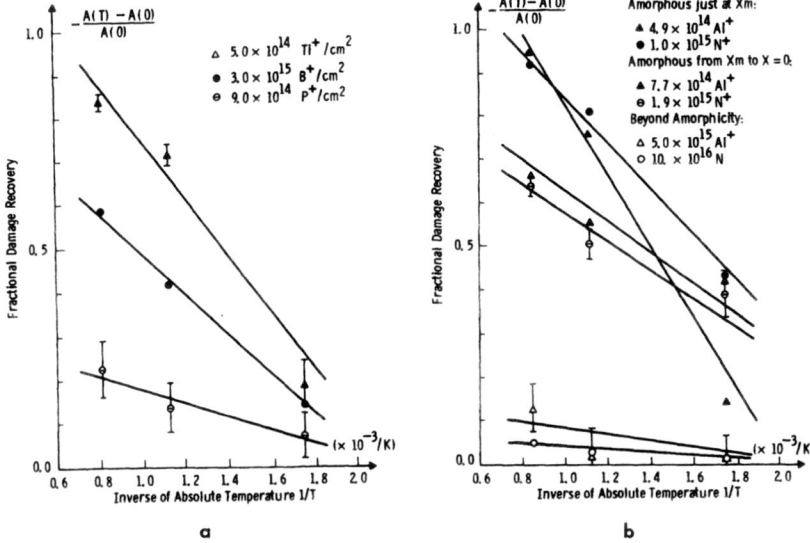

Figure 2. Temperature dependence of damage recovery calculated from integration over the area under the Si direct backscattering peak: a) B^+, P^+ and Ti^+ implantation; b) Al^+ and N^+ implantation [12]

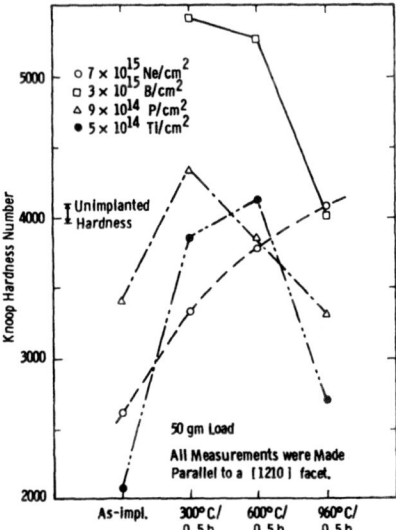

Figure 3. Knoop microhardness measurements of Ne^+, B^+, P^+ and Ti^+ implanted and annealed 6H SiC

As discussed elsewhere [13] creation of an amorphized damage zone tends to soften the near surface region. Thus, on the basis of the RBS/channeling results, it was anticipated that the phosphorous implanted sample would have a lower KHN value than the titanium implanted crystal. The converse is true. Moreover, with post-implantation annealing, both the P^+ and Ti^+ doped samples first recover to the unimplanted hardness level then soften again at the highest annealing temperature. It is concluded that the near-surface mechanical properties, as reflected in indentation response, depend on the chemical/electronic nature of the implant species as well as on the number of atomic displacements.

SUMMARY

1. The critical deposited damage energy to produce a metaminct (randomized or amorphous) state in SiC is 2×10^{21} keV/cm^3 for N, Al, P, (and Ne) ions. Higher values are required for B and Ti which are lighter or heavier than C and Si, respectively.

2. Annealing and regrowth rates depend on the number of atomic displacements for 300 K implantation. They are relatively insensitive to the chemical/electronic characteristics of the implanted ions with the possible exception of boron.

3. Microhardness effects are complex and vary with the chemical/electronic nature of the implanted ion.

ACKNOWLEDGMENTS

This work was partially supported under ONR Contract No. N00014-85-C-0021 and by the NSF under Contract No. DMR 84-03596.

REFERENCES

1.   W. J. Choyke and Lyle Patrick, Phys. Rev. B4, 1843 (1971).
2.   V. V. Makarov, T. Tuomi, K. Naukkarinen, M. Luomajavi, and M. Riihonen, Appl. Phys. Lett. 35, 922 (1979).
3.   D. A. Thompson, M. C. Chan, and A. B. Campbell, Can. J. Phys. 54, 626 (1976).
4.   V. V. Makarov, Sov. Phys. Solid State 13, 1974 (1972).
5.   R. R. Hart, H. L. Dunlap, and O. J. Marsh, Rad. Eff. 9, 261 (1971).
6.   A. Addamiano, G. W. Anderson, J. Comas, H. L. Hughes, and W. Lucke, J. Electrochem. Soc. 119, 1355 (1972).
7.   R. B. Wright and D. M. Gruen, Rad. Eff. 33, 133 (1977).
8.   R. B. Wright, R. Varma, and D. M. Gruen, J. Nucl. Mater. 63, 415 (1976).
9.   A. B. Campbell, J. B. Mitchell, J. Shewchun, D. A. Thomson, and J. A. Davies, Silicon Carbide 1973, Proc. 3rd Int. Conf. on SiC, edited by R. C. Marshall, J. W. Faust, Jr., and C. E. Ryan (South Carolina Press, 1974), p. 486.
10. A. B. Campbell, J. Shewchun, D. A. Thomson, and J. B. Mitchell, in Ion Implantation in Semiconductors, edited by S. Namba (Plenum, New York, 1975).
11. H. G. Bohn, J. M. Williams, C. J. McHargue, and G. M. Begun, J. Mater. Res. 2, 107 (1987).
12. J. A. Spitznagel, Susan Wood, W. J. Choyke, N. J. Doyle, J. Bradshaw, and S. G. Fishman, Nucl. Instr. and Methods B16: 237-243 (1986).
13. Susan Wood, J. A. Spitznagel, W. J. Choyke, J. Bradshaw, J. Greggi, Jr., and N. J. Doyle, Mat. Res. Symp. Proc., Vol. 60, 459 (1986).
14. R. M. More and J. A. Spitznagel, Radiation Effects, 60: 27 (1982).

IN–SITU REFLECTANCE MEASUREMENTS OF SEMICONDUCTORS DURING ION IMPLANTATION

PIETER L. SWART, BEA M. LACQUET AND MICHAEL F. GROBLER
Sensors Sources and Signal Processing Research Group, Faculty of Engineering, Rand Afrikaans University, Johannesburg 2000

ABSTRACT

Damage introduced during ion implantation of semiconductor materials coalesce at a certain critical dose, for a particular energy and ion species. After this threshold dose rapid changes occur in the reflectance. This may be used for studying the amorphization process, or it may be applied as a non–destructive dosimeter and uniformity tool.

An automated reflectometer was developed for studying reflectance during ion implantation of semiconductors with various ion species. Results on argon and phosphorous implants into silicon at energies ranging from 50 to 240 keV are presented.

1. INTRODUCTION

Ion implantation is a well established technique for the introduction of dopants into semiconductors. Crystalline silicon can be transformed into amorphous Si by ion implantation [1]. High–dose implantations produce a heavily damaged or amorphous layer near the surface. For device manufacture it is usually necessary to remove the damage by an annealing process in order to achieve electrical activation [2] of the implanted dopants. Therefore the mechanism of amorphous layer formation and recrystallization are of considerable interest. Diverse experimental techniques have been employed to study the nature of the damage, amorphization and recrystallization associated with ion implantation and subsequent annealing.

Optical techniques provide the possibility of in–situ measurements of reflectance or transmittance during implantations or during annealing. Changes in the optical reflectivity of polished Si wafers due to ion implantation were reported in 1969 by Kurtin, et al [3], and Hart, et al [4]. There exists general agreement now that these changes occur as result of a modification in the optical constants of a thin layer near the crystal surface, leading to interference effects [3–15].

The recent increased interest in this subject is related to the possibility of in–situ measurements of reflectance during implantation. An automated reflectometer was developed for studying the reflectance of ion implanted silicon.

2. REFLECTOMETER

The reflectometer is shown schematically in fig. 1. It employs a laser (currently a He–Ne laser with $\lambda = 632.8$ nm) as the optical excitation. The optics consist of an attenuator, mirrors k_1, k_3, k_5, beam splitter k_8, quartz windows k_2 and k_4 covered by anti–reflective coatings and Si P–I–N photodiodes D_1 and D_2.

The sample is mounted in a Faraday cup at an angle of approximately 7° with respect to the ion beam. The angle of incidence of the laser beam is less than 5° with respect to the surface normal, justifying the assumption of near–normal incidence. Provision has also been made for secondary electron suppression. The implantation area is defined by an aperture with a diameter of 55 mm.

The system is computer controlled with facilities for automatic calibration, adjustable data capture rate, data storage, signal processing and graphics.

Mat. Res. Soc. Symp. Proc. Vol. 147. ©1989 Materials Research Society

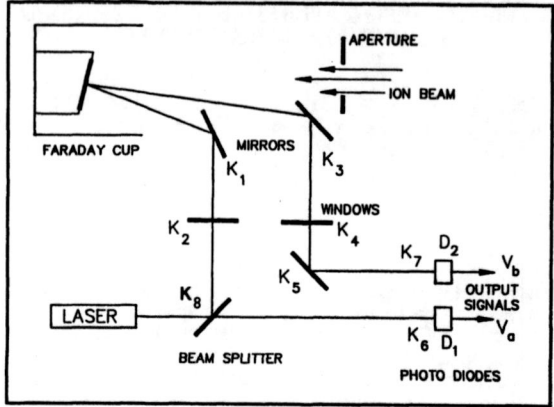

SCHEMATIC DIAGRAM OF
THE REFLECTOMETER.

Fig. 1 Schematic diagram of the reflectometer

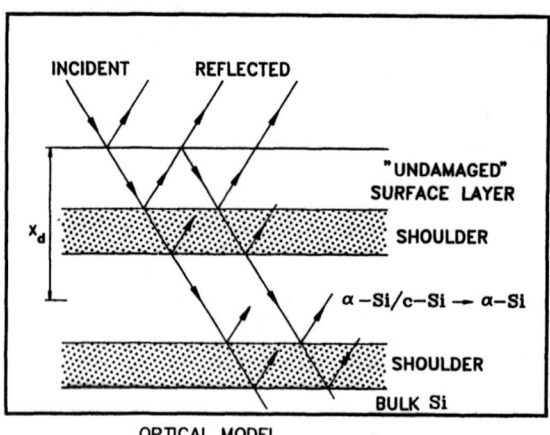

OPTICAL MODEL

Fig. 2 Four layer optical model

3. MODEL

3.1 Damage profile and boundaries.

The damage profile in ion–implanted semiconductors is approximately of Gaussian shape for fluence less than the threshold for amorphization [6,16,17].

$$D(x) = \hat{D}\exp[-(x - X_d)^2/2\Delta X_d^2] \qquad (1)$$

where X_d = range of the damage profile
and ΔX_d = standard deviation of the damage profile.

Sigmund and Sanders [16] have shown that there is a one–to–one correspondence between $(X_d, \Delta X_d)$ and the projected range and standard deviation of the profile for the implanted ions.

$$X_d = aR_p \qquad (2)$$

and
$$\Delta X_d = b\Delta R_p \qquad (3)$$

The a and b are dimensionless constants with values depending on the mass ratio of the projectile and the target atoms.

From profiling data obtained using various techniques such as Rutherford back scattering [18], ESR [19], and optical reflectometry in conjunction with layer removal [6,7,17], it appears that the peak value of the disorder $\hat{D}$ is an increasing function of ion fluence. However, above a certain critical dose, which is a function of the ion species, ion energy and implantation temperature, the damage saturates. This leads to a flat–topped profile with approximately Gaussian shoulders. The material constituting the flat–topped portion is assumed to be totally amorphous. If the ion energy is high enough, this damage layer is buried beneath the crystal surface, sandwiched between a layer of single crystal material and the bulk crystal.

The damage is related to the ion fluence through the theoretical work of Kinchin and Pease [20]. One can estimate the number of atoms N_t that will be displaced by a projectile as it comes to rest. This estimate is based on the assumption that displacements result from a strong Coulomb interaction between the projectile and the target atom. By employing the Gaussian shaped profile as proposed by Sigmund and Sanders [16] and the number of displaced atoms N_t, one can relate the damage $D(x)$ to the fluence ϕ.

$$D(x) = \frac{N_t\phi}{\sqrt{2\pi}\,\Delta X_d}\exp\left[-(x - X_d)^2/2\Delta X_d^2\right] \qquad (4)$$

It is therefore possible to solve for the boundaries of the totally amorphous layer.

$$x_1, x_2 = X_d \pm \sqrt{2}\,\Delta X_d\left[\ln\left(\frac{N_t\phi}{\sqrt{2\pi}\,\Delta X_d D_{th}}\right)\right]^{1/2} \qquad (5)$$

where D_{th} is the threshold for amorphization.

3.2 Optical model.

It has been established experimentally that the refractive index of ion implanted semiconductor material is different from that of undamaged single crystals [eg. 9–14]. It can therefore be expected that the optical constants change in a continuous fashion throughout the damaged layer. Following damage saturation, the structure consists of a uniform amorphous layer, flanked by shoulders consisting of a mixed phase [21] with optical constants that change progressively from the values for single crystal material to those corresponding to the amorphous material. The optical model is presented in fig. 2.

4. EXPERIMENTAL

The experiments were performed using CZ–grown electronic grade single crystal silicon substrates with the following specification: (111)–orientation, 76.2 mm diameter, $380 \pm 20~\mu m$ thick, mirror–shine polished on one side, matt finish on back–side, p–type, boron–doped, $8\Omega - cm \pm 15\%$.

Directly prior to implantation, the wafers were chemically cleaned using a standard cleaning procedure. We performed reflectivity vs dose experiments by implanting $^{31}P^+$ and $^{40}Ar^+$ into the Si with a dose range of $1 \times 10^{16}~cm^{-2}$. The beam current (5 μA), the tilting angle (7°), and implantation aperture (55 mm diameter) were kept constant throughout the experiments.

5. RESULTS AND DISCUSSION

The reflectance versus ion dose was measured for samples implanted with $^{40}Ar^+$ and $^{31}P^+$ between 50 keV and 240 keV at a dose rate of 0.2 $\mu A/cm^2$.

5.1 Argon.

The experimentally determined curves for argon implantations at 50, 100, 150 and 200 keV, are presented in figs. 3(a) − (d), respectively. The R vs. N curves for the various energies are qualitatively the same.

One can observe a pre–amorphous zone at low dose where only small changes in the reflectance are noticeable. Interference maxima and minima may occur in this dose range, especially for the higher energy implants (>100 keV).

In the intermediate dose range, the reflectance increases rapidly with increasing dose. The amorphous threshold has been exceeded, and the internal layer of amorphous material becomes continuous. The higher refractive index of amorphous silicon causes the rapid increase in reflectance. With the proper compensation for the interference effect, one may deduce the threshold dose for amorphization from this region of the curve. It is obtained from the intersection of the pre–implanted reflectance and the asymptote to the curve.

In the third section of the curve, interference maxima and minima are observed as a result of the increasing thickness of the damage layer. The interference pattern is damped as a result of the relatively large extinction coefficient of the amorphous silicon. The refractive index (n,k) of the α − Si may be obtained by doing a curve–fit in this region. The experimental values of the threshold dose for amorphization for Ar in Si range from 1.4×10^{14} ions/cm² at 50 keV to 1×10^{15} ions/cm² at 200 keV.

A very interesting feature of the argon implanted silicon, is the appearance of sharp peaks in the reflectance at high dose. This may be seen in fig. 3(a), but the effect is especially evident in fig. 3(c) (150 keV). The origin of this effect has not conclusively been identified, but it is believed to be stress related. This effect is presently under investigation.

5.2 Phosphorous.

The experimentally determined curves for phosphorous implantations at 100, 160 and 240 keV, are presented in figs. 4(a) − (c), respectively. These curves are qualitatively the same as those for argon. The threshold for amorphization, however, has shifted to higher fluence. This is to be expected since the phosphorous ion is lighter [1]. The experimental values of the threshold dose for amorphization for P in Si range from 3×10^{14} ions/cm² at 50 keV to 6×10^{15} ions/cm² at 240 keV. Detailed model calculations and curve fits had been performed previously for 50 keV implantations [22]. From those results it is evident that the model yields values of the refractive index of the

amorphous layer ($\tilde{n}_2 = 4.4 - i0.67$) which compare well with electron beam evaporated amorphous silicon ($n = 4.3 - 4.4$ and $k = 0.7 - 0.8$ at 632.8 nm) [23]. Ellipsometry also gives comparable results [14].

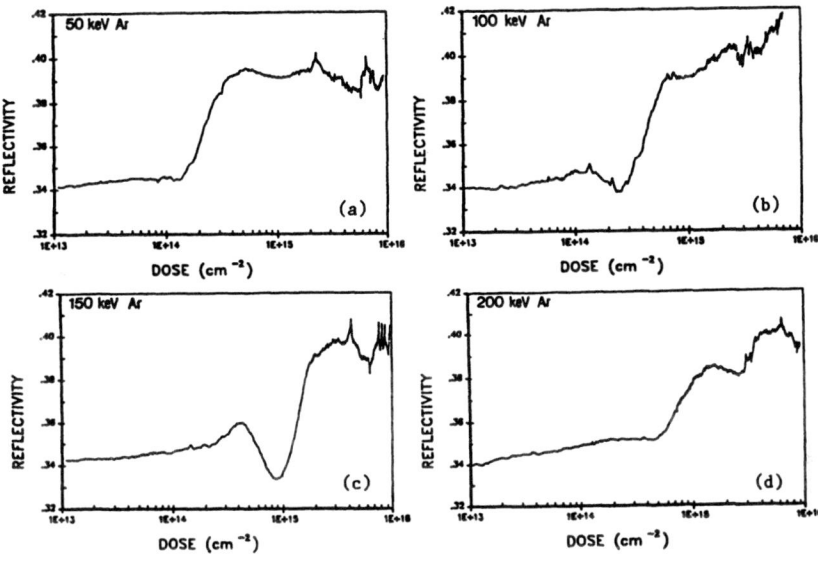

Fig. 3   Monochromatic reflectance of silicon during implantation with Ar

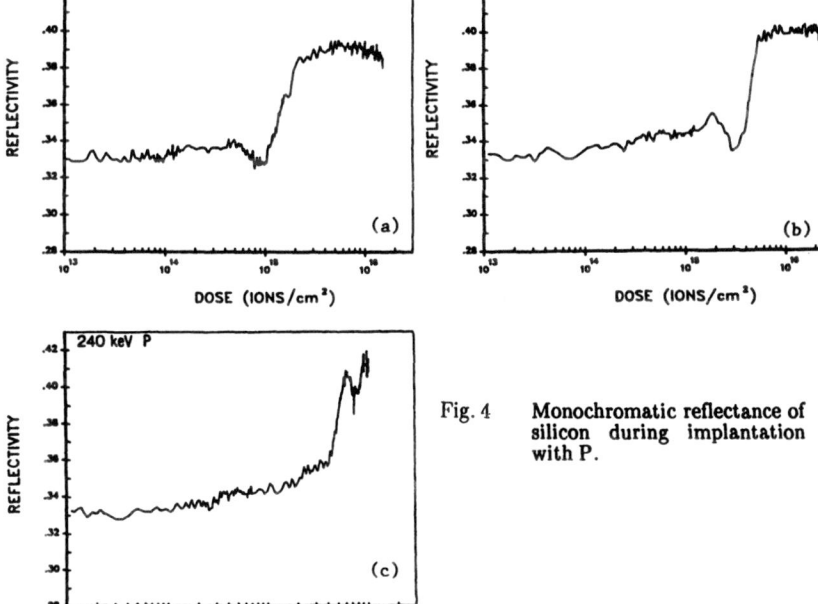

Fig. 4   Monochromatic reflectance of silicon during implantation with P.

6.   CONCLUSION

A dose dependent increase in the optical reflectivity of ion implanted Si is observed during ion implantation as a result of the fact that the crystal is damaged.   The dose and energy dependence of the reflectivity is a complicated function of the depth distribution of the damage and the parameters of the damage layer.

Reflectance measurements for argon and phosporous implantations into Si in the energy range 50–240 keV and dose up to 1×10^{16} cm^{-2} were obtained using an automated in–situ reflectometer.   The measured reflectance agrees qualitatively with the conceptual model.   At high Ar dose sharp peaks were observed in the reflectance curves.   This effect is attributed to stress.

ACKNOWLEDGEMENT

This work was supported by the Foundation for Research Development, Pretoria.

REFERENCES

1.   J.F. Gibbons, Proc. IEEE 60, 1062 (1972).
2.   B.L. Crowder, J. Electrochem. Soc. 117, 671 (1970).
3.   S. Kurtin, G.A. Shifrin and T.C. McGill, Appl. Phys. Lett. 14, 223 (1969).
4.   R.R. Hart and O.J. Marsh, Appl. Phys. Lett. 14, 225 (1969).
5.   T.C. McGill, S.L. Kurtin and G.A. Shifrin, J. Appl. Phys. 41, 246 (1970).
6.   E.T. Yen, B.J. Masters and R. Kastl, in *Ion Implantation in Semiconductors and Other Materials*, edited by S. Namba (Pergamon Press, NY & London, 1974) p. 501.
7.   M. Miyao, T. Miyazaki and T. Tokuyama, Japan J. Appl. Phys. 17, 955 (1978).
8.   A.H. Kachare and W.G. Spitzer, J. Appl. Phys. 45, 2938 (1974).
9.   K. Watanabe, M. Miyao, I. Takemoto and N. Hashimoto, Appl. Phys. Lett. 34, 518 (1979).
10.   K. Nakamura and M. Kamoshida, Rad. Eff. 42, 29 (1979).
11.   K. Nakamura, T. Gotoh and M. Kamoshida, J. Appl. Phys. 50, 3985 (1979).
12.   K. Watanabe, T. Motooka, N. Hashimoto and T. Tokuyama, Appl. Phys. Lett. 36, 451 (1980).
13.   T. Motooka and K Watanabe, J. Appl. Phys. 51, 4125 (1981).
14.   M. Delfino and R.R. Razouk, J. Appl. Phys. 52, 386 (1981).
15.   M. Fried, T. Lohner, E. Jároli, Gy. Vizkelethy, G. Mezey, M. Somogyi, H. Kerkow and J. Gyulai, Thin Solid Films 116 (1984).
16.   P. Sigmund and J.–B. Sanders, in *Applications of Ion Beams to Semiconductor Technology*, edited by P. Glotin (Edition Ophrys, France, 1967) p. 215.
17.   M. Miyao, N. Yoshihiro, T. Tokuyama and T. Mitsuishi, J. Appl. Phys. 49, 2573 (1978).
18.   J.W. Mayer, L. Eriksson and J.A. Davies: *Ion Implantation in Semiconductors*, (Academic Press, NY, 1970).
19.   B.L. Crowder, R.S. Title, M.H. Brodsky and G.D. Pettit, Appl. Phys. Lett. 16, 205 (1970).
20.   G.H. Kinchin and R.S. Pease, Rep. Prog. Phys. 18, 2 (1955).
21.   D.M. Wood and N.W. Ashcroft, Phil. Mag. 35, 269 (1977).
22.   P.L. Swart, H. Aharoni, B.M. Lacquet, Nucl. Instr. Meth. Phys. Res. B6 365 (1985).
23.   D. Beaglehole and M. Zavetova, J. Non–Cryst. Sol. 4, 272 (1970).

PART III

Focused Ion Beams

FOCUSED ION BEAM INDUCED DEPOSITION

JOHN MELNGAILIS AND PATRICIA G. BLAUNER
Research Laboratory of Electronics, M.I.T., 77 Massachusetts Avenue,
Cambridge, MA 02139

ABSTRACT

Focused ion beam induced deposition is already in use commercially for the repair of clear defects in photomasks, where missing absorber is added. Research is being carried out to extend this technique to the repair of x-ray lithography masks and to the restructuring and repair of integrated circuits, particularly in the prototype phase. In this technique a local gas ambient is created, for example, by aiming a small nozzle at the surface. The gas molecules are thought to adsorb on the surface and to be broken up by the scanned focused ion beam. A deposit is formed with linewidth equal to the beam diameter which can be below 0.1 μm. At small beam diameters and low currents (50-100 pA) the time to deposit 1 μm^3 is in the vicinity of 10-20 sec. If the gas is a hydrocarbon, the deposit is largely carbon, which is useful for photomask repair. On the other hand, if the gas is a metal halide or a metal organic, the deposit is metallic. The deposits have substantial concentrations of impurities due to the atoms in the organometallic, to the ion species used, or to the ambient in the vacuum chamber. Thus the resistivities of the "metal" films deposited typically range from 150 to 1000 μΩcm which is usable for some repairs. (Pure metals have resistivities in the range 2.5 to 12 μΩcm.) We have deposited gold from dimethyl gold hexafluoro acetylacetonate and have achieved linewidths down to 0.1 μm, patches of 1 μm thickness with steep side walls and in some cases, resistivities approaching the bulk value. Other workers have reported deposits of Al, W, Ta, and Cr. We will review previous work in the field and present some of our own current results.

INTRODUCTION

The patterning of thin films is a key step in the fabrication of almost all microelectronic devices. In conventional processing, conductors, for example, are created by (1) depositing a thin film of metal using evaporation or sputtering, (2) covering the surface with radiation sensitive polymer, (3) patterning the polymer using incident radiation (lithography), and then (4) removing the metal from the areas not covered by the polymer. This multistep process treats the entire sample the same, i.e. the remaining film is the same thickness everywhere but patterned into, perhaps, millions of features.

In some cases, such as pattern repair, device customization, or in-situ fabrication, a beam writing process may be desired. Such a process combines deposition and patterning into a single step but is necessarily slower since the writing is done serially, point-by-point. Photon beams (usually lasers), electron beams, and ion beams have been used. An appropriate gas ambient, often a metal organic or a metal halide, is created over the sample and a chemical reaction is produced at the point where the beam is incident. By far the most widely studied of these processes is laser deposition,[1] which has been demonstrated for local wiring of integrated circuits.[2] Ablation is used in conjunction with the deposition to cut conductors or to make via holes in insulating layers.[2] These processes are generally limited to dimensions well above 1 μm.

Electron beams and ion beams, on the other hand, can operate in the submicrometer regime and thus are of interest for repair processes as dimensions of microelectronic devices shrink. With electron beams a number of materials have been deposited with submicrometer dimensions including: Cr[3][4], W[4][5], Au[5], and Fe[6]. Tungsten deposition by e-beams has been proposed for x-ray mask repair.[7] The removal of material by e-beam in a controlled fashion is more difficult although stimulated etching of Si with XeF_2 at 0.5 μm dimensions has been reported.[8]

This paper will focus on ion induced deposition. This is of particular interest because focused ion beam systems have recently been developed. (For a review of the subject see Ref. 36.) These ion beams operate with a bright liquid metal (usually Ga) ion source and usually produce beams focused to a spot below 0.1 μm at energies in the 20-70 keV range. Ion beams can be used to induce deposition at submicrometer dimensions. In addition, when no gas ambient is present, the same beams can remove material by sputtering with control of dimensions down to 0.1 μm. The focused ion beam can also be used for imaging when secondary electrons (sometimes also ions) are collected. This works very much like a scanning electron microscope, except that the surface is eroded. By frame storage techniques this erosion can be kept to about a monolayer.

The combination of deposition, milling and imaging has been exploited commercially in focused ion beam repair of photomasks.[9-15] See Fig. 1. This is likely to be the preferred technique over laser repair as dimensions shrink below 1 μm (Ref. 16). Similarly focused ion beams are also expected to play a role in microelectronic circuit repair or restructuring as dimensions shrink. Cutting of conductors has already been used, as well as voltage contrast imaging (see Ref. 17 and 18 for a review of this field and references to the literature). Recently two metal lines have been joined by milling through the passivation layers and depositing a tungsten connection from one line to another. [22] $W(CO)_6$ gas was used, and the deposited material had a resistivity of 30 to 50 times higher than pure bulk W. The resistance of the connection was nevertheless acceptably low.

Another application which is motivating research on focused ion beam induced deposition is x-ray lithography mask repair. For X-ray lithography, the masks used have the same dimensions as the final samples. Demagnification which is sometimes used in optical lithography is not available. Thus x-ray masks for the next generation of devices will have minimum features in the range of 0.25 to 0.3 μm and metal absorber thickness of 0.3 to 0.7 μm, depending on the x-ray wavelength used. The repair of such high-aspect ratio features of small dimensions presents special challenges. In milling off unwanted features redeposition effects have to be considered[19] and in the deposition of absorber, sidewall steepness and material density are important issues. [20]

In this paper we will discuss ion induced deposition, review the results obtained so far, speculate on the mechanism of the process, and consider the applications.

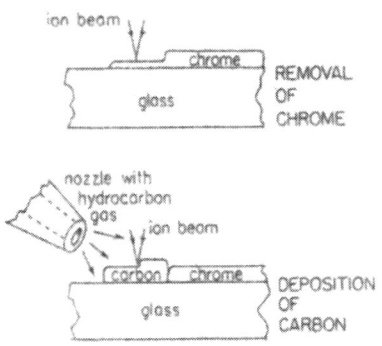

FOCUSED ION BEAM PHOTOMASK REPAIR

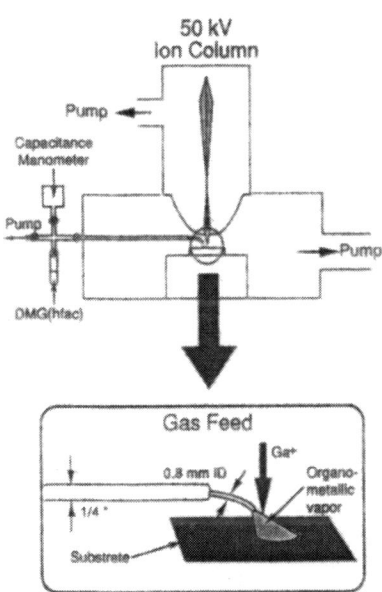

Fig. 1: Schematic of the photomask repair process. The absorbing film deposited is shown to be carbon, but a metal could equally well be used.

Fig. 2: Schematic of the focused ion-beam-induced-deposition apparatus. The inset shows the details of the gas feed. The ion beam is scanned over the area of the surface where the flux of gas is incident.

ION INDUCED DEPOSITION

a)   Apparatus

As discussed above, what is of current technological interest is <u>focused</u> ion beam induced deposition. The apparatus used commercially usually consists of a focused ion beam column which is differentially pumped and a gas feed system which points a fine nozzle at the sample surface where the ion beam is incident.[10-15] In some cases the "gas" is a solid at room temperature e.g. $Cr(CO)_6$ (Ref. 21), or $W(CO)_6$ (Ref. 22). Then all of the gas feed system has to be heated. The apparatus used at MIT which does not include a heated feed system is shown schematically in Fig. 2 (Refs. 23 & 24). The gas pressure on the sample is a function of nozzle height above the sample. The pressure on the substrate at the point where the ion beam writing is carried out was measured as a function of nozzle height by replacing the sample with a "stagnation tube" as shown in Fig. 3. Using these results as calibration one can perform deposition experiments at various gas pressures by simply varying the tube height above the sample.

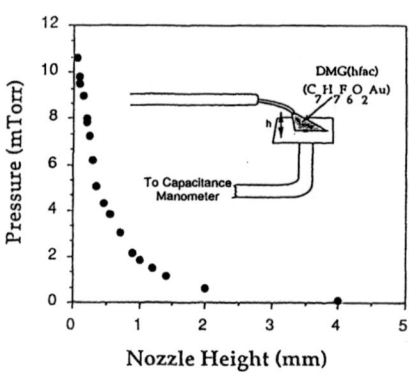

Fig. 3: The pressure as a function of nozzle height above the sample (measured for dimethyl gold hexafluoro acetylacetonate (DMG (hfac)). A 1 mm diameter hole is in the sample and is connected to a capacitance manometer which reads the pressure. From Ref. 24.

Another scheme for creating a local gas ambient consists of enclosing the sample in a box (as shown in Fig. 4) which has a hole in it through which the ion beam enters. Except in cases where the sample is close to the hole, this arrangement provides a measurable, uniform pressure on the sample. This scheme was used in the earliest experiments.[25,26]

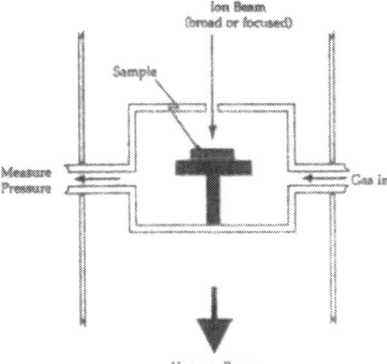

Fig. 4: For some studies of ion induced deposition the sample is enclosed in a cavity where a uniform gas pressure is maintained and into which the gas enters through a small hole.

In order to study the fundamental process by which an ion beam causes adsorbed gas molecules to break apart, a broad ion beam such as from an implanter can be more convenient than a focused beam. Large area deposits can be formed, noble gas ions can be used, and the mass and energy can be varied easily.[25,27,28,30] The results should be directly comparable to focused ion beam induced deposition in those cases where the focused beam is scanned rapidly enough so that the deposition rate is independent of scan rate.[29] This is generally the region in which one would wish to operate with a focused ion beam since at slow scan speeds the deposition rate drops because the supply of adsorbed molecules is depleted.[24,29]

The ion induced deposition process can be affected by the temperature of the sample. Accordingly, heated and cooled stages have been used.[28,30] In addition, in-situ measurement of adsorption rates and deposition rates can be readily obtained by using a quartz crystal microbalance as the "sample" (Ref. 31,32).

b) Films deposited

Ion induced deposition has been demonstrated with a number of metal halides and metal organic gases, as well as with hydrocarbons. In many instances, high concentrations of impurities are observed in the film in addition to the desired metal. There are three main sources of impurities: the parent molecule, e.g. films made from metal organics usually contain carbon; the incident ion, if Ga is used it will, of necessity, be implanted in the film; and the ambient, if the vacuum level is 10^{-6} torr then a monolayer can form in about 1 sec, partly oxidizing a reactive metal film during growth.

Under ideal conditions, namely, a metal halide gas, an ultrahigh vacuum chamber, and the incident ion of the same species as the metal to be deposited or perhaps a noble gas ion, high purity metal films should be achievable. Depositions under such conditions have as yet not been reported. The results reported so far have been summarized in Tables 1 and 2.

Most films are seen to contain high concentrations of impurities and to have resistivities much higher than bulk metal values. The exceptions are the gold depositions carried out on samples at elevated temperatures (100-160° C) using dimethyl gold hexafluoro acetylacetonate, $(C_7H_7O_2F_6Au)$, abbreviated as DMG(hfac), and the tungsten deposits formed in well pumped vacuum systems. For the gold films, the higher temperature presumably permits the hydrocarbon products of the dissociated DMG(hfac) to desorb before they are in turn broken up by the incident ions.[30,33]

The gold films deposited at room temperature using an Ar ion beam have carbon concentrations of 30-50%. Transmission electron microscope studies of these films indicate that initially the gold grows in unconnected islands of about 40-60 nm size.[30] Since the electron diffraction pattern of this film is that of pure gold, the carbon must be between the islands. Cross-sectional TEM indicates that, as these films thicken, a columnar structure results.[30] The film deposited at elevated temperatures on the other hand shows the continuous polycrystalline structure expected for e-beam evaporated films (see Fig. 5). Films deposited at room temperature with a Ga focused ion beam have similar island structure when examined by TEM.[24]

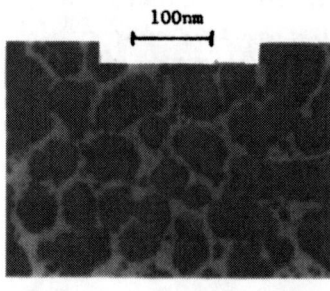

(a)

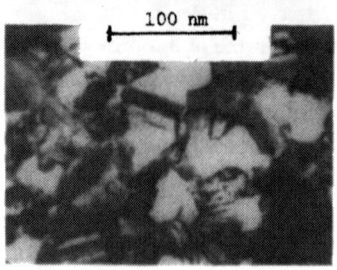

(b)

Fig. 5: Transmission electron micrographs of gold films grown by ion-induced deposition at two substrate temperatures: (a) Room temperature, nominal thickness 60 nm, 70 keV Ar^+ ions at 0.7 $\mu A/cm^2$, total dose $1X10^{16}/cm2$. The film is seen to be made up of unconnected gold islands. This discontinuous columnar structure was observed by cross sectional TEM to persist even to 250 nm thicknesses. (b) 160°C, otherwise same conditions, thickness 100 nm. The film is seen to be continuous and has the microstructure typical of conventionally evaporated films (from Ref. 30).

Table 1. Ion Induced Deposition

Gas (Reference)	Ion, Energy	Current Density (A/cm^2)	"Yield" (atoms/ion)	Deposit Composition	Resistivity (μΩcm)
Al(CH3)3 (1)	Ar$^+$ 50 keV		13	Al:O:C 14:70:16 59:39:2	
Ta(OC2H5)5 (2)	Ar$^+$ 50 keV		27	Ta:O:C 56:27:17	
WF6(RT) (2)	Ar$^+$ 50 keV		31	W:O 75:25	
WF6 (80K) (2)	Ar$^+$ 50 keV		3800	high oxygen content	
WF6 (3)	Ar$^+$ 750 eV	4×10^{-4}	5	W:O:C 90:5:5	350 (Bulk = 4.6)
WF6 (7)	Ar$^+$ 500 eV	1×10^{-6}		W:F:C 93.3:4.4:2.3	10 (Bulk = 4.6)
W(CO)6 (4)	Ga$^+$ 25 keV		2	W:C:Ga:O 75:10:10:5	150 - 225
W(CO)6 (8) (40° C)	Ga/In/Sn 16 keV			W:C:O 50:40:10	100
Cr(CO)6 (5)	Au$^+$ 50 keV	0.7		Cr:C:Au:O 40:40:16:4	
Styrene (6)	Ga$^+$ 20 keV	0.5	3.6	C:O 65:30	

(1) K. Gamo et al., Japan J. Appl. Phys. 23 (1984) L293.
(2) K. Gamo et al., Microelectronic Engineering 5, 163 (1986).
(3) H. Lezec, MIT (unpublished).
(4) D.K. Stewart, L.A. Stern, and J.C. Morgan, SPIE (1989).
(5) W.P. Robinson, SPIE (1989).
(6) L.R. Harriott and J. Vasile, JVST B6, 1037 (1988).
(7) Z. Xu, K. Gamo and S.Namba, Riken Conf., Mar. 1989.
(8) Y. Madokoro, T. Ohnishi, and T. Ishitani, Riken Conf., Mar. 1989.

Table 2. Ion Beam Induced Au Deposition Using DMG (hfac)

| System | Ion | | | | Deposit Characteristics | | | |
	Energy	Species	Current Density	Substrate Temp.	Yield (Au Atoms/ion)	Composition (Atomic %)	Resistivity (μΩ-cm)	Ref.
Focused Beam	40 keV	Ga+	4-1000 μA/cm² *	Room Temp.	~3†	50% Au 35% C 15% Ga	500 - 1,500	1
	40 keV	Ga+	200 μA/cm² *	100-125 °C	~3†	80% Au 10% C 10% Ga	3 - 20	2
	20 keV	Ga+	--	--	25-100	-	–	3
Broad Beam	70 keV	Ar+	0.7-7 μA/cm²	Room Temp.	15-25†	50-65% Au 50-35 % C	10,000	4
	70 keV	Ar+	0.7μA/cm²	100-160 °C	50-60†	>90% Au	3-10	4
	5 keV	Ar+	5 μA/cm²	0-15 °C	3-12	50-65% Au 50-35% C	–	5

† Measured assuming a film density is equal to that of Au.

*Average current density of a rastered beam. The instantaneous current density was 0.3-3 A/cm .

1. P.G. Blauner, J.S. Ro, Y. Butt, and J. Melngailis, to be published J. Vac. Sci. Technol. B (1989).
2. P.G. Blauner, J.S. Ro, Y. Butt, and J. Melngailis, submitted for publication.
3. A. Wagner and J.P. Levin, Ion Implantation Conference, Kyoto, Japan, 1988.
4. J.S. Ro, A.D. Dubner, C.V. Thompson, and J. Melngailis, Mat. Res. Soc. Proc. 101, 255 (1988).
5. A.D. Dubner and A. Wagner, to be published, J. Appl. Phys. (July 15, 1989).

c) Mechanism

A model for the deposition process has been formulated in terms of the adsorbate coverage and the ion and gas fluxes[24,31,34] which is similar to the model proposed for electron beam induced deposition.[5] Such a model would be useful for predicting how a given deposition system could be optimized. However, as yet the fit with experimental data is uncertain and optimum conditions can best be determined by measurement. For example, in gold deposition consider the case where the focused ion beam is scanned fast enough to avoid any transient, local adsorbate depletion effects. Then the rate of film growth will increase with average current density until the average adsorbate density becomes depleted. At that point the deposition rate falls rapidly and in fact would become negative as milling takes over (Fig. 6). These results indicate that, for a given focused ion beam current and gas flux, there is a minimum area which can be continuously scanned to achieve maximum growth rate and scanning a smaller area will lead to a reduced rate and, eventually, milling. Thus for a 100 pA beam a 7 μm x 7 μm area can be scanned, but for smaller areas the beam would need to be blanked off for part of the time.

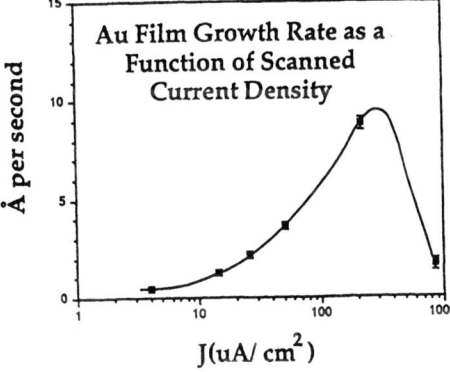

Fig. 6: The deposition rate as a function of average current density (Ref. 24). The pressure of the DMG (hfac) gas is constant at 10 mtorr.

As more data becomes available the type of models discussed up to now[24,31,34] can no doubt be refined to describe ion induced deposition. However, the underlying mechanism responsible for coupling the energy of the scanning ion to the molecules adsorbed on the surface remains unexplained. The quartz crystal microbalance experiments permit the density of adsorbed molecules to be measured and at the same time also the deposition yield (number of atoms deposited per incident ion). The cross section is then estimated to be 2×10^{-3} cm^2 for 5 keV Ar ions. [31] This means that on the average all molecules are decomposed within a 2.5 nm radius of the point of impact. At 5 keV the yield at pressures of ~5 mtorr is about 4 atoms/ion while for 70 keV Ar ions[30] the yield is 60. The density of adsorbed molecules was not measured in the 70 keV experiment, but since the dependence of density on pressure is very slow in the range of 1-20 mtorr, the coverage is likely to be similar. Thus the cross section is about a factor of 15 higher, and we can say that the incoming ion decomposes the molecules within a 10 nm radius.

How does a molecule sitting on the surface 10 nm from the point of ion impact receive enough energy to break apart? Three mechanisms can be imagined: electromagnetic radiation, electronic excitation, and atomic motion e.g. collision cascades or lattice vibrations. The fact that heavier ions under the same conditions have a significantly higher yield tends to favor the atomic motion, since lighter ions would be expected to produce more electronic or electromagnetic excitation. The limited data available[28] for the deposition from WF$_6$ is summarized in Table 3, and compared to the energy loss mechanisms for the incident ion. Clearly neither electronic energy loss rate of the incident ion nor the loss rate to substrate nuclei correlates perfectly with the yield. For what it is worth, we note that the square root of the total loss rate correlates reasonably well with the yield. More data is very much needed for yield as a function of ion mass and ion energy to shed light on the mechanism.

The scanning tunneling microscope may be a useful tool for observing the size and shape of deposited material due to individual ions. In fact, the reverse process, craters created by incident ion sputtering, has been observed by scanning tunneling microscopy.[37] For 8 keV Kr ions incident on a PbS (001) surface the area surrounding the point of incidence over which atoms are displaced from equilibrium is seen to be 3 to 5 nm in diameter.[37]

APPLICATIONS

a) ### Photomask repair.
In the case of photomask repair most of the systems designed use carbon deposition to replace missing absorber material,[9-15] although Cr is also used.[21] One preferred carbon source that has been reported is styrene[35] (Table I). The deposition time can be estimated from Table 4.

Table 3. Mechanisms of Ion Induced Deposition

(Film deposited from WF_6 at 20 mtorr, Ions 50 keV 1 $\mu A/cm^2$
by Gamo et al. ref. 28)

Deposition Ion Yield (a)	Rate of Energy Loss to Electrons (b) keV/μm	Rate of Energy Loss to Nuclei (b) keV/μm	Ion Range in W R_p (μm)	Range Straggle R_p (μm)
He 5.9	225	15	.110	.090
Ar 14.1	770	690	.015	.014
Xe 27.9	830	2900	.008	.005

Notes on table:

a) Calculated from data of K. Gamo et al. Microcircuit Engineering 5, 163 (1986) assuming pure tungsten is deposited.

b) Calculated using the TRIM program, original developed by J.P. Biersack and J.F. Ziegler, see for example, Ion Implantation Techniques, Eds. H. Ryssel and H. Glawischnig, Springer-Verlag, (1982), p. 122.

Table 4. Time to deposit 1 μm^3

(or 5 x 10^{10} atoms)

Beam Current (nA)	Beam diameter (assuming 4A/cm^2) (μm)	Yield (atoms/ion)	Time (sec)
0.1	0.06	3	28
0.1	0.06	10	8.3
1.0	0.18	3	2.8
1.0	0.18	10	0.83

Clearly to deposit large areas with high resolution will take considerable time, but then most repairs are not expected to exceed a few square micrometers.

b)   X-ray lithography mask repair.

The repair of clear defects (missing absorber) in x-ray masks will require high aspect ratio structures to be deposited. If the x-ray wavelength is in the range of 0.9 to 1.4 nm, then the absorber, say gold or tungsten, has to be about 0.3 to 0.5 μm thick. Thus minimum circuit features of 0.3 μm will lead to aspect ratios in excess of 1:1. In addition, the features will need to have nearly vertical side walls. Fig. 7 shows a photo of a 3.4 x 3.4 μm patch of gold 1 μm thick with almost vertical side walls. Since the gold contains a substantial percentage of carbon, the x-ray transmission is no doubt increased. In practice, this can be compensated for by increasing film thickness. Such patches are now being evaluated on x-ray masks at the Fraunhofer Institute in Berlin. Tungsten patches produced by ion induced deposition have also shown x-ray opacity.[20]

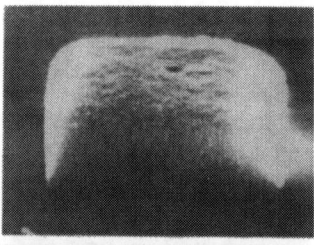
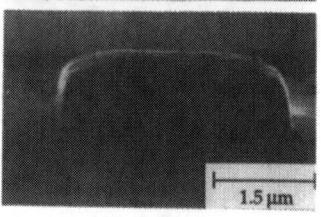

Fig. 7: SEM of a 3.4 x 3.4 m deposit shown in perspective (top) and in profile (bottom). This feature was written at a deposition rate of 11 Å/s. The height (~1 μm) and high aspect ratio of the edges of this deposit are of particular interest for application in x-ray lithographic mask repair. Ref. 23.

1.5 μm

c)   Circuit repair.

For restructuring or repair of the conductors on integrated circuits the deposited film must be conducting. In many cases short distances are covered and the deposited film can be thick so that resistivities up to 100 times higher than the bulk metal value may still be acceptable. In addition the contact resistance between the existing metal and the deposited metal must be low. Low resistance "jumpers" of W deposited using $W(CO)_6$ have been used to connect adjacent metal lines.[22] No evidence of contact resistance was reported.

We have deposited gold films with linewidths down to 0.1 μm (see Fig. 8). In some experiments thin lines were deposited across 4 fingers of Al, nichrome, or Au. By comparing 4 point and 2 point measurements we found that the contact resistance was negliable [24,33] As shown in Table 2, we have succeeded in depositing gold films with resistivity nearly equal to the bulk metal value. This should make repair of integrated circuits more convenient in that thinner films can be used.

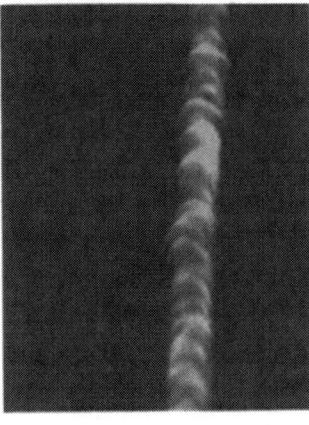

Fig 8: SEM of deposited line. This line was written by repeatedly and continuously scanning of 50 pA ion beam over 25 μm at 300 cm/s for 30 s.

0.1 μm

In any repair process the ability of the focused ion beam to image the area to be altered is critical as well as the ability to mill out via holes through the passivation layer and to cut conductors where needed.

In-situ deposition of contacts on III-V's, for example, is another potential area of application, particularly if lower energy ions can be used.

CONCLUSIONS:

Focused ion beam induced deposition is a novel process which is as yet not understood in detail. A number of metals have been deposited with submicrometer control of dimensions and some of the deposition parameters have been studied. The field is motivated by applications in photomask repair, x-ray mask repair and by circuit restructuring and repair.

ACKNOWLEDGEMENTS:

The authors wish to acknowledge the contributions of their colleagues at MIT to this work, in particular Jae Sang Ro, Andy D. Dubner, Yousaf Butt, Carl V. Thompson, and Henry I. Smith. Ion induced deposition at MIT is funded by the Army Research Office contract no. DAAL03087-K-0126 and by Draper Laboratory.

REFERENCES:

(1)   For reviews of laser-microchemical processing see, for example, D.J. Ehrlich and J.Y. Tsao, J. Vac. Sci. Technol. B4, 299 (1986) or D. Bauerle, Chemical Processing with Lasers (Springer, Berlin 1986).

(2)   J.G. Black, S.P. Doran, M. Rothschild and D.J. Ehrlich, Appl. Phys. Lett. 50, 1016 (1987).

(3)   S. Matsui and K. Mori, Jpn. J. Appl. Phys. 23, L706 (1984).

(4)   S. Matsui and K. Mori, J. Vac. Sci. Technol. B4, 299 (1986).

(5)   H.W.P. Koops, R. Weiel, D.P. Kern and T.H. Baum, J. Vac. Sci. Technol. B6, 477 (1988).

(6)   R.R. Kunz and T.M. Mayer, Appl. Phys. Lett. 50, 962 (1987).

(7)   W. Brünger, Microcircuit Engineering to be published (proceedings of Microcircuit Engineering Conference, Vienna, 1988).

(8)   S. Matsui and K. Mori, Appl. Phys. Lett. 51, 1498 (1987).

(9)   A. Wagner, Nucl. Instr. and Methods in Physics Research 218, 355 (1983).

(10)   J.R.A. Cleaver, H. Ahmed, P.J. Heard, P.D. Prewett, G.J. Dunn, H. Kaufman, Microelectronic Engineering 3, 253 (1985).

(11)   M. Yamamoto, M. Sato, H. Kyogoku, K. Aita, Y. Nakagawa, A. Yasaka, R. Takasawa, O. Hattori, SPIE vol. 632, 97 (1986).

(12)   H.C. Kaufman, W.B. Thompson, and G.J. Dunn, SPIE vol. 632 (1986).

(13)   B.W. Ward, D.C. Shaver, M.L. Ward, SPIE 537, 110 (1985).

(14)   N.P. Economou, D.C. Shaver, B. Ward, SPIE 773, 201 (1987).

(15)   W.P. Robinson and N.W. Parker, SPIE vol. 773, 216 (1987).

(16)   P.J. Heard, P.D. Prewett, and R.A. Lawes, Microelectronic Engineering 6, 597 (1987).

(17)   P. Sudraud, G. Benassayag and M. Bon, Microelectronic Engineering 6, 583 (1987).

(18)   L.R. Harriott, Applied Surface Science 36, 432 (1989).

(19)   K.P. Müller, U. Weigmann, and H. Burghause, Microelectronic Engineering 5, 481 (1986).

(20)   H.C. Petzold, H. Burghause, R. Putzar, U. Weigmann, N.P. Economou, and L.H. Stern, to be published SPIE Proceedings "Electron Beam, X-Ray and Ion Beam Technologies: Submicrometer Lithographies VIII (San Jose, Calif., Mar, 1989).

(21)   W.P. Robinson, to be published SPIE Proceedings "Electron Beam, X-Ray and Ion Beam Technologies: Submicrometer Lithographies VIII" (presented San Jose March 1989).

(22)   D.K. Stewart, L.A. Stern and J.C. Morgan, to be published SPIE Proceedings "Electron Beam, X-Ray and Ion Beam Technologies: Submicrometer Lithographies VIII" (presented San Jose March 1989).

(23) P.G. Blauner, J.S. Ro, Y. Butt, C.V. Thompson and J. Melngailis, Materials Research Society Proceedings, vol. 129 (1989) to be published.

(24) P.G. Blauner, J.S. Ro, Y. Butt, C.V. Thompson and J. Melngailis, J. Vac Sci. Technol. $\underline{B}$, to be published (July/Aug 1989).

(25) K. Gamo, N. Takakura, N. Samoto, R. Shimizu, and S. Namba, Jpn. J. Appl. Phys. $\underline{23}$, L293 (1984).

(26) K. Gamo, N. Takakura, D. Takehara, and S. Namba, Extended Abstract 16th International Conference on Solid State Devices and Materials (Kobe, Japan, 1984), p. 31.

(27) K. Gamo and S. Namba, in Proceedings of Symposium on Reduced Temperature Processing for VLSI (Electrochemical Society, Pennington, NJ, 1986), vol. 86-5.

(28) K. Gamo, D. Takehara, Y. Hamamura, M. Tomita and S. Namba, Microelectronic Engineering $\underline{5}$, 163 (1986).

(29) G.M. Shedd, A.D. Dubner, H. Lezec and J. Melngailis, Appl. Phys. Lett. $\underline{49}$, 1584 (1986).

(30) J.S. Ro, A.D. Dubner, C.V. Thompson, J. Melngailis, Mat. Res. Soc. Symp. Proc. vol. 101, p. 255 (Mat. Res. Soc. 1988).

(31) A.D. Dubner and A. Wagner, J. Appl. Phys. to be published, July 15, 1989.

(32) A.D. Dubner and A. Wagner, J. Appl. Phys. to be published, May 1, 1989.

(33) P.G. Blauner, Y.Butt, J.S. Ro, C.V. Thompson, J. Melngailis, submitted for publication.

(34) F.G. Rudenauer, W. Steiger, and D. Schrottmayer, J. Vac. Sci. Technol. $\underline{B6}$, 1542 (1988).

(35) L.R. Harriott and M.J. Vasile, J. Vac. Sci. Technol. $\underline{B6}$, 1035 (1988).

(36) J. Melngailis, J. Vac. Sci. Technol. $\underline{B5}$, 469 (1987).

(37) I.H. Wilson, N.J. Zheng, V. Knipping, and I.S.T. Tsong, Appl. Phys. Lett. $\underline{53}$, 2039 (1988).

LATTICE DEFECTS GENERATED BY ION IMPLANTATION
INTO SUBMICRON Si AREAS

M. TAMURA, S. SHUKURI and Y. KAWAMOTO

Central Research Laboratory, Hitachi, Ltd., Kokubunji, Tokyo 185, Japan

ABSTRACT

Cross-sectional transmission electron microscopy observations have been carried out to clarify two-dimensional depth distributions of lattice defects generated in high-dose (5×10^{15} ions/cm^2), P, As, BF$_2$ and B implanted, annealed submicron Si areas as a function of implantation areas. Monte Carlo simulation is also adapted for ion-implantation into submicron Si through fine mask patterns to predict the effect of mask size on spatial damage and impurity profiles. Simulation results predict that the above profiles have a strong mask size dependence for regions below the critical size, where the dopant concentration decreases and damage depth moves toward the surface-side with a reduced implantation area. Some experimental results support simulation results, although most defects, mainly in the P and As implantation, are confined within the original amorphized layers, independent of mask size. However, in BF$_2$ and B implantation, unexpected defect behavior such as variations in defect distribution from one implanted layer to another is found to occur in submicron regions doped by implantation.

INTRODUCTION

Continuous demands for the miniatuarization of individual devices in Si LSIs have led to the development of finer and shallower junctions formed by ion implantation and annealing. With the reduction in junction size, it has become more important in designing junctions to precisely know the two-dimensional distributions of primary and secondary defects and impurities in implanted and annealed layers. The following parameters mainly determine the two-dimensional distributions of the ion-implanted area including impurities and defects: (1) mask-pattern dimensions, (2) lateral straggling of implanted ions, and (3) subsequent thermal processes. There have been a few reports on the spatial distributions of impurities and lattice defects in ion-implanted submicron regions after annealing [1 - 4]. So far, we have reported some characteristic features of defects both in focused ion beam implanted layers [2, 3] and in regions implanted through fine mask patterns by a broad beam [4].

The present paper reports the two-dimensional distributions of damage in ion-implanted submicron Si areas obtained by simulation and experiments. First, we discuss changes in the spatial distribution of damage with a reduction in pattern sizes into which implantations are carried out. These changes in spatial distribution are based on Monte Carlo simulation results. Then, the two-dimensional defect distributions obtained by cross-sectional transmission electron microscope (XTEM) observations are reported on high-dose P, As, BF$_2$ and B implanted submicron (100) Si areas.

MONTE CARLO SIMULATION

Simulation program

The Monte Carlo program, developed by Ishitani et al. [5], was used to follow the three-dimensional trajectory of the energetic ions in amorphous Si substrates as well as to calculate the primary displacement energy transferred to the lattice atoms. In this program, the two-body collision model is

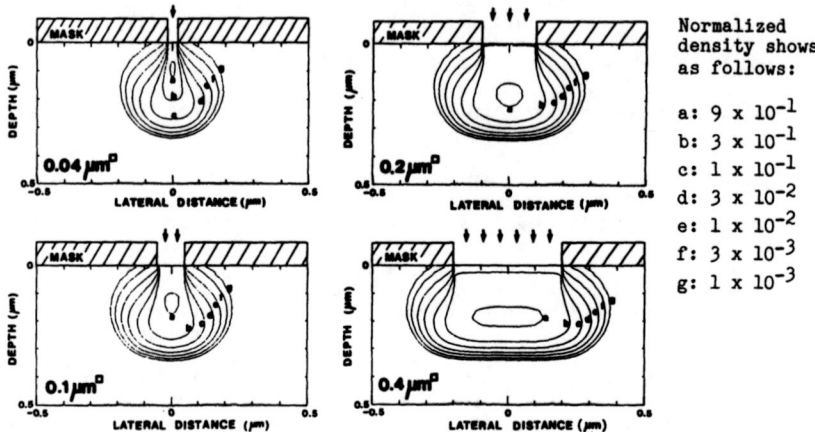

Normalized density shows as follows:

a: 9×10^{-1}
b: 3×10^{-1}
c: 1×10^{-1}
d: 3×10^{-2}
e: 1×10^{-2}
f: 3×10^{-3}
g: 1×10^{-3}

Fig. 1 Two-dimensional Frenkel pair density contours normalized by a peak concentration for 70 keV, B implantation as a function of mask opening size.

employed and the two-body interaction potential is the Moliere type [6]. In the calculation, 4×10^5 incident ions were used with a fixed incident energy to simulate the three-dimensional distribution of primary damage due to ion implantation below the surface.

Primary damage calculation

The distribution of Frenkel pairs was calculated using the Kinchin-Pease equation. The total number of Frenkel pairs, ν, produced by primary nuclear energy transfer was determined from $\nu = 1$ for $E_d < E_\nu < 2.5 E_d$, $\nu = 0.8 E_\nu/2 E_d$ for $2.5 E_d < E_\nu$, where E_ν and E_d are the transfer energy of the nuclear collision and displacement energy (approximately 25 eV), respectively.

Figure 1 shows a typical example of calculation results indicating two-dimensional Frenkel pair density contours normalized by a peak concentration for 70 keV, B ion implantation. Mask pattern sizes are between 0.04 and 0.4 μm. Results clearly show that the maximum concentration depth of Frenkel pairs approaches the surface-side with the decrease in mask size.

EXPERIMENTS

Conventional P, As, BF_2 and B ion implantation was carried out through striped windows in 400 nm poly-Si on 20 nm SiO_2 deposited on (100) Si. Window opening size varied mainly between 0.1 and 1.0 μm. The implantation with a dose of 5×10^{15} ions/cm^2 was done through 10 nm thick SiO_2 layers for all ions used here. Implantation energies were 50 and 140 keV for P, 25, 80 and 150 keV for As, 25 and 70 keV for BF_2 and 10, 70 and 100 keV for B. After implantation, 15 min isochronal annealing was performed in a dry N_2 atmosphere. The two-dimensional depth distribution of defects formed in implanted layers was then observed by XTEM.

RESULTS AND DISCUSSION

Prediction by simulation

Monte Carlo simulation results predict some specific features of mask size dependence of both defect and impurity distributions in submicron regions implanted through fine mask patterns. Figure 2 shows the mask size dependence of depths of the maximum Frenkel pair concentrations normalized by depths of the maximum impurity concentrations for various implantations indicated in the figure. The results clearly show the Frenkel pair depths move toward the surface with the decrease in mask size when the mask size diminishes below the threshold value for each implantation. It can be clearly understood from the figure that the threshold size becomes smaller with an increase in implanted ion mass: that is, it is 0.4 μm for 70 keV B, 0.1 μm for 100 keV P and 0.01 μm for 100 keV As. This result is mainly due to the magnitude of the lateral straggling of each ion.

The simulation program used here does not take into account the knock-on effect caused by interactions between host atoms for damage accumulation. This is to keep the calculation simple. Therefore, the damage depth for heavy ion, particularly As, should be shifted to deeper regions in real situations than in simulated ones.

Similar mask size effect on damage depth is also observed for the same ion species used for implantation, if the implantation energy is varied. This feature is typically shown in Fig. 3 in the B implantation. In the 10 keV B implantation case, the critical mask size for making the defect distribution change is below 0.1 μm, about 1/4 as small as that for 70 keV B implantation. This is also due to the degree of lateral straggling.

Another typical situation showing the mask size effect on the behaviors of implanted impurities and generated defects in submicron areas can be seen in Fig. 4. Here, we see that the maximum concentration of defects and impurities decreases with a decrease in mask size below the threshold value. The critical size is about 0.4 μm for B, 0.2 μm for P and 0.1 μm for As. This result suggests that the impurity concentration doped by implantation will vary for each implanted region, if various submicron sizes below critical ones are included in the doping area for implantation.

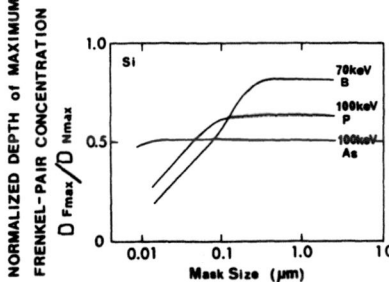

Fig. 2 Mask size dependence of depths of maximum Frenkel pair concentrations, D_{Fmax}, normalized by depths of maximum impurity concentrations, D_{Nmax}, for various implantations.

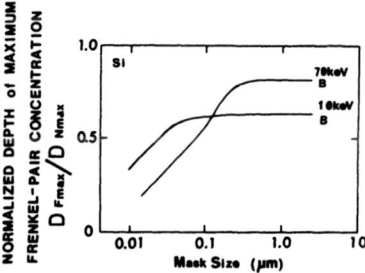

Fig. 3 Mask size dependence of D_{Fmax} normalized by D_{Nmax} for B implantation with different implantation energies.

In addition to these phenomena, it is important to note the lateral spread of implanted ions outside patterned areas. The ratio of the lateral spread of B concentration contours, D, to the mask opening size, d, is shown as a function of mask size and implantation energy in Fig. 5. Two different curves are shown for ion concentration contours of 10^{-1} and 10^{-3} from the maximum value.

The ratio of the lateral spread of implanted ions to the mask opening size increases sharply for mask sizes below 0.3 µm for 70 keV and 0.1 µm for 10 keV implantation, as the figure clearly shows. These results strongly indicate that interactions between impurities and /or defects will occur severely between adjacent different implanted areas with submicron sizes smaller than a 0.2 ∿ 0.3 µm mask opening, although low energy implantation, such as 10 keV, can shift the critical size into smaller values for this phenomenon to occur, as suggested from the figure.

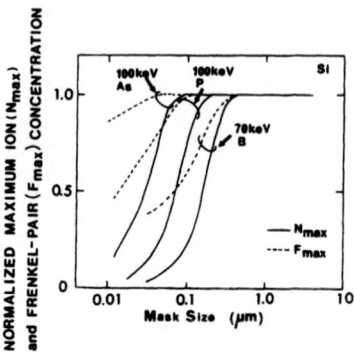

Fig. 4 Mask size effect on concentrations of N_{max} (maximum ion) and F_{max} (maximum Frenkel pair) for various implantations.

Damage formation and its annealing behavior
P and As implantation

In this section, characteristic features of defects generated in high-dose (5×10^{15} ions/cm^2) P and As implanted and annealed layers through sub-micron mask patterns are described. XTEM micrographs showing two-dimensional distributions of defects generated in submicron areas with different mask sizes by 50 keV, P implantation are summarized in Fig. 6. In as-implanted layers, continuous amorphous Si (a-Si) layers are formed from the surface to a depth of 107 nm in each submicron implanted layer, irrespective of mask size. The angle of the poly-Si mask edge to the substrate surface is approximately 70°. Therefore, the poly-Si masks are partially amorphized even for surface regions near the substrate surface. After 800°C annealing, the well

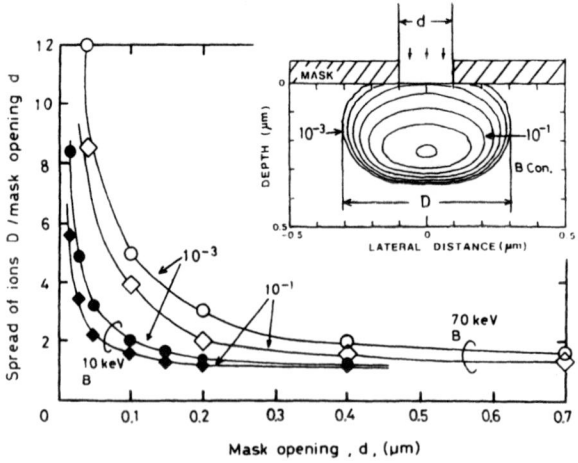

Fig. 5 Ratio of lateral spread of B concentration contours, D, to mask opening size, d, as a function of mask size. Two B implantation cases with 10 and 70 keV implantation enegies are compared.

known dislocation loop formation occurs just below the original amorphous/crystalline (a/c) interface for all the implanted layers. Together with this loop formation, dislocation loops are also formed near the range, R_p ($\sim$60 nm) for samples with mask opening sizes larger than 0.3 µm, as shown in the figure.

The dislocation loops at the R_p are considered to be originating from precipitation due to the high-concentration P atoms at R_p, where P atoms ($\sim$8 x 10^{20} atoms/cm^3 at R_p) exceed solid solubility limits ($\sim$4 x 10^{20} atoms/cm^3) at 800°C. However, interestingly, no loop formation at R_p is detected for any samples with mask opening sizes below 0.1 µm, as recognized from the figure. This fact may reflect that the maximum impurity concentration slightly decreases below the solubility limit at 800°C for implanted layers below 0.1 µm regions, as discussed in the previous section in connection with Fig. 4.

Annealing at 1000°C in Fig. 6, results in complete elimination of all the dislocation loops formed at 800°C. However, defects just under the mask corners remain in all the implanted regions. These defects exist separately under two mask edges for samples with mask sizes above 0.1 µm, but interactions between these defects occur in implanted regions with smaller mask sizes than 0.1 µm, as shown in the figure. Such a mask edge defect formation process has already been discussed for high-dose As implantation by considering two-dimensional SPE recovery process of implantation-induced amorphous layers with circular profiles under the mask edges [7].

Figure 7 shows XTEM results for a 140 keV P implantation case. Here, a continuous amorphous layer of 270 nm, thicker than that of the 50 keV implantation, is equally formed in individual implanted layers. We also note that this thick amorphous layer formation results in the 0.2 µm extension into the Si substrate surface under the poly-Si mask edge. Thus, although the two-dimensional amorphous shapes in the 140 keV P implantation are different from those in the 50 keV P implantation, the structure and nature of secondary defects generated after 800 and 1000°C are nearly the same as those observed in the 50 keV P implantation except for the following two slight differences: One is that dislocation loops at the R_p depth are not generated in all the submicron areas mainly due to the decrease in the maximum impurity concentration ($\sim$3.6 x 10^{20} atoms/cm^3) at R_p by the increase of implantation energy. Another is that the mask edge defect generation is clearly observed even after 800°C annealing. No special size effects on defect generation and distribution after annealing are also detected in 140 keV implantation.

In As implantation, the shapes of amorphous regions and the nature of post–annealing-induced defects in submicron regions showed nearly the same features as in the P implantation, although the survival or elimination of R_p defects after high-temperature annealing varied according to implantation energies. This phenomenon is attributed to the difference in interactions between recoiled oxygen atoms and R_p defects. We have already shown in 80

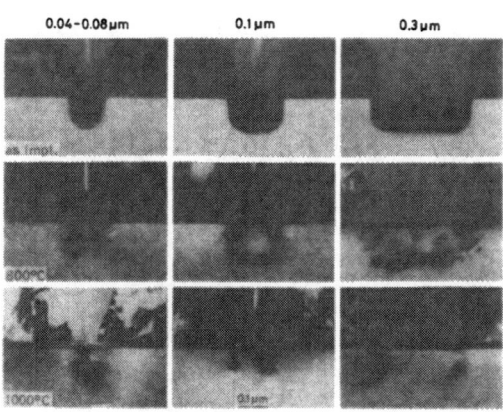

50 keV, P, 5 x 10^{15}ions/cm^2

Fig. 6 XTEM micrographs showing two-dimensional distributions of defects observed in 50 keV, P implanted submicron areas. Mask opening size and annealing temperatures are indicated.

148

keV As implantation that oxygen
atoms of about 2×10^{19} atoms/cm^3
have a strong pinning effect on R_p
defects [8]. Whereas, oxygen
above 2×10^{19} atoms/cm^3 has the
reverse effect, i.e., dissolution,
on these defects [8]. One typical
example is shown in Fig. 8 for 25
keV As implantation. Here, no R_p
depth (∿22 nm) defects remain in
the original amorphized regions
which have an equal thickness of
about 40 nm in individual implanted submicron areas, after 800°C
annealing.

The maximum concentration of
As atoms at an R_p depth under 25
keV, 5×10^{15}/cm^2 implantation
condition is about 2.5×10^{21}
atoms/cm^3, well above the solid
solubility (∿2×10^{20} atoms/cm^3
[9]) at 800°C. Nevertheless, the
fact that precipitate-related R_p
defects are completely eliminated
clearly suggests that the interaction between high-concentration
knocked-on O and As atoms after
800°C annealing suppressed the As
clustering phenomenon in these
submicron regions. In this connection, such R_p depth defects
were severe in both 80 and 150 keV
implanted layers even after 1000°C
annealing. This is considered to
be because of sharp decrease in

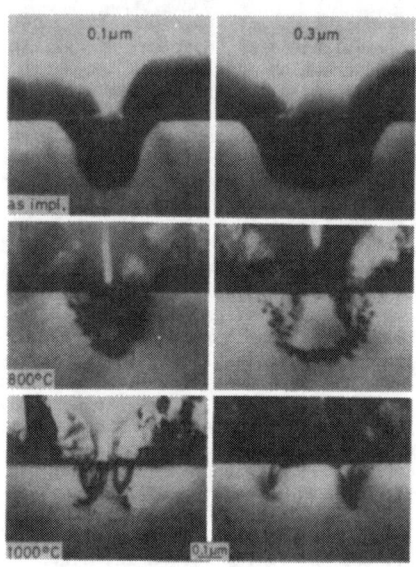

140 keV, P, 5×10^{15} ions/cm^2

Fig. 7 XTEM micrographs showing two dimensional defect distributions observed
in 140 keV, P implanted submicron areas.
Mask opening size and annealing conditions are indicated. Note severe defects under mask edges after annealing
at temperatures above 800°C.

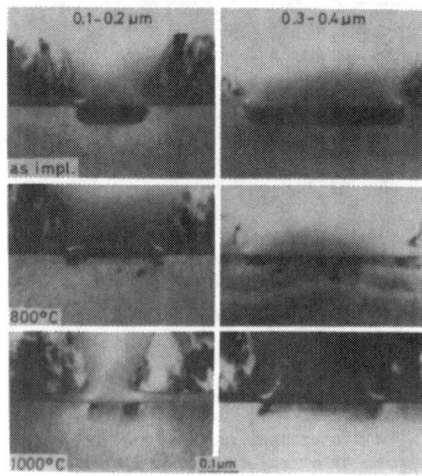

25 keV, As, 5×10^{15} ions/cm^2

Fig. 8 XTEM micrographs showing
two-dimensional defect distributions observed in 25 keV, As implanted submicron areas. Mask
opening size and annealing temperatures are indicated.

knocked-on O concentration at the Rp depth for both 80 and 150 keV, resulting in the stabilization of Rp defects by pinning them.

From the above results, we can conclude for high-dose P and As implantation into submicron Si where continuous amorphous layers form that (1) amorphous depth variation was not observed among samples with implanted submicron areas under the present experimental conditions; (2) all the defects generated after annealing were confined within the original a-Si regions; (3) common defects observed in all the submicron regions were the mask edge defects which were generated by a complex recovery process of mask edge amorphous layers. They survived in all the submicron regions even after 1000°C annealing.

BF$_2$ and B implantation

In this section, characteristic features of defects in submicron regions with high-concentration p-type layers doped by BF$_2$ and B implantation are described. In BF$_2$ layers implanted at 25 and 70 keV, amorphous layers were formed from the surface to 40 nm and 110 nm depths, respectively. These depths did not change among samples with different submicron regions. The annealing sequence results of defects by XTEM observations are shown for 25 keV implanted layers in Fig. 9. At 800°C annealing, two high-density defect bands remain. One exists near the surface region and another along the original a/c interface. The former defects are composed of microtwins, as already revealed in 110 keV BF$_2$ implantation by Nieh and Chen [10]. The latter, a/c dislocation loops, are decorated by F atoms, as clearly shown by the F profile results in BF$_2$ implanted layers [11]. Accordingly, the loop-band width is three or four times larger than that observed in P and As implanted layers.

As a result, in implanted layers with a 0.1 μm mask opening size, some a/c interface defects were observed to be grown, extending from the a/c interface to about a 50 nm depth into the substrate. These severe defects almost disappear at 1000°C annealing, as seen from the figure. Since F atoms at 1000°C move to the substrate surface [12], the pinning effect of F atoms on defects vanished, resulting in an a/c interface dislocation loop dissolution. Only a low density of microtwins was observed in the top surface region about 30 nm in thickness from the surface, but no bubble formation [10] was observed. These defects disappearance results were independent of mask size.

On the other hand, in 70 keV BF$_2$ implantation, the decoration effect on a/c dislocation loops by F atoms at 800 °C weakend in the micrograph in comparison with the 25 keV implantation. This may be due to the separation between the Rp depth of F and the a/c interface. Therefore, defects (twins near the surface and a/c dislocation loops) were almost confined within the original a/c regions after 800°C annealing. However, in some submicron regions, rodlike defects were found to exist outside the a/c interface at a depth of 100 - 150 nm below the a/c interface. The same result has already been observed in 110 keV BF$_2$ implanted and 700 - 900°C

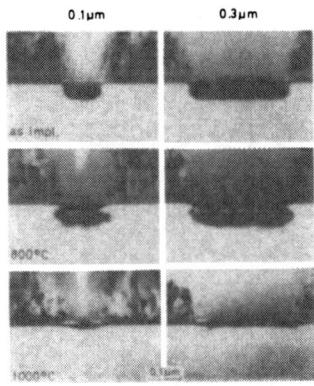

Fig. 9 XTEM micrographs showing two-dimensional defect distributions observed in 25 keV, BF$_2$ implanted submicron areas. Mask opening size and annealing temperatures are indicated.

150

annealed Si for blanket implantation [10]. Such
a phenomenon as where defects laterally distribute
under the mask material beyond the a/c interface
was not observed.

(a)

However, contrary to the results of the 25
keV implantation, the generation of severe defects
extending into the area beneath the original a-Si
layer was observed in 1000°C annealed submicron
regions, although rodlike defects were completely
absent. Typical examples are shown in Fig. 10
for the samples with 0.3 μm mask opening size.

Figure 10(a) shows an example where defects
are confined within the original a-Si layers, and
10(b) and (c) show examples where they grow from
the a-Si areas into outside ones. Thus, the
grown defects' two-dimensional distributions
varied from sample to sample with submicron implanted layers. The figure also clearly shows
that defect types such as long-elongated dislocations about 0.4 μm in length along the <211>
direction (Fig. 10(c)) and large loops (Fig. 10
(b)) differ in these respective implanted regions. Such behavior of residual defects after
1000°C annealing markedly contrasts with that
observed in the P and As implanted layers, where
all the defects remained within the original a-Si
layers, even if the situation is the same as the
continuous amorphous layer formation is similarly
created in the submicron regions. It is not
clear at the moment whether or not this difference can be attributed to the existence of
high-concentration F atoms in the BF_2 implantation.

(b)

(c)

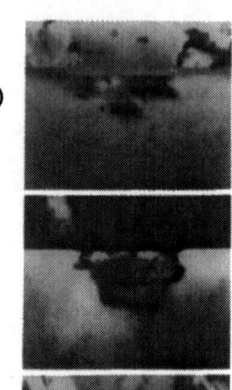

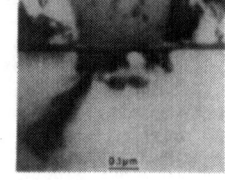

70 keV, BF_2, 5 x 10^{15} ions/cm^2, 1000°C

Fig. 10 XTEM micro-
graphs showing defect
behavior variations
from sample to sample
observed in 70 keV,
BF_2 implanted and
1000°C annealed sub-
micron areas of 0.3
μm size.

It is understood that no clear mask edge
defects are generated in BF_2 implanted layers by
comparing Figs. 6, 7 and 8 with Figs. 9 and 10.
This result shows another big contrast between P
and As, and BF_2 implantation. The SPE recovery
process of a-Si layers in the BF_2 implantation
(110 KeV, 5 x 10^{15} ions/cm^2) is reported [12].
According to the results, the regrowth speed of a BF_2 implanted layer is
strongly affected by the concentration of F atoms. The average regrowth
speed of a BF_2 implanted layer is roughly 2 nm/min, about 1/20 that of P and
As implanted layers. Another interesting point is that (111) faceting occurs at the growing a/c interface, resulting in the formation of microtwins
near the surface region. These two results suggest that the orientation dependence of SPE regrowth speed is weakened in BF_2 implanted layers. This
will result in missing the clear formation of mask edge defects as a consequence of nearly the same slow growth rate of mask edge amorphous layers.
The detailed observations on the mask edge amorphous recovery process will be
reported elsewhere [13].

Next, characteristic of defects in high-dose B implanted submicron areas
are reported. Figure 11 shows XTEM results for a 10 keV B implantation case.
In the present experiment, no amorphous layers were created in the implanted
layers. This is a different situation from P, As and BF_2 implantation.
However, as shown in the figure, a strong black contrast region showing the
damage accumulation phenomenon is detected at about a 0.8 Rp (Rp ≈ 33 nm)
depth in 10 keV implanted layers with sizes larger than 0.3 μm. This damage
accumulation area is seen to slightly move deeper in the substrate in a
0.1 μm region. After 800°C annealing, the agglomeration of small dislocation

0.1 µm   0.3 µm   ≥ 0.5µm

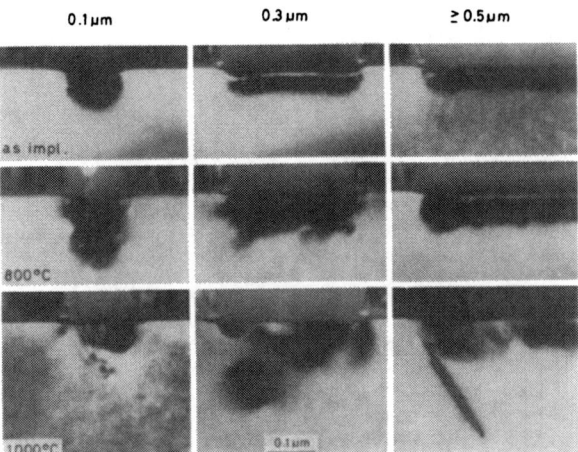

10 keV, B, 5 x 10^{15} ions/cm^2

Fig. 11 XTEM micrographs showing two-dimensional defect distributions observed in 10 keV, B implanted submicron areas. Mask size and annealing temperatures are indicated.

loops is formed at about Rp depth in the region with mask sizes above 0.3 µm, while this defect position in the 0.1 µm is about 2 times deeper than in the larger mask sizes. However, particular extension of defects into the substrate under the mask material was not observed. After 1000°C annealing, both dislocation loops and stacking faults are generated in the implanted layers. Some of these defects are grown to extend into a region of 0.2∿0.3 µm depth as shown in the figure. They are elongated along the <211> directions in the micrograph. These defect configuration and formation varied in the individual implanted submicron areas. This situation is nearly the same as that described for 70 keV BF$_2$ implantation results in Fig. 10. However, amorphization and the effect of F atoms on defect formation are excluded in the B implantation case, when we consider the defect expansion process. The same phenomenon has already been observed in focused 70 keV B ion beam implantation with a 7 x 10^{14} ions/cm^2 dose, where defects remaining after 1000 °C annealing showed remarkably different distributions and shapes in each 0.2 µm implanted layer [3].

When there is no amorphous layer formation, point defects which will form secondary defects, move freely and condense on {111} planes. This forms residual defects of different shapes, sizes and types in each implanted region after high-temperature annealing such as 1000°C. This result is also believed to be affected by the statistical fluctuation in the growth of lattice defects originating from condensed-point defects existing in small dimensions.

In 70 and 100 keV B implantation, nearly the same features of defect growth and distributions were observed after annealing at temperature above 800 °C. Examples of the nature and distribution of such defects are illustrated in Fig. 12 for 70 keV implantation.

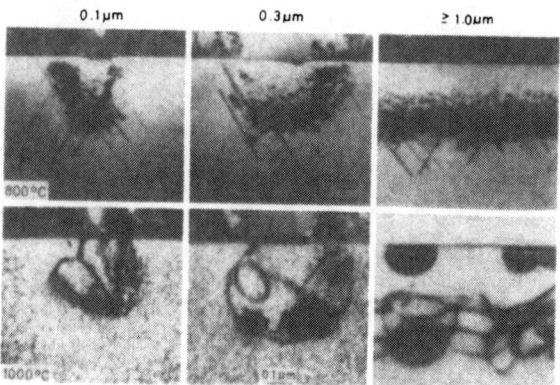

70 keV, B, 5 x 10^{15}ions/cm^2

Fig. 12 XTEM micrographs showing two-dimensional defect distributions observed in 70 keV, B implanted submicron areas. Mask size and annealing temperatures are indicated.

The figure shows that defects in 800°C annealed samples consist of buried agglomeration of small dislocation loops near Rp depth ($\sim$ 217 nm) together with rodlike defects extending from the lower edge of the above dislocation loop band into the substrate. The defects extend about two times Rp below the surface, independent of mask size. Such defects reach to the surface from Rp depth under the mask edge region by reflecting a 70° inclination of the mask edge to the substrate surface. 1000°C annealing results reveal the following defect charactersitics: in implanted layers with mask sizes above 1.0 μm, two different types of defects are present. One is faulted loops near the surface region. The other is dislocation lines distributed in deeper regions. The dislocation lines are believed to develop from dislocation loops under the strong strain field induced by the high-concentration B distribution gradient, although they were not observed in the 10 keV implantation.

Table 1 Maximum defect depths in B implanted submicron areas. Implartation conditions and mask opening size are indicated.

		0.1 μm	0.3 μm	≧1.0 μm
70 keV	800°C	330 nm	330 nm	330 nm
	1000°C	330 - 400 nm	400 - 530 nm	530 nm
100 keV	800°C	400 nm	400 nm	400 nm
	1000°C	400 - 500 nm	470 - 530 nm	600 nm

In contrast, in layers implanted through mask sizes below 0.3 μm, dislocation lines distribute throughout the entire region from the surface to depths which are dependent on the mask sizes. The observed depths in which defects exist are summarized in Table 1.

From Table 1, we note that dislocations generated after the 1000°C anneal are distributed at deeper regions with the increase in mask size. This feature is nearly the same as dislocations for 70 and 100 keV implantation. This is in contrasts with the result of 800°C in which defects distribute at the same depths independent of mask size. This fact may have a strong connection with the B diffusion process; that is, with the movement of B atoms to greater depths, dislocations will simultaneously move to relax the strain field induced by solute lattice contraction of B atoms. This phenomenon suggests that the more B atoms are included in implanted regions, the deeper the diffused areas are. Therefore, as suggested by simulation results in Fig. 4, one can speculate that the effective maximum B concentration decreases in submicron implanted regions below 0.3 μm size.

Finally, the spread of defects from the mask edges into the substrate was compared between calculation and observation results for 10 and 70 keV B implantation. Results are shown in Fig. 13. Experimentally, the spread of defects was evaluated as the lateral extension from the mask corner by neglecting the effect of mask edge inclination to the substrate surface. Nevertheless, qualitative agreement between simulation and experiment is obtained. The results indicate that interactions of defects (or impurities) between adjacent submicron regions including implanted B atoms become quite important with the increase in implantation energy, for example, from 10 to 70 keV.

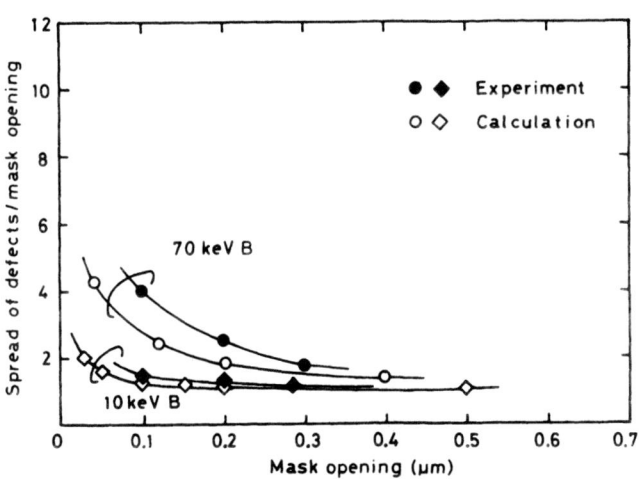

Fig. 13 Comparison of ratio of lateral spread of defects to mask opening size between calculation and experiment in 10 and 70 keV B implantation into submicron areas. The definition of the ratio is the same as in Fig. 5.

CONCLUSION

The annealing behavior of two-dimensional damage distribution in sub-micron Si areas implanted through fine mask patterns was investigated as a function of mask opening size. This was done mainly through cross-sectional TEM observations for samples implanted with high-dose (5×10^{15} ions/cm^2), 50 and 140 keV P, 25, 80 and 150 keV As, 25 and 70 keV BF$_2$ and 10, 70 and 100 keV B ions. Post-annealed, induced defects in amorphized Si layers by both P and As implantation were confined within the original amorphized Si, independent of mask size. However, severe defects other than implantation-induced ones remained under the mask edges as a consequence of the compli-cated process for recovering the mask edge amorphous layer. In contrast, in the BF$_2$ implantation with the formation of amorphous regions, there was no clear mask edge defect generation, although typical phenomena such as defect extension into deeper regions from amorphous layers in the substrate, were detected in submicron regions.

In B implanted layers without amorphous formation, defect distributions varied from one implanted layer to another, indicating that statistical fluctuations played an important role for growth from condensed-point de-fects to lattice defects in very limited areas. Monte Carlo simulation also suggested that spatial damage and impurity profiles in submicron areas in-troduced by implantation had a strong mask size dependence.

REFERENCES

1. E. Miyauchi, H. Arimoto, Y. Bamba, A. Takamori, H. Hashimoto and T. Utsumi, Jpn. J. Appl. Phys. 22, L243 (1983).

2. S. Shukuri, T. Ishitani, M. Tamura and Y. Wada, 18th Symp. Ion Impl. and Submicron Fabrication (Riken, March 1987) p. 73.

3. M. Tamura, S. Shukuri and M. Madokoro, J. Vac. Sci. Technol. B6,996 (1988).

4. M. Tamura, M. Horiuchi and Y. Kawamoto, Nucl. Instr. and Meth. B37/38, 329 (1989).

5. T. Ishitani, A. Shimase and S. Hosaka, Jpn. J. Appl. Phys. 22, 329 (1983).

6. G. Moliere, Z. Nat. A2, 133 (1947).

7. M. Horiuchi and M. Tamura, Nucl. Instr. and Meth. B37/38, 285 (1989).

8. M. Tamura and M. Horiuchi, Jpn. J. Appl. Phys. 27, 2209 (1988).

9. A. Lietoilas, J. F. Gibbons and T. W. Sigmon, Appl. Phys. Lett. 36, 765 (1980).

10. C. W. Nieh and L. J. Chen, J. Appl. Phys. 60, 3114 (1986).

11. M. Y. Tsai, D. S. Day, B. G. Streetman, P. Williams and C. A. Evans, Jr. J. Appl. Phys. 50, 188 (1979).

12. C. W. Nieh and L. J. Chen, J. Appl. Phys. 60, 7546 (1986).

13. M. Tamura and M. Horiuchi, to be published in J. Crystal Growth.

IN-SITU MEASUREMENT

OF ION BEAM INDUCED DEPOSITION OF GOLD

A.D. DUBNER* and A. WAGNER**
* Research Laboratory of Electronics, Massachusetts Institute of Technology,
  Cambridge, MA 02139
** IBM Research Division, T.J. Watson Research Center, Yorktown Heights,
  NY 10598

Work performed at: IBM Research Division, T.J. Watson Research Center
                      Yorktown Hts., NY 10598

ABSTRACT

A mechanistic model of the ion beam induced deposition (IBID) process is proposed, and an experimental system for measuring IBID is described. In IBID, the kinetic energy of the incident ion is deposited in the substrate surface resulting in the ejection of atoms (sputtering), and the decomposition of adsorbed molecules. Net deposition occurs when the number of atoms added through decomposition exceeds the number removed by sputtering.

Gold films were deposited by decomposing dimethyl gold hexafluoroacetylacetonate ($C_7H_7F_6O_2Au$) molecules with a 5 keV argon ion beam. The beam current density was 5 $\mu A/cm^2$. These films were deposited on quartz crystal microbalances (QCM). The QCM's provide an in-situ measurement of the deposition rate as a function of ion dose, dose rate, gas pressure, and substrate temperature. The mass of the depositing film increases with increasing ion dose. The deposition yield, or mass deposited per incident ion, is independent of the power input of the ion beam and therefore, the decomposition is not due to macroscopic heating. The deposition yield is shown to depend on the sputter yield of the substrate. The yield increases with increasing gas pressure and decreasing substrate temperature. This increase is proportional to the adsorption of the organometallic gas.

INTRODUCTION

Deposition of thin films is an important step in microelectronics fabrication. In the past several years, a number of techniques which use lasers [1], or electron beams [2,3,4] for localized film deposition have been described. In 1984, Gamo et. al [5] described a method of depositing thin metal films called ion beam induced deposition (IBID). In IBID, metal atoms are deposited on a substrate by decomposing organometallic molecules with an ion beam. Deposition occurs only where the ion beam is incident on the substrate, and localized deposition is achieved using a focused ion beam [6]. Thin films of aluminum [5], tungsten [7], tantalum [7], and gold [6] have been deposited using this technique.

An experimental system has been built to study the IBID process [8]. The system uses a quartz crystal microbalance (QCM) to continuously measure the mass of the deposited film. The deposition rate is measured as a function of ion current, gas pressure, and substrate temperature. The QCM is sensitive enough to measure the adsorption of less than one monolayer of the organometallic precursor. Thus, this experimental system is also used to measure gas adsorption on the substrate as a function of gas pressure and substrate temperature.

MODEL

Figure 1 illustrates a model of the IBID process. A substrate is surrounded by a few millitorr of an organometallic gas. Some of the gas molecules adsorb onto the substrate surface. Simultaneously, a beam of ions is incident on the substrate. The kinetic energy of each incident ion is deposited in the substrate surface through both nuclear and electronic excitation. This deposited energy results in the ejection of atoms (sputtering), and the decomposition of adsorbed molecules. Decomposition of the molecules results in the non-volatile component (the metal atom) being deposited and the volatile components, such as oxygen and fluorine, leaving the surface. Net deposition of metal occurs when the number of atoms added through decomposition exceeds the number of atoms removed by sputtering, i.e.,

$$Y_{net} = Y_D - S \tag{1}$$

where Y_{net} is the number of atoms deposited per incident ion, Y_D is the number of molecules decomposed by the ion, and S is the number of atoms sputtered by the ion.

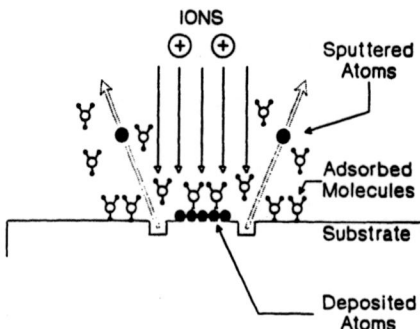

Figure 1: Schematic of the ion beam induced deposition process. Three elements are involved: the beam of ions, the organometallic gas, and the substrate. Atoms are deposited on the substrate surface by decomposing organometallic molecules with the ion beam. Net deposition occurs when the number of atoms added through decomposition exceeds the number of atoms removed by sputtering.

Note that Y_{net} is the experimentally measured quantity since the QCM measures a net mass change. The sputter yield (S) can be measured by removing the gas from the system which reduces decomposition (Y_D) to zero. The dependence of the net deposition yield (Y_{net}) on the sputter yield (S) in equation 1 was demonstrated in a previous publication [8]. The number of molecules decomposed per ion (Y_D) should depend on the number of molecules adsorbed and the ion decomposition cross section as will be discussed later.

RESULTS

The experimental system and measurement technique are described in reference 8, and will be reviewed only briefly here. The deposited mass is measured as a function of time by recording the resonant frequency of the QCM every two seconds. The gas pressure surrounding the QCM and the ion current hitting the QCM are recorded as well. The data are stored in an IBM PC/XT and can be uploaded and processed by the host IBM 3090.

Figure 2a shows an example in which gold films were deposited by decomposing dimethyl gold hexafluoroacetylacetonate $C_7H_7F_6O_2Au$, (abbreviated (DMG(hfac)) with 5 keV argon ions at two different dose rates (i.e. current densities). The gas pressure and substrate temperature were held constant during the entire run at 1 millitorr and 5°C. Negative frequency change is plotted verses time so that a positive slope corresponds to a mass increase. The mass of the deposit is shown on the right. The sloped regions of the curve (800-1600 seconds and 2000-2700 seconds) are where deposition took place. The horizontal regions are where the beam was blocked and no growth occurred. The current densities during the two depositions were ~5 $\mu A/cm^2$ and ~1.25 $\mu A/cm^2$ respectively. The reduction in dose rate drastically reduces the deposition rate.

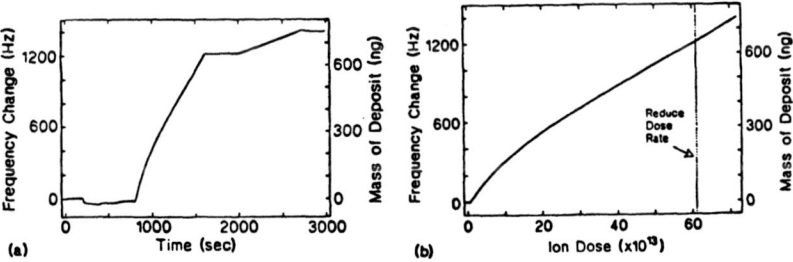

Figure 2: (a) A plot of frequency vs. time for a gold deposition run in which the dose rate was varied. The mass of the deposit is shown on the right. (b) A plot of frequency (and mass) vs. ion dose for the same deposition run. The reduction in dose rate does not affect the slope of the curve (i.e. yield is unaffected). Thus, the deposition process is not due to macroscopic heating.

158

Figure 2b shows the same deposition run plotted as a function of ion dose. The dose is calculated from the ion current which is recorded as a function of time during deposition. The reduction in dose rate does not affect the slope of the curve, i.e. the yield (in units of mass deposited per incident ion). This indicates that the deposition process is not due to macroscopic heating. When the power input of the ion beam is reduced by a factor of four the yield is not affected. The change in yield between 800 and 1400 seconds (or 0 to ~45x10[13] ions) is due to sputtering [8].

Figure 3a shows another deposition run. The deposition was performed with 5 keV argon ions at a current density of 5 μA/cm[2], and DMG(hfac) pressures of 5, 4, 3, 2, 1, 0.5, and 0.25 millitorr. The QCM temperature was held at 15°C. The slope of the curve corresponds to the deposition rate in nanograms per second. The deposition yield (mass deposited per incident ion) can be calculated by dividing the deposition rate by the ion impingement rate. Figure 3b shows a plot of deposition yield versus time for the data in figure 3a. The units have been converted from nanograms per incident ion to atomic mass units per incident ion. An approximate yield of gold atoms deposited per incident ion is shown on the right of figure 3b. It was calculated by assuming that all of the deposited mass is gold. While the film contains ~35 to 50 atomic percent carbon [8], the percentage by weight of gold in the film is very large (>90%) and dominates the measured mass change. Therefore, the assumption that all of the deposited mass is gold is not unreasonable. In figure 3b, each step corresponds to a deposition performed at a different pressure (5, 4, 3, 2, 1, 0.5, and 0.25 millitorr). The scatter in the data is due to scatter in the current reading.

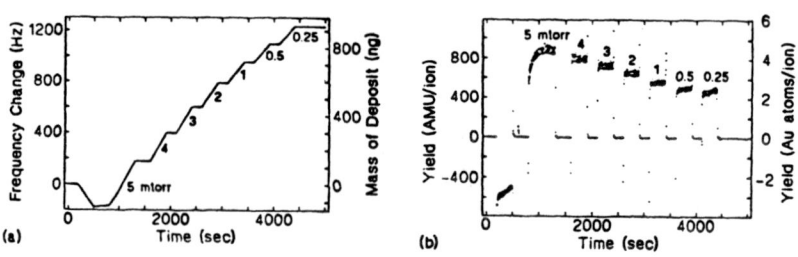

Figure 3: (a) A plot of frequency and mass vs. time for a run in which the temperature was held a 15°C and the gas pressure was varied. The slope of the curve corresponds to the deposition rate. (b) A plot of yield vs. time for the data in (a). The yield was calculated by dividing the deposition rate by the ion impingement rate.

From experimental runs like that shown in figure 3, a curve of yield verses pressure at a single temperature can be generated. By repeating the procedure at different temperatures a family of yield verses pressure curves can be generated. The deposition yield increases with increasing gas pressure and decreasing substrate temperature [9].

The QCM is sensitive enough to measure the adsorption of DMG(hfac). Adsorption isotherms for DMG(hfac) were measured at temperatures and pressures which correspond to the family of yield curves. The adsorption increases with increasing pressure and decreasing temperature [9]. Figure 4 shows the result of plotting the deposition yield curves against the DMG(hfac) adsorption at the same temperatures and pressures. The dashed line is a least squares fit to the data.

Figure 4 demonstrates that the deposition yield is proportional to the adsorption of the organometallic gas (as was postulated earlier). Thus, the number of molecules decomposed per ion (Y_D) depends on the number of molecules adsorbed and the ion decomposition cross section, i.e.,

$$Y_D = \Theta \Omega_D \tag{2}$$

where Θ is the number of molecules/cm² adsorbed on the surface, and Ω_D is the decomposition cross section in cm². Combining equations 1 and 2, the net yield depends on adsorption, the cross section for decomposition, and sputtering, i.e.,

$$Y_{net} = \Theta \Omega_D - S \tag{3}$$

This is the equation for the line in figure 4. The cross section is equivalent to the slope of the line. For a 5 keV Ar⁺ ion, the decomposition cross section is about 2×10^{-13} cm². In general, the decomposition cross section will depend strongly on the ion/solid interaction (i.e. ion mass and energy).

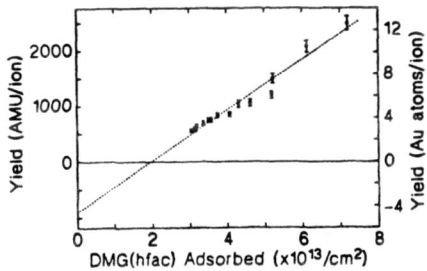

Figure 4: A plot of deposition yield verses DMG(hfac) adsorption for deposition of gold with 5 keV argon ions. The yield is proportional to the gas adsorption.

SUMMARY

Gold films were deposited by decomposing DMG(hfac), $(C_7H_7F_6O_2Au)$, molecules with a 5 keV argon ion beam. The mass of the depositing film increases with increasing ion dose, but is independent of ion dose rate (at the current densities used, i.e. $\leq 5 \ \mu A/cm^2$). Thus, the deposition yield, or mass deposited per incident ion, is independent of the power input of the ion beam. Therefore, the decomposition is not due to macroscopic heating. The deposition yield increases with increasing gas pressure and decreasing substrate temperature. This increase is proportional to the adsorption of the organometallic gas. The decomposition process (addition of atoms to the surface) competes with sputtering (ejection of atoms from the surface). The following equation can be written and has been verified experimentally: $Y_{net} = \Theta\Omega_\eta - S$.

REFERENCES

1. D.J. Ehrlich and J.Y. Tsao, J. Vac. Sci. Technol., B1 (4), 969 (1983).

2. R.R. Kunz and T.M. Mayer, J. Vac. Sci. Technol., B6 (5), 1557 (1988).

3. S. Matsui and K. Mori, J. Vac. Sci. Technol., B4 (1), 299 (1986).

4. H.W.P. Koops, R. Weiel, D.P. Kern, and T.H. Baum, J. Vac. Sci. Technol., B6 (1), 477 (1988)

5. K. Gamo, N. Takakura, N. Samoto, R. Shimizu, and S. Namba, Jpn. J. Appl. Phys., 23, L293 (1984).

6. K. Gamo, D. Takehara, Y. Hamamura, M. Tomita, and S. Namba, Microelectronic Engineering, 5, 163 (1986).

7. G.M. Shedd, H. Lezec, A.D. Dubner and J. Melngailis, Appl. Phys. Lett., 49, 1584 (1986).

8. A.D. Dubner and A. Wagner, J. Appl. Phys., 65 (9), 3636 (1989).

9. A.D. Dubner and A. Wagner, J. Appl. Phys., July 15, 1989.

BROAD AND FOCUSED ION BEAM Ga+ IMPLANTATION DAMAGE IN THE FABRICATION OF p+-n Si SHALLOW JUNCTIONS

A.J. STECKL*, C-M. LIN**, D. PATRIZIO***, A.K. RAI*** and P.P. PRONKO***

*University of Cincinnati, Cincinnati, OH 45221-0030 (author to whom correspondence should be addressed)
**Intel Corporation, Santa Clara, CA 95052-8121
***Universal Energy Systems, Dayton, OH 45232

ABSTRACT

The use of focused and broad beam Ga+ implantation for the fabrication of p+-n Si shallow junctions is explored. In particular, the issue of ion induced damage and its effect on diode electrical properties is explored. FIB-fabricated junctions exhibit a deeper junction with lower sheet resistance and higher leakage current than the BB-implanted diodes. TEM analysis exhibits similar amorphization and recrystallization behavior for both implantation techniques with the BB case generating a higher dislocation loop density after a 900°C anneal.

1. INTRODUCTION

The fabrication of Si shallow p+-n junction by implantation of Ga and/or other heavy Column III ions has recently been reported [1-3]. In these prior reports we have concentrated on using conventional (i.e. Broad Beam) Ga+ ion implantation and have achieved junctions with depths less than 100nm. We have also reported [4] the use of focused ion beam (FIB) implantation of the same species to fabricate p-n junctions and thin layers.

FIB technology is a particle beam technology which is proving to be extremely versatile [5,6]. FIB applications include optical and x-ray mask repair, circuit restructuring, micromilling, direct-write lithography, direct-write doping, etc. In particular, the direct-write doping is foreseen to have many attractive features: maskless and resistless processing, locally customized implant conditions (such as energy, dose, multiple implants, lateral profiling, etc.). To properly utilize the potential FIB advantages one must understand the nature of ion-solid interactions which take place at a current density 10^4-10^5 greater than conventional implantation. We have previously found that FIB-implanted Ga profiles exhibit a longer tail and a higher activation percentage than BB-implanted Ga [4]. In this paper, we report on an investigation of the damage generated by the two techniques and the effect on certain electrical properties.

We have used (100) Si n-type wafers with a carrier concentration of $5x10^{15}$/cm^3. Prior to implantation, an 8nm thermal oxide was grown. The Ga implantation energy and dose ranges were 50-75keV and 10^{13}-10^{15}/cm^2, for both methods. The FIB Ga+ current and current density were 200pA and 80mA/cm^2 while the corresponding values for the BB Ga+ were 75µA and 0.8µA/cm^2. Some of the samples were pre-implanted with Si (80keV, $2x10^{15}$/cm^2) to generate an amorphous surface layer. Post-implantation anneals were performed at either 600°C for 30sec, 900°C for 10sec, or 1,000°C for 10sec. The FIB Ga beam was mass analyzed and found to contain primarily Ga+. In the BB implantations the Ga beam was obtained by mass separation.

2. EXPERIMENTAL RESULTS

The Ga *atomic* concentration depth profiles of FIB and BB implantation obtained from SIMS analysis are shown in Fig. 1 for Ga-implanted at 75keV with a dose of $1x10^{15}$cm^{-2}. The implants were performed into crystalline and into amorphous Si substrates. The peak concentration occurs at a depth of 46nm in all cases, which is very close to the calculated projected range of 46.5nm. For implants into c-Si a long Ga tail is evident for both FIB and

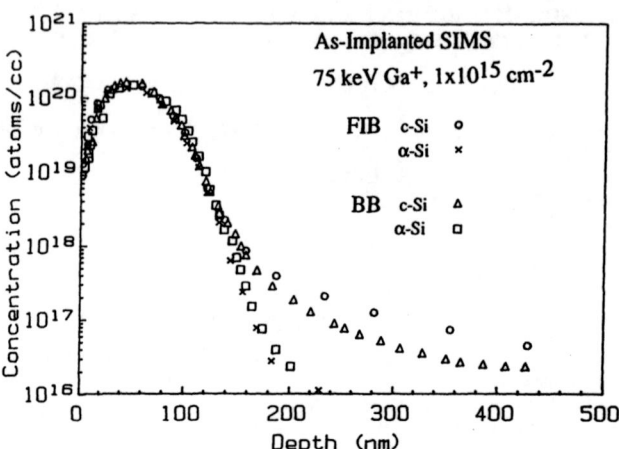

Fig. 1 SIMS Depth Profiles of 75keV Ga FIB and BB As-Implanted p+-n Junctions.

BB, with the former being more pronounced. With the Si top layer preamorphized by self-implantation, the Ga depth profiles begin to approach the ideal Gaussian distribution. The corresponding carrier concentration depth profiles obtained from spreading resistance (SRP) measurements performed after a 600°C anneal for 10sec are shown in Fig. 2. For the preamorphized case, a junction depth of 130nm is obtained for both BB and FIB samples. For implants into c-Si, the BB Ga-implanted *carrier* concentration profile does not change significantly, while the FIB depth profile exhibits higher peak activation and a deeper tail. This results in a junction depth for the FIB case of 220nm, as compared to 140nm for the BB implant. In general, the Ga activation appears greater for the FIB implants in all regions: R_p, end-of-range and tail. This is probably due to the greater damage, and hence larger number of Ga-accepting vacancies, created by the FIB implant.

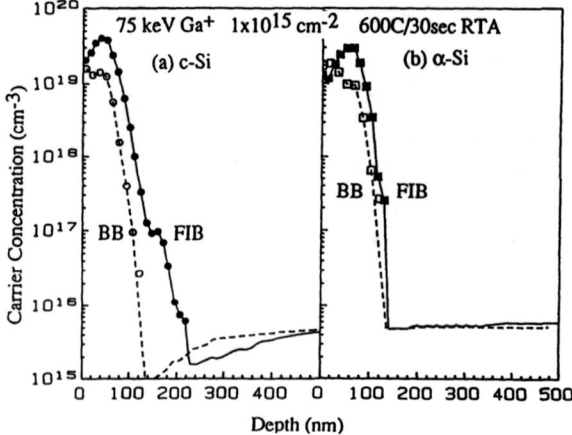

Fig. 2 Post-Anneal SRP Depth Profiles of 75keV Ga FIB- and BB-Implanted p+- Junctions.

The electrical properties of BB- and FIB-implanted p-n junctions produced by implantation with 50 and 75keV Ga for a $1 \times 10^{15}/cm^2$ dose and annealed at 600°C for 10sec are summarized in Table 1. The sheet resistance and the junction depth of the p+ layer were obtained from the SRP measurements. Diode leakage current density values were obtained from 5V reverse bias measurements. Comparing the properties of FIB- and BB-implanted diodes, a pattern emerges where in general the FIB diodes have a lower sheet resistance, deeper junction depth and higher leakage current density than their BB counterparts. The lower R_S is likely due to the higher Ga activation and the longer X_J is caused by a combination of the more pronounced tail in the atomic depth profile and its higher level of activation. Upon anneal at 900°C, the non-preamorphized samples resulted in a 25-50% increase in X_J, while those with the Si self-implant showed only a minor increase. The reason for the higher J_R in FIB diodes is not clear, but must be related to a greater level of defects in the depletion layer of the reverse-biased junction. In our p+-n junctions, the depletion layer extends primarily into the lightly doped n-Si substrate.

RBS analysis in channeling mode was performed with a 2MeV He+ ion beam on the BB-implanted samples only. From the aligned spectra of as-implanted samples shown in Fig. 3, amorphous layers of 175 and 120nm are measured to have been produced by the 75keV Ga implantation, with and without Si preimplant, respectively. The amorphous layers of both samples were regrown by RTA at 600°C for 30sec. The activated Ga percentage is estimated to be about 75%, after anneal. A small backscattering peak at the surface implies that the regrowth was not totally completed. A second peak, at 0.72MeV, is associated with disorder in the end-of-range implanted region. BB Ga-implanted samples annealed at 900°C for 10sec indicated reduced surface damage and somewhat greater end-of-range disorder. The 25 and 50keV Ga-implanted BB samples examined by RBS exhibited amorphous layers of 45 and 85nm, respectively, prior to annealing. With an anneal at 600°C for 30sec, the samples show similar behavior as in the 75keV Ga-implanted case.

Cross-sectional TEM analysis was performed using a Hitachi H600 scanning transmission electron microscope. Only 75keV Ga BB and FIB samples implanted into c-Si with $1 \times 10^{15}/cm^2$ dose have been analyzed to date. Fig. 4 shows TEM cross-sections of BB-implanted samples, as implanted and annealed. In the as-implanted case, an SiO_2 top layer is clearly visible. Next, a uniform amorphized Si region of ~91-92nm is measured. This is followed by an end-of-range region (11-12nm) showing heavy accumulation of point defects. Finally, the crystalline Si substrate follows. After the 600°C/30sec anneal, the amorphized region has recrystallized (to a thickness of 98nm) and a line of defects is now observed in the end-of-range. The BB-implanted sample annealed at 900°C for 10sec indicates that the individual defects have coalesced into dislocation loops.

Species	Implant Energy (keV)	Ion Implant Technique	Sheet Resistance (ohm/sq)	Junction Depth (nm)	$J_R(-5V)$ nA/cm^2)
Ga	50	FIB	1010	200	24.4
		BB	1660	130	2.5
	75	FIB	437	220	28.0
		BB	972	140	4.0
Si+Ga	50	FIB	950	90	9.13×10^7
		BB	974	110	5×10^6
	75	FIB	499	130	2.29×10^8
		BB	722	130	1.8×10^6

Table I. Electrical Properties of FIB-and BB-Implanted p+-n Junctions.

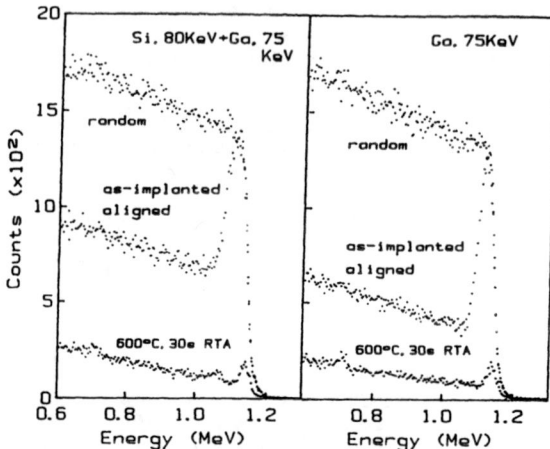

Fig. 3 RBS/Channeling Spectra for BB-Implanted p+-n Junctions.

A very similar behavior is observed for the FIB-implanted samples, as shown in Fig. 5. The as-implanted sample shows a uniformly amorphized layer of ~113nm, followed by an end-of-range defect region of 11-12nm. Upon anneal at 900°C for 10sec, a 98nm recrystallized region is measured along with similar dislocation loops as in the BB-implanted case. An estimate of the dislocation loop (area) density was obtained from these cross-sectional samples: $5.5 \times 10^{19}/cm^2$ for the BB case and $2.1 \times 10^{19}/cm^2$ for the FIB case.

3. DISCUSSION

From the combination of depth profiles (SIMS and SRP), TEM analysis and diode electrical measurements some (but not all) of the p-n FIB and BB Ga-implanted junction characteristics can be understood.

For the *non-preamorphized* samples, the implanted dose ($1 \times 10^{15}/cm^2$) was higher than the critical dose for amorphizing the Si crystal, which is $2 \times 10^{14}/cm^2$ for Ga implantation in Si. As confirmed by TEM, amorphous layers were therefore formed during implantation and recrystallized in the RTA process leading to an activated impurity concentration higher than the solid solubility at the anneal temperature. From the SRP analysis, peak carrier concentrations higher than $2 \times 10^{19}/cm^3$ were observed in the samples with 600°C RTA for 30sec, indicating the activated Ga impurities exceed their solid solubility in Si at these temperatures ($\leq 5 \times 10^{18}/cm^3$, maximum solid solubility of Ga in Si is $4 \times 10^{19}/cm^3$ at 1200°C). The thickness of the amorphized region was found to be 100-120nm from TEM and RBS measurements for both BB and FIB diodes. This is also consistent with a kink observed in the depth profiles which is normally located at the amorphous-crystalline interface.

The TEM analysis indicated that, in so far as it was observable, the damage due to FIB and BB implantation behaves similarly. Indeed, the first order measurements performed indicate that the dislocation loop density generated by the broad beam is higher than that due to the focused ion beam. No defects below the end-of-range were observed in the TEM cross-sections of either case. Unfortunately, this prevents us from being able to explain the leakage current density differences between the two cases, as the dislocation loops lie well inside the p+-side of the junction. Therefore, further analytical studies are necessary to explain the difference between FIB and BB implantation.

The authors wish to thank S. Hashimoto for the RBS measurements, G. Smith for the SIMS profiles and T.P. Chow for many useful discussions.

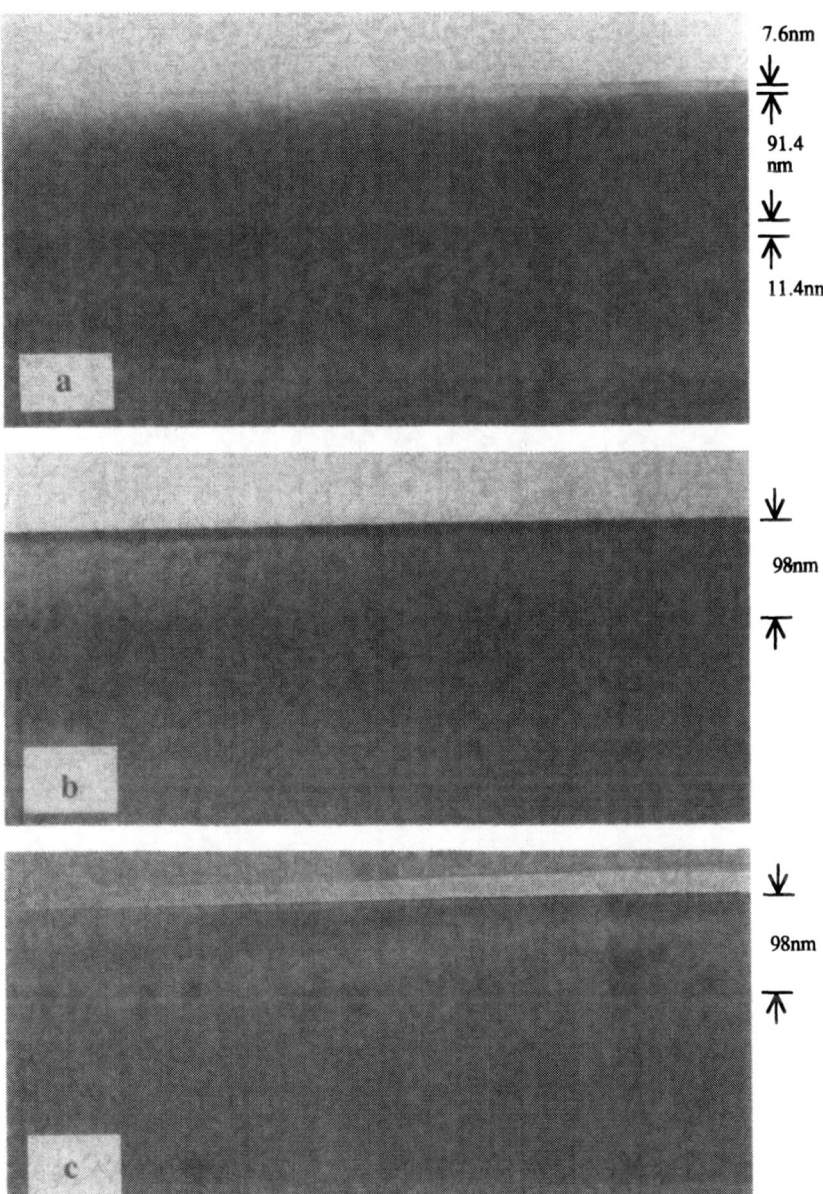

7.6nm

91.4 nm

11.4nm

a

98nm

b

98nm

c

Fig. 4 Cross-Section TEM Analysis of 75keV Ga BB-Implanted Samples: (a) as implanted (262,500x); (b) after 600°C/30sec anneal (153,000x); (c) after 900°C/10sec anneal (153,000x).

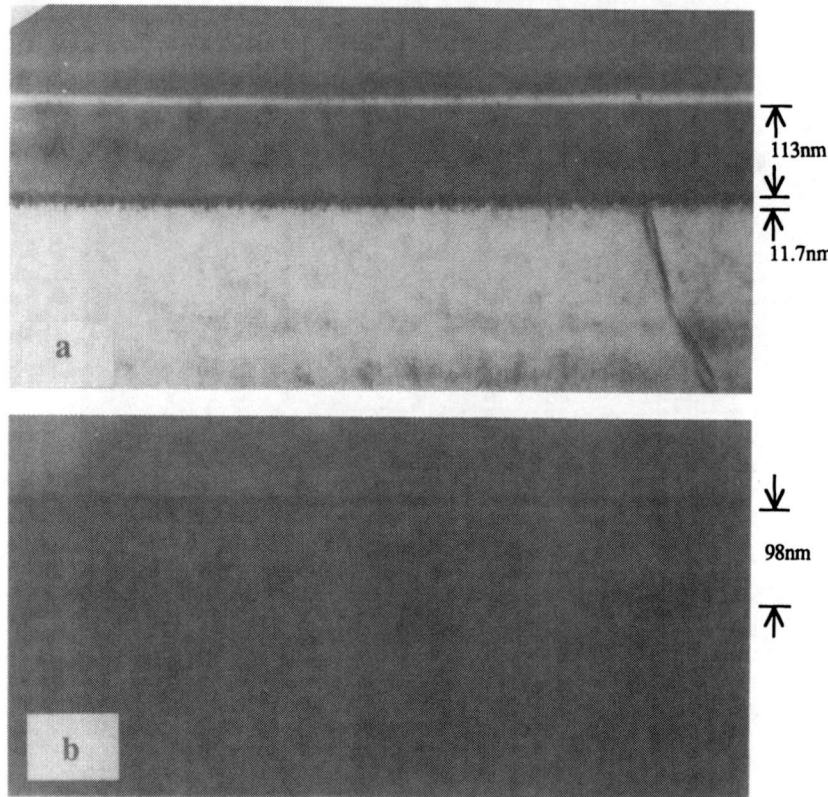

Fig. 5 Cross-Sectional Analysis of 75keV Ga FIB-Implanted Samples: (a) as implanted
(150,000x); (b) after 900°C/10sec anneal (153,000x).

REFERENCES

[1]   C.-M. Lin, A.J. Steckl and T.P. Chow, "Si p⁺-n Shallow Junction Fabrication Using
       On-Axis Ga⁺ Implantation," Applied Physics Letters, Vol. 52, pp. 2049-2051, June
       1988.

[2]   C.-M. Lin, A.J. Steckl and T.P. Chow, "Electrical Properties of Ga-Implanted Si p⁺-n
       Shallow Junctions Fabricated by Low Temperature Rapid Thermal Annealing," IEEE
       Electron Device Letters, Vol. EDL-9, pp. 594-597, November 1988.

[3]   C.-M. Lin, A.J. Steckl and T.P. Chow, "Sub-100nm p⁺-n Shallow Junctions
       Fabricated by Group III Dual Ion Implantation," Applied Physics Letters, Vol. 54, pp.
       1790-1792, May 1989.

[4]   C.-M. Lin, A.J. Steckl and T.P. Chow, "Thin Layer p-n Junction Fabrication Using Ga
       and In Focused Ion Beam Implantation," J. Vac. Sci. Tech., Vol. B6, pp. 977-981,
       May/June 1988.

[5]   A.J. Steckl, "Prospects for Particle Beam and In-Situ Processing of Integrated
       Circuits," Proceedings of IEEE, Vol. 74, pp. 1753-1774, December 1986.

[6]   J. Melngailis, "Focused Ion Beam Technology and Applications," J.Vac. Sci. Tech.,
       Vol. B5, pp. 469-495, March/April 1987.

PART IV

MeV Implantation

DAMAGE GROWTH IN Si DURING SELF-ION IRRADIATION: A STUDY
OF ION EFFECTS OVER AN EXTENDED ENERGY RANGE

O. W. HOLLAND, M. K. EL-GHOR, AND C. W. WHITE, Solid State Division, Oak Ridge
National Laboratory, Oak Ridge, TN 37831

ABSTRACT

Damage nucleation/growth in single-crystal Si during ion irradiation is discussed. For MeV
ions, the rate of growth as was well as the damage morphology are shown to vary widely along
the track of the ion. This is attributed to a change in the dominant, defect-related reactions as the
ion penetrates the crystal. The nature of these reactions were elucidated by studying the interaction of MeV ions with different types of defects. The defects were introduced into the Si crystal
prior to high-energy irradiation by self-ion implantation at a medium energy (100 keV). Varied
damage morphologies were produced by implanting different ion fluences. Electron microscopy
and ion-channeling measurements, in conjunction with annealing studies, were used to characterize the damage. Subtle changes in the predamage morphology are shown to result in markedly
different responses to the high-energy irradiation, ranging from complete annealing of the damage to rapid growth. These divergent responses occur over a narrow range of dose $(2-3 \times 10^{14}$
cm^{-2}) of the medium-energy ions; this range also marks a transition in the growth behavior of the
damage during the predamaging implantation. A model is proposed which accounts for these
observations and provides insight into ion-induced growth of amorphous layers in Si and the role
of the amorphous/crystalline interface in this process.

INTRODUCTION

The mechanisms responsible for damage accumulation in Si during heavy ion irradiation
have been the subject of many investigations. Obviously many of the studies were motivated by
the importance of ion-related technologies in the manufacture of Si-integrated circuits. Also, the
commercial availability of low-defect density, high-purity Si single crystal makes it an attractive
material for studying such phenomena. While much progress has been made, no comprehensive
model exists which can predict the nature of the ion-solid interaction over an extended range of
ion mass and energy and substrate temperature. Part of the problem has been in identification of
the many different defects which form during ion irradiation. The Frenkel defect (vacancy-
interstitial pair), which is produced directly when ion scattering in the lattice results in atomic
displacements, is mobile at RT (and below). These point-defects will participate in the formation
of other defects; ones which are stable at the irradiation temperature. While much progress has
been made in defect classification [1], little is known about their mutual interaction (which
obviously plays a role in the evolution of the damage morphology during irradiation) or their
formation kinetics.

The understanding of ion-related damage processes is further complicated by an uncertainty
in the role of the collision cascade. The passage of an energetic ion through a solid will initiate a
sequence of displacement events. If atoms are displaced with sufficient energy, their interaction
with other lattice atoms can lead to further displacements. Frenkel pair production is therefore a
cascade process in which the incident ion is simply the primary damage-producing particle. The
total of all such events is commonly referred to as the collision cascade of the ion. One model
envisions that stable damage is nucleated within these dense cascades during the time that it takes
for the energy spike to dissipate [2,3] (i.e., the time for cascade quenching, which is of the order
of 10^{-12}s). Such a model has been referred to as heterogeneous because the nucleation of damage is localized in the vicinity of the collision cascade. An alternate model considers damage
nucleation and growth to occur homogeneously within the lattice as a result of random interaction
between the point defects generated by different collision cascade [4]. Obviously in such a
model, the point defects are assumed to be long-lived relative to the quench time of the cascade so
that they become homogeneously distributed throughout the irradiated region without any spatial
correlation to the cascade volume.

In this paper, damage nucleation and growth in single-crystal Si during self-ion irradiation
over a wide range of ion energy is discussed. MeV ions were used because they offer the

possibility of studying ion-impact phenomena over an extended range of nuclear and electronic stopping. The nature of the damage growth varied over the penetration depth of the high-energy ions. Two distinct regions were identified, and the growth behavior in each is discussed. Insight into the mechanisms responsible for the different behaviors was gained from studying the interaction of the MeV ions with preexisting damage in the Si lattice. Predamaging of the lattice, prior to the MeV-ion irradiation, was accomplished by 100-keV self-ion irradiation. Different fluences were used to vary the predamage morphology. Characterization of the damage at the lower energy was done to better understand its interaction with the MeV ions. However, this characterization yielded interesting results which, when correlated with the higher-energy results, leads to a better understanding of ion-induced amorphization of Si and the role of the amorphous/crystal interface in this process. Analytical techniques used in the study included Rutherford backscattering/channeling spectroscopy (RBS) and cross-sectional transmission electron microscopy (TEM).

EXPERIMENTAL

Czochralski-grown, p-type Si (100) wafers with resistivities between 6–8 Ω-cm were used in this study. High-energy implantations (>1.0 MeV) were done with a raster-scanned beam of $^{28}Si^+$-ions from a 1.7 MeV General Ionex tandem accelerator. An energy of 1.25 MeV and an average current density <0.4 $\mu A/cm^2$ were generally used. The use of a tandem accelerator insures that the $^{28}Si^+$ beam will be free of any contaminants, such as molecular ions of the same mass-to-charge ratio, as might be encountered using a single-ended machine. For lower energy implantations of the Si^+ self-ion (which were done with a single-ended, mass-analyzed implanter), isotope ^{30}Si was used to minimize the possibility of contaminant beams. Post-implantation annealing was carried out in a conventional, quartz-tube furnace in flowing, dry nitrogen gas. Structural characterization of cross-sectionally thinned samples was done using a Philips EM400T transmission electron microscopy. RBS with 2.75-MeV-He$^+$ ions was used to measure damage profiles.

HISTORICAL PERSPECTIVE

Heterogeneous Model

Brinkman [5] argued that, when the nuclear stopping of an ion is sufficiently high to displace at least one atom per lattice plane, there will be a net outward motion of the target atoms along the incident ion trajectory. Such correlated motion will produce a vacancy-rich core surrounded by an interstitial-rich region and is known as a displacement spike. He suggested that nucleation of extended defects such as dislocations can result from such a process. It is clear that the high concentration of defects localized within dense collision cascades offer other possibilities for damage nucleation. Swanson [6] studied the ion-induced, crystalline-to-amorphous (c-a) transformation in Si and concluded that it occurs spontaneously when the free energy of the defective crystalline regions becomes equal to or greater than that of the amorphous phase. If the defect density (deposited energy density) is sufficiently high then it was suggested that the c-a transformation can occur directly within the volume of the collision cascade. For less-dense cascades, it was envisioned that overlap of the damage regions produced by different ions was needed to achieve the critical defect density to stimulate the phase change. Discrete damaged regions associated with collision cascades of ions have been observed by TEM [7–9]. Figure 1(a) shows the cascade damage observed by plan-view TEM in (110) Si after implantation at 80K with 100-keV Bi$^+$ ions to a dose of 1.0×10^{12} cm^{-2}. The cascades appear as black/white spots with an average size of 50 Å. The image reversal in the figure occurs across a thickness contour in the sample. The absence of <110> chains of atoms in the central region of the cascade damage is evident in the high-resolution micrograph in Fig. 1(b). Howe et al. [7] showed that the efficiency of creating such damaged regions was <0.1 when the deposited energy density was <0.1 eV/atom and was >0.7 when the density exceeded 0.4 eV/atom. It is tempting to conclude from such studies that the formation of a continuous amorphous layer proceeds by a coalescence or overlapping of such regions. Gibbons [10] incorporated this idea into a model which allowed for amorphization to occur either within a single cascade or as a result of cascade overlap. In the model, the fractional amorphous area is given as

$$\frac{A_A}{A_o} = 1 - \left(\sum_{k=0}^{m-1} \frac{(A_i\Phi)^k}{k!} \exp(-A_i\Phi) \right)$$

(1)

where m is the number of ions required to damage a region sufficiently to form the amorphous phase, Φ is the ion fluence, and A_i is the cross-sectional area of the amorphous region formed around the track of the ion. Somewhat surprising, Dennis and Hale [11] found that, for a wide range of ions (from Li^+ to Kr^+) and implantation temperature (80–300K), the m value determined from fitting their data to Eqn. (1) was always greater than 1 and was within the 2–3 range. This is suggestive that some other mechanism for amorphization other than heterogeneous growth must be considered. This was further confirmed by the work of Ruault et al. [9] who observed very dense cascades in Si by TEM. The cascades were found to have an amorphous core, and their number density correlated well with the implanted fluence. However, it was shown that this type of damage could not account for the formation of the continuous amorphous layer observed at higher fluences. Rather, it was found that nucleation of amorphous Si occurred randomly in the regions between the damaged regions of the collision cascade. This growth mechanism appeared to be more dominant than the cascade mechanism, and produced an amorphous phase which had recrystallization kinetics similar to that of a continuous amorphous layer, while the amorphous cascade regions annealed at anomalously low temperatures.

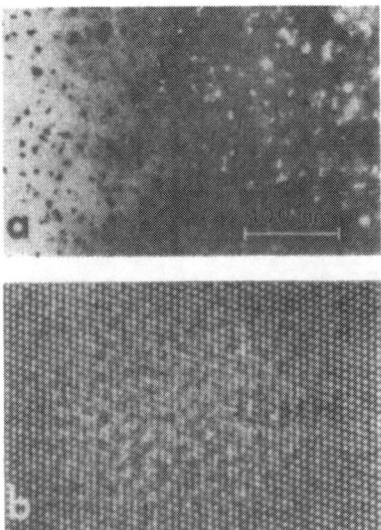

Fig. 1. (a) Bright-field electron micrograph in plan view of discrete damage produced by Bi^+ (1.0×10^{12} cm^{-2}, 100 keV, at LN$_2$) implantation of Si (110), and (b) high-resolution image from a damaged region showing the amorphous structure.

Homogeneous Model

In contrast to the heterogeneous model, is the homogeneous model in which damage accumulation is envisioned to occur as a result of outdiffusion of the Frenkel defects from the cascade volume where they interact with defects from other cascades to nucleate stable complexes. Damage is therefore nucleated homogeneously, rather than locally within the region of the cascade. Clearly, this model assumes that Frenkel defects formed within an ion cascade are long-lived so that any spatial correlation with the cascade volume is lost prior to their interaction. While it is reasonable to assume that this model is more appropriate to those cases where the ion-solid combination results in low-density cascades, the observations discussed above for dense cascades suggest that the homogeneous mechanism may be important over a larger range. While this model differs from the heterogeneous one, amorphization of the lattice during irradiation is still thought to occur by a spontaneous phase transformation when the local defect density exceeds the critical value. However, the onset of amorphization is expected to occur only after an incubation period (during which damage levels increase to the critical concentration), and there-after the growth will occur abruptly over a narrow range in dose. This is in sharp contrast to the heterogeneous model [given in Eqn. (1)] and therefore provides a stringent test to differentiate between the two models.

RESULTS

1.25 MeV Si+ Ion Irradiation

Aligned and random spectra from Si after implantation with 1.25 Si+ ions are shown in Fig. 2. Spectra from two samples which had been implanted with 0.25×10^{16} and 1.0×10^{16} cm^{-2}, respectively, are compared. The deviation of the aligned scattering yield at any point within the implanted samples from that in the virgin sample indicates the presence of ion-induced damage. In the spectrum from the sample implanted with the high-dose (HD), the scattering yield is seen to be at the random level in a region centered on the ions' end of range, EOR, (~1.3 μm). This is consistent with the formation of a buried amorphous layer. By comparison, the amount of damage ahead of the EOR and extending to the surface is seen to be much reduced. The results of computer simulation using TRIM [12] of 1.25-MeV Si+ ions in Si, normalized to an implant fluence of 1×10^{15} cm^{-2}, are shown in Fig. 3. The density of Frenkel defects

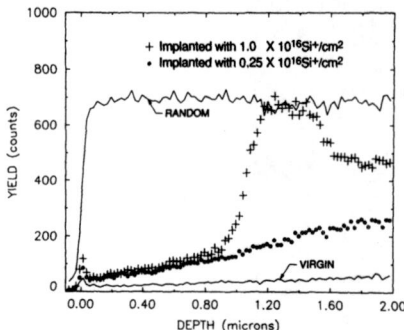

Fig. 2. <100> aligned spectra from Si single crystals implanted with 1.25-MeV Si+ ions.

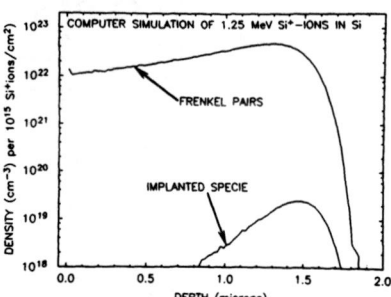

Fig. 3. TRIM calculation of the distribution of Frenkel defects and the implanted specie for 1.25-MeV Si+ ions in Si.

produced along the path of the ion is seen to increase near the ions' EOR, but the variation is small and clearly does not directly correlate with corresponding damage levels observed in the HD sample. The scattering yield in the lower dose (LD) sample is seen to rise slowly in a regular manner with depth until near the EOR where there is a small but perceptible change in the rate of increase. This suggests that a discrete change in the damage concentration or type occurs at this depth similar to that observed at the higher fluence. A comparison of the aligned spectra from the two implanted samples surprisingly shows that the yield from the top ~1.0 μm is the same in both samples. Apparently, there is some critical level for damage in this region after which no further growth occurs even though substantial numbers of Frenkel defects are produced within the region during subsequent irradiation. Therefore, two distinct regions are observed: in the vicinity of the ions' EOR, damage increases monotonically with dose until the lattice turns amorphous, while ahead of this region damage growth increases initially with dose but saturates at a low level. The damage remains saturated until further growth of the buried amorphous layer encroaches on the region. This is seen in Fig. 4, which compares aligned spectra from three samples which were implanted with 1.25-MeV Si+ ions at slightly different doses. The formation of a buried layer is clear in all samples. The damage (i.e., scattering yield) ahead of the buried layer is seen to remain constant, even though there the layer grows with increasing dose. The abrupt change in the scattering yield in the vicinity of the a-c interface in all samples indicates a sharp transition between the two phases and is consistent with a layer-by-layer growth mechanism for the a-layer.

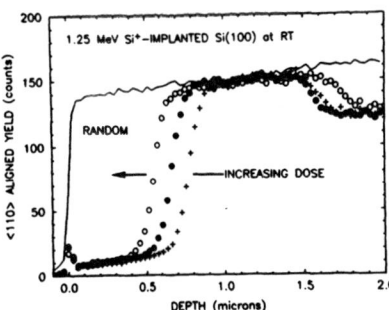

Fig. 4. Comparison of aligned spectra from Si single crystals implanted with different doses of 1.25-MeV Si+ ions.

Cross-sectional TEM micrographs from both the HD- and LD-implanted samples are presented in Fig. 5. The morphology determined from analysis of the micrographs is consistent with the RBS results above. The formation of the buried layer in the HD sample is clearly seen in Fig. 5(b). Extended defects and loops are evident both beyond the buried layer and in a narrow region ahead of the layer. However, the near-surface extending over a range of ~1.0 μm is remarkably free of any defects. It is clear that nucleation of extended defects such as dislocations is suppressed in this region. The absence of these types of defects plays an important role (which is discussed later) in the saturation of damage in this region. It is worth noting that the growth of the buried amorphous layer during continued irradiation occurs only at the a-c interface. The presence of dislocation loops in proximity to the a-c interface was characteristic in all samples examined in which an amorphous layer was formed. The morphology of the LD-implanted sample is shown in Fig. 5(a). No extended defects are seen over the top ~1.0 μm, while a band of defects is located near the EOR region. These defects were stable to temperatures in excess of 600°C and are thought to be small dislocation loops.

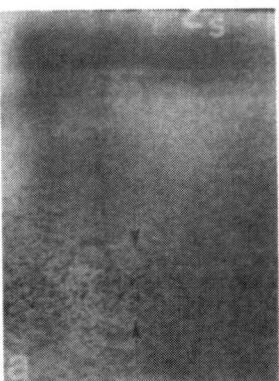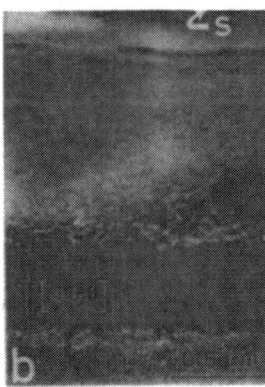

Fig. 5. Cross-sectional TEM micrograph from Si samples implanted with 1.25-MeV Si+ ions at a dose of (a) 0.25 × 10^{16} and (b) 1.0 ×10^{16} cm^{-2}.

The absence of extended defects in the near-surface after high-energy ion-irradiation suggests that the damage determined by RBS consists of small defect clusters. Since defects anneal with a characteristic activation energy, they are often mobile in different time-temperature domains and can therefore be identified by their annealing behavior. Aligned spectra, before and after annealing at 200°C for 1 h, from the HD Si⁺-implanted sample are compared in Fig. 6. This temperature is too low to anneal either dislocations or amorphous damage. It can be seen from the spectra that, while no regrowth of the buried amorphous layer occurred during the anneal, most of the damage in the near surface was removed. The characteristic annealing time (e.g., the time for annealing half of the damage) at 200°C for the defects in this region was 30 min. This is similar to that for the divacancy which is known to form in Si under certain irradiation conditions [13]. Therefore, the dominant defect in the region where damage saturation occurs appears to be the divacancy.

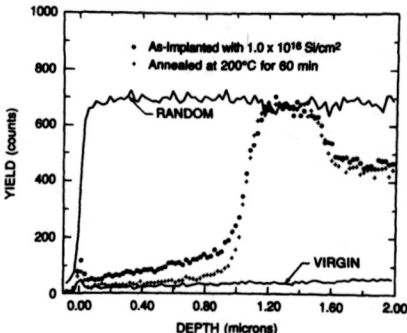

Fig. 6. Aligned spectra from a Si⁺ (1.25 MeV, 1.0 × 10¹⁶ cm⁻²) implanted Si sample before and after annealing at 200°C for 1 h.

Observations presented to this point are thought to be the result of a homogeneous nucleation and growth mechanism for damage in Si. Insight into details of the model which account for damage saturation is gained by analysis of the defect reactions using chemical rate equations. Rather than starting from first principles, only those reactions which are consistent with observation will be allowed. This eliminates any reactions which lead to nucleation or growth of extended defects such as dislocations. The reason for the absence of these kinds of defects in the saturated region is unclear at this point. Reactions involving the Frenkel defects are therefore limited to (a) interstitial/vacancy recombination, (b) divacancy formation by vacancy pairing, and (c) divacancy annealing by combination with an interstitial. This reaction leads to the following rate equations for growth of the different defect populations:

$$\frac{dn_i}{dt} = G - k_1 n_i n_v - k_2 n_i n_o,$$

$$\frac{dn_v}{dt} = G - k_1 n_i n_v - k_3 n_v^2 + k_2 n_i n_o,$$

$$\frac{dn_o}{dt} = \frac{1}{2} k_3 n_v^2 - k_2 n_i n_o,$$

(2)

where n_i, n_v, n_o are the respective concentration of the interstitial, vacancy and divacancy, and the k's are the reaction rate constants. The homogeneous generation rate of Frenkel defects by the ions is given by G. It is clear that there exists a steady-state solution of these nonlinear differential equations. This occurs as a result of a detailed balancing between the annihilation and creation rates for each type of defect. This balancing is responsible for the saturation of the damage and clearly would not be possible if unsaturable sinks, such as dislocations, were present in the reaction volume. A square root dependence of the saturation level on dose rate is predicted by the steady-state solution. This dose-rate dependence was confirmed using different current density implants (0.4 and 0.1 μA/cm^2). However, the dependency was found to be slightly weaker than predicted, varying approximately as the cube root of the dose rate.

<u>Medium-Energy (100-keV) Si$^+$ Ion Irradiation</u>

Further investigations into damage effects which occur during high-energy ion irradiation were done using lower-energy (100-keV) ions to predamage the near-surface region. The morphology of the damage formed during the 100-keV ion irradiation depended on the ion fluence used. The subsequent interaction of high-energy ions with the predamage was then investigated. However, before this interaction could be studied it was necessary to characterize the nature of the damage produced by the 100-keV ions and the growth behavior. Different annealing stages were studied to determine the nature of the damage present in the implanted region. Well defined annealing stages have been observed in ion-irradiated Si at 250 and 600°C. As previously discussed, the lower temperature stage is associated with the divacancy defect, while 600°C is sufficiently high to remove amorphous damage [14]. It should be cautioned that the identification of the divacancy as the defect annealed at 250°C is somewhat restrictive. Higher-order vacancy complexes and simple defect clusters which act as trapping sites during the anneal can be involved in this stage. Extended defects (such as dislocations, staking faults, and twins) generally require temperatures significantly above 600°C to anneal. The distribution of damage was determined after each anneal stage from ion channeling spectra. The profiles could then be manipulated to yield the distribution of a particular defect. For example, the difference in the as-implanted damage and that remaining after 250°C annealing yields the divacancy profile.

The variation of the different defects with implantation dose in a depth range of 80–110 nm is shown in Fig. 7. This depth range brackets the EOR of the 100-keV ions. Only the distribution of the divacancy (i.e., simple defect clusters) and amorphous damage are shown. While small dislocation loops were observed over a portion of the fluence range, their contribution to the total damage fraction was very small (~0.02). Total damage is seen to increase with dose in a very nonlinear manner; initially growing sublinearly with dose, but afterwards increasing quit rapidly. Such behavior for ion-implanted Si has been reported by others [4,15]. The transition between these two behaviors is seen to occur at a critical dose of ~2 × 10^{14} cm^{-2}. The range of dose below this transition point will be referred to as region I, while region II will reference the range above. Not only does the behavior of the total damage change at this critical point but that of the different defects is also seen to diverge. The damage produced in region I consists almost entirely of the divacancy defect, while amorphous damage is dominant in region II. The critical dose marks the onset of the growth of amorphous damage and also a marked increase in the divacancy population which eventually disappears when the region is completely amorphized. The divergent behavior of both the total damage, and the divacancy component is seen to correlate with the onset of amorphous growth.

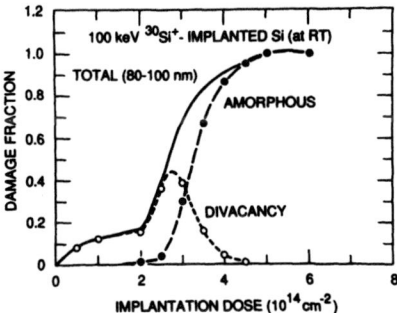

Fig. 7. Dose dependence of damage (and its components) produced during 100-keV Si$^+$ ion irradiation of Si (100) single crystal.

It is clear that different defect processes must dominate in the two growth regimes. During the initial sublinear regime, simple defects are formed and their growth greatly slows near the transition. These defects (in region I) and their growth behavior is similar to that which occurs in the near-surface region ahead of the EOR in Si irradiated with the megaelectron volt ions. The constrained growth in the present case is therefore thought to be the result of a mechanism similar to that which produces damage saturation during MeV ion irradiation. It should be recalled that constrained growth was attributed to the detailed balancing between simple defect reactions in the irradiated volume.

The transition between the two growth regimes must coincide with some nucleation phenomena which precipitates the rapid growth of damage [4]. The presence of dislocation loops in both regions seems to discount their role in this process, although they may contribute to damage growth in region I. TEM showed that the loop size and density increased slightly through the transition between the two regions, but the change was continuous and seemed unrelated to the divergent behavior at the critical dose. Rather, the onset of amorphous damage growth at the critical point seemed to initiate the different growth behavior. It is believed that the formation of amorphous damage within the irradiated volume significantly alters the balance between the defect reactions which is established during the initial growth, as is evidenced by the rapid increase in the divacancy defect at the critical point. While the details of the process are not clear, the amorphous Si is thought to trap interstitial-type defects which in turn removes the constraint on the divacancy growth in the surrounding crystalline regions, permitting it to increase until it reaches the critical value for amorphization. Therefore, the ion-induced c-a phase transition in Si is considered to be a cooperative process in which its onset triggers a rapid growth of other defect-types which then leads to further amorphization. It should be mentioned that this cooperative mechanism is consistent with the growth behavior of the buried amorphous layer which occurs during high-energy ion irradiation. The growth of the layer is recalled to occur only within the the vicinity of the a-c interface, while ahead of the layer damage remained saturated at a low level. The presence of the a-c interface appears to alter the balance between the various defect reactions in its vicinity leading to an unconstrained growth of damage which results in further amorphization.

Since damage nucleation during both medium- and high-energy ion irradiation has been assumed to be homogeneous in nature (rather than heterogeneous at the site of the collision cascade), it is of interest to compare the 100 keV results to the predictions of the cascade model given by Eqn. (1). The curves given for different values of m (the number of cascade overlaps required to produce the amorphous transition) are shown as the solid lines in the Fig. 8. It is clear that, for heterogeneous growth, the amorphous transition becomes more abrupt with dose as the value of m increases. However, it is seen that the actual a-c transition which occurs in Si is more abrupt than that predicted by Eqn. (1), even for m=12. It therefore seems clear that heterogeneous nucleation does not dominate for self-ion implantation in Si.

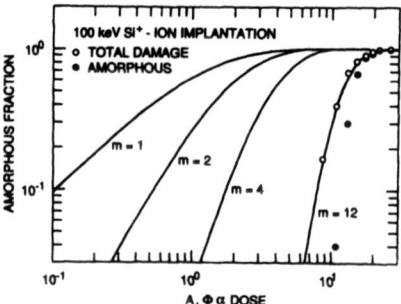

Fig. 8. The variation of amorphous damage with ion dose. Gibbon's model predictions are given for different values of m by the solid curves. Experimental values for both total and amorphous damaged produced by 100-keV Si+ ions are given by the data points.

<u>Interaction of MeV-Ions with Preexisting Damage</u>

The interaction of MeV ions with different predamage morphologies is illustrated in Fig. 9. Aligned spectra in Figs. 9(a) and (b) are from samples predamaged with 2×10^{14} and 3×10^{14} cm^{-2} implants at 100 keV, respectively. A scattering peak is seen at ~0.09 μm in each of the (100 keV) as-implanted spectra indicating a peak in the damage distribution at this location. According to Fig. 7, the damage at the lower dose consists predominantly of simple defects (clusters), while substantial amounts of both amorphous damage and divacancy defects are present at the higher dose. The effect of high-energy ions on each of these different damage morphologies is seen in the figure by comparing the scattering yield in the as-implanted spectra (over a region extending to ~0.3 μm) with that from samples subsequently irradiated with 1.25-MeV Si$^+$ ions at a fluence of 10^{16} cm^{-2}. It is immediately clear that the interaction of the high-energy ions with the two predamage morphologies results in strikingly different results. In Fig. 9(a), it is seen that the predamage was annealed by high-energy irradiation, while in Fig. 9(b) substantial damage growth occurs as evidenced by the large scattering yield in the near-surface which approaches the amorphous level. The large scattering yield from deep in the MeV ion-irradiated samples is obviously the result of damage produced near the EOR of the 1.25-MeV Si$^+$ ions. TEM examination of the samples confirmed the RBS results. Implantation with 2×10^{14} Si$^+$/cm^2 at 100 keV produces a band of dislocation loops centered near ~0.09 μm which were observed to be slightly coarsened after high-energy, ion irradiation. Therefore, the ion-induced annealing in this sample is thought to be the result of a reduction in the divacancy concentration.

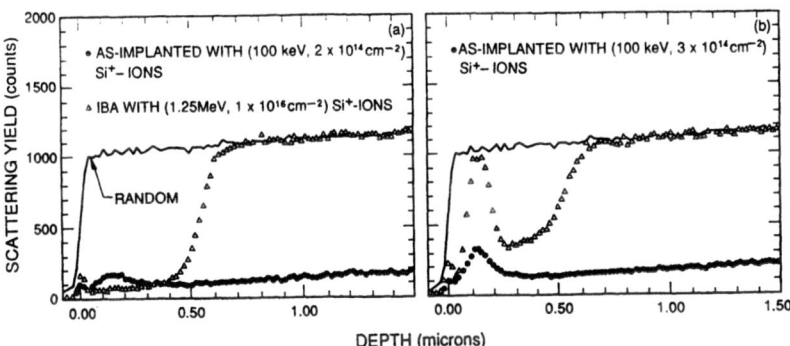

Fig. 9. Comparison of aligned spectra showing the effect of 1.25-MeV Si$^+$ ions on samples predamaged with 100-keV Si$^+$ ions at (a) 2×10^{14} and (b) 3×10^{14} cm^{-2}.

At the higher fluence (3×10^{14} cm^{-2}), a band of small dislocation loops was also observed along with microdiffraction ring patterns consistent with amorphous Si. Damage growth which occurred during the subsequent MeV ion irradiation was found to result in the formation of a continuous, buried amorphous layer in the predamaged region. The markedly different nature of the interaction of the MeV ions on either side of the critical point (which marks the transition in the medium-energy, ion-induced damage growth) illustrates the importance of the defect morphology in this process. Predamaging at fluences below the critical point (in region I) leads to an interaction which results in substantial annealing of the damage, while above this value (in region II) rapid damage growth occurs. The reason for such behavior is readily discerned. Remember that, in the absence of any predamage, only divacancy defects are observed in the near-surface after the passage of MeV ions, and their concentration saturates during irradiation at a very low value. Predamaging in region I creates a very similar damage morphology, although the concentration of the divacancy defect is higher. This high concentration represents a supersaturation of these defects relative to the dynamical conditions established during high-energy ion irradiation and as such is expected to decrease during irradiation to a value consistent with it steady-state value. However, amorphous damage is formed over the entire region II. This type of damage has been shown to disrupt the the dynamical balancing which constrains damage growth. Therefore, it is not surprising to find that the interaction of the MeV ions with the amorphous damage results in rapid damage growth.

CONCLUSION

The dose dependence of damage production during 100-keV self-ion irradiation of Si was determined. Two distinct regimes were observed: an initial regime in which growth is constrained and only simple defects are formed, followed by a regime of rapid, unconstrained growth which results in complete amorphization of the lattice. Damage nucleation over an extended range of energy for self-ion irradiation of Si was shown to be dominated by a homogeneous mechanism. In this model, the rapid growth during 100-keV ion irradiation is precipitated by the onset of amorphization. This is envisioned to occur as a result of a cooperative mechanism in which the formation of amorphous damage leads to further amorphization by promoting growth of other types of defects. The homogeneous model was also shown to account for the saturation of damage in the surface region which occurs during high-energy bombardment, as well as the layer-by-layer growth of the buried amorphous layer formed near the EOR of the ions. Markedly different responses were observed in predamaged Si during irradiation with high-energy ions, varying from damage annealing to rapid growth. These different responses were shown to be consistent with the damage growth model.

ACKNOWLEDGMENTS

Research was sponsored by the Division of Materials Sciences, U.S. Department of Energy under contract DE-AC05-84OR21400 with Martin Marietta Energy Systems, Inc.

REFERENCES

1. See Review by J. W. Corbett, J. P. Karins, and T. Y. Tan, *Nucl. Instrum. and Methods* **182/183**, 457 (1981).

2. See Review by D. A. Thompson, *Radiat. Eff.* **56**, 105 (1981).

3. L. Eriksson, J. A. Davies, N. G. E. Johansson, and J. W. Mayer, *J. Appl. Phys.* **40**, 842 (1969).

4. L. T. Chadderton, *Radiat. Eff.* **8**, 77 (1971).

5. J. A. Brinkman, *J. Appl. Phys.* **25**, 961 (1954).

6. M. L. Swanson, J. R. Parsons, and C. W. Hoelke, *Radiat. Eff.* **9**, 249 (1971).

7. L. M. Howe and M. H. Rainville, *Nucl. Instrum. and Methods* **182/183**, 143 (1981).

8. J. Narayan, O. S. Oen, D. Fathy, and O. W. Holland, *Mater. Lett.* **3**, 67 (1985).

9. M. O. Ruault, J. Chaumont, J. M. Penisson, and A. Bourret, *Philos. Mag.* **50**, 667 (1984).

10. J. F. Gibbons, *Proceedings of the IEEE* **60**, 1062 (1972).

11. J. R. Dennis and E. B. Hale, *J. Appl. Phys.* **49**, 1119 (1978).

12. J. P. Biersack, *Nucl. Instrum. and Methods* **174**, 257 (1980).

13. F. L. Vook and H. J. Stein, *Radiat. Eff.* **2**, 23 (1969).

14. G. L. Olson, A. Kokorowski, J. A. Roth, and L. D. Hess, in *Laser-Solid Interactions and Transient Thermal Processing of Materials*, ed. by J. Narayan, W. L. Brown, and R. A. Lemons (North-Holland, New York, 1983) p. 141.

15. F. H. Eisen and B. Welch, *Radiat. Eff.* **7**, 143 (1971).

EPITAXIAL EXPLOSIVE CRYSTALLIZATION OF AMORPHOUS SILICON LAYERS BURIED IN A SILICON (100) AND (111) MATRIX

P.A. STOLK, A. POLMAN and W.C. SINKE
FOM-Institute for Atomic and Molecular Physics
Kruislaan 407, 1098 SJ Amsterdam, the Netherlands
C.W.T. BULLE-LIEUWMA and D.E.W. VANDENHOUDT
Philips Research Laboratories, P.O. Box 50000, 5600 JA Eindhoven, the Netherlands

ABSTRACT

420 nm thick amorphous Si layers buried in a Si (100) or Si (111) matrix, produced by 350 keV Si-implantation, were irradiated using a pulsed ruby laser. Time-resolved reflectivity measurements show that melting can be initiated buried in the samples at the crystalline-amorphous interface. Melting is immediately followed by explosive crystallization of the buried amorphous layer, which is started from the crystalline top layer. The velocity of this self-sustained crystallization process is determined to be 15.0 ± 0.5 m/s for Si (100) and 14.0 ± 0.5 m/s for Si (111). RBS and cross-section TEM reveal that epitaxially grown crystalline Si, containing a high density of twin defects, is formed in both the Si (100) and the Si (111) sample.

INTRODUCTION

The recently developed technique of high-energy (MeV) ion implantation [1] has uncovered new areas in semiconductors research. Implantation of high-energy ions in single-crystal silicon (c-Si) enables the formation of buried amorphous silicon (a-Si) layers, embedded in a crystalline matrix. Several well-established techniques can be applied to achieve annealing of the implantation-damage, such as thermal treatment or laser annealing.

In the past ten years pulsed lasers have been employed to induce rapid crystallization phenomena in a-Si. Using laser pulses of high energy density (ED), recrystallization of a-Si surface layers can be achieved via melting of the complete a-Si layer, followed by epitaxial crystal growth from the c-Si substrate. Furthermore, it has been shown that for low ED irradiation 'explosive crystallization' can occur in the amorphous material [2-5]. Under suitable conditions, pulsed laser heating of a-Si surface layers results in formation of a thin liquid silicon (l-Si) layer at the surface. Since the melting temperature of a-Si (T_{ma}) is approximately 225 K below that of c-Si (T_{mc}) [6], this liquid is undercooled with respect to T_{mc} and formation of crystalline Si from the melt is thermodynamically favourable. Since a-Si is metastable with respect to c-Si [7], the transition of a-Si to c-Si via the l-Si state will lead to an effective heat-release. Deeper lying a-Si is melted by this heat and will, again, crystallize. In this way a self-sustained or 'explosive' crystallization process (XC) can take place. XC yields fine grained polycrystalline silicon (fgp-Si) [2-5]. Until now, the nature of the processes that trigger and sustain XC is still under debate [8-10].

In this paper studies on pulsed laser-induced crystallization phenomena in ≈ 420 nm thick a-Si layers buried in a Si (100) and Si (111) matrix are presented. It is shown that *epitaxial* XC can be initiated buried in the sample at the c-Si/a-Si interface. The velocity at which XC proceeds in both the Si (100) and the Si (111) sample is determined, and the microstructure of the crystalline material formed at these velocities is studied in detail.

EXPERIMENTAL

350 keV ^{28}Si$^+$ ions were implanted into both Si (100) and Si (111) substrates, in a random direction close to the surface normal. The total implantation dose amounted to $4*10^{15}$ cm^{-2}. At this dose an amorphized region is formed buried beneath a heavily damaged crystalline top layer containing implantation-induced point defects and amorphous clusters. The surface-damage was annealed by a thermal treatment of 15 min at 490 °C in vacuum (10^{-6} mbar). Under these conditions, no significant solid phase epitaxy of the buried a-Si layer occurs (i.e. < 10 nm) [11]. Samples were irradiated with a single pulse from a Q-switched ruby laser (λ=694 nm) having a pulse length of 32 ns full width at half-maximum. The ED was varied between 0.15 J/cm^2 and

180

1.4 J/cm². Time-resolved reflectivity (TRR) measurements [12] were performed during irradiation using a continuous wave AlGaAs laser, operating at a wavelength of 825 nm. The total system response-time was ≈ 1 ns. After pulsed laser irradiation the microstructure of the samples was analyzed by means of Rutherford backscattering spectrometry (RBS) and cross-section transmission electron microscopy (TEM). RBS spectra were obtained by using 2 MeV He⁺ and a scattering angle of 110°.

RESULTS

RBS data of the Si (100) sample are given in Fig. 1a. The aligned spectrum of the sample after implantation and thermal annealing shows a peak at the Si surface backscattering energy, followed by a low yield region, showing that a c-Si top layer is present. Furthermore, a region at random height is observed, indicating the presence of a buried amorphized layer. At low backscattering energies the yield is decreased below the random yield, due to channeling in the c-Si substrate. The thickness of the c-Si top layer is determined to be 130 nm, whereas the thickness of the buried a-Si layer amounts to about 420 nm. These values are in agreement with cross-section TEM data of the Si (100) sample before irradiation (not shown) [13].

Figure 2 shows TRR transients recorded during irradiation of the Si (100) sample at various energy densities. The reflectivity level before irradiation is 35 % in all cases. For irradiation with an ED below 0.20 J/cm² no change in reflectance is measured. At 0.22 J/cm² a sudden decrease in reflectance is observed, followed by a maximum and a minimum. Upon raising the ED to 0.25 J/cm² 8 extrema can be observed. At 0.27 J/cm² the last three extrema are superimposed on a broader peak appearing in the TRR-trace. For 0.29 J/cm² this peak develops to a plateau at 75 %, which equals the reflectivity level of an optically thick l-Si surface layer. With increasing ED (up to 1.4 J/cm², not shown) this high reflectivity plateau (HRP) lengthens,

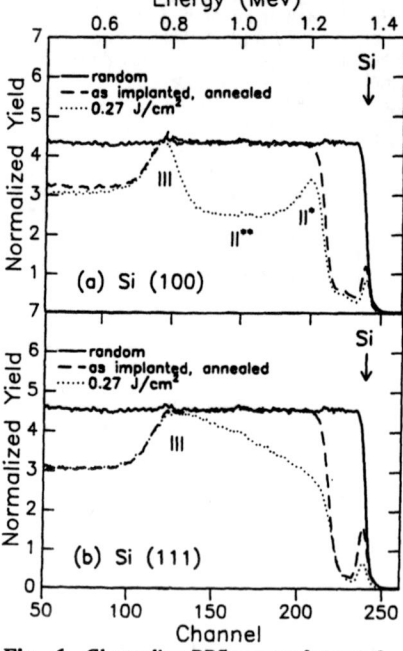

Fig. 1 Channeling-RBS spectra for samples after ion implantation and thermal annealing, and after subsequent laser irradiation at 0.27 J/cm². Random spectra are shown for reference. (a) Si (100), (b) Si (111).

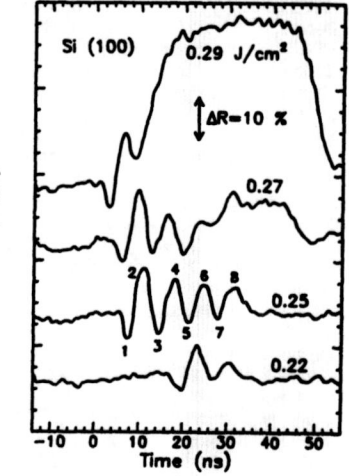

Fig. 2 Time-resolved reflectivity traces of the Si (100) samples obtained during laser irradiation at several energy densities (indicated in the figure). Reflectivity traces are shifted for clarity and the arrow indicates the reflectivity scale. The peak of the laser pulse is taken as the origin of the time-axis.

but in all cases it is preceded by at least one minimum in reflectance. TRR measurements on Si (111) show qualitatively the same behaviour for all energy densities.

The RBS spectrum in Fig. 1a shows that after irradiation at an ED of 0.27 J/cm^2, the yield from the original a-Si layer is decreased considerably. A plateau at about 60 % of the random height, indicated in the figure by II**, is observed, lying between two peaks II* and III.

Figure 3a shows a typical (011) cross-section TEM micrograph of the Si (100) sample after irradiation at 0.35 J/cm^2. Four distinct regions are present, numbered I to IV in the picture. A 130 nm thick crystalline surface layer (I) is visible on top of an approximately 360 nm thick crystallized region (II). After irradiation a 60 nm thick a-Si layer (III) is left, lying on top of the crystalline substrate (IV). The top 100 nm of the surface layer (I) contains a low density of dislocations and micro-twins. High-resolution lattice imaging reveals that a large density of small micro-twins has formed in the 30 nm thick interface area with region II. These micro-twins are only a few interplanar distances (d$_{111}$) thick. Region II consists of columnar structures of c-Si. The columns appear to be separated by boundaries perpendicular to the surface (indicated by arrows in the image) and contain [100]-oriented monocrystalline Si, as was determined by electron diffraction. Large twin regions (marked 'T') extend from the column-boundaries into the c-Si. No fgp-Si or amorphous clusters are present in the 360 nm thick crystallized region.

Figure 1b shows RBS results for Si (111). The data confirms formation of a buried amorphous layer after implantation and thermal annealing. The thickness of the buried a-Si layer is determined to be ≈ 420 nm, that of the crystalline top layer 120 nm, in good agreement with values obtained by cross-section TEM (not shown) [13]. Upon laser irradiation at 0.27 J/cm^2, the yield from the initially amorphous layer is changed drastically. These changes in the RBS spectrum are distinctly different than those observed in the Si (100) sample. Near the c-Si surface region the backscattering yield is decreased to about 50 % of the random height. For lower backscattering energies the yield increases, reaching the random height at point III.

(011) cross-section TEM micrographs of the sample after laser irradiation at 0.35 J/cm^2 (Fig. 3b) show that ≈ 300 nm of the initial a-Si layer is converted into c-Si (II). Besides [111]-oriented columns of c-Si, a high density of twin lamellae, inclined at 70° to the surface, is observed. These lamellae (marked 'T' in the micrograph) extend from the crystalline top layer (I) down to the remaining amorphous region (III) and are oriented parallel to {111} planes. The micrograph also shows the characteristic boundaries, separating [111]-oriented columns (indicated '→ '). Large twin lamellae extend from the column-boundaries into the c-Si. Near the 130 nm thick a-Si layer (III) large pockets of micro-twins are present. High-resolution imaging and micro-diffraction did not show amorphous or micro-crystalline areas in the 300 nm thick c-Si region.

DISCUSSION

1) Melting and solidification scenario

From the TRR, RBS and TEM data the following melting and solidification scenario is derived. Amorphous Si has a lower melting temperature [6] and a higher optical absorption coefficient than c-Si. Therefore, during laser irradiation, melting of a-Si near the c-Si/a-Si interface can occur, while the temperature of the relatively transparent c-Si surface layer remains below Tmc. Upon melting of a-Si a buried l-Si layer is formed, having a temperature near Tma. Hence, the buried melt is undercooled with respect to Tmc and will tend to crystallize. The non-molten c-Si surface layer acts as a seed for crystallization. Heat release upon solidification [7] will result in melting of deeper lying a-Si. In this way XC [2-5] can take place; a thin self-propagating melt moves through the a-Si, leaving behind c-Si.

The optical reflectivity of a sample containing a buried liquid layer was calculated using a Fresnel matrix calculation for an isotropic planar multilayer system [13]. Calculations show that variation of the depth of a buried liquid layer gives rise to minima and maxima in the reflectivity of the sample. These extrema are caused by changing interference conditions for light reflected at the surface and at the buried c-Si/l-Si interface. The calculations strongly indicate that the oscillations in reflectivity observed in the TRR measurements on Si (100) (Fig. 2) are the result of the inward motion of a buried melt starting at 130 nm depth. The 1/4-wavelength distance between two consecutive interference-extrema amounts to 52 nm [13]. From the observation of eight extrema in the reflectivity transient at 0.25 J/cm^2 (Fig. 2) it is inferred that the self-propagating melt reached a total depth of approximately 500 nm, leaving a 50 nm thick a-Si layer unaffected. At this depth the process of XC stops, probably because the heat release by crystallization of l-Si is insufficient to melt the remaining a-Si, since this a-Si is strongly cooled

182

by the well-conducting crystalline substrate.

For energy densities above 0.26 J/cm^2 the laser pulse supplies enough energy to raise the surface temperature above T_{mc}. In the ED range between 0.26 and 0.29 J/cm^2 lateral inhomogenieties in the laser spot result in formation of a non-uniform and optically thin (i.e. <10 nm at λ=825 nm [14]) surface melt. This melt does not completely shield the oscillations in reflectance caused by XC. In the ED range of 0.29 J/cm^2 upto 1.4 J/cm^2 an optically thick l-Si surface layer is formed, giving rise to an HRP in the TRR-traces (Fig. 2). In all cases at least one minimum in reflectance is observed, indicating that melting of c-Si at the surface is always

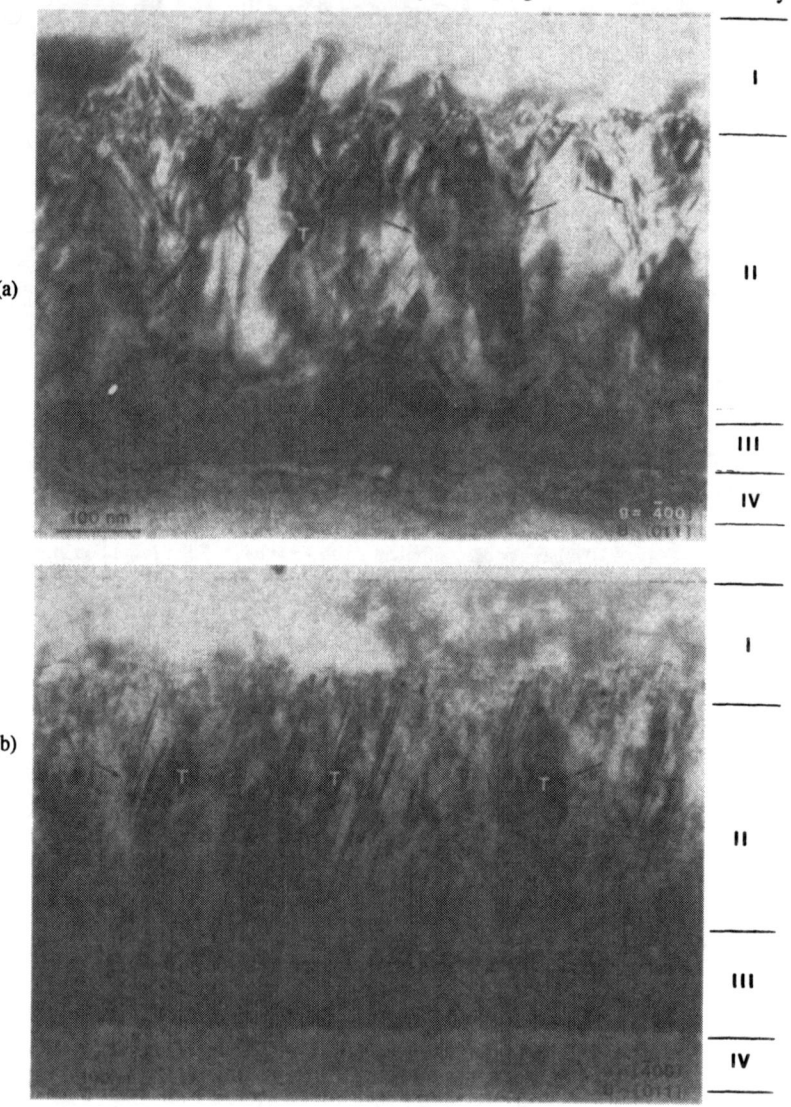

Fig. 3   *(011) cross-section [400] bright-field TEM micrographs after irradiation at 0.35 J/cm^2. (a) Si (100),(b) Si (111).*

preceded by buried melting in the a-Si layer.

TEM (Fig. 3a) shows that most of the crystalline silicon formed by XC in Si (100) samples has the same orientation as the surface layer. This implies that crystal growth during XC has taken place epitaxially in the [100] direction; crystallization occurs at the trailing c-Si/l-Si interface. The twin regions, present in the crystallized material, are formed by twinning about {111} planes during crystal growth.

Scattering from the large number of micro-twins near the c-Si surface region results in peak II* in the RBS-spectrum (Fig. 1a). Both dechanneling from these twin defects and the presence of column-boundaries and twin regions in the epitaxially grown Si (100) may result in the relatively high channeling plateau II** in the RBS spectrum. Peak III at random height in the RBS spectrum (Fig. 1a) and region III in the TEM picture (Fig. 3a) indicate that a thin amorphous layer remained unaffected after irradiation at 0.27 J/cm² and 0.35 J/cm² respectively. This corroborates the idea that XC quenches within the a-Si near the c-Si substrate, as was also derived from the TRR measurements at 0.25 J/cm².

TEM reveals that columns of epitaxially grown c-Si, containing a high density of thin (i.e. < 30 nm) twin lamellae, are formed in the Si (111) sample. Direct scattering and dechanneling from the mixture of differently oriented c-Si result in an increase with depth of the channeling yield of the crystallized layer in the RBS spectra, as can be seen in Fig. 1b.

2) High-speed crystal growth

From the calculated position of the interference-extrema [13] and the time differences between the consecutive extrema in the TRR-trace of 0.25 J/cm², the XC velocity is determined as a function of depth (Fig. 4) for the Si (100) sample. It is found that the process proceeds at a mean velocity of 15.2 m/s and is stopped abruptly. By averaging over several experiments, the velocity of the self-propagating melt was found to be constant over depth at 15.0 ± 0.5 m/s for Si (100). This analysis shows that the epitaxial crystal growth proceeds at a speed near the maximum value at which any growth of Si (100) from l-Si can take place [15]. It has been demonstrated that amorphization occurs if the crystallization velocity exceeds 15.0 ± 1.5 m/s [15]. Twin defects have been shown to form during crystal growth at high (15 m/s) velocities [16], which tallies with the presence of micro-twins in the Si (100) grown by epitaxial XC.

Analyzing the TRR measurements for Si (111), an average XC-velocity of 14.0 ± 0.5 m/s was determined. In the past, high-speed melting and recrystallization processes in Si have been studied using ultra-short (1 ns) laser pulses [16]. On the basis of computer modeling of these experiments, the amorphization threshold for Si (111) has been estimated to be 12 m/s. Furthermore, breakdown of perfect epitaxial growth on Si (111) has been shown to occur above 5 to 8 m/s, due to formation of twin defects [16]. In comparison with epitaxial growth on Si (100), single crystal growth on (111) planes is hampered by a relatively high barrier to plane nucleation. Crystal growth is facilitated if twinning about {111} planes occurs, since nucleation sites become available at facets in the growing surface. XC was measured to proceed at a macroscopic velocity of 14.0 ± 0.5 m/s. This implies that for the present experimental conditions the amorphization threshold lies well above the estimated value of 12 m/s [16]. As the crystallization process is influenced by the density of nucleation sites, the apparent discrepancy between the maximum growth velocities in the two experiments may be related to differences in twin density.

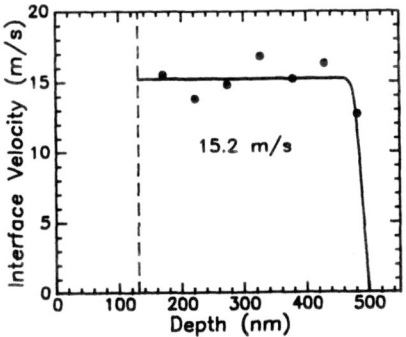

Fig. 4 *The c-Si/l-Si interface velocity during epitaxial XC as a function of depth for the Si (100) sample.*

184

3) Comparison to conventional XC

Tsao et al. [8] have suggested that 'conventional' XC [2-5] of amorphous top layers is triggered and sustained by nucleation at the leading l-Si/a-Si interface of the buried melt, providing an explanation for the formation of fgp-Si by this type of XC. Roorda et al. [10] have proposed that nucleation occurs in the solid a-Si layer during the heating phase prior to melting. The nuclei formed would grow upon exposure to undercooled l-Si. It appears that neither of these nucleation processes play a role in epitaxial XC, since fgp-Si is absent in the c-Si formed in both Si (100) and Si (111). This will be discussed in detail in a forthcoming publication [17].

CONCLUSIONS

Amorphous layers buried in crystalline Si can be crystallized using pulsed laser irradiation. Under specific irradiation conditions melting can be induced in the buried layer, whereafter epitaxial explosive crystallization is triggered from the crystalline surface layer. The crystallization velocity in Si (100) and Si (111) has been determined using real time reflectivity measurements. Explosive crystal growth in Si (100) proceeds at a velocity of 15.0 ± 0.5 m/s, yielding epitaxially grown c-Si columns. Growth on a Si (111) top layer proceeds at a macroscopic speed of 14.0 ± 0.5 m/s, indicating that under the present experimental conditions the amorphization threshold for crystal growth on Si (111) is significantly higher than the maximum value of 12 m/s obtained in earlier experiments [16]. The crystal growth rate for Si (111) may be enhanced by inclined twin defects. No polycrystalline grains have formed during the crystallization processes in both Si (100) and Si (111). This implies that heterogeneous nucleation at the leading l-Si/a-Si melt front [8] or solid state nucleation [10] do not play a role under the circumstances of epitaxial explosive crystallization.

We would like to acknowledge A.G. Cullis (RSRE, Malvern, UK) for helpful discussions. The work at the FOM-Institute is part of the research program of the Stichting voor Fundamenteel Onderzoek der Materie (FOM) and was made possible by financial support from the Stichting voor Technische Wetenschappen (STW) and the Nederlandse Organisatie voor Wetenschappelijk Onderzoek (NWO).

REFERENCES

1 A. Polman et al., Nucl. Instr. and Methods B37/38, 935 (1989)
2 J. Narayan and C.W. White, Appl. Phys. Lett. 44, 35 (1984)
3 W.C. Sinke and F.W. Saris, Phys. Rev. Lett. 53, 2121 (1984)
4 D.H. Lowndes, G.E. Jellison, S.J. Pennycook, S.P. Withrow and D.N. Mashburn, Appl. Phys. Lett. 48, 1389 (1986)
5 J.J.P. Bruines, R.P.M. van Hal, H.M.J. Boots, A. Polman and F.W. Saris, Appl. Phys. Lett. 49, 1160 (1986)
6 M.O. Thompson, G.J. Galvin, J.W. Mayer, P.S. Peercy, J.M. Poate, D. Jacobson, A.G. Cullis and N.G.Chew, Phys. Rev. Lett. 52, 2360 (1984)
7 E.P. Donovan, F. Spaepen, D. Turnbull, J.M. Poate and D.C. Jacobson, Appl. Phys. Lett. 42, 698 (1983)
8 J.Y. Tsao and P.S. Peercy, Phys. Rev. Lett. 58, 2782 (1987)
9 D.H. Lowndes, S.J. Pennycook, G.E. Jellison, Jr., S.P. Withrow and D.N. Mashburn, J. Mater. Res. 2, 648 (1987)
10 S. Roorda and W.C. Sinke, Appl. Surf. Sci. 36, 188 (1989)
11 G.L. Olson and J.A. Roth, Mater. Sci. Rep. 3 (1988)
12 D.H. Auston, C.M. Surko, T.N.C. Venkatesan, R.E. Slusher and J.A. Golovchenko, Appl. Phys. Lett. 33, 437 (1978)
13 A. Polman, D.J.W. Mous, P.A. Stolk, W.C. Sinke, C.W.T. Bulle-Lieuwma and D.E.W. Vandenhoudt (unpublished)
14 K.M. Sharev, B.A. Baum and P.V. Gel'd, High Temperature 15, 548 (1977)
15 M.O. Thompson, J.W. Mayer, A.G. Cullis, H.C. Webber, N.G. Chew, J.M. Poate and D.C. Jacobson, Phys. Rev. Lett. 50, 896 (1983)
16 A.G. Cullis, N.G. Chew, H.C. Webber and D.J. Smith, J. Cryst. Growth 68, 624 (1984)
17 A. Polman, S. Roorda, P.A. Stolk and W.C. Sinke (unpublished)

MeV IMPLANTATION OF GALLIUM ARSENIDE

H. KANBER, J. C. CHEN, AND M. J. BARGER
Hughes Aircraft Company, Microwave Products Division, 3100 W. Lomita Blvd., Torrance, CA 90509

ABSTRACT

We investigated n-type MeV implants into semi-insulating GaAs to fabricate fully implanted varactor diodes for monolithic millimeter-wave applications. Single and multiple implants of Si from 6 to 2.5 MeV and 6 MeV S implants in GaAs were studied before and after capless furnace annealing and rapid thermal annealing by chemical profiling using secondary ion mass spectrometry (SIMS) and electrochemical and traditional differential capacitance-voltage profiling. A buried active layer at a depth of 2.8 μm with a peak carrier concentration of 2.5×10^{18} cm^{-3} was achieved by using 6 MeV Si + 4 MeV Si + 6 MeV S implants and rapid thermal annealing them at a temperature 1050°C for 10 seconds. The chemical profiles are correlated to the electrical profiles to determine activation and annealing behavior of MeV Si and S implants in semi-insulating GaAs.

INTRODUCTION

Selective ion implantation is one of the primary fabrication techniques for monolithic integration of different types of GaAs devices on monolithic microwave and millimeter-wave integrated circuits (MMICs). Through the years, selective ion implantation has been used to develop a very sophisticated FET-based IC technology in GaAs. Advanced future circuits will require monolithic fabrication on the same wafer of FETs and other types of GaAs devices, which at present can only be made in hybrid form. An example is the family of devices that requires a buried n+ layer: mixer diodes, vertical varactors, and vertical PIN diodes. The monolithic integration of varactor diodes and GaAs FETs is a critical fabrication technology, since these devices are key components in V-band 60-GHz control circuits based on semi-insulating GaAs. The requirement of a buried n+ layer 2.5 to 5 μm deep below the GaAs surface led us to decide on using 6 MeV Si and 6 MeV S implants. The projected range R_p and the implant profiles of deep MeV Si implants in GaAs are areas of very active research in several laboratories [1-3] and are not fully established. High activation of these deep implants requires furnace annealing at high temperatures, above 850°C, or using relatively novel rapid thermal annealing at elevated temperatures. We present in this paper optimization of rapid thermal annealing cycles for combined multiple Si and S MeV implants in GaAs. Atomic concentration and electron concentration profiles were measured for as-implanted, furnace annealed, and rapid thermal annealed GaAs wafers.

EXPERIMENTAL

The wafers used in this study were semi-insulating GaAs grown by the high-pressure liquid encapsulated Czochralski (LEC) technique. They were implanted with ^{28}Si and ^{32}S ions using single and multiple implant schedules, as shown in Table I. The MeV implantation was performed at Universal Energy Systems at Dayton, Ohio, using a 1.7 MV Tandetron accelerator. Using triply charged ions, energies up to 6 MeV were obtained with 1.5 MV on the terminal. To reduce channeling effects, the wafers were mounted with a 7-degree tilt and with

TABLE I
HIGH ENERGY Si + S IMPLANTATION FOR FULLY
IMPLANTED VARACTOR DIODES

Wafer #	Species	Dose (cm^{-2})	Energy MeV	CAT Furnace Anneal	Rapid Thermal Anneal
16	^{28}Si	1E14	6.0	900°C/30 m 950°C/30 m	950°C/10 s 1000°C/10 s 1050°C/10 s
17, 18	^{28}Si	1E14	6.0	900°C/30 m 950°C/30 m	950°C/10 s 1000°C/10 s 1050°C/10 s
	^{28}Si ^{28}Si	6.6E13 6.3E13	4.0 2.5		
19, 20, 21	^{28}Si	1E14	6.0	900°C/30 m 950°C/30 m	950°C/10 s 1000°C/10 s 1050°C/10 s
	^{28}Si ^{32}S	6.6E13 1E14	4.0 6.0		

the major flat rotated 20 degrees with respect to the horizontal plane of the beam. Wafer 16 was implanted with 6 MeV ^{28}Si only to study the implant profile and its annealing behavior. Wafers 17 and 18 were implanted with ^{28}Si at 6.0 MeV, 4.0 MeV, and 2.5 MeV to produce a square-shaped buried n+ layer. We also used ^{32}S implantation at 6 MeV to raise the carrier concentration above the 1×10^{18} cm^{-3} level, since Si is known to be amphoteric in GaAs. At these higher energies, the maximum carrier concentration that can be obtained from Si implantation alone is only 7 to 8×10^{17} cm^{-3} [2]. Thus, wafers 19, 20, and 21 were implanted with both 6.0 MeV and 4.0 MeV ^{28}Si and 6 MeV ^{32}S ions. We used both capless furnace annealing by the controlled atmosphere technique (CAT) [4] and rapid thermal annealing (RTA) to activate the implants, as shown in Table I. The capless furnace annealing was done with an AsH$_3$/H$_2$ atmosphere-supplied arsenic overpressure; the RTA was carried out in a Heatpulse 410 system in a face-to-face configuration with a bare GaAs wafer in a pure Ar atmosphere.

The atomic concentration profiles were obtained by secondary ion mass spectrometry (SIMS) analysis at C. Evans and Associates, Redwood City, CA. The electrical profiles were obtained by Polaron electrochemical profiling. Traditional differential capacitance-voltage (C-V) profiling was also used to study the near-surface electrical profiles by metallized Pt-Au Schottky dots and mesa etching 1.94 to 2.8 μm for the back contact.

RESULTS AND DISCUSSION

Atomic Concentration Profiles of MeV Implants

The SIMS profiles of the Si and S MeV implants in GaAs are shown in Figures 1 through 4. Figure 1 shows the 6 MeV Si chemical concentration as a function of depth into GaAs after implantation, after furnace annealing, and after rapid thermal annealing. The SIMS-determined experimental projected range for the 6 MeV implant is at 2.98 μm, with a peak Si

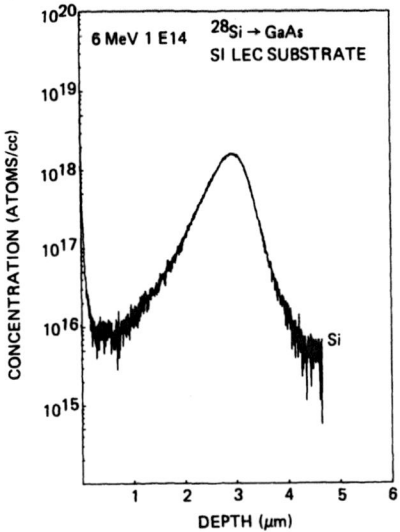

Figure 1   SIMS profile of 6 MeV Si→GaAs after implant, after furnace annealing, and after RTA.

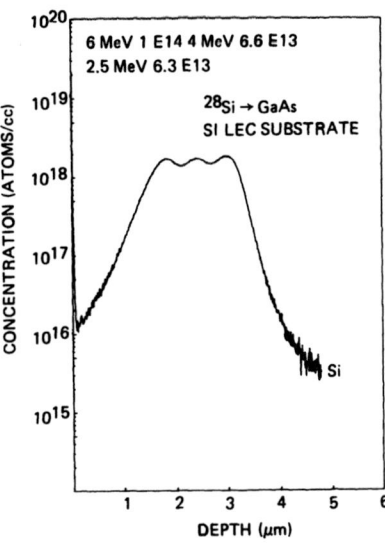

Figure 2   SIMS profile of multiple MeV Si implants in GaAs after implant, after furnace annealing, and after RTA.

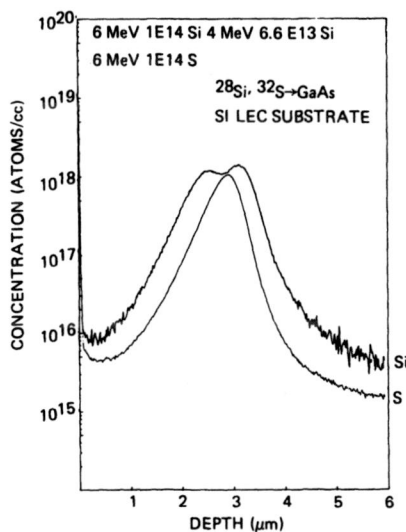

Figure 3   SIMS profiles of 6 MeV Si + 4 MeV Si, and 6 MeV S implants in GaAs after implant.

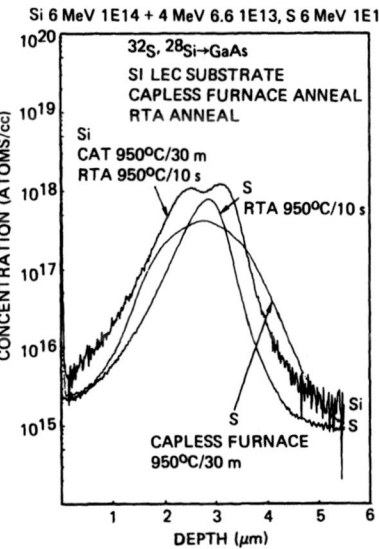

Figure 4   SIMS profile of 6 MeV Si + 4 MeV Si and 6 MeV S implants after annealing.

concentration of 1.6×10^{18} cm^{-3}. As expected, Si does not diffuse in semi-insulating GaAs after either method of annealing, and the SIMS profiles confirm this. Figure 2 shows triple Si implants of 6 MeV, 4 MeV, and 2.5 MeV that produced a flat-topped buried n + layer between 1.8 and 3 μm. Again, there was no diffusion or movement of the Si after both furnace annealing and RTA. Figure 3 shows the SIMS profiles of the combined 6 and 4 MeV double-Si implants and the 6 MeV S implant after implantation. Since ^{32}S is slightly heavier than ^{28}Si, the S peak at its projected range $R_p (= 2.88$ μm) with a peak atomic concentration of 1.05×10^{18} cm^{-3} is slightly shallower than the 6 MeV Si peak at a depth of 3.12 μm. Figure 4 shows the same profiles of Figure 3 after capless furnace annealing at 950°C for 30 minutes and rapid thermal annealing at 950°C for 10 seconds. Comparing Figures 3 and 4, the Si profile is the same as the as-implanted profile. However, the S diffuses considerably during furnace annealing both at 900°C (not shown) and at 950°C. An advantage of rapid thermal annealing is that it inhibits the fast S diffusion by the interstitial mechanism. The RTA-annealed S profile shown in Figure 4 is the same as the as-implanted S profile shown in Figure 3. These data indicate that RTA is a suitable technique for annealing of MeV S implants.

<u>Electrical Profiles of MeV Implants</u>

The annealing schedule for Si and S MeV implants into semi-insulating GaAs was described in Table I. Each wafer was divided into five pieces; two pieces were annealed by capless furnace annealing with As overpressure, and the remaining three pieces were rapid thermal annealed in a face-to-face configuration with a pure Ar atmosphere. A comparison of the furnace-annealed Polaron profile with the RTA-annealed profile of the 6 MeV Si implant is shown in Figure 5. An anomalous result of p-type conductivity near the surface of the implanted GaAs wafer was seen in all the furnace-annealed wafers. The causes and possible impurities contributing to the p-type conductivity, whether originating from the substrate or from the furnace environment, are currently under investigation. We have previously seen impurity redistribution both in furnace annealing [4] and rapid thermal annealing [5]. There was negligible diffusion noted in the Polaron profile upon annealing the single 6 MeV Si implant at the higher temperature of 1050°C for 10 seconds. Figure 6 shows the maximum activation of the triple Si MeV implants for RTA conditions of 950°C for 10 seconds. The peak carrier concentration of 8.6×10^{17} cm^{-3} was obtained at a depth of 2.87 μm. Addition of the 6 MeV S implant to the double 6 MeV + 4 MeV Si implant enabled us to raise the carrier concentration above the 1×10^{18} cm^{-3} level. We obtained a maximum carrier concentration of 2.0×10^{18} cm^{-3} at a depth of 2.8 μm for the furnace annealed (950°C for 30 minutes) multiple Si and S implant with the chemical profiles shown in Figure 4. Figure 7 shows the electrical profile of this multiple implant after RTA annealing at 1050°C for 10 seconds. A maximum carrier concentration of 2.5×10^{18} cm^{-3} was measured at a depth of 2.74 μm for the multiple Si and S MeV implants. Comparing Figures 4, 5, and 7, it can be seen that RTA annealing is advantageous for the combined Si and S MeV implants since it sharpens the electrical profile and achieves the highest carrier concentration measured.

Our goal for this study was to optimize the implant and annealing schedules to be able to fabricate fully implanted varactor diodes for V-band phase shifters. The fabrication sequence would require additional shallower implants to create an n- region and a very shallow p + region near the surface. Thus, we studied the near-surface region after MeV implantation and annealing by traditional Schottky metallization and differential C-V profiling using a Miller Feedback C-V system. The detailed electrical surface profiling involved photolithography,

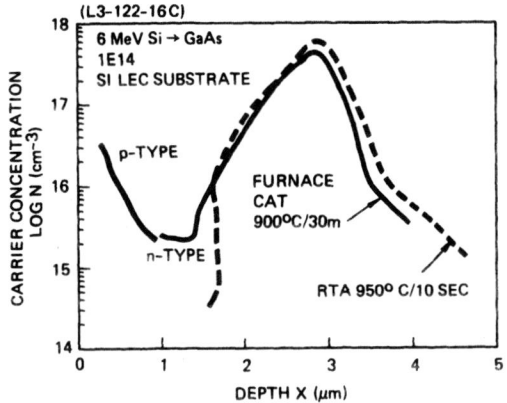

Figure 5 Electrical profile of 6 MeV Si in GaAs after annealing.

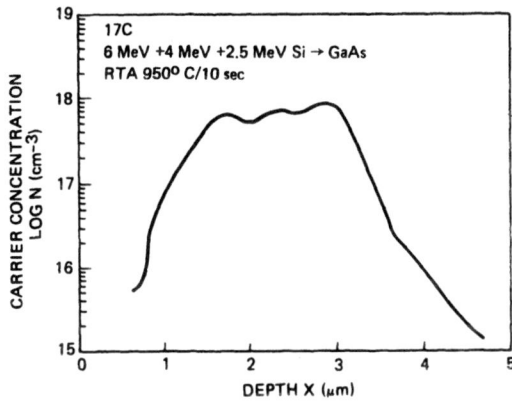

Figure 6 Electrical profile of multiple MeV Si implants in GaAs after rapid thermal annealing.

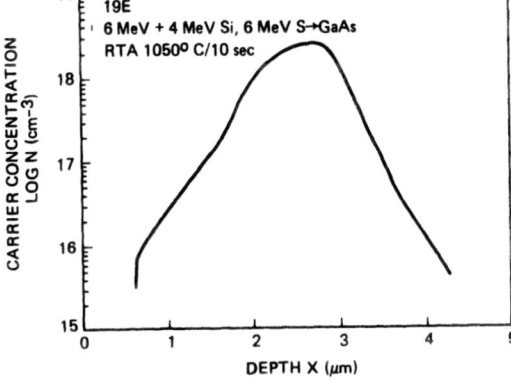

Figure 7 Electrical profile of Si and S MeV implants in GaAs after rapid thermal annealing.

metallization of Schottky dots using Pt-Au alloy, and mesa etching 1.94 to 2.8 μm for the back n+ contact. Overall, our surface profiling of the near-surface region, which is shallower than 0.6 μm, indicated that the carrier concentration was less than 1×10^{16} cm^{-3} between the surface and 0.6 to 0.8 μm from the surface for all MeV implant and anneal conditions studied in this paper. There is a gradual rise in the n-type carrier concentration profile from 0.6 μm (where n < 10^{16} cm^{-3}) to 1.2 μm (where n = 6 to 7 x 10^{16} cm^{-3}) following the near-surface tail of the MeV implants. We verified that the leading tail of the MeV implants near the surface was abrupt after implant annealing and that the carrier concentrations near the surface regions of these GaAs wafers were low enough that we would fabricate the shallower layers required for the varactor diodes by further ion implantation. These device fabrication and device results will be described in another publication.

CONCLUSIONS

Chemical SIMS profiles and electrical carrier concentration profiles have been obtained for Si and S implants between 2.5 and 6 MeV in semi-insulating GaAs substrates. Both furnace annealing between 900 and 950°C for 30 minutes and rapid thermal annealing between 950 and 1050°C for 10 seconds were used to activate the MeV implants. No redistribution of Si was observed using either annealing technique. On the other hand, for S implants, RTA seems advantageous over furnace annealing because it prevents the redistribution of the fast-diffusing S ions. A buried active layer at a depth of 2.8 μm with a peak carrier concentration of 2.5 x 10^{18} cm^{-3} was achieved by using 6 MeV Si + 4 MeV Si + 6 MeV S implants and rapid thermal annealing them at a temperature 1050°C for 10 seconds. This implant and anneal optimization study leads to the capability to fabricate fully implanted varactor diodes for monolithic millimeter wave circuit applications.

ACKNOWLEDGEMENTS

This work was partially supported by the U.S. Air Force Avionics Laboratory, WPAFB, under Contract Number F33615-86-C-1069.

REFERENCES

1. P. E. Thompson, H. B. Dietrich, M. Spencer, and D. C. Ingram, SPIE 530, 35 (1985).

2. P. E. Thompson and H. B. Dietrich, in *GaAs and Related Compounds*, Las Vegas 1986, Inst. Phys. Conf. Ser. No. 83, edited by W. T. Lindley (Institute of Physics, Bristol, 1987), pp. 271-276.

3. P. E. Thompson, R. G. Wilson, D. C. Ingram, and P. P. Pronko in *Materials Modification and Growth Using Ion Beams*, edited by U. Gibson, A. E. White, and P. P. Pronko (Mater. Res. Soc. Proc. 93, Pittsburgh, Pa 1987), pp. 73-77.

4. H. Kanber, M. Feng, and J. M. Whelan, *J. Appl. Phys.*, 55, 347(1984).

5. H. Kanber and J. M. Whelan, *J. Electrochemical Society*, 134, 2596 (1987).

INFLUENCE OF THE TEMPERATURE OF IMPLANTATION ON THE MORPHOLOGY OF DEFECTS IN MeV IMPLANTED GaAs.

G. BRAUNSTEIN, SAMUEL CHEN, S.-TONG LEE AND G. RAJESWARAN.
Corporate Research Laboratories, Eastman Kodak Company, Rochester, NY 14650.

ABSTRACT

We have studied the influence of the temperature of implantation on the morphology of the defects created during 1-MeV implantation of Si into GaAs, using RBS-channeling and TEM. The annealing behavior of the disorder has also been investigated.

Implantation at liquid-nitrogen temperature results in the amorphization of the implanted sample for doses of 2×10^{14} cm^{-2} and larger. Subsequent rapid thermal annealing at 900°C for 10 seconds leads to partial epitaxial regrowth of the amorphous layer. Depending on the implantation dose, the regrowth can proceed from both the front and back ends of the amorphous region or only from the deep end of the implanted zone. Nucleation and growth of a polycrystalline phase occurs concurrently, limiting the extent of the epitaxial regrowth. After implantation at room temperature and above, two distinct types of residual defects are observed or inferred: point defect complexes and dislocation loops. Most of the point defects disappear after rapid thermal annealing at temperatures $\geq$ 700°C. The effect of annealing on the dislocation loops depends on the distance from the surface of the sample. Those in the near surface region disappear upon rapid thermal annealing at 700°C, whereas the loops located deeper in the sample grow in size and begin to anneal out only at temperatures in excess of 900°C. Implantation at temperatures of 200 - 300°C results in a large reduction in the number of residual point defects. Subsequent annealing at 900°C leads to a nearly defect-free surface region and, underneath that, a buried band of partial dislocation loops similar to those observed in the samples implanted at room temperature and subsequently annealed.

INTRODUCTION

The recrystallization behavior of GaAs, ion-implanted in the keV energy regime, is a complex process which depends strongly on implantation and processing parameters. Under implantation conditions that induce the formation of a continuous amorphous layer (low temperature of implantation, high dose), thermal annealing can result in polycrystal formation, epitaxial regrowth, formation of stacking faults and microtwins, nucleation of dislocation loops and other phenomena, depending on the temperature and duration of the annealing stage.[1,2] However, when the implantation conditions are such that the amorphous layer is not formed (implantation at or above room temperature, low dose), appropriate thermal annealing can result in very low levels of residual defects.[1] Furthermore, when the temperature of implantation is near or above room temperature, significant dynamic annealing of the implantation damage occurs, reducing the amount and influencing the nature of the residual lattice defects.[1,3] More specific details about the recovery of GaAs as a function of implantation and annealing conditions in the keV energy regime can be found in references 1, 2, and 3, and references therein.

A rather complex recrystallization behavior can also be expected in MeV implantation of GaAs. It has been shown recently that the lattice structure of GaAs is highly preserved after implantation of 1-MeV Si ions, at room temperature, to doses as high as 1×10^{16} cm^{-2}.[4,5,6] The intense dynamic annealing which takes place during the MeV irradiations not only prevents the amorphization of the lattice but also induces the formation of certain well-defined lattice defects. Indeed, only perfect and partial dislocation loops have been identified by transmission electron microscopy (TEM) analysis of the as-implanted samples. In addition, the presence of discrete point defect complexes was inferred as a result of ion-channeling studies. Most of the point defects annealed out upon rapid thermal annealing at 500 - 700°C for 10 seconds. Concurrently, the near surface region became increasingly defect free while, deeper into the sample, the dislocation loops transformed into the imperfect type. With increasing annealing temperature the loops grew in size while decreasing in number, but did not completely disappear, even after annealing at 1050°C. In the present work we have studied the influence of the temperature of implantation on the build-up, and subsequent annealing behavior, of defects in GaAs implanted with 1-MeV Si ions over an extended implantation temperature range varying from -198°C to 300°C.

EXPERIMENTAL

Semi-insulating, (100) GaAs samples were implanted in a nonchanneling direction with 1-MeV Si ions at temperatures between -198°C and 300°C. Doses ranging from 1×10^{13} cm^{-2} to 1×10^{16} cm^{-2} were employed. In order to evaluate beam heating effects a very thin thermocouple was glued to a test sample, which was subsequently irradiated at various temperatures. It was found that when the beam current density was kept below 0.75 μamp/cm^2, the temperature of the test sample would not rise more than 20 - 30°C above the temperature of the copper block used as sample holder, irrespective of the implantation temperature. Consequently, a beam current of ~ 0.7 μamp/cm^2 was employed throughout the present study. Rapid thermal annealing (RTA) was carried out under flowing nitrogen gas at temperatures of 500°C, 700°C and 900°C for periods of 10 to 40 seconds using GaAs proximity capping. The residual disorder, after implantation and after subsequent annealing, was investigated using 2-MeV He^{++} channeling and cross-sectional TEM.

RESULTS

Figure 1 shows aligned channeling spectra of GaAs implanted with 1-MeV Si ions, to a dose of 5×10^{15} cm^{-2}, at various temperatures. The aligned and random spectra of unimplanted GaAs are also shown. The aligned spectrum of the sample implanted at -98°C coincides with the random spectrum over the entire implanted region, revealing the formation of a continuous amorphous layer. However, significant recovery of implantation damage, revealed by the reduction in scattering yield particularly in the near-surface region, is evident when the implantation temperature is increased to -52°C. The aligned spectrum of the sample implanted at room temperature (24°C) shows further reduction of residual disorder, and is characterized by a gradual increase in yield with increasing depth. A similar dechanneling depth dependence, but with a uniformly reduced yield, is observed in the sample implanted at 204°C.

The different characteristics of the ion-channeling spectra, shown in Figure 1, suggest that, by varying the temperature of implantation, not only the amount of residual defects can be changed but, more significantly, the nature of these defects can be altered. In order to test this assumption we have investigated the recovery, upon RTA, of the radiation damage produced by implantation at various temperatures (i.e., the annealing behavior of different types of defects).

Figure 2 shows channeling spectra of GaAs after implantation with 1×10^{15} Si/cm^2 at -198°C and after subsequent annealing at 900°C for 40 seconds, illustrating the annealing behavior of a continuous amorphous region. The aligned spectrum of the annealed sample exhibits the characteristic surface peaks for As and Ga, indicating almost complete recovery of crystallinity in the near surface. Below that, there is a very steep rise in backscattering yield reflecting the existence of a rather sharp interface between a crystalline region in the near surface and a disordered region underneath. The aligned yield, corresponding to the disordered region, does not reach the random level, indicating that it is a buried band of polycrystalline, or partially crystalline, material that retained some alignment with respect to the underlying lattice structure. The deeper end of this defective band is much shallower than the original amorphous-crystalline interface in the as-implanted sample, revealing that lattice recovery has also proceeded from the underlying unimplanted crystal.

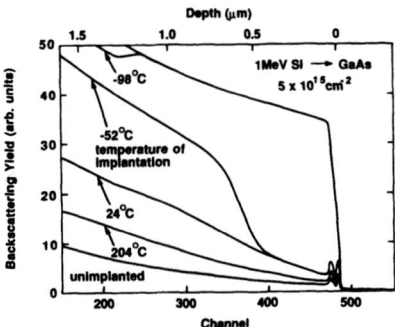

Figure 1: Ion-channeling spectra of GaAs implanted with 1-MeV Si ions to a dose of 5×10^{15} cm^{-2} at various temperatures.

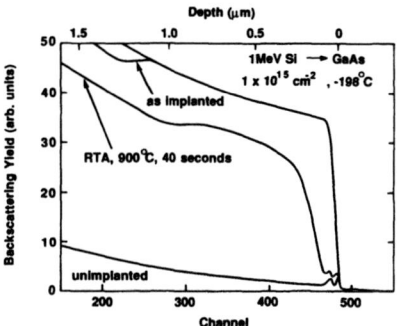

Figure 2: Ion-channeling spectra of GaAs implanted at -198°C with 1-MeV Si ions to a dose of 1×10^{15} cm^{-2} and annealed at 900°C for 40 seconds.

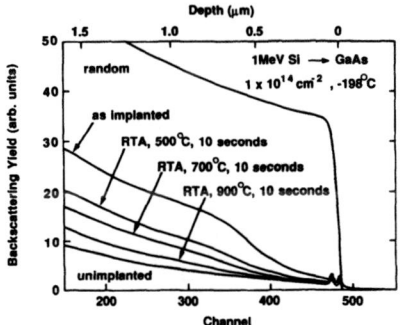

Figure 3: Ion-channeling spectra of GaAs after implantation at -198°C with 1-MeV Si ions to a dose of 1×10^{14} cm^{-2} and after RTA at various tempreatures.

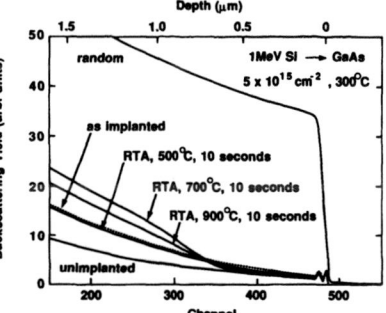

Figure 4: Ion-channeling spectra of GaAs after implantation at 300°C with 1-MeV Si ions to a dose of 5×10^{15} cm^{-2} and after RTA at various temperatures.

Figure 3 presents the annealing behavior of frozen-in lattice defects, formed after implantation of GaAs with 1×10^{14} Si/cm^2 at -198°C. A disorder peak is noticeable centered at about 0.8±0.1 μm in good agreement with TRIM[7] calculations. The dechanneling yield is observed to decrease gradually with increasing annealing temperature. Upon RTA at 900°C for 10 seconds the yield of the aligned channeled spectrum is only slightly higher than the spectrum of unimplanted GaAs, indicating that most of the defects have been annealed out.

Finally, Figure 4 shows channeling spectra of GaAs after implantation with 5×10^{15} Si/cm^2 at 300°C and after subsequent RTA at 500°C, 700°C and 900°C for 10 seconds, illustrating the annealing behavior of the defects created in high-temperature implantations. The aligned spectrum of the as-implanted sample shows a gradual increase in backscattering yield with increasing depth similar to that of the GaAs implanted at 204°C in Figure 1. No major change is observed after RTA at 500°C. The spectra corresponding to the 700°C and 900°C anneals exhibit a peculiar sigmoid-like shape, already found in samples implanted at room temperature and rapid thermally annealed at 700 - 900°C.[5,6] In the near-surface region the yield coincides with that of unimplanted GaAs, indicating complete recovery of implantation damage. However, deeper into the sample, a pronounced rise of the backscattering yield reveals the presence of residual defects. In references 5 and 6 we have shown that the sigmoid-like spectrum reflects the formation of a buried band of partial dislocation loops.

DISCUSSION

The results of the present work indicate that in MeV-implantation of GaAs the temperature of implantation has a significant role in determining the nature of the radiation-induced defects. At about liquid-nitrogen temperature defects have very low mobilities. Thus, when the implantation is performed in this temperature regime, lattice displacements remain essentially frozen-in. For 1-MeV Si implanted into GaAs at -198°C to the relatively low dose of 1×10^{14} cm^{-2} radiation damage most likely consists of vacancies, self-interstitials, point-defect complexes and possibly small amorphous clusters. As the implantation dose is increased, the amount of disorder increases until a continuous amorphous layer throughout all the implanted region is formed. When the temperature of implantation is raised, migration of self-interstitials and vacancies becomes possible due to increased thermal activation as well as radiation-enhanced mobility. Mobile point defects have larger probabilities to annihilate by recombination or by agglomeration into extended defects, thus inhibiting amorphization and giving rise to the process of dynamic annealing. In fact, a competition takes place between the accumulation of lattice disorder and the dynamic annealing process, which is particularly evident in the sample implanted at -52°C. In the near surface region of this sample, dynamic annealing predominates and the aligned yield is close to that of samples implanted at higher temperatures. Deeper into the sample, where the nuclear energy deposition is higher, accumulation of lattice disorder prevails and the aligned channeling spectrum shows a damage peak which appears to track the nuclear-energy-loss depth distribution. A TEM study to determine the nature of the defects in this region is underway.

When the GaAs samples are implanted at or above room temperature, dynamic annealing prevails. No amorphicity could be detected by TEM analysis in 1-MeV Si, room-temperature implanted GaAs, even when the dose was as high as 1×10^{16} cm^{-2}. Indeed, in previous works[5,6] we have shown that the lattice structure of GaAs, implanted at room temperature with 1-MeV Si ions, consists of isolated partial and perfect dislocation loops as well as discrete point-defect complexes dispersed in perfectly crystalline material. We have also observed that a large fraction of the point defects created in room temperature MeV implantations were annealed out after RTA at 500°C for 10 seconds. Correspondingly, the aligned channeling spectrum of the sample annealed at 500°C showed a uniformly reduced yield with respect to the as-implanted sample, much in the same way as the sample implanted at 204°C in the present study shows a uniformly reduced dechanneling yield with respect to the sample implanted at room temperature. This similarity suggest that dislocation loops are the predominant residual defect after implantation of GaAs with 1-MeV Si ions at 204°C.

Having established the nature of the implantation-induced lattice defects as a function of implantation temperature and dose, we proceed to discuss the annealing behavior ot the defects.

The recrystallization of the continuous amorphous region, formed by high dose implantation at -198°C, appears to be the result of two competing processes: epitaxial regrowth, proceeding from both the surface and the deep amorphous-crystalline interface, versus grain nucleation and growth in the central zone. A similar phenomenon was observed by F. Xiong et al.[8] in their study of amorphization and recrystallization of MeV-implanted InP. Our interpretation is further supported by the cross-sectional TEM micrographs, shown in Figures 5a and 5b, of a sample implanted at -198°C with 1-MeV Si ions to a dose of $1x10^{16}$ cm^{-2} and subsequently rapid thermally annealed at 900°C for 10 seconds. In the micrograph of the as-implanted sample we observe a uniform amorphous region starting at the surface and ending, in a rather sharp amorphous-crystalline interface, at a depth of about

Figure 5: Cross-sectional TEM micrographs of GaAs implanted at -198°C with 1-MeV Si ions to a dose of $1x10^{16}$ cm^{-2}; (a) as implanted and (b) after RTA at 900°C for 10 seconds.

1.3 μm. The micrograph of the annealed sample shows the formation of a polycrystalline structure which is narrower than the original amorphous region. In the polycrystalline region the grains closer to the surface are much larger than those closer to the buried interface, although no epitaxial regrowth proceeding from the surface is observed in this sample. It could be that, for samples implanted with doses lower that 1×10^{16} cm^{-2} like the one in figure 2, small crystallites remain after implantation which serve as seeds for epitaxial recrystallization upon annealing.

The rest of the discussion of the dependence of annealing behavior on the temperature of implantation is based on the results of our previous works[5,6] which were briefly summarized in the introduction section. The disorder created by 1×10^{14} Si/cm^2 implanted at -198°C gradually decreases with increasing annealing temperature and is almost completely annealed out after RTA at 900°C for 10 seconds. We interpret this behavior as proof that the implantation-induced defects were mainly point defects or small point defect aggregates since more complex extended defects, such as dislocations, or continuous amorphous regions, as described above, exhibit a very different recovery behavior in this temperature regime. We have already argued that after implantation at 204 - 300°C most point defects have been annealed out and only dislocation loops remain. The annealing behavior of the sample implanted at 300°C supports our assumption. Little change is observed in the channeling spectrum after RTA at 500°C, the temperature at which recovery of point defects appears to be the dominant annealing process. Furthermore, the sigmoid-like channeling spectra observed after RTA at 700°C and 900°C has been shown to represent a signature for the transformation of perfect into partial dislocation loops. Annealing at a temperature of 1050°C for 10 seconds was not sufficient to completely anneal out these dislocation loops.

ACKNOWLEDGMENTS

We appreciate the help of J. Madathil with the ion implantations, J. deJohn with the TEM sample preparation and S. Wilson and L.R. Zheng with the RTA. We also thank J.S. Williams for the preprint of reference 3.

REFERENCES

1) D.K. Sadana, Nucl. Instrum. Methods, B7/8, 375 (1985).

2) W.G. Opyd, J.F. Gibbons, J.C. Bravman, and M.A. Parker, Appl. Phys. Lett., 49, 974 (1986).

3) J.S. Williams, R.G. Elliman, S.T. Johnson, D.K. Sengupta, and J.M. Zemansky, in: Processing and Characterization of Materials Using Ion Beams, Mat. Res. Soc. Symp. Proc., in press.

4) S.-Tong Lee, G. Braunstein, and Samuel Chen, Mat. Res. Soc. Symp. Proc. 126, 183 (1988).

5) Samuel Chen, G. Braunstein, and S.-Tong Lee, in: Characterization of Structure and Chemistry of Materials, Mat. Res Soc. Symp. Proc., in press.

6) G. Braunstein, L.-R. Zheng, Samuel Chen, S.-Tong Lee, D.L. Peterson, K.Y. Ko, and G. Rajeswaran, in: Processing and Characterization of Materials Using Ion Beams. Mat. Res. Soc. Symp. Proc., in press.

7) J.P. Biersack and L.G. Haggmark, Nucl. Instrum. Methods, 174, 257 (1980).

8) F. Xiong, C.W. Nieh, D.N. Jamieson, T. Vreeland Jr., and T.A. Tombrello, Mat. Res. Soc. Symp. Proc., 100, 105 (1988).

DEFECTS PRODUCED BY HIGH ENERGY OXYGEN IONS IMPLANTED IN SILICON

A. GROB*, J.J. GROB*, A. PERIO**, P. THEVENIN* and P. SIFFERT*
*Centre de Recherches Nucléaires (IN2P3), Laboratoire PHASE (UPR du CNRS n°292), 23 rue du Loess, F-67037 Strasbourg Cedex, France
**Centre National d'Etudes des Télécommunications, BP 98, F-38243 Meylan Cedex, France

ABSTRACT

Rutherford backscattering/channeling analysis was used to study the damage creation processes occuring during 1 MeV O^+ implantation in silicon. The target temperature was varied from RT to 500°C using beam heating. The corresponding damage profiles and dechanneling behaviour were studied. Several implantations were performed at 77K for comparison. Transmission electron microscopy observations were connected to the dechanneling measurements in order to determine the dominant kind of defect in each case.

For 77K implants, the defects are mainly interstitials distributed according the energy deposition in elastic collisions, extending up to the surface. At 500°C, the defects are imperfect dislocations confined in a narrow band around the mean range of oxygen ions. We demonstrate that dechanneling in samples implanted at intermediate temperatures results from a mixing of point defects and distorsion centers. The relative importance of the two kind of defects is followed as a function of implantation temperature.

INTRODUCTION

High energy implantation is one of the most promising new areas of implantation technology for Si integrated circuits. Several advantages are offered by MeV implantation, particularly the possibility to dope below the surface [1] or through overlayers [2] or to form buried insulating [3] or conductive [4] layers. Most of the physical mechanisms known in the traditional energy range of ion implantation ($\leq$ 200 keV) are not basically modified at high energy. However, there are several differences which lend themselves to new applications and/or lead us to reconsider some particular points. Damage creation mechanisms are particularly concerned since they can be modified by the combined effect of the reduction in the primary defects production rate, increased distance from surface and beam heating. Surprisingly, there have been few papers dealing with defects produced by MeV implantation and their annealing. We only know that "there are distinct regions in which different damage nucleation and growth mechanisms are dominant" [5] and that secondary defects exist which cannot be completely annealed out even after long high temperature heating cycles [6].

The aim of this work is to collect further informations about the damage production mechanisms at high energy with a particular attention devoted to beam power (i.e. temperature) related effects like dynamic annealing, agglomeration of single defects and nucleation of secondary (extended) defects. The nature, number and distribution of damages created by 1 MeV oxygen ions in silicon have been investigated as a function of dose and using different beam power densities. These results have been compared to the damage parameters measured at 77K. An attempt is made to describe the evolution of each kind of defects as implantation temperature increases.

EXPERIMENTAL

1 MeV oxygen ions were implanted in single crystal silicon wafers using a 4 MV Van de Graaff accelerator. During implantation, the samples were slightly tilted ($\approx$ 10°) relatively to the <100> axis in order to minimize channeling. The beam current was varied from 0.1 up to 7-8 µA and the total dose from 10^{16} to 10^{17} cm^{-2}. The temperature reached during implantation strongly depends on experimental conditions : wafer size, wafer holding, scanning system, scanned area relatively to total (radiating) area, surrounding, ... [7]. In our case, the samples were thermally insulated and the O^+ beam was electrostatically scanned over the area of 1 cm^2. The target temperature was calibrated and often checked as a function of the beam power P_b using a test sample of the same size in which a thermocouple was embedded and cemented. The

198

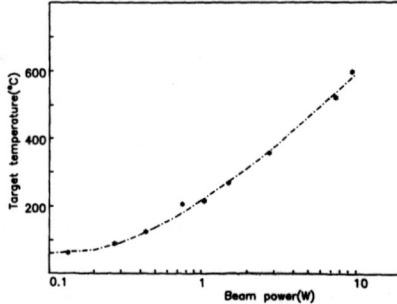

Fig. 1 : Calculated (dashed line) and measured equilibrium target temperature as a function of incident beam power.

calibration curve is shown in Fig. 1. The wafer temperature varies as $(P_b)^{1/4}$ indicating that ion beam power and radiative losses are in equilibrium. The latter is reached after a time corresponding to a dose of about 10^{15} cm^{-2}. In these conditions, even with the moderate O$^+$ current which can be obtained with our accelerator, temperatures as high as 500°C can be achieved. It must be emphasized that higher temperatures could be reached using conductive but also reflective wafer holders [7].

A different set-up was used for implantation at 77K. In this case, smaller samples were placed on a goniometric holder which can be cooled by a flow of liquid nitrogen. They were implanted at lower doses in order to obtain similar defect concentrations and analysed in-situ using the same accelerator.

The disorder created by the O$^+$ bombardment was mainly examined by the Rutherford backscattering (RBS)/channeling technique [8,9]. The energy of backscattered particles was measured with a surface barrier detector located at 160° with respect to the incident beam. Random and <100> aligned spectra were recorded on both implanted and unimplanted part of the samples. Attention was paid to frequently change the beam spot location in order to avoid the damage produced during the analysis. The profile of displaced atoms was extracted from direct scattering events and that of more complex defects by estimating or measuring the dechanneling level as a function of depth. Additional informations on the nature of the defects have been obtained by varying the energy of the analysing particles (He$^+$ or protons) [10]. Transmission electron microscopy (TEM) observations (plan views) have been connected to these energy-dependent dechanneling measurements in order to determine the dominant type of defects in each case.

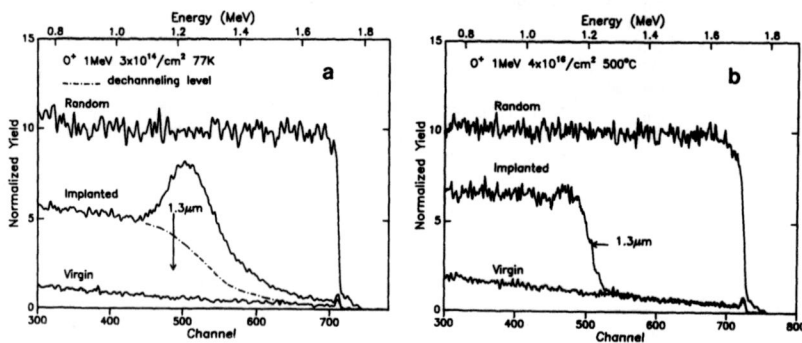

Fig. 2 a and b : Backscattered spectra of 3 MeV He$^+$ random and <100> aligned on Si before and after bombardment with 1 MeV O$^+$ at (a) 77K - 3 x 10^{14} cm^{-2} and (b) 500°C - 4 x 10^{16} cm^{-2}. The dechanneling level in (a) (dashed line) was calculated using a plural scattering cross-section. The mean projected range of O$^+$ (1.3 μm) is marked by an arrow.

RESULTS AND DISCUSSIONS

The dependence of both production and depth distribution of defects on implantation temperature is evidenced in Fig. 2a and b. At 77K (Fig. 2a), a dose of 3.10^{14} cm^{-2} is nearly sufficient to amorphize the crystal and the damaged region extends up to the surface. At 500°C (Fig. 2b) a dose of 4.10^{16} cm^{-2} does not produce any displacement in the first μm of silicon. A sharp increase of the dechanneling level occurs at 1.3 μm which is exactly the mean projected range (R_p) of 1 MeV O$^+$ in silicon [11].There is practically no direct scattering observed in this case.

In a first stage, we will try to demonstrate that the nature of defects is very different in these two extreme situations. By measuring the dechanneling fractions of aligned particles when passing through the disordered region as a function of their energy E, we can distinguish between obstruction and distorsion defects [12] :

- obstruction centers refer to randomly distributed isolated or clustered point defects or even small amorphous zones [10]. They do not induce any lattice deformation around them. In the case of axial channeling, the dechanneling cross section (σ_{d1}) (see definition in [10]) on this first kind of damage is given by the Rutherford cross section (or by a multiple scattering cross section in the case of a large concentration of defects), integrated over all angular deviations greater than the critical value [10]. In such a case, σ_{d1} varies as E$^{-\alpha}$ where $0.5 \leq \alpha \leq 1$ ($\alpha = 1$ for small point defect concentrations n_{d1} and $\alpha = 0.5$ for large n_{d1} [13]). More refined calculations of the dechanneling cross section lead to the same conclusion [13,14];

- distorsion centers refer to stress-induced lattice deformations around more complex defects like dislocations or precipitates. The maximum curvature of atomic rows as a function of the distance from the defect center can be estimated from elastic theory and the dechanneling cross section σ_{d2} calculated from the condition that the force due to the repulsive potential of the row must be greater than the centrifugal force of the particle [15]. If the strain extends infinitevely as around a perfect edge dislocation σ_{d2} varies as $\sqrt{E}$ [15]. A similar result was obtained using a Monte Carlo simulation and was often experimentally verified [16,17]. If the strain is limited as around dislocation loops, σ_{d2} varies as $\sqrt{E}$ up to a threshold energy and becomes constant higher [18].

Assuming an additivity rule of the contributions to dechanneling coming from the two kinds of defects we can define the dechanneling parameter $P_d(x)$ directly proportional to the dechanneling cross sections σ_{d1} and σ_{d2} and related to the experimentally measured dechanneling fractions in damaged (χ_D) and virgin (χ_V) silicon [9] :

$$P_d(x) = \sigma_{d1} \int_0^x n_{d1}(z)\, dz + \sigma_{d2} \int_0^x n_{d2}(z)\, dz = P_{d1}(x) + P_{d2}(x) \qquad (1)$$

$$\text{and} \quad P_d(x) = \ln \frac{1 - \chi_V(x)}{1 - \chi_D(x)} \qquad (2)$$

$n_{d1}(x)$ and $n_{d2}(x)$ are the local concentrations of obstruction and distorsion defects respectively. The values of the dechanneling parameter P_d measured behind the damage peak are reported in Fig. 3. as a function of the ratio E/Z_1 (Z_1 is the atomic number of the particles), for two samples damaged at 77K using 2×10^{14} O$^+$/cm^2 and 3×10^{14} O$^+$/cm^2. These results show that P_d varies as E$^{-0.5}$, demonstrating that the defects are randomly distributed interstitials. Their mean production rate is estimated to be ≈ 5000 per incident ion, in agreement with the model proposed by Thompson and Walker [19].

Once known the dominant defect in these samples, we are able to separate dechanneling events and direct scatterings using the well known iterative procedure [8] and a plural scattering cross section [20]. This gives the dechanneling level as plotted in Fig. 2a which connects with the experimental spectrum beyond the damage peak. The difference between this level and the total backscattering yield gives the depth distribution of displaced atoms. This profile is shown to correspond to that of energy deposited into elastic collisions [21].

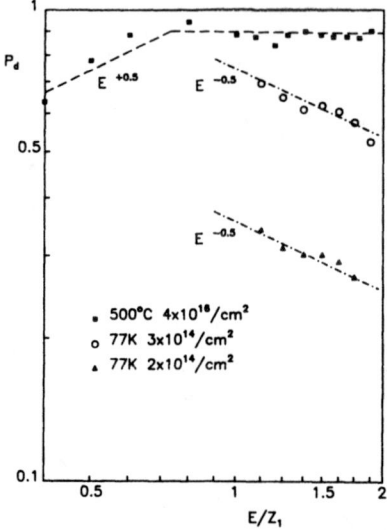

Fig. 3 : Dechanneling parameter P_d as a function of the ratio E/Z_1 of the incoming particles energy to their atomic number, for samples damaged at 77K and 500°C.

A similar investigation was performed on samples implanted at 500°C (Fig. 3). P_d first increases as $\sqrt{E}$ and becomes constant at higher energies. As mentioned above, this behaviour indicates that the dominant species are extension limited strain defects. More precise informations on the morphology of these defects may be drawn from TEM observations (Fig. 4a and b).

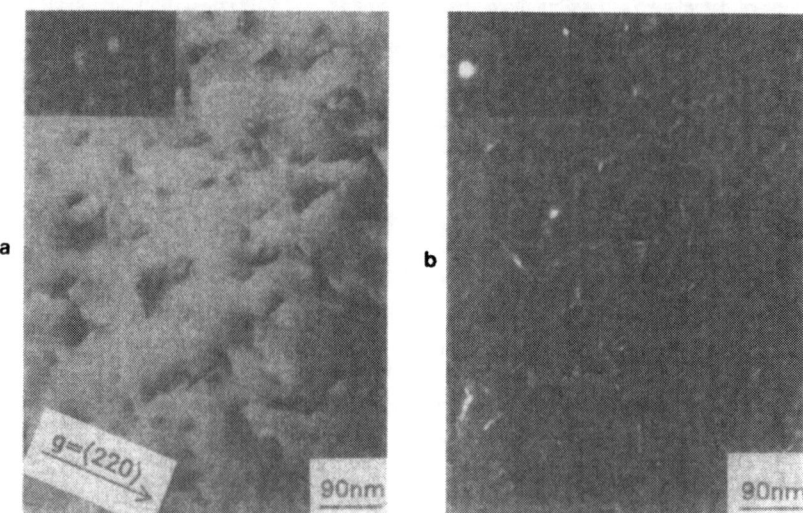

Fig. 4 a and b : Plan view TEM micrographs of the Si sample damaged with 4×10^{16} O$^+$/cm^2 at 500°C : (a) bright field ; (b) weak beam.

The bright field picture (Fig. 4a) shows an important density (5.10^{10} cm^{-2}) of linear dislocations whose length ranges from 20 to 100 nm. The weak beam micrograph (Fig. 4b) reveals that they are dissociated. Because of the narrowness and depth of these defects, it was not possible to evidence neither the stacking faults contrast nor the habit plane. However, these results strongly suggest that the drastic increase of dechanneling observed in Fig. 2b is mainly due to imperfect dislocations. Their profile can be determined using the relation (1) in which the first contribution (P_{d1}) is neglected :

$$n_{d2}(x) = \frac{1}{\sigma_{d2}} \frac{d}{dx} \left\{ \ln \left[\frac{1 - \chi_V(x)}{1 - \chi_D(x)} \right] \right\} \qquad (3)$$

σ_{d2} is estimated knowing the total concentration of dislocations from the TEM micrograph. Fig. 5 shows that the dislocations form a narrow band (FWHM ≈ 100 nm) around the mean projected range of oxygen ions.

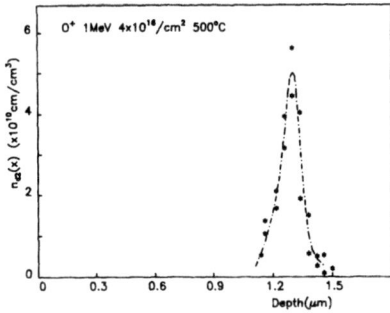

Fig. 5 : Concentration profile of the dislocations produced during the implantation of 4 x 10^{16} O$^+$/cm^2 at 500°C.

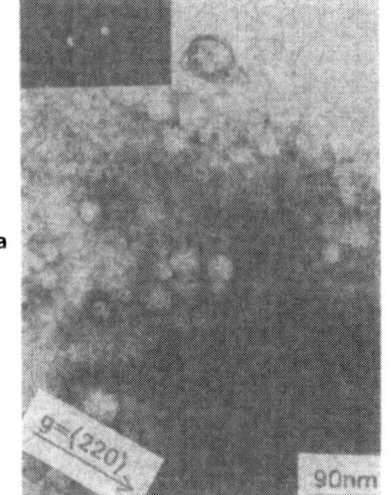

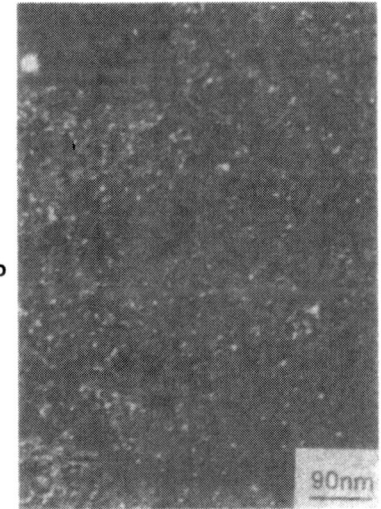

Fig. 6 a and b : TEM micrographs of the sample damage with 4 x 10^{16} O$^+$/cm^2 at 280°C : (a) bright field ; (b) weak beam.

Between the two extreme situations described above, one can reasonnably think that, as implantation temperature increases, the defect accumulation progressively changes from the simple agglomeration of point defects to the nucleation and growth of extended defects. Such an intermediate situation is shown in Fig. 6a and b. A large variety of defects are present in the crystal. The bright field micrograph (Fig. 6a) shows local variations of contrast due to elastic strains around defects. The weak beam observation (Fig. 6b) reveals aggregates, probably amorphous, embedded in the crystalline material and a large density (a few 10^{10} cm^{-2}) of small diameter (3-4 nm) spherical precipitates (probably SiO_2). The RBS spectra also show a mixing of different defects since simultaneously direct scattering events and a high dechanneling level is observed (Fig. 7). Furthermore, the latter cannot be connected to the experimental level beyond the damage peak using only a plural scattering treatment. We demonstrate that in all these intermediate cases, the variation of the dechanneling parameter P_d with He$^+$ energy can be fitted using the two contributions :

$$P_d = P_{d1} + P_{d2} = \frac{A}{\sqrt{E}} + B\sqrt{E} \qquad (4)$$

In other words, the main contributions to dechanneling are point defects and unlimited strains. The dislocations observed at 500°C are not actually formed but distorsions exist. A similar behaviour was already observed for B$^+$ implanted silicon at lower energies [22].

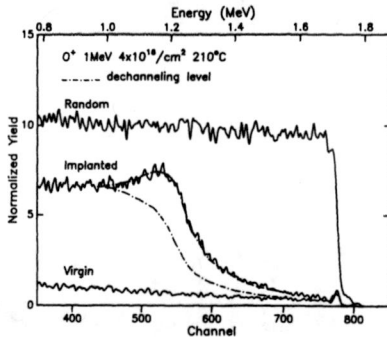

Fig. 7 : Backscattering spectra of 3 MeV He$^+$ random and <100> aligned on a Si sample before and after bombardment with 4 x 10^{16} O$^+$/cm^2 at 210°C. The dechanneling level (dashed line) was calculated taking into account the contributions of both kinds of defects.

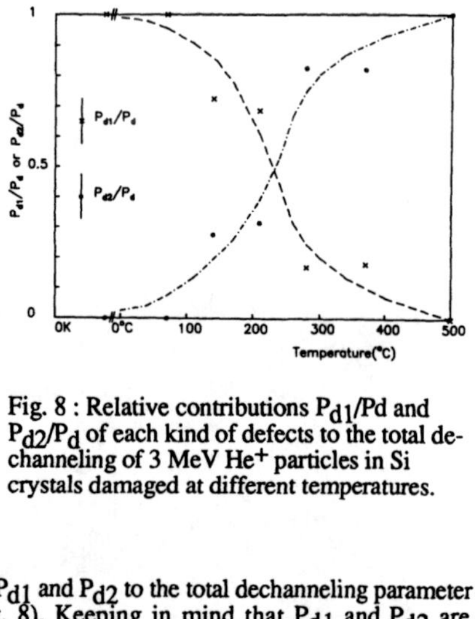

Fig. 8 : Relative contributions P_{d1}/Pd and P_{d2}/P_d of each kind of defects to the total dechanneling of 3 MeV He$^+$ particles in Si crystals damaged at different temperatures.

We have studied the relative importance of P_{d1} and P_{d2} to the total dechanneling parameter as a function of implantation temperature (Fig. 8). Keeping in mind that P_{d1} and P_{d2} are proportional to the total number of each type of defects, Fig. 8 shows that the production of interstitials progressively diminishes (at 280°C there are only 16 displacements per incident ion) while that of distorsions increases. The dashed lines in Fig. 8 are only to guide the eye ; presently we are not able to distinguish between a continuous or a step by step transformation from point defects to dislocations. It must be pointed out that the mean transition temperature (here ≈ 250°C) between type 1 and type 2 defects should depend on Frenkel pair production rate (i.e. on both mass and energy of implanted ions). Furthermore, in the precision limits of our measurements, it seems that the partition between the two kinds of defects is not dose dependent except in three particular situations :
- at low dose ($\leq 5.10^{15}$ O$^+$/cm^2) where distorsions are difficult to bring out : either the sensitivity of our experiment is not sufficient or there is a minimum dose required for their apparition ($\approx 10^{16}$ O$^+$/cm^2). The non-linear behaviour of the production of distorsions was often observed in other experimental situations [22, 23, 24] ;

- at very high dose ($\geq 5.10^{16}$ O$^+$/cm^2) and low temperature ($< 200°$C) when a critical point defect concentration is achieved and amorphization occurs. In such a case, the term P_{d1} increases drastically ;

- at very high dose and high temperature ($> 300°$C) where we have observed that the production of distorsion centers (term P_{d2}) saturates.

Finally, assuming that the depth distribution of point defects follows that of the energy deposited in elastic collisions and that strains have a gaussian profile centered at 1.3 µm, we have reconstructed several aligned RBS spectra of crystal damaged at different temperatures. The only fitting parameter was the width of the strain distribution which was varied between that measured at 500°C and that of the O$^+$ distribution. The profile of point defects was normalized knowing that the total number of diplaced atoms extracted from P_{d1} and using the dechanneling cross section measured at 77K. Such calculations are shown to fit exactly the experimental spectra (Fig. 7). This result confirms our assumptions concerning the presence of two dominant kinds of defects and that they are mainly located in two distinct regions : point defects spread roughly from surface up to R_p and distorsions are confined in a narrow band around R_p. Both production and annealing of type 1 defects seems to be only related to collision parameters and target tempeature and hence is insensitive to the implanted ion specie [5] while distorsions should depend on the formation of precipitates (i.e. on impurity concentration).

CONCLUSIONS

In summary, we demonstrate that the dechanneling behaviour of light particles in silicon crystals damaged by MeV O$^+$ ions using different beam powers results from the presence of two main kinds of defects : point defects and distorsions. Isolated or clustered point defects are the dominant species up to an implantation temperature of 200-250°C while distorsions, probably located around SiO$_2$ precipitates, become the major dechanneling centers at higher temperatures. The two kinds of damage are produced in two distinct regions : point defects extend from surface up to R_p, distorsions are centered on R_p.

ACKNOWLEDGMENTS

We are grateful to A. Golanski for many fruitful discussions and to A. Inard for his help in preparing the samples for TEM observations. The authors wish to acknowledge L.R. Doolittle for providing the computer program RUMP used to analyse the RBS data.

REFERENCES

[1] A. Kosta and S. Kalbitzer in Ion Implantation in Semiconductors, edited by S. Namba (Plenum New York, 1975) p.689.

[2] D. Pramanik and A.N. Saxena, Nucl. Instr. and Methods, B21, 116 (1987).

[3] J.J. Grob, A. Grob, P. Thevenin, P. Siffert, C. d'Anterroches and A. Golanski, MRS 1988 Fall Meeting, Boston (december 1988) ; to be published in Journal of Material Research.

[4] A.E. White, K.T. Short, R.C. Dynes, R. Hull and J.M. Vandenberg, Nucl. Instr. and Methods B39, 253 (1989).

[5] O.W. Holland, M.K. El-Ghor and C.W. White, Appl. Phys. Lett. 53 (14), 1282 (1988).

[6] M. Tamura, N. Natsuadi, Y. Wada and E. Mitami, Nucl. Instr. and Methods B21, 438 (1987).

[7] P.D. Parry, J. Vac. Sci. Technol. 13 (2), 622 (1976).

[8] W.K. Chu, J.W. Mayer and M.A. Nicolet, Backscattering Spectrometry (Academic Press, New York, 1978).

[9] L.C. Feldman, J.W. Mayer and S.T. Picraux, Material Analysis by Ion Channeling (Academic Press, New York, 1982) Chapter 5.

[10] Ibid. Chapter 4.

[11] J.J. Grob, A. Grob, P. Thevenin and P. Siffert, Nucl. Instr. and Methods B30, 34 (1988).

[12] Y. Quéré in "Session d'Etudes sur la Canalisation des Particules" (Ed. CEA-INSTN, 1974) p. 99.

[13] K. Gärtner, K. Hehl and G. Schlotzhauer, Nucl. Instr. and Methods B4, 55 (1984).

[14] J. Mory in "Session d'Etudes sur la Canalisation des Particules" (Ed. CEA-INSTN, 1974) p. 251.

[15] Y. Quéré, Phys. Stat. Sol. 30, 713 (1968).

[16] D.V. Morgan and D.V. Van Vliet in Atomic Collision Phenomena in Solids, edited by D.W. Palmer, M.W. Thompson and P.D. Townsend, (North Holland, 1970) p. 477.

[17] S.T. Picraux, E. Rimini, G. Foti and S.U. Campisano, Phys. Rev. B18 (5) 2078 (1978).

[18] G.G. Bentini, M. Bianconi and M. Servidori, Nucl. Instr. and Methods B18, 145 (1987).

[19] D.A. Thompson and R.S. Walker, Radiation Effects 36, 91 (1978).

[20] L. Meyer, Phys. Status Solid. 44, 253 (1972).

[21] D.K. Brice, Ion Implantation Ranges and Energy Deposition Distributions, Vol. 1 (Plenum, New York, 1975).

[22] J.J. Grob and P. Siffert, Nucl. Instr. and Methods 209, 413 (1983).

[23] C. Prunier, E. Ligeon, A. Bourret, A.C. Chami and J.C. Oberlin, Nucl. Instr. and Methods B17, 227 (1986).

[24] S. Coffa, L. Calgagno, C. Spinella, S.U. Campisano, G. Foti, E. Rimini, Nucl. Instr. and Methods B39, 357 (1989).

PART V

High-Dose Implantation

PARTIALLY IONIZED BEAM DEPOSITION OF THIN FILMS

T.-M. Lu
Center for Integrated Electronics and Physics Department
Rensselaer Polytechnic Institute, Troy, NY 12180

ABSTRACT

It has been shown recently that using less than a few percent of self-ions, i.e., ions derived from the deposition material itself, (partially ionized beam (PIB) deposition), one can dramatically modify thin film properties. In this paper we discuss phenomena such as surface cleaning, enhanced surface ordering, and ion-induced surface damage associated with PIB-surface interactions. PIB metal deposition has been emphasized.

Introduction

It is known that ion bombardment during film growth can dramatically modify the compositional, structural, and electronic properties of the films[1]. One of the most important features in the ion-assisted techniques is the fact that additional deposition parameters related to the ion bombardment are provided to give additional degrees of freedom to control the film properties. Very often, inert gas ions such as Ar^+ are employed to assist the film growth. Examples can be found in many plasma-assisted and ion-beam-assisted deposition techniques. Sometimes reactive gas ions such as O^+ are used to study compound formation. Examples are reactive plasma-assisted (or activated reactive) and reactive ion-beam-assisted deposition techniques [1].

Recently, a number of researchers [2-9] have employed self-ions, which are derived from the deposition material itself, as the source of ions to control the properties of elemental thin films. Very often, only a small amount of self-ions are employed in the deposition. The ion/atom ratio in the beam is normally less than a few percent. A bias potential of 0-5kV is applied to the substrate during deposition. This technique has been generally called the partially ionized beam (PIB) deposition. In this paper, we shall describe some of the important characteristics involving PIB deposition.

PIB Sources

The design of a PIB deposition system is very similar to that of ion plating technology [10] except that no foreign gas such as Argon is used. A wide variety of PIB sources have been used to deposit metal, semiconductor, insulator, and polymer thin films. A conventional PIB source involves first evaporation of the deposition material and then ionization of the vapor stream. A bias potential is applied to accelerate the charged species to the substrate. Evaporation of the material can be achieved by several means: resistance heating, inductive heating, and electron-bombardment heating. The vapor stream is then partially ionized by electron impact ionization. An example of the conventional PIB source design is shown in Fig. 1. The deposition material is vaporized in a high purity graphite crucible by electron bombardment heating. The electrons are drawn from a hot filament and are accelerated towards the crucible, heating it upon impact. The evaporated vapor stream is then partially ionized in an ionization unit located a few centimeters above the crucible. In the ionization unit, a hot filament provides electrons for the impact ionization of the vapor stream. Also falling in this class of design is the ionized cluster beam (ICB) deposition technique [4]. The only difference is that in the ICB deposition the crucible is covered with a cap containing a small orifice so that the vapor stream undergoes a supersonic expansion through it for cluster formation.

208

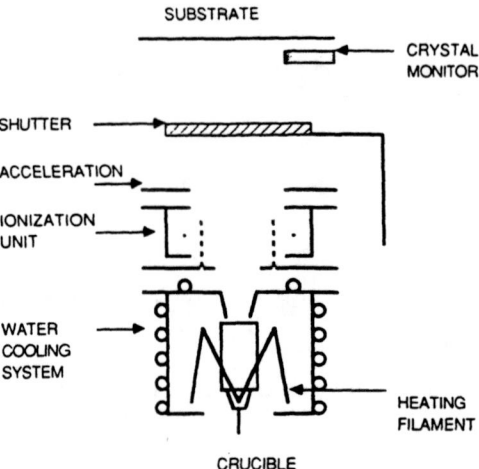

SUBSTRATE

CRYSTAL
MONITOR

SHUTTER

ACCELERATION

IONIZATION
UNIT

WATER
COOLING
SYSTEM

HEATING
FILAMENT

CRUCIBLE

Fig. 1   A schematic diagram of the conventional partially ionized beam deposition system.

The ion/atom ratio of the vapor stream received at the substrate depends on the potential and the current used for the impact ionization as well as the size of the substrate and the source to substrate distance. For a two-inch substrate and a source to substrate distance of one foot, the maximum value of the ion/atom ratio one can achieve with this design is about 0.1%. The total ion beam current is a fraction of a mA. However, recent studies [11-14] of the PIB deposition indicated that very often a dramatic change of the film properties occurs with an ion/atom ratio ranging from 0.1 to 2.0%. It is therefore desirable to have a PIB source design such that a higher ion/atom ratio value can be achieved without a severe restriction on the size of the substrate.

One way is to install a conductance-limited ionization unit [3,15] so that a high vapor pressure can be confined in the ionization unit for more efficient ionization. Another way is simply to ionize the vapor stream closer to the crucible opening where the vapor density is higher [2,9]. Fig. 2 shows the later design [2]. In this scheme, the ionization unit is completely removed. A helical coil or a set of U-shape grids is placed on top of the crucible. Electrons from the filament are drawn not only to bombard the crucible to heat it, but also are drawn toward the region above the crucible by the coil or grids to ionize the vapor stream coming out of the crucible. That is, a common filament is used to provide electrons for both the crucible heating and the ionization of the vapor stream. The ion/atom ratio can be easily controlled by varying the vertical position of the crucible. Depending on the distance between the source and the substrate, an ion/atom ratio of 50% can be achieved. Typically, for a source-substrate distance of 30cm, one can obtain an ion/atom ratio up to 8%, or about 10mA of the total beam current. Very uniform ion beam current distribution can be obtained over four-inch diameter area with this source to substrate distance.

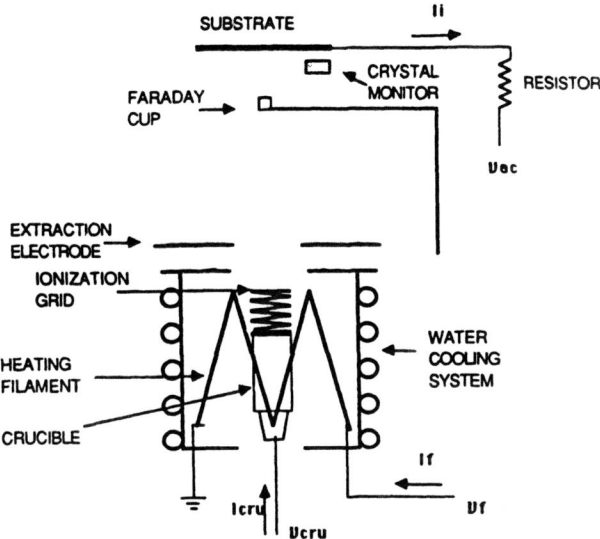

Fig. 2 A schematic diagram of the partially ionized beam deposition system with a high ion/atom ratio.

There are other ways to obtain a high ion/atom ratio. In the conventional e-beam evaporation technique in which an energetic electron beam is used to heat the charge directly without an excessive heating of the boat that contains the charge, one can easily get a few percent of self-ions [1]. Unfortunately, it is not possible to vary the ion/atom ratio. An extreme case of the high ion/atom ratio source is the mass-separated pure ion source for ion-beam deposition [16]. Here a hundred percent ions are used in the deposition. We shall not discuss this case here.

PIB Induced Surface Cleaning

One of the remarkable achievements in the PIB technology is the ability to deposit epitaxial films at lower substrate temperatures in a conventional vacuum condition without in-situ cleaning of the substrate prior to deposition [7,9,17,18]. To form an epitaxial layer one must have a clean surface free of contaminants. The observation of the epitaxial growth of thin films without in-situ cleaning of the substrate therefore implies that the PIB technique possesses a self-cleaning capability.

Before the substrate is loaded into the chamber for deposition, it is normally cleaned by some chemical means. After the chemical cleaning, except for certain substrates such as graphite, the surface immediately becomes oxidized. A thin layer of oxide, together with some contaminants such as hydrocarbons would inevitably exist at the surface. After the substrate is loaded into the chamber, the substrate must be cleaned in the vacuum prior to deposition for subsequent epitaxial growth of the overlayers. A routine surface cleaning technique is to use Ar or Xe ions of several kV to sputter the surface [19]. After the removal of the contaminants, the surface is normally badly damaged, and a thin layer of clean, but disordered material is formed at the surface. The surface is then annealed at an elevated temperature to recrystallize the damaged layer for subsequent epitaxial growth. Another way to clean the surface in vacuum is simply to heat the substrate to high temperatures so that the impurities would be desorbed from the surface. For example, for the Si substrate, an oxygen impurity can be desorbed at 850°C and a carbon impurity at 1200°C. After the surface is cleaned, it may be re-contaminated quickly if the vacuum is less than 10^{-9} Torr. Therefore, for an in-situ cleaning process to be effective, one must operate at an ultra-high vacuum condition.

In the PIB case the process of removing impurities may be quite different. The film deposition is occurring at the same time as the ion bombardment. With a substrate bias of 1kV, the penetration depth of metal ions into the Si substrate, for example, is less than $30\AA$ [20]. For a deposition rate of $15\AA/s$, the interface becomes buried in a few seconds. Assuming that the ion/atom ratio used in the experiment is less than 2%, the self-ions which are responsible for the removal of the contaminants have been estimated to be less than $5 \times 10^{+14}/cm^2$ [21]. This dose is at least one order of magnitude lower than that used in the conventional sputter-cleaning of surface impurities using inert gas ions [19].

The impurities, including hydrogen, oxygen, and carbon, at the interface between the PIB deposited metal film and the substrate have been measured using secondary ion mass spectrometry [21] and $^1H(^{15}N, \alpha\gamma)^{12}C$ nuclear resonance reaction [22] techniques. Fig. 3 is a plot of the hydrogen concentration at the metal/Si interface as a function of the metal ion flux arriving at the substrate in the first 10 monolayers of PIB deposition. The plot includes both Al/Si and Cu/Si interfaces. The hydrogen concentration at the Al/Si and Cu/Si interfaces was obtained using the $^1H(^{15}N, \alpha\gamma)^{12}C$ nuclear resonance reaction and secondary ion mass spectrometry techniques, respectively. The Al film was deposited at a bias potential of 1.5kV and the Cu film at 1kV. The hydrogen concentration at the Cu/Si interface for a Cu^+ flux of $\sim 2.5 \times 10^{+14}$ ions/cm² (equivalent to a 2% ion/atom ratio) shown in the figure is actually equal to the background hydrogen concentration in the Si substrate. That is, for this ion flux the additional hydrogen impurity located at the substrate surface due to contamination prior to deposition has been completely removed during the deposition.

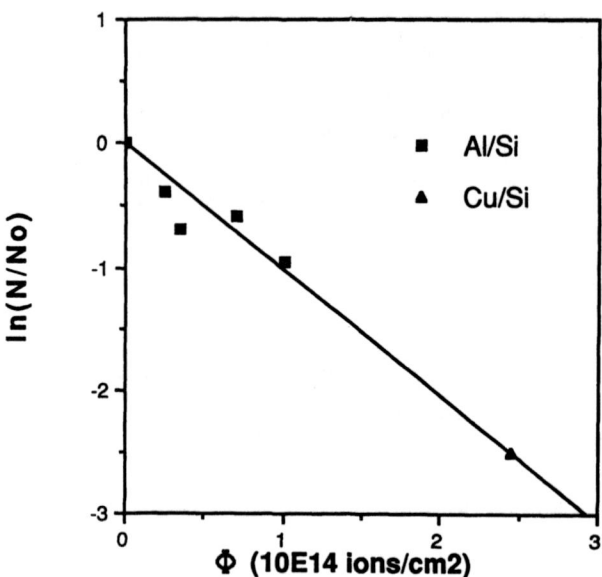

Fig. 3   The relative hydrogen concentration, $\ln(N/N_o)$, as a function of the metal ion flux Φ. N_o is the hydrogen concentration at the metal/Si interface when no ionization is applied to the beam, i.e., the zero ion flux case. The Al/Si and Cu/Si data were obtained by the nuclear resonance reaction[22] and secondary ion mass spectrometry[21] techniques, respectively.

The hydrogen desorption cross section, calculated from the slope of the plot shown in Fig. 3, is about $10^{-14}cm^2$. Similar desorption cross sections have been found for oxygen and carbon impurities. These are unusually large desorption cross sections. Conventional inert gas ion beam sputtering processes gives a desorption cross section of the order of $10^{-15}cm^2$ or less. In the sputter-cleaning of the substrate using inert gas, it is believed that direct collisions between the ions and the impurities located at the surface, together with the usual cascade process are significant processes responsible for the removal of the impurities [19]. In the PIB surface cleaning, these processes alone do not seem to be able to give such a high desorption cross section. It has been suggested [22] that the "thermal spike" mechanism [23,24] may also play an important role in the desorption of surface impurities. In this process, a very high surface transient temperature can be induced at the surface locally to desorb a number of impurities per impinging ion. Assuming that the transient "thermal spike" generated by each ion can heat (and therefore "clean") a surface area having a diameter of ~10$\mathring{A}$, the whole surface can be covered with the spikes during the initial stage of deposition with less than 1% of self-ions. However, there has not been a strong experimental demonstration of the existence of the thermal spike.

For effective surface cleaning the bias voltage has to be equal to or larger than 1kV. The optimum bias potential for cleaning the surface impurities should depend somewhat on the material deposited and the type of impurities to be removed.

If the ionization is turned on during the subsequent film growth, the surface cleaning continues in the growing film itself [21]. High purity thin films can therefore be grown at less stringent vacuum conditions.

PIB Induced Substrate Defects

It has been shown that the optimum bias potential that leads to metal/Si epitaxial growth is around 1kV. This voltage is sufficiently high to cause the desorption of surface impurities which is necessary for epitaxial growth, but not high enough to cause severe damage in the substrate. This is because the ion dose is small. The interface is buried by the growing film in two seconds for a deposition rate of 15$\mathring{A}$/sec and the ions cannot penetrate to the substrate during the subsequent growth. For example, high quality Al/Si contacts can be formed with a bias potential of 1kV and an ion/atom ratio ranging from 0.1 to 1.6%. Our recent cross section transmission electron microscopy study [25] shows no observable ion-induced dislocations or other extended defects near the substrate surface. Unusually uniform and stable Al/Si Schottky diodes with barrier heights as high as 0.77eV can be obtained [26,27].

For a bias potential larger than 1kV, surface damage occurs and the quality of the interface deteriorates. Good quality metal/Si contacts cannot be formed for a bias potential larger than 2kV [21,27,28]. No epitaxy has been observed for potentials larger than 2kV. (The situation may be different for PIB semiconductor deposition. For examples, single crystal Si/Si [7] and Ge/Si [5] can be formed at 5kV and 10kV, respectively.) At these bias potentials, the Schottky barrier height reduces and the leakage current increases. An anomalous frequency dependence in the C-V characteristics was observed for the Al/Si film deposited with a bias potential of 2.5kV [28]. More interestingly, at this bombardment voltage, an inverted n-type surface layer on an otherwise p-type substrate was observed in samples of ultrathin (10-50$\mathring{A}$) Al/Si films [29].

DLTS (deep level transient spectroscopy) has been used to study levels in the band gap induced by the energetic ion bombardment near the substrate surface during the growth of the Al/Si thin films [30]. Fig. 4 shows the DLTS spectra on Al/Si Schottky contacts deposited with the bias potentials of 0, 0.5, 1, and 3kV. The ion/atom ratio has been fixed at 0.1%. These films were deposited at room temperature. No deep levels were created for bombardment voltages less than 1kV. At 1kV, one deep level was observed. At 3kV, five deep levels were detected. Some of these levels have been identified. Level E_1, with the activation energy of 0.23eV, corresponds to a singly charged divacancy. Level E_3 has an energy of 0.38eV and is either due to a neutral state divacancy or the E-center. All these levels can be completely eliminated after annealing the sample at 400°C for an hour. The bias potential range for the creation of deep levels agrees qualitatively with the structural and electrical measurements regarding the ion-induced substrate damage.

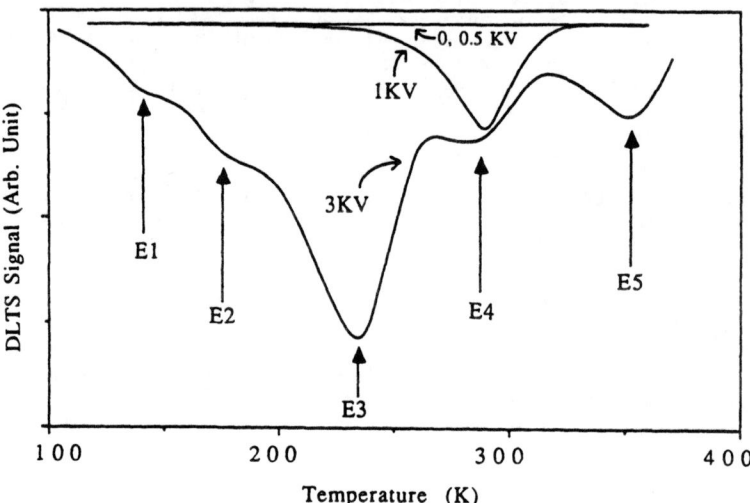

Fig. 4   DLTS spectra obtained from the Al/Si films deposited by the PIB technique with the bias potential of 0kV, 0.5kV, 1kV, and 3kV.

PIB Thin Film Modifications

Similar to many ion-assisted deposition techniques, the PIB technique can be used to modify and control many thin film properties. Examples are: promotion of adhesion, reduction of stress, increased density, and control of orientation and epitaxy. Many desirable electrical and optical properties are a result of these structural modifications. The extra kinetic energy provided by the energetic ions can improve the dynamics of film formation by increasing the mobility of the atoms arriving at the surface. For an ion/atom ratio of 1%, the average energy per atom arriving at the surface is 10eV. This is two orders of magnitude higher than the energy provided by thermal means.

Many low temperature metal and semiconductor epitaxies have been reported using the PIB deposition technique [5-9,17,18]. An example is the room temperature epitaxy of the very large lattice mismatch Al(111)/Si(111) system at a conventional vacuum condition without in-situ cleaning[17,18]. A high-resolution electron transmission microscopy study shows that the interface appears to be incommensurate [18], and therefore not strained. This observation opens the possibility of growing a dislocation free and strain free thin film which cannot be achieved in small lattice mismatch systems.

A dramatic change of the texture of metal films deposited on an amorphous surface can be obtained using the PIB technique [11,31]. Fig. 5 is a summary of X-ray 2θ diffraction measurements performed to study the effect of the ion/atom ratio on the Al thin film (5000Å thick) texture deposited by the PIB technique on an oxidized Si substrate. Both X-ray Al(111) and Al(200) peaks are plotted as a function of the ion content (ion/atom ratio) in the beam. The bias potential was kept at 2kV. The results demonstrate the existence of an optimum condition for growing a highly oriented thin film. For the 2kV bias case, the optimum condition occurs at an ion/atom ratio of 0.7%. This value increases as the bias potential is reduced. The existence of an optimum ion content for growing an oriented film implies at least a twofold role of the ion bombardment during the film formation. First, ion bombardment supplies the energy required for atoms to overcome an activation energy barrier for the formation of the low energy (111) plane. Second, too high an ion content or bombardment energy may create severe extended defects so that the texture growth is inhibited. The existence of two competing factors guarantees an optimum condition for the texture growth. Highly textured grains can enhance the electromigration resistance of the films.

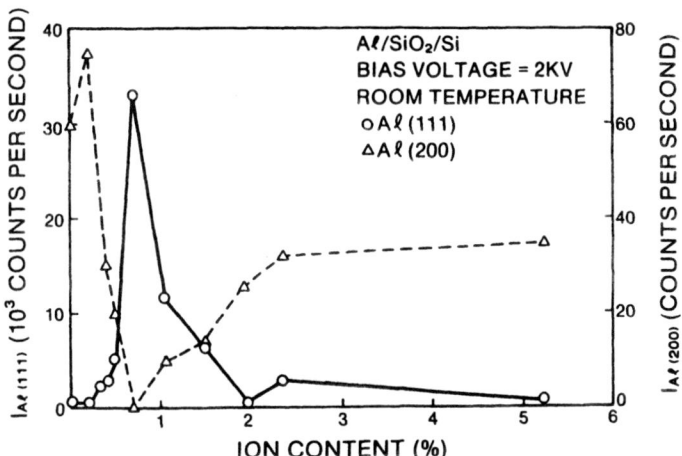

Fig. 5   The measured X-ray 2θ diffraction intensity of the Al films deposited on SiO_2 surface by the PIB technique is plotted as a function of the ion content (ion/atom ratio) in the beam. The bias potential was kept at 2kV. Note the difference in scale for the (111) and (200) peaks.

PIB can also be used to control the morphology of thin films during growth. Both "Al plugs" [13] and a planarized Al layer [32] can be deposited on deep oxide vias or trenches. Thick Al films ($\sim$4-5μm) can be produced without the usual columnar structure [14]. Control of thin film structure and morphology is an important area of research for multilevel interconnect and packaging applications.

Discussions

One main advantage of the PIB deposition technique over many other ion-assisted deposition techniques is that no foreign gas is used during the film growth and therefore no gas is incorporated in the film. Since the self-ions have the same mass as the surface atoms during the growth, energy transfer of the ions to the surface atoms via collisions is most effective. The surface mobility of the growing film is expected to be higher. The self-sputtering of the deposited material in the PIB deposition is not considered to be a serious problem due to the small amount of ions used in the deposition. The sputtering yield of the deposited material due to the ion bombardment can be high, but the amount that gets sputtered away represents only a small fraction of the total deposited material.

In cases where in-situ cleaning is not required, it may in principle be possible to operate the PIB deposition at a lower bias potential (lower than 1kV) to obtain similar thin film modifications described in this paper by using a higher ion/atom ratio. The extreme case would be to use a 100% self-ions [16]. In practice it is much harder to generate such a metal or semiconductor ion beam with very high current for a high rate of deposition. Even if one can generate such a high current beam, the density of the beam may still be limited by the space charge effect.

Acknowledgement: This work is supported in part by the Semiconductor Research Corporation and IBM Corporation. I thank members of the RPI Center for Integrated Electronics Ion Beam Processing group for their contributions in this work. I also thank all my collaborators for their interest and support in this work.

References

[1] For a review, see S. M. Rossnagel and J. J. Cuomo, in Materials Research Society Bulletin Vol.XII, No. 2 (1987), p. 40.

[2] S.-N. Mei and T.-M. Lu, J. Vac. Sci. Technol. A6, 9 (1988).

[3] J. E. Greene, Solid State Technology 30 (4), 115 (1987); M. A. Hassan, A. S. Barnett, J. E. Sundgren, and J. E. Greene, J. Vac. Sci. Technol. A5, 1883 (1987).

[4] I. Yamada, H. Takaoka, H. Usui, and T. Takagi, J. Vac. Sci. Technol. A4, 722 (1986); and references therein.

[5] G.-R. Yang, X.-M. Wang, and F.-G. Qin, Chinese Journal of Semiconductors 6, 203 (1985); Journal of Instrument Materials 15, 37(1984).

[6] T. Narusawa, S. Shimuzu, and S. Komiya, J. Vac. Sci. Technol. B2, 306 (1984).

[7] K. Mameno, H. Hanafusa, J. Nishino, M. Nakao, and T. Nakakado, Proc. 9th Symp. on ISIAT' 85, ed. T. Takagi, Tokyo (1985), p. 329.

[8] T. Ito, T. Nakamura, M. Muromachi, and T. Sugiyama, Jpn. J. Appl. Phys. 15, 1145 (1976); T. Itoh and T. Nakmura, ibid., 16, 553 (1977).

[9] Y. Namba and T. Mori, J. Vac. Sci. Technol. 13, 693 (1976).

[10] D. M. Mattox, in "Deposition Technologies for Films and Coatings", ed. R. F. Bunshah, Noyes Publications, Park Ridge, NJ (1982), p. 244.

[11] A. S. Yapsir, L. You, T.-M. Lu, and M. Madden, J. Mater. Res. 4, 343 (1989).

[12] C.-H. Choi, R. Ramanarayanan, S.-N. Mei, and T.-M. Lu, Mat. Res. Soc. Symp. Proc. Vol. 93, 267 (1987).

[13] S.-M. Mei, T.-M. Lu, and S. Roberts, IEEE Electron Device Lett. EDL-8, 503 (1987).

[14] R. Selvaraj, S.-N. Yang, J. F. McDonald, and T.-M. Lu, Proc. IEEE VLSI Multilevel Interconnection Conference (Electron Device Society, New York, 1987), p. 440.

[15] A. Rockett, S. A. Barnett, and J. E. Greene, J. Vac. Sci. Technol. B2, 306, (1984).

[16] B. R. Appleton, R. A. Zuhr, T. S. Noggle, N. Herbots, S. J. Pennycook, and G. D. Alton, Materials Research Society Bulletin Vol. XII, No. 2, 52 (1987); and references therein.

[17] C.-H. Choi, R. A. Harper, A. S. Yapsir, and T.-M. Lu, Appl. Phys. Lett. 51, 1992 (1988); A. S. Yapsir, C.-H. Choi, S.-N. Yang, T.-M. Lu, M. Madden, and B. Tracy, Mat. Res. Soc. Symp. Proc. Vol. 116, 465 (1988).

[18] T.-M. Lu, A. S. Yapsir, P. Bai, P.-H. Chang, and T. J. Shaffner, Phys. Rev. B, in press.

[19] H. F. Winters and E. Taglauer, Phys. Rev. B 35, 2174 (1987).

[20] P. Bai and T.-M. Lu, unpublished.

[21] G.-R. Yang, P. Bai, T.-M. Lu, and L. W. M. Lau, to be published.

[22] A. S. Yapsir, T.-M. Lu, and W. A. Lanford, Appl. Phys. Lett. 52, 1962 (1988).

[23] M. Szymonski and A. Poradzisz, Appl. Phys. A 28, 175 (1982).

[24] K. H. Muller, J. Vac. Sci. Technol. A 4, 184 (1986).

[25] K. Rajan and T.-M. Lu, unpublished.

[26] A. S. Yapsir, P. Bai, and T.-M. Lu, Appl. Phys. Lett. 53, 905 (1988).

[27] J. Wong, S.-N. Mei, and T.-M. Lu, Appl. Phys. Lett. 50, 679 (1987).

[28] J. Wong, C. Lam, and T.-M. Lu, Mat. Res. Soc. Symp. Vol. 101, 189 (1988).

[29] R. Srinivasan, S. P. Murarka, and T.-M. Lu, J. Appl. Phys. 65, 1198 (1989).

[30] W. I. Lee, J. Wong, J. M. Borrego, and T.-M. Lu, J. Appl. Phys. 64, 2206 (1988).

[31] D. B. Knorr and T.-M. Lu, Appl. Phys. Lett., in press.

[32] P. Bai, T.-M. Lu, and S. Roberts, Proc. IEEE VLSI Multilevel Interconnection Conference (Electron Devices Society, New York, 1988), p. 446.

FABRICATION OF HIGH QUALITY SILICIDE LAYERS BY ION IMPLANTATION

Karen J Reeson*, Ann De Veirman°, Russell Gwilliam*, Chris Jeynes*, Brian J Sealy*, J Landuyt°, Udo Bussmann*, J K N Lindner* and E H te Kaat*

*   Dept Electronic and Electrical Engineering, University of Surrey, Guildford, Surrey, GU2 5XH, United Kingdom

°   University of Antwerp (RUCA), Groenenborgerlaan 171, B2020, Antwerpen, Belgium

♦   Dept Physics, University of Dortmund, POB 500 500, 4600 Dortmund 50 FRG

ABSTRACT

Buried layers of $CoSi_2$ have been successfully fabricated in (100) single crystal silicon by implanting 350 keV Co^+ to doses in the range $2 - 7 \times 10^{17}$ cm^{-2} at a temperature of ~550°C. For doses $\geq 4 \times 10^{17}$ $^{59}Co^+$ cm^{-2}, a continuous buried layer of $CoSi_2$ grows epitaxially, during implantation. After annealing (1000°C 30 minutes) continuous layers of stoichiometric $CoSi_2$, which are coherent with the matrix are produced for doses $\geq 4 \times 10^{17}$ $^{59}Co^+$ cm^{-2}. For doses of $\leq 2 \times 10^{17}$ $^{59}Co^+$ cm^{-2}, discrete octahedral precipitates of monocrystalline $CoSi_2$ are observed. Isochronal annealing (for 5s) at temperatures in the range 800-1200°C, shows that at temperatures ≥ 900°C there is significant redistribution of the Co from B-type or interstitial sites → substitutional A-type lattice sites. As the anneal temperature is increased there is a corresponding improvement in the crystallinity and coherency of the Si and $CoSi_2$ lattices. This shows that at a given temperature much of the Co redistribution takes place within the first 5s of the anneal.

INTRODUCTION

Recently it has been demonstrated that Ion Beam Synthesis (IBS) is a viable technique for producing both surface and buried silicide layers [1-12]. This involves the implantation of high doses of energetic transition metal ions into (100) or (111) silicon substrates, which are subsequently annealed at 1000°C for 30 minutes. Silicide layers fabricated using IBS have possible applications as metal bases for bipolar transistors [13-14] and as surface/ buried interconnects or contacts [15]. IBS silicides offer several distinct advantages over conventionally grown surface silicide films. These include some of the lowest reported resistivities for thin film silicides [1-2,9]. Additionally single crystal aligned $CoSi_2$ films can readily be grown on (100) silicon using IBS, whereas, it is very difficult to grow this type of film using evaporation or MBE growth techniques [11,16]. In this paper we study the effects of dose, annealing temperature and time on the evolution of the $CoSi_2$ layers. This is in order to gain a better understanding of the physical processes which operate during their formation and so optimise the implantation and annealing parameters.

EXPERIMENTAL

3 inch device grade (100) 20-30 Ωcm (B doped) silicon wafers were implanted, over an area of 6.25 cm² with 350 keV $^{59}Co^+$ to doses of $2 - 7 \times 10^{17}$ $^{59}Co^+$ cm^{-2}. During implantation, the silicon substrate was maintained at a temperature of ~ 550°C using ion beam heating. Subsequent to implantation, the implanted area was cut into smaller specimens, one specimen from each wafer was retained in its 'as implanted' state and the others were annealed. The annealing was carried out either in a conventional nitrogen flow furnace at 1000°C for 5 or 30 minutes or in a dual graphite strip annealer [17] where isochronal 5s anneals at 800°C, 900°C, 1000°C, 1100°C and 1200°C were used. Specimens are identified by the dose of Co implanted. Thus specimens 1, 2 and 3 were implanted with doses of 2×10^{17} $^{59}Co^+$ cm^{-2}, 4×10^{17} $^{59}Co^+$ cm^{-2} and 7×10^{17} $^{59}Co^+$ cm^{-2} respectively. The specimens were analysed by Rutherford Backscattering (RBS), Cross Sectional Transmission Electron Microscopy (XTEM), High Resolution Electron Microscopy (HREM) and spreading resistance measurements.

Mat. Res. Soc. Symp. Proc. Vol. 147. ©1989 Materials Research Society

RESULTS AND DISCUSSION

a) As Implanted Structure

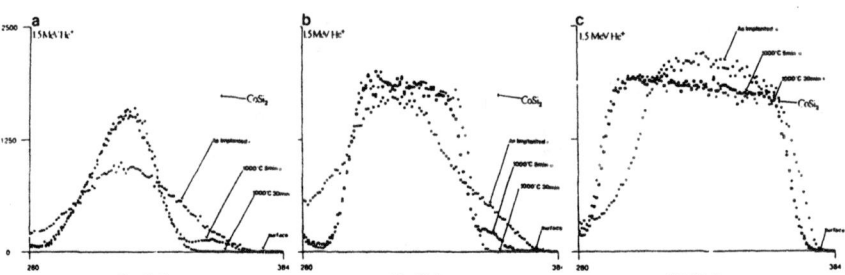

Fig 1: Non channelled RBS spectra for specimen 1 (fig 1a, 2×10^{17} $^{59}Co^+$ cm^{-2}, 350 keV), 2 (fig 1b, 4×10^{17} $^{59}Co^+$ cm^{-2}, 350 keV) and 3 (fig 1c, 7×10^{17} $^{59}Co^+$ cm^{-2}, 350 keV) as implanted ⊕, annealed 1000°C 5 min ⊖ and annealed 1000°C 30 min ⊙.

Figs 1a, b and c show the non-channelled RBS spectra, for all three specimens, in the region of the spectrum (channels 280 -375) corresponding to the cobalt depth distribution, before and after annealing at 1000°C for 5 and 30 minutes. The cobalt distributions, prior to annealing, are essentially gaussian with the tail of the distribution extending up to the silicon surface. At the peak of the distribution, as the dose increases, x in the formula CoSi$_x$ decreases from 4.7 (specimen 1) to 2.1 (specimen 2) to 1.8 (specimen 3).

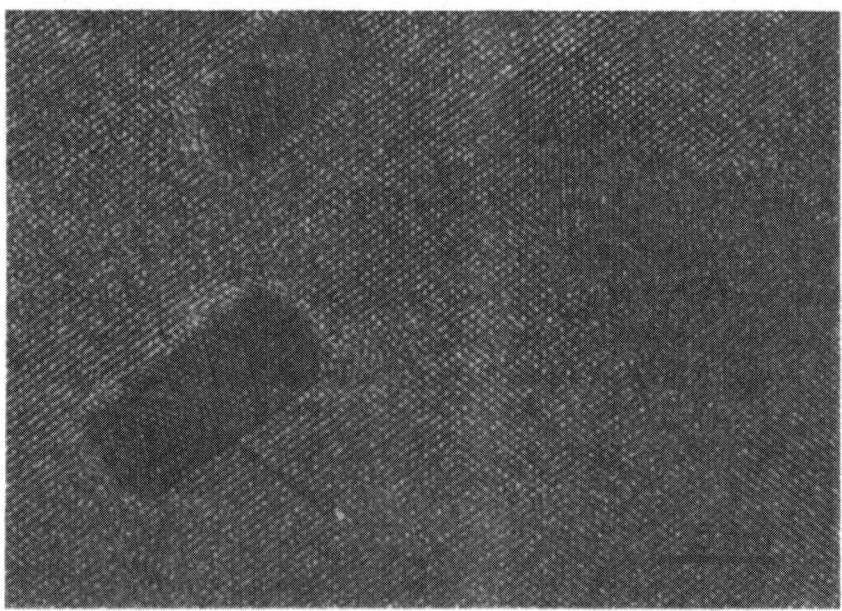

Fig 2: HREM micrograph of specimen 1, showing both A- and B- type CoSi$_2$ precipitates.

A HREM image of specimen 1 (fig 2) shows discrete precipitates of both A- and B- type $CoSi_2$ at the peak of the implanted distribution. For a dose of 4 x 10^{17} $^{59}Co^+$ cm^{-2} XTEM micrographs (fig 3a) [10] reveal that a continuous layer of $CoSi_2$, which is coherent with the matrix is formed during implantation. Above and below this buried layer precipitates of both A- and B- type $CoSi_2$ are observed.

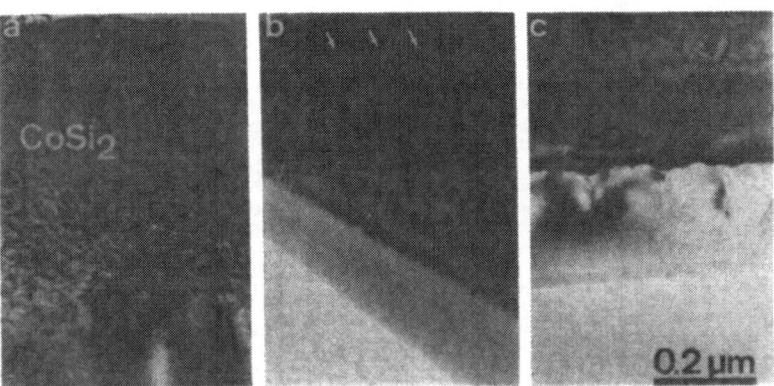

Fig 3: XTEM micrographs of specimen 2 (4 x 10^{17} $^{59}Co^+$ cm^{-2}, 350 keV) a) as implanted, b) annealed 1000°C 5 min and c) annealed 1000°C 30 min.

The channelled and non-channelled RBS spectra for specimen 3, prior to annealing, are shown in fig 4. The spectra can be divided into four distinct regions: Region I; Cobalt depth distribution from the surface (channel 375), Region II; Silicon overlayer, Region III; Loss in the silicon yield, due to the presence of the cobalt atoms, Region IV; Silicon substrate. Comparison of the channelled and non channelled spectra in regions I and III, shows a significant degree of dechannelling. Since the channelled spectrum was taken with the specimen aligned to facilitate channelling in the (100) single crystal silicon substrate, the reduction in the channelled yield within the implanted layer indicates that it is coherent with the silicon. Comparison of the channelled and non channelled spectra in region I shows that close to the silicon surface (z) (channels 342 - 356) 79% of the implanted Co is in substitutional sites. The percentage of cobalt in substitutional sites falls at the peak of the implanted distribution (y) to 37%. This is consistent with the formula of $CoSi_{1.8}$ derived from the RBS spectrum and implies that the excess Co atoms are accommodated in interstitial sites. Beyond the peak of the distribution (x) the amount of cobalt in substitutional sites increases to 50%. At the interfaces between the synthesised layer and the very thin silicon overlayer (region II) and silicon substrate (region IV) there is a slight rise in the cobalt dechannelled yield. This increase can be correlated to both the presence of B-type $CoSi_2$ precipitates (in addition to A-type) above and below the buried layer and to damage introduced by the ion beam, during implantation. Below the synthesised layer this damage is in the form of 311 defects which result from a supersaturation of silicon self interstitials [10-12, 18].

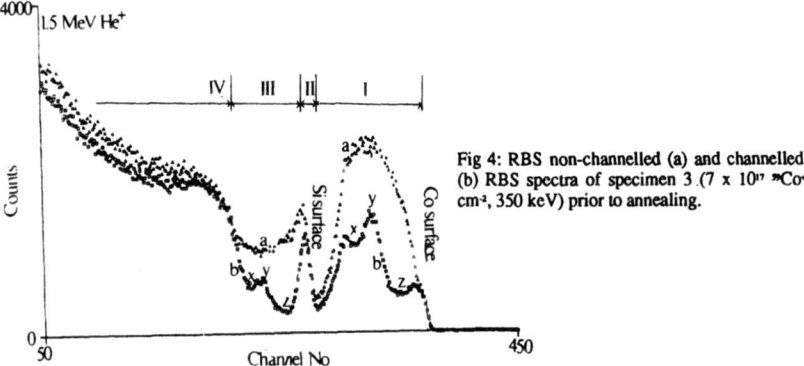

Fig 4: RBS non-channelled (a) and channelled (b) RBS spectra of specimen 3 (7 x 10^{17} $^{59}Co^+$ cm^{-2}, 350 keV) prior to annealing.

b) Furnace Anneals at 1000°C

Figs 1a, 1b and 1c also show the cobalt distributions, for specimens 1, 2 and 3 after annealing at 1000°C for 5 and 30 minutes. After 5 minutes annealing the cobalt distribution has become more rectangular and for the two highest doses saturates at a level commensurate with $CoSi_2$. A tail in the cobalt distribution is still seen extending up to the silicon surface in figs 1a and 1b (indicated by →). After 30 minutes annealing the profiles are very similar in shape to those after 5 min annealing, however the cobalt tail is no longer apparent. XTEM micrographs of specimen 2, before and after annealing at 1000°C for 5 and 30 minutes are shown in figs 3a, 3b and 3c respectively. The structure of the as implanted specimen (fig 3a) has already been discussed in the previous section. After annealing at 1000°C for 5 min, a continuous layer of $CoSi_2$, which is aligned with the matrix is observed. Precipitates of A-type $CoSi_2$ are also observed in the silicon overlayer (shown by → in fig 3b). These precipitates account for the tail in the cobalt distribution observed in the RBS spectrum. After annealing at 1000°C for 30 min (fig 3c) the structure is very similar to that after 5 minutes annealing, however, the silicon overlayer now appears to be devoid of $CoSi_2$ precipitates. This indicates that the $CoSi_2$ layer grows by gettering Co released from the dissolution of smaller precipitates, in the wings of the distribution. Consequently as the anneal time is increased the silicon overlayer and substrate become progressively denuded of Co as the buried layer grows. For specimens 1 and 3 a similar type of behaviour is observed. However, for specimen 3, the excess cobalt at the peak of implanted distribution also migrates to the wings where it reacts with silicon to form the disilicide phase. For this specimen most of the migration appears to be towards the lower $CoSi_2$/Si interface as most of the silicon overlayer has been consumed during implantation. For specimen 1, the volume concentration of cobalt, at the peak of the distribution, is insufficient, even after annealing for 30 min to synthesise a continuous buried layer and so discrete A-type octahedral precipitates of $CoSi_2$ are formed. After annealing for 5 and 30 min defects are observed in both the silicon overlayer and substrate. These are thought to result from stresses introduced during the annealing process. Such stresses may arise from the differences in the thermal expansion coefficients of $CoSi_2$ and Si (10.44 and 3 ppm/°C respectively [19]). The Si/$CoSi_2$ interfaces have been examined by HREM [10]. This shows that in some places the interfaces are planar and atomically abrupt, where as in others steps of up to 150Å occur [10].

c) Spreading Resistance Measurements

Depth resolved spreading resistance (R_s) measurements similar to those described by Lindner et al [6,9] were carried out on specimen 1. After implantation the R_s value at the profile minimum is found to be six orders of magnitude below that of the bulk silicon substrate. An R_s profile after annealing at 1000°C for 30 min is shown in fig 5. Due to the finite probe volume contributing to each point of measurement the thickness of the synthesised layer appears broadened with respect to XTEM and RBS results. R_s is now reduced by seven orders of magnitude. Taking into account the substrate resistivity of 20-30Ωcm this indicates a very low resistivity of the synthesised layer comparable to the best reported values for UHV thin films [eg. ref 20].

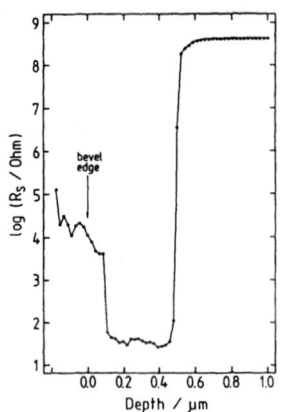

Fig 5: Spreading resistance profile of specimen 1 (2×10^{17} ^{59}Co$^+$ cm^{-2}, 350 keV) after 1000°C 30 min anneal.

d) Rapid Thermal Isochronal Annealing

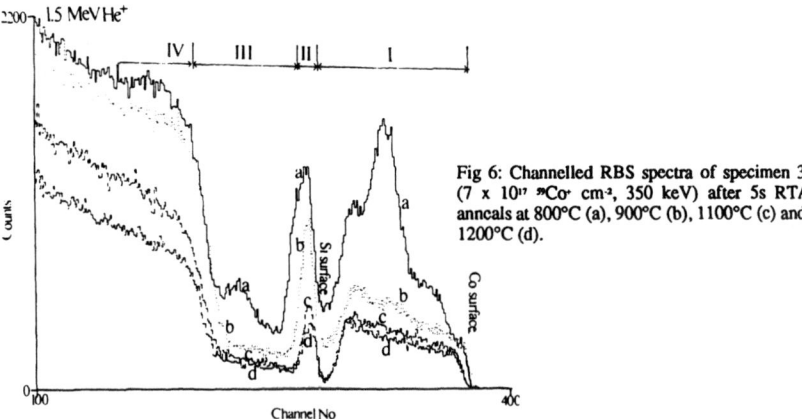

Fig 6: Channelled RBS spectra of specimen 3 (7 x 10¹⁷ ⁵⁹Co⁺ cm⁻², 350 keV) after 5s RTA anneals at 800°C (a), 900°C (b), 1100°C (c) and 1200°C (d).

Fig 6 shows channelled RBS spectra for specimen 3 after annealing, for 5s, at 800°C (curve a) 900°C (curve b) 1100°C (curve c) and 1200°C (curve d). For reasons of clarity the channelled spectrum after annealing for 5s at 1000°C is not shown. The regions labelled I-IV in fig 6 can be correlated with the areas discussed in fig 4. As the anneal temperature is increased, there is a reduction in the dechannelled yield in all four regions. This can be correlated with the migration of cobalt from interstitial → substitutional A-type sites and the annealing of the implantation induced damage. The reduced channelled yield in region III also indicates that the coherency between the synthesised layer and the matrix is increasing with increasing anneal temperature. In region I, the channelled spectrum, after annealing at 800°C for 5 s (curve a) is very similar in profile to that after implantation (fig 4). However, after annealing at 900°C for 5s (curve b) the channelled yield has been significantly reduced and the dechannelled fraction within the synthesised layer is approximately constant. The shapes of the channelled spectra b-d, in region I, are similar, however, the dechannelled yield decreases with increasing annealing temperature, as progressively more cobalt atoms move into substitutional A-type lattice sites. These results show, that even for very short annealing times (5s) at temperatures ≥900°C there is rapid migration of the excess interstitial Co from the peak to the wings of the distribution. For the highest anneal temperature (1200°C 5s) the resulting structure is comparable to structures annealed at 1000°C 30 min.

CONCLUSIONS

We have shown that by implanting cobalt into silicon at temperatures of ~550°C it is possible to fabricate layers of $CoSi_2$ which have a high degree of coherency with the matrix. For doses ≥ 4 x 10¹⁷ ⁵⁹Co⁺ cm⁻² at 350keV a continuous buried layer of $CoSi_2$ grows during implantation, above and below this layer precipitates of A- and B- type $CoSi_2$ are observed. For a dose of 2 x 10¹⁷ ⁵⁹Co⁺ cm⁻² the synthesised layer is discontinuous and consists of discrete precipitates of both A- and B- type $CoSi_2$. After annealing at 1000°C for 5 min the distributions become more rectangular and for the two highest doses saturate at a level commensurate with stoichiometric $CoSi_2$. Some cobalt in the form of discrete A-type $CoSi_2$ precipitates is also present in the silicon overlayer giving rise to a tail in the cobalt distribution which extends up to the silicon surface. After annealing at 1000°C for 30 min $CoSi_2$ precipitates are no longer observed in the silicon overlayer as all the Co has been gettered by the buried layer. For the lowest dose the volume concentration of Co is insufficient to form a continuous layer, even after annealing, and so the synthesised layer consists of discrete octahedral precipitates of $CoSi_2$. 5s anneals carried out on the highest dose specimen, at temperatures of 800-1200°C indicate that as the anneal temperature is increased there is a corresponding improvement in the crystallinity of the synthesised layer, the silicon overlayer and the substrate. This can be correlated with a migration of Co atoms from interstitial to A-type substitutional sites, and the annealing of implantation induced damage. The coherency between the synthesised layer and the matrix also improves accordingly.

ACKNOWLEDGEMENTS

The authors would like to thank the staff of the D R Chick laboratory, University of Surrey for their technical assistance during ion implantation and ion beam analysis. ADV is indebted to the Belgian fund for Scientific Research (IIKW) for her fellowship. The work is supported, in part, by the UK

Science and Engineering Research Council (SERC).

REFERENCES

1 A E White, K T Short, R C Dynes, J M Gibson and R Hull Materials Research Society Symposia Proceedings **107**, 3 (1988).

2 A E White, K T Short, J L Batstone,D C Jacobson, J M Poate and K W West Applied Physics Letters **50**, 95 (1987).

3 A E White, K T Short, R C Dynes,J P Garno and J M Gibson Materials Research Society Symposia Proceedings **74**, 481 (1987).

4 J C Barbour, S T Picraux and B L Doyle Materials Research Society Symposia Proceedings **107**, 269 (1988).

5 P Madakson, G J Clark, F K Legoues, F M d'Heurle and J E E Baglin Materials Research Society Symposia Proceedings **107**, 281 (1988).

6 J K N Lindner and E H te Kaat Materials Research Society Symposia Proceedings **107**, 275 (1988).

7 A E White Ion Beam Modification of Materials-88, Tokyo, Japan, June (1988), to be published Nuclear Instruments and Methods B.

8 A H van Ommen, J J M Ottenhiem, A M L Theunissen and A G Mouwen Appl Phys Lett **53**, 669 (1988).

9 J K N Lindner and E H te Kaat J Materials Research 3, 1238 (1989).

10 K J Reeson, A De Veirman, R Gwilliam, C Jeynes, B J Sealy and J Van Landuyt Oxford Conference on Microscopy of Semiconducting Materials, April (1989) to be published in Inst of Phys Conf Ser .

11 C W T Bulle-Lieuwma, A H Van Ommen and L J Van Ijzendoom Appl Phys Lett **54**, 244 (1989).

12 C W T Bulle-Lieuwma, A H Van Ommen Oxford Conference on Microscopy of Semiconducting Materials, April (1989) to be published in Inst of Phys Conf Ser.

13 J C Hensel, A F J Levi, R T Tung and J M Gibson Appl Phys Lett **41**, 151 (1985).

14 A F J Levi, R T Tung, J L Batstone and M Anzlowar Materials Research Society Symposia Proceedings **107**, 259 (1988).

15 S P Murarka in Silicides for VLSI Applications, Academic Press, New York, (1983).

16 C W T Bulle-Lieuwma, A H van Ommen and J Hornstra Materials Research Society Symposia Proceedings **102**, 377 (1988).

17 R Gwilliam, R Bensalem, B J Sealy and K G Stephens, Physica **129B**, 440 (1985).

18 A Bourret in Microscopy of Semiconducting Materials Inst Phys Conf Ser **87**, 39 (1987)

19 G V Samsanov and I M Vinitskii in Handbook of Refractory Materials IFI/Plenum, New York, 1980

20 R T Tung, J M Poate, J C Bean, J M Gibson and D C Jacobson Thin Solid Films **93**, 77 (1982).

LATERAL CONFINEMENT OF SILICIDE LAYERS SYNTHESIZED WITH HIGH DOSE IMPLANTATION AND ANNEALING

ALICE E. WHITE, K. T. SHORT, S. D. BERGER, H. A. HUGGINS, AND D. LORETTO
AT&T Bell Laboratories, Murray Hill, NJ 07974

ABSTRACT

Using mesotaxy, a technique which involves high dose implantation followed by high temperature annealing, we have created narrow wires of $CoSi_2$ buried beneath the surface of a silicon wafer. The implantation masks are fabricated directly on the silicon substrate using high resolution electron beam lithography in combination with reactive ion etching. TEM analysis shows that the wires are single-crystal and oriented with the substrate with very abrupt interfaces. The electrical continuity of the wires has been confirmed with electron-beam-induced current measurements.

INTRODUCTION

We have already demonstrated that, after implantation of Co into silicon and annealing at 1000°C, the Co in the implant distribution coalesces to form a buried single-crystal layer of $CoSi_2$ [1]. In order to explore whether this process operates laterally as well as in the direction of the ion beam, electron beam (e-beam) lithography and reactive ion etching were used to create an ion implantation mask to fabricate narrow wires. The advantage of this approach is that these wires naturally form with very abrupt interfaces and are free of the residual damage effects which ordinarily result from etching structures of this type.

MESOTAXY

Details of the mesotaxy process for $CoSi_2$, a cubic silicide with -1.2% mismatch with silicon, are given in ref. 1. Briefly, a high dose of Co ($\sim 3 \times 10^{17}/cm^2$) is implanted at $\sim$ 200 keV into a Si wafer which is held at 350°C in order to promote dynamic annealing. This results in a roughly gaussian profile of Co buried beneath the surface of the Si. In the as-implanted sample, 50% of the Co occupies substitutional sites and there is no amorphous material. When this sample is annealed at 600°C for 1 hr. and 1000°C for 1/2 hr., the Co coalesces to a well-defined layer of $CoSi_2$ and the crystallinity of the Si overlayer improves dramatically. TEM examination shows that these layers are single-crystals aligned with the substrate and they have atomically abrupt interfaces in both (100) and (111) Si.

The structural integrity of the $CoSi_2$ layers is reinforced by their electrical characteristics [2]. The residual resistivities of the mesotaxy $CoSi_2$ layers in (100) Si are $\sim 1 \mu\Omega$-cm, lower by a factor of two than the best UHV-deposited $CoSi_2$ [3]. This implies that the electron mean free path (λ) in these layers is greater than 1000Å, so we have the hope of making a metal structure with dimensions that are smaller than λ in order to study quantum size effects [4].

Mat. Res. Soc. Symp. Proc. Vol. 147. ©1989 Materials Research Society

NANOSTRUCTURE FABRICATION

The mesotaxy layers can be made thinner by reducing the implant dose [2], so we thought it would be interesting to try confining them laterally as well. This procedure is outlined in Fig. 1. A Si wafer with 5000Å of thermally-grown SiO_2 on it is coated with ~2000Å of PMMA. Then a wire pattern is exposed in the resist using a JEOL JBX-5D e-beam lithography system with an ~150Å probe size. After development, a 200Å layer of chromium is evaporated onto the wafer surface at an oblique angle so as to coat only the resist surfaces. The chromium then serves as an etch mask for transferring the pattern into the SiO_2 using reactive ion etching with CHF_3 gas. The PMMA and chromium are then removed in acetone and the sample is mounted in the implant chamber, heated to 350°C, and implanted with 170 keV Co to a dose of $2\times10^{17}/cm^2$. The oxide thickness was chosen to insure that all the Co ions hitting the masked region would be stopped in the mask. Following the 600°C anneal, a CVD oxide cap is deposited to protect the Si surface during the 1000°C anneal. Both the oxide cap and the masking oxide are removed by etching in buffered HF after the two-step anneal is complete.

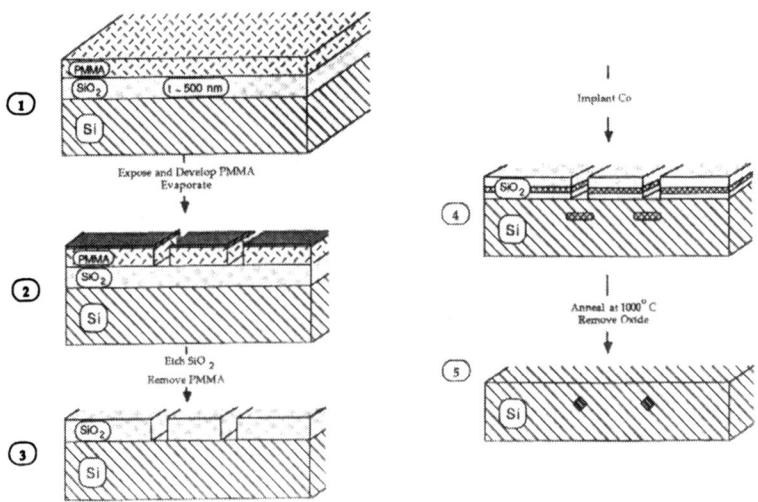

Fig. 1:   Fabrication process for $CoSi_2$ nanostructures.

STRUCTURAL AND ELECTRICAL CHARACTERIZATION

Arrays of wires lying along the 110 direction with linewiths from 1μm to 1500Å were fabricated for study with TEM. Fig. 2 shows the electron diffraction pattern from a single wire. The presence of sharp 200 diffraction spots confirms that the wire consists of single-crystal $CoSi_2$ aligned with the Si. When a single wire is imaged in plan view using the 200 reflection which originates solely from the silicide (Fig. 3), it looks remarkably similar to the cross-sectional view of a mesotaxy layer indicating that the mesotaxy process operates laterally as well as vertically. This wire has a linewidth of $\sim$1500Å with the same abrupt interfaces that were observed for the layers. Reductions of greater than 20% in the actual wire linewidth from the size of the opening in the mask were obtained because of the lateral coalescence of the Co to form the silicide. The plan view image also shows dense bands of dislocations running parallel to the wire on either side, although they are difficult to see because of the choice of imaging condition.

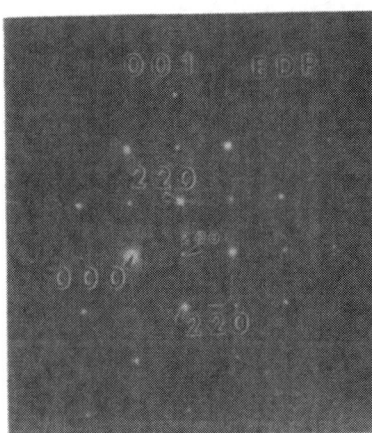

Fig. 2: Electron diffraction pattern from a single wire.

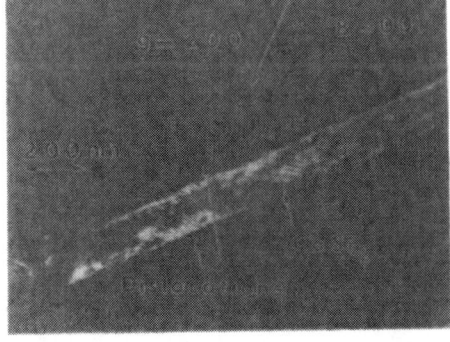

Fig. 3: Plan view dark field image of a $CoSi_2$ wire.

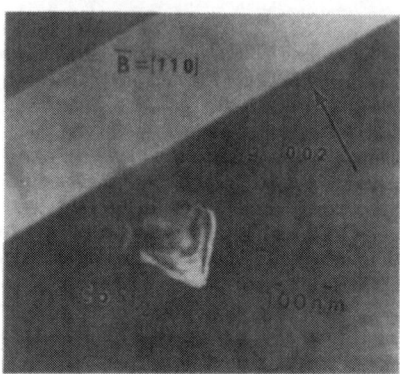

Fig. 4: Dark field image of the cross-section of a CoSi$_2$ wire.

Examination of one of the wires in cross-section reveals that the CoSi$_2$/Si interfaces prefer to form on (111) planes (Fig. 4). The dislocation bands can be seen more clearly in this view and appear to run from the CoSi$_2$ wire to the Si surface also along (111) planes. These dislocations form during the anneal, presumably to relieve the stess in the wire. Although the top surface of this particular wire appears to be a bit rough (probably because of irregular etching of the oxide mask), the other interfaces are of very high quality and have been formed *totally without etching*.

In addition to the wire arrays, a bridge structure was fabricated in order to investigate the electrical characteristics of these wires. This consisted of several wires of varying widths, all 10μm long, stretched between two contacts. The measurements were complicated by the fact that the CoSi$_2$ was completely surrounded by low resistivity Si. However, we were able to check the continuity of the wires using electron-beam-induced current (EBIC). This was accomplished by contacting the buried metal structure and scanning the e-beam across the sample. Electron-hole pairs are generated by the beam, and these carriers are collected by the Schottky contact providing contrast in the EBIC signal. This is illustrated in Fig. 5 which shows both the backscattered electron image and the EBIC image of one of the wires. Since the wire appears bright in the EBIC image, it is electrically continuous. All 23 wires that were measured at room temperature were conducting over the entire 10μm length.

We were also able to measure the residual resistivity of a 1μm wire by taking advantage of the superconducting properties of CoSi$_2$. Measuring the resistance change between the two contacts as the silicide went through the superconducting transition (at $\sim$1.3K), made it possible to separate the contribution of the silicide from the contribution of the silicon. The result for this wire was $\rho = 0.5\mu\Omega$-cm at 1.3K.

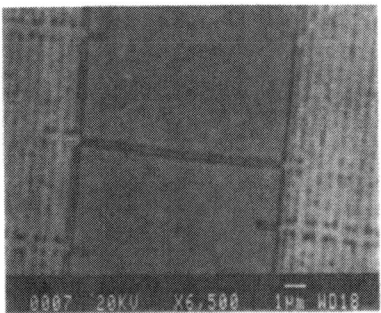

Backscattered Electron Image

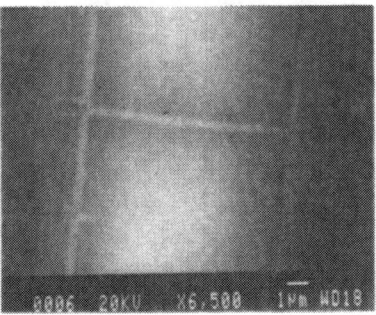

Fig. 5:   Backscattered electron (top) and
            EBIC (bottom) images of a 1μm wire.

SUMMARY

We have fabricated buried single-crystal CoSi$_2$ wires in Si using high dose implantation through oxide masks. These wires have sharp interfaces and cross-sectional areas as small as 2×10^{-10}cm^2. They are electrically continuous over micrometers of length and preliminary transport measurements show that they have resistivities that are comparable to the mesotaxy CoSi$_2$ layers. Since the implantation masks are produced by e-beam lithography, there is considerable flexibility concerning the size, shape, and crystallographic orientation of the final nanostructures. In this way, we hope to study both the growth mechanisms and the transport properties of these structures as a function of these variables.

ACKNOWLEDGEMENTS

We would like to thank J. M. Valles and R.C. Dynes for helpful discussions and the use of the low temperature facility.

REFERENCES

[1] Alice E. White, K. T. Short, R. C. Dynes, J. P. Garno, and J. M. Gibson, Appl. Phys. Lett. *50*, 95 (1987).

[2] Alice E. White, K. T. Short, R. C. Dynes, J. M. Gibson, and R. Hull, Mater. Res. Soc. Symp. Proc. *100*, 3 (1988).

[3] J. C. Hensel, R. T. Tung, J. M. Poate, and F. C. Unterwald, Appl. Phys. Lett. *44*, 913 (1984).

[4] Z. Tesanovic, M. V. Jaric, and S. Maekawa, Phys. Rev. Lett. *57*, 2760 (1986).

FORMATION OF BURIED IRIDIUM SILICIDE LAYER IN SILICON BY HIGH DOSE IRIDIUM ION IMPLANTATION

K. M. Yu, B. Katz, I. C. Wu, and I. G. Brown
Center for Advanced Materials, Materials and Chemical Sciences Division
Lawrence Berkeley Laboratory, Berkeley CA 94720

ABSTRACT

We have investigated the formation of $IrSi_3$ layers buried in <111> silicon. The layers are formed by iridium ion implantation using a metal vapor vacuum arc (MEVVA) high current metal ion source at room temperature with average beam energy ≈130 keV. Doses of the Ir ions ranging from $2x10^{16}$ to $1.5x10^{17}/cm^2$ were implanted into <111> Si. The formation of $IrSi_3$ phase is realized after annealing at temperatures as low as 500°C. A continuous $IrSi_3$ layer of ≈200 Å thick buried under ≈400 Å Si was achieved with samples implanted with doses not less than $3.5x10^{16}/cm^2$. Implantated doses above $8x10^{16}/cm^2$ resulted in the formation of an $IrSi_3$ layer on the surface due to excessive sputtering of Si by the Ir ions. The effects of implant dose on phase formation, interface morphology and implanted atom redistribution are discussed. Radiation damage and regrowth of Si due to the implantation process was also studied.

INTRODUCTION

Recently, several studies have been published describing silicide formation by high dose implantation of transition metal ions. Silicides which have been formed by this process include $CoSi_2$ [1-6], $TiSi_2$ [1,7,8], $CrSi_2$ [3,4,9], $FeSi_2$ [9], and $ViSi_2$ [10]. White et.al. [2-4] have fabricated epitaxial buried layers of $CoSi_2$ and $CrSi_2$ by implanting metal ions into a heated Si substrate (~350-450°C) with subsequent annealing at ~1000°C. In fact, buried silicide formation has only been observed when implantation was carried out at elevated substrate temperature (300°C) [2,7,9]. This is thought to be the result of diffusion of the metal ions away from the surface so that the net implantation profile becomes deeper than that expected by theoretical calculations [2].

We have chosen to study the formation of $IrSi_3$ by Ir ion implantation because $IrSi_3$ has several interesting structural and electrical properties which make it a unique silicide system. In the last ten years, only a few reports were published on the solid-state reactions between thin film Ir on Si [11-14]. Among the metal silicides, $IrSi_3$ has the highest silicon concentration (75%) and therefore requires the least amount of metal atoms for its formation. When $IrSi_3$ is formed by implantation with Ir ions, this means less sputtering of the Si and shorter implantation time due to the relatively small dose of Ir ions needed. Solid-state reactions between Ir thin films and Si substrates indicate that the formation of $IrSi_3$ phase is initiated at an annealing temperature around 1000°C [11], higher than the formation temperature of any other metal silicides. Because of its high formation temperature, $IrSi_3$ formation does not follow the diffusion controlled kinetics as in most cases since at such high temperature diffusion of the Ir and Si are very fast in the layer. Instead, $IrSi_3$ formation was governed by nucleation controlled kinetics [11] where the growth of the silicdie did not follow a layer-by-layer process.

According to Ishiwara et.al. [15], lattice mismatch between $IrSi_3$ (hexagonal with a=4.354 Å and c=6.628 Å) and <111> Si is 1.87%, so that epitaxial growth of $IrSi_3$ on <111> Si is very likely. Recently Chu et.al. [14] investigated the epitaxial growth of $IrSi_3$ on <111> and <100> Si and observed large grains of ~30μm-40μm of epitaxial $IrSi_3$ can be formed on <111> Si. Formation of epitaxial buried $IrSi_3$ layer is therefore very possible with high dose ion implantation. In addition to the interesting structural properties, Schottky barriers formed between $IrSi_3$ and n-Si exhibit barrier heights=0.94eV, the highest in any silicide/silicon contact structure. In this work we investigate the formation of buried layer of $IrSi_3$ in <111> Si by direct Ir ion implantation into Si at room temperature.

EXPERIMENTAL

The implantation was performed by MEVVA (metal vapor vacuum arc) high current ion source [16]. Briefly, in this ion source the highly ionized metal plasma that is created in the metal vapor vacuum arc discharge to provide the "feedstock" from which the ion beam is extracted. For this work beam extraction voltage was 50kV. The mean charge state of the iridium ions produced was measured and was ≈2.6. Since no energy analysis was carried out the mean implantation energy for this work was ≈130kV. The wafer was cooled by thermal conduction of the support structure. The beam pulse repetition rate was limited to 5 pulse/second to avoid overheating of the target. No intentional heating of the wafer was applied and hence the implantation was considered to be carried out at room temperature. The ion current density was ~20-30μA/cm². Implantation doses varied from $2x10^{16}$ to $1.5x10^{17}$/cm².

Annealing was carried out in the temperature range of 400-900°C for time duration of 20 min. to 5 hr. in flowing N_2 with the samples covered by a blank Si wafer. Another set of samples has undergone rapid thermal annealing (RTA) in Ar ambient at 820°C for 20 sec.

Rutherford backscattering spectrometry (RBS) with 1.8 MeV $^4He^+$ ion beams was used to study the ion dose, depth profiles, atomic composition and interface sharpness of the structures. The backscattering angle θ was 165° and the samples were tilted with the beam to a surface normal angle =60-65° to improve depth resolution to ≤50Å. X ray diffraction (XRD) in the Seeman-Bohlin geometry was carried out for high sensitivity phase identification. Phase formed in a ≤100Å thick layer can be identified by this technique. Microstructures and interface morphology were studied by transmission electron microscopy (TEM) in both plane-view and cross-sectional modes.

RESULTS AND DISCUSSION

Figure 1 shows a series of RBS spectra for Ir implanted samples with different doses. The total Ir doses in these samples as measured by RBS are 3.5, 7.0 and $8.0x10^{16}$/cm² while the intended doses were ≈3.5, 7.0 and $10x10^{16}$/cm². Therefore, the maximum retention doses for 130 keV Ir implanted into Si is ≈$8x10^{16}$/cm². Note that the thickness of surface Si layer decreases and the signal moves to the surface as the implanted dose increases due to sputtering of Si by the Ir ions. For low dose implants (<$3x10^{16}$/cm²), the Ir profile in Si shows a Gaussian shape with range straggling ≈190 Å while the calculated value according to the LSS theory for 130 keV implant is ≈100 Å. The broadened Gaussion distribution comes from the multiple charge states of the Ir ions in this work. For high dose samples (>$7x10^{16}$/cm²), the implant profile becomes flat-top with a low energy tail.

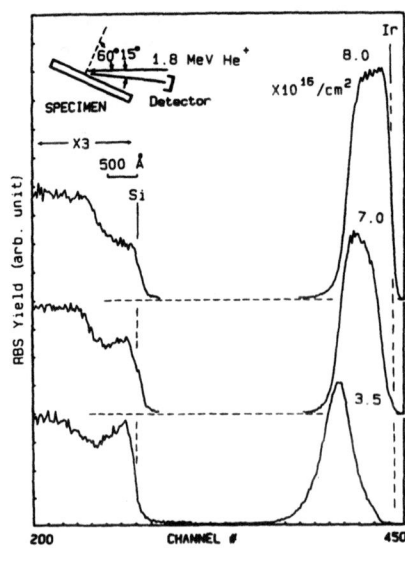

Fig. 1. RBS spectra of as-implanted samples for various doses. The doses indicated are the retained doses measured by RBS. Note that the Ir signal moves toward the surface as the dose increases.

The surface Si layer thickness as measured by RBS is plotted against the total dose in Figure 2(a). When extrapolating the curve to high dose, we observe that $\sim 10^{17}/cm^2$ of implant dose is needed in order to sputter away all the Si on the surface. This is consistent with the RBS results in Figure 1 where we observe that all the surface Si is sputtered away in the sample with a retained dose of $8 \times 10^{16}/cm^2$. The projected range can also be estimated to be ≈ 530 Å by extrapolating this curve to zero dose. This is in good agreement with the calculated value of 550 Å for 130 keV Ir ions implanted into Si from the LSS theory. In Figure 2(b), the peak concentration of the Ir atoms are plotted against the total measured dose. A linear dependence up to 30 atomic percent of Ir in Si is observable in this figure. For the stoichiometric $IrSi_3$, Ir content in the silicide is 25 at.%. From 2(b), we can estimate that a total dose of $\sim 6.3 \times 10^{16}/cm^2$ is needed so that the peak Ir concentration reaches 25 at.% which we call the stoichiometric dose level $(Nt)_{st}$. XRD analysis of samples with retained Ir doses varying from 1.5-$8 \times 10^{16}/cm^2$, reveals that the $IrSi_3$ phase appears for all doses following annealing at temperatures above 500°C while the as-implanted samples exhibit an amorphous phase.

The RBS spectra for samples with low dose, $3.5 \times 10^{16}/cm^2$ are shown in Figure 3(a) both for as-implanted and after annealing at 700°C for 1.5 hr. The spectrum for the as-implanted sample indicates that the original Ir profile is Gaussian with Ir peak concentration ≈ 15 at.% percent buried underneath ≈ 400 Å of Si. The spectrum for the sample after annealing shows that the Ir atoms diffuse toward the peak of the profile from both sides forming a close to rectangular distribution. The buried layer formed after annealing has a nominal composition of $IrSi_{3.4}$ with thickness of ≈ 200 Å. However, it is also observed that in the annealed sample ≈ 2 at.%

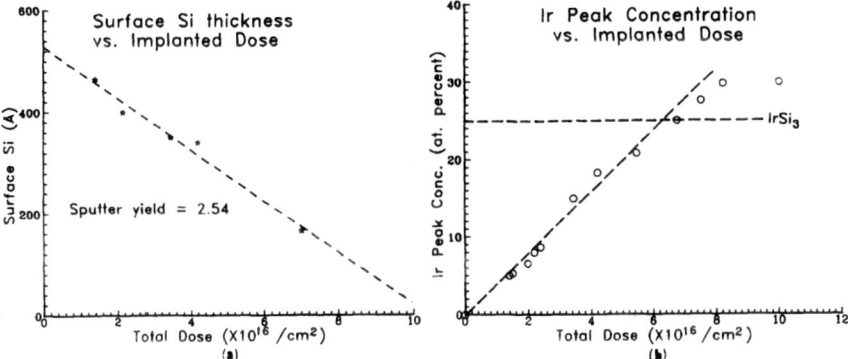

Fig. 2. (a) A plot of surface Si layer thickness measured by RBS as a function of total implanted Ir: (b) A plot of Ir peak concentration in Si as a function of implanted Ir dose.

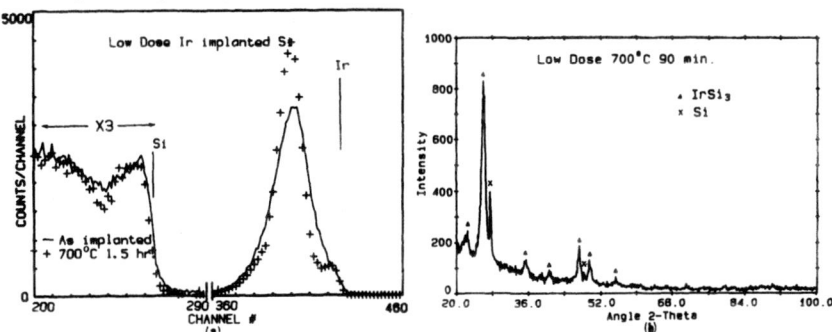

Fig. 3. (a) RBS spectra of low dose sample, dose=$3.5 \times 10^{16}/cm^2$ as-implanted and annealed at 700°C for 1.5 hr. (b) XRD spectrum for the annealed sample in (a) indicating the presence of $IrSi_3$ and polycrystalline Si phases.

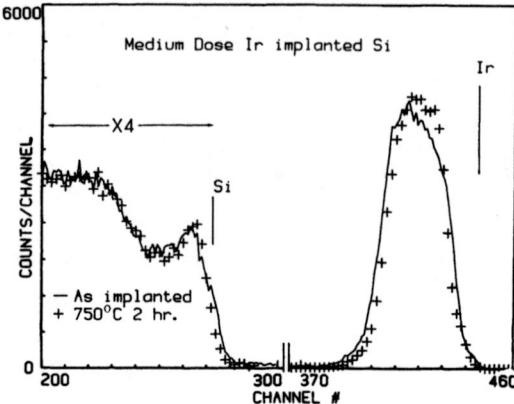

Fig. 4. RBS spectra of samples with medium dose, $7 \times 10^{16}/cm^2$, as-implanted and annealed at 750°C for 2 hr.

Fig. 5. Cross-sectional TEM micrograph of the sample with the same dose as those in Fig. 4 but after RTA. Note that the interfaces between the buried silicide layer and the surface and substrate Si are very sharp.

of Ir is distributed uniformly in the surface Si layer. The XRD analysis of the same sample shown in Figure 3(b) reveals patterns of IrSi$_3$ and polysilicon phases. The polysilicon diffraction pattern in Figure 4, however, has peaks shifted to lower Bragg angle as compared to the Si standard in the XRD data file indicating an expansion in the Si lattice, $\Delta a/a \approx 0.04$. This is probably due to the incorporation of Ir atoms in the surface layer as observed in the RBS results. Electron diffraction study on the surface Si of the annealed sample shows clearly ring pattern of polycrystalline Si. From the RBS, XRD and TEM results, we conclude that a continuous buried IrSi$_3$ layer is formed at a retained Ir dose of $\geq 3.5 \times 10^{16}/cm^2$ (with as-implanted Ir peak concentration ≈ 15 at.%).

RBS spectra for samples with retained dose $\approx 7 \times 10^{16}/cm^2$, close to $(Nt)_{st}$, as-implanted and furnace annealed at 750°C for 2 hr. are shown in Figure 5. The spectrum for the annealed sample shows sharpening of the Ir profile forming a layer ≈ 360 Å thick with nominal composition of IrSi$_3$ buried under ≈ 250 Å of Si. Similar structure is observed for sample annealed with RTA at 820°C for 20 sec. XRD on the furnace annealed sample shows diffraction peaks of the IrSi$_3$ and polysilicon phases indicating that the surface Si layer is polycrystalline. Cross-sectional TEM micrograph of the sample after RTA is shown in Figure 5. The TEM micrograph shows a continuous IrSi$_3$ buried layer with atomically sharp Si/IrSi$_3$/Si interfaces. Dark field plan-view TEM image on the same sample shows polycrystalline rod-shaped IrSi$_3$ grains of ~5000-1000 Å in length formed in Si.

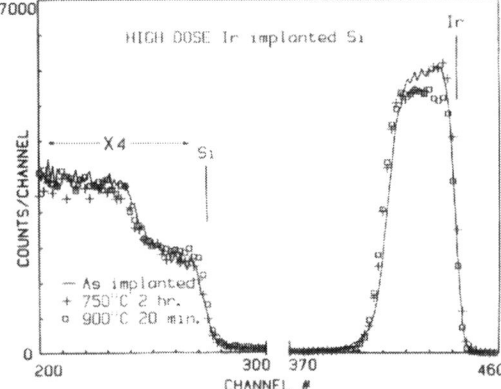

Fig. 6. RBS spectra of samples with high dose, $8 \times 10^{16}/cm^2$, as-implanted and annealed at 750°C for 2 hr. and 900°C for 20 min. Note that the redistribution of the Ir signals as the samples are annealed.

Ir implanted Si to a dose well above $(Nt)_{st}$ results in peak Ir concentration above 25 at.%. Figure 6 shows RBS spectra for samples with retained dose $=8 \times 10^{16}/cm^2$ as-implanted and annealed at 700°C for 2 hr. and 900°C for 20 min. The as-implanted sample has a Ir-Si layer of ≈ 400 Å thick on the surface with [Ir]:[Si] ratio $\approx 1:2.34$. The RBS spectrum for the sample annealed at 700°C for 2 hr. shows layered structure of (170 Å) $IrSi_{2.34}/(250$ Å) $IrSi_3$ on Si. A uniform layer of $IrSi_3$ on Si is achieved after annealing at 900°C for 20 min. Samples annealed at lower temperatures (500-700°C) only show a sharpening in the Ir-Si layer/Si interface but the [Ir]:[Si] ratio in the layer remains $\approx 1:2.34$. However, XRD shows only diffraction pattern of $IrSi_3$ phase for samples annealed at temperatures above 500°C.

From the RBS and XRD results, we conclude that for samples with doses higher than $(Nt)_{st}$, annealing at temperatures in the range of 500-700°C results in the formation of $IrSi_3$ phase with nominal composition of $IrSi_{3-x}$. The excess Ir atoms in the layer diffuse into the Si substrate when the samples are annealed at higher temperatures (>700°C) forming a thicker layer of stoichimetric $IrSi_3$. RTA at 820°C for 20 sec. on this sample also results in a sharp layer of stoichiometric $IrSi_3$ on Si.

Ion channeling measurements on the high dose samples shows no reduction in the RBS yield in the layer for channeling orientation indicating small grain polycystalline $IrSi_3$ in the layer. A damage layer of ≈ 1000 Å Si below the $IrSi_3$ layer is observed for the sample which had been annealed at 500°C for 1 hr. This sample after RTA shows epitaxial regrowth of this Si layer leaving <200 Å of damaged Si. Complete regrowth of the damaged layer is achieved after annealing at 800°C for 40 min.

SUMMARY AND CONCLUSIONS

We have successfully fabricated a buried $IrSi_3$ layer in Si by implantation of Ir ions into Si at room temperature at an average energy of 130 keV. The critical minimum dose for the formation of a buried continuous layer of $IrSi_3$ in Si is $\approx 3.5 \times 10^{16}/cm^2$. This critical dose is much lower than those observed for other buried silicides studied mainly because of the fact that $IrSi_3$ has the lowest metal concentration in any metal-silicide. $IrSi_3$ phase is formed after annealing at temperature ≥ 500°C for doses as low as $1.5 \times 10^{16}/cm^2$.

For samples with doses below $(Nt)_{st}$, upon annealing the Ir atoms diffuse to the center of the implant from the tails of the profile. For samples with doses higher than $(Nt)_{st}$, annealing at temperature below 700°C results in $IrSi_3$ phase with higher Ir content ($IrSi_{3-x}$). Redistribution of the Ir atoms into the substrate occurs during annealing at temperatures above 700°C forming a thicker layer of stoichiometric $IrSi_3$. Complete regrowth of the damaged Si below the $IrSi_3$ layer due to implantation is observed after annealing at 800°C for 40 min.

However, no epitaxial layer of $IrSi_3$ is formed in this study. It is likely that a buried epitaxial $IrSi_3$ in Si can be formed by Ir implantation in a heated Si substrate where in-situ annealing is performed.

234

ACKNOWLEDGEMENT

This work was supported by the Director, Office of Energy Research, Office of Basic Energy Sciences, Materials Science Division, of the U.S. Department of Energy under Contract No. DE-AC03-76SF00098.

REFERENCES

1. F. H. Sanchez, F. Namavar, J. I. Budnick, A. Fasihudin and H. C. Hayden, Mater. Res. Soc. Symp, Proc. **51**, 439 (1986).
2. Alice White, K. T. Short, R.C. Dynes, J. P. Garno, and J. M. Gibson, Appl. Phys. Lett. **50**, 95 (1987).
3. Alice White, K. T. Short, R. C. Dynes, J. M. Gibson, and R. Hull, Mater. Res. Soc. Symp. Proc. **100**, 3 (1988).
4. Alice White, K. T. Short, R. C. Dynes, R. Hull and S. M. Vandenberg, Nucl. Instrum. Meth. **339**, 253 (1989).
5. A. H. Van Ommen, J. J. M. Ottenheim, A. M. L. Theunissen, and A. G. Mouwen, Appl. Phys. Lett. **53**, 669 (1988).
6. K. Kohlhof, S. Mantl, B. Stritzker and W. Jager Nucl. Instrum. Meth. **B39**, 276 (1989).
7. P. Madakson, G. J. Clark., F. K. Leguoues, F. M. d'Heurle and J. E. E. Baglin, Mater. Res. Soc. Fall Meeting 1987, Symp. H, Boston, MA, Nov. 30-Dec. 5. 1987.
8. V. P. Salvi, S. V. Vidwan, A. A. Rangwala, B. M. Arora, Kuldeep and A. K. Jain, Nucl. Instrum,. Meth. **B28**, 242 (1987).
9. F. Namavar, F. H. Sanchez, J. I. Budnick, A. H. Fasihudin and H. C. Hayden, Mater. Res. Soc. Spring Meeting 1987, Symp. C, Anaheim, CA, April 21-23, 1987.
10. Subhaga Vidwans, V. P. Salvi, A. A. Rangwala, B. M. Arora, A. K. Jain and Kuldeep., Nucl. Instrum. Meth. **B39**, 280 (1989).
11. S. Petersson, J. Baglin, W. Hammer, F. D'Heurle, T. S. Kuan, I. Ohdomari, J. de Sousa Pires, and P. Tove, J. Appl. Phys. **50**, 3357 (1979).
12. I. Ohdomari , T. S. Kuan, and K. N. Tu, J. Appl. Phys. **50**, 7020 (1979).
13. M. Wittmer, P. Oelhafen, and K. N. Tu, Phys. Rev. **B35**, 9073 (1987).
14. J. J. Chu, L. J. Chen, and K. N. Tu, J. Appl. Phys. **63**, 1163 (1988).
15. Hiroshi Ishiwarsa, Shuichi Saitoh and Kohki Hikosaka, Jpn. J. Appl. Phys. **20**, 843 (1981).
16. I. G. Brown, J. E. Galvin, B. F. Gavin and R. A. MacGill, Rev. Sci. Instrum. **57**, 1069 (1986).

BACK CHANNEL DEGRADATION AND DEVICE MATERIAL IMPROVEMENT BY Ge IMPLANTATION

F. NAMAVAR, B. BUCHANAN, E. CORTESI, AND P. SIOSHANSI
Spire Corporation, Patriots Park, Bedford, MA 01730

ABSTRACT

Because of potential "back channel" leakage problems in silicon-on-insulator (SOI) metal-oxide-semiconductor (MOS) devices, especially n-channel MOS devices which must operate in an ionizing radiation environment, it is desirable to produce Separation by IMplantation of OXygen (SIMOX) wafers which have a layer of poor quality silicon near the Si/buried SiO_2 interface. At the same time, these wafers must have low defect, high quality silicon near the wafer surface for device fabrication.

We have demonstrated that with Ge ion implantation and solid phase epitaxy regrowth, the surface region of the silicon top layer of the SIMOX wafer is improved and the region adjacent to the buried SiO_2 is degraded. These results have been observed by RBS/channeling, XTEM, and plane view TEM.

INTRODUCTION

The fabrication of silicon-on-insulator (SOI) material with the Separation by IMplantation of OXygen (SIMOX) process has gained increasing importance in the fabrication of radiation hard devices. However, one of the obstacles to more widespread use of SIMOX material is the high density of dislocations in the silicon top layer.

Standard SIMOX wafers typically have a silicon surface layer about 1500 angstroms thick and a buried layer about 4000 angstroms thick. The interfaces are generally smooth and sharp, but there are some silicon islands in the buried oxide layer. The density of defects in this material has typically been observed at a level of about 10^8 to 10^{10} dislocations/cm^2 [1-3].

Recently, substantial progress has been made in reducing the density of threading dislocations [4-8]. Although low defect SIMOX material is a prerequisite for fabricating high performance devices, a degraded region at the Si/buried SiO_2 interface may be advantageous for radiation hard device applications.

We have studied the effect of Ge implantation into the Si top layer of SIMOX material followed by solid phase epitaxy regrowth to reduce the density of threading dislocations. At the same time, we produced many end-of-range defects [9] near the Si/buried SiO_2 interface.

There are a variety of ways [10] to reduce "back channeling" effects in metal-oxide-semiconductor (MOS) devices fabricated in SIMOX, but preliminary indications are that the Ge implantation is more practical and achieves both a higher quality surface Si region and a degraded Si region near the Si/buried SiO_2 interface with one low dose, room temperature implantation.

BACK CHANNEL LEAKAGE

An MOS device fabricated in SOI material may have an unwanted parasitic MOS device at the silicon/buried insulator interface after being exposed to ionizing radiation [10,11]. The passage of ionizing radiation through an insulator such as SiO_2 generates free electrons and holes. It is well known [12,13] that the electrons are more mobile and are swept out of the oxide and the holes tend to be trapped near the silicon interface. The trapped positive charge induces a negative charge in the silicon at the insulator interface, which can cause a change in the electrical characteristics of component MOS devices and can create inversion layers in p-type silicon (Figure 1) [11]. If the inversion layer connects n-type regions of different potentials, a parasitic NMOS can be turned on by ionizing radiation, creating a permanent leakage path (Figure 1).

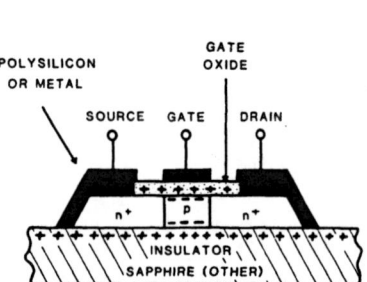
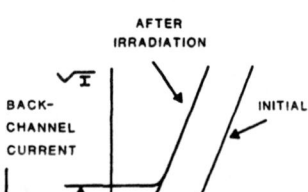

Figure 1.   Cross section of NMOS transistor on insulating substrate illustrating trapped positive charge induced parasitic back-channel and effect on transfer characteristics [11].

Thus, it is desirable to produce a silicon top layer which consists of a high quality layer at the surface for device fabrication, adjacent to a damaged layer. The high quality silicon surface layer is used for fabrication of devices and the damaged silicon layer serves to reduce the back channel current because of reduced mobility in this layer.

IMPROVING THE CRYSTALLINE QUALITY OF SOI MATERIAL BY IMPLANTATION

It has been demonstrated [14] that the crystalline quality of the silicon layer grown on sapphire substrates (SOS) by chemical vapor deposition (CVD) can be improved by implantation of silicon ions and subsequent thermal annealing. This improvement can be achieved because defects in SOS material are planar (twins and stacking faults) [15]. Only amorphization and solid phase epitaxy regrowth is required to improve the crystalline quality of the silicon. On the other hand, defects in SIMOX material are threading dislocations, which are linear and align along the (110) direction [1]. Partial amorphization of the silicon top layer is necessary but not sufficient to reduce the density of defects in SIMOX since the threading dislocations will regrow with the Si during solid phase epitaxy.

Threading dislocations are terminated by interfaces and in SIMOX material these interfaces are normally at the buried layer and the wafer surface. It has been observed that epitaxially grown layers produced by CVD may also exhibit the interface effect if the new layer is sufficiently strained relative to the substrate [16]. We anticipated that implantation of Ge would not only produce an amorphized silicon layer (as with Si implantation but with lower dose) but would also create a strained layer because of the larger lattice constant of silicon alloyed with germanium as opposed to that of pure silicon. Thus, the strained layer would create an artificial interface and stop the propagation of threading dislocations during solid phase epitaxy regrowth.

EXPERIMENTAL RESULTS

A four-inch p-type silicon wafer was implanted with a dose of 1.6×10^{18} O^+/cm^2 at 160 keV and then annealed for six hours at 1300°C in N_2. Pieces of this wafer were implanted with Ge with nominal doses of 1×10^{14}, 5×10^{14}, and 1×10^{15} Ge^+/cm^2 at 100 and 150 keV. During the Ge implant, the sample was tilted about seven degrees with respect to the (100) direction to avoid channeling. The Ge-implanted samples were annealed for 0.5 hours at 850°C in N_2.

An as-Ge-implanted Si control sample (implanted with a nominal dose of 5 $\times 10^{14}$ Ge^+/cm^2 at 150 keV) was studied by RBS/channeling measurements. The aim of these measurements was to determine the depth of damage resulting from Ge implantation and also to measure Ge concentration. As shown in Figure 2a, the damaged silicon layer is about 1400 angstroms thick. The spectrum shown in this figure is only partially channeled and was obtained without the use of a goniometer. Figure 2b shows the concentration profile of Ge as a function of depth. The concentration of Ge at the peak is about 0.08 atomic percent. It is clear from the RBS data that the depth of damage is almost equal to the maximum Ge depth penetration.

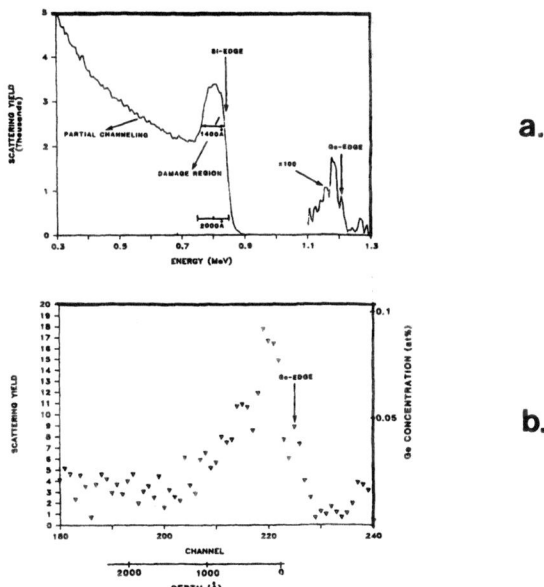

Figure 2.   (a) 1.5 MeV He^+ RBS spectrum of a Si sample implanted with a nominal dose of 5×10^{14} Ge^+/cm^2 at 150 keV. Partial channeling was obtained by tilting the sample. (b) Ge concentration depth profile.

Figure 3 is an XTEM of the SIMOX sample implanted with a dose of 5×10^{14} Ge^+/cm^2 at 150 keV and annealed for 0.5 hours at 850°C in N_2. There are two regions of the silicon top layer. The surface region is about 1200–1400 angstroms thick and is high quality silicon. Only one defect that reaches the surface is visible in this micrograph. The lower region is about 400–500 angstroms thick and contains a large number of defects. Comparing RBS and XTEM results, one can see that the width of the silicon region damaged by Ge implantation (for the as-implanted sample) corresponds to the width of the high quality surface region of the annealed sample. It appears that the lower (degraded) region is out of the range of the implanted Ge.

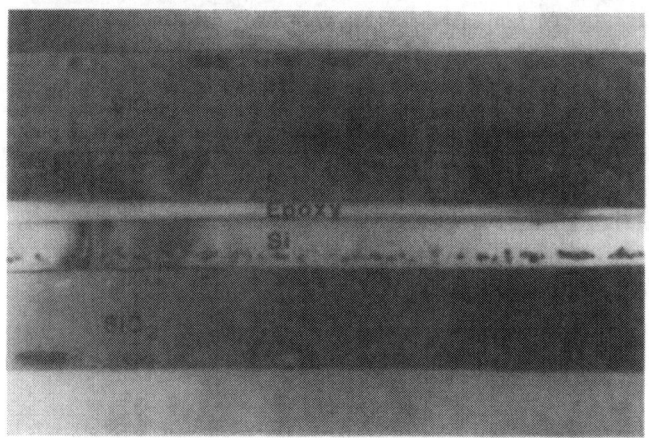

0.5um

Figure 3. XTEM of a Ge-implanted SIMOX sample (after solid phase epitaxy regrowth) shows high quality Si in the surface region of the wafer and damaged Si near the Si/SiO_2 interface. Only one defect reaches the surface of the sample in this micrograph. (See also Figures 2 and 4).

Figure 4a shows a plane view TEM of the silicon top layer. In this case, all defects were observed, namely those present in the surface region and in the lower degraded region. The sample was then thinned from the lower part of the sample and observed by plane view TEM. Figure 4b shows the plane view TEM of only the surface region of the silicon top layer. As can be seen, no defects were observed in the region implanted with Ge. In the region that was out of range of the implanted Ge (lower degraded region), two types of defects were observed: threading dislocations and dislocation loops. In Figure 4a we observed the presence of several threading dislocations. The length of these defects is reduced as compared to unimplanted SIMOX material. At this time, we cannot conclude that the apparent shortening of the threading dislocations is due only to their stopping at the interface with the Ge-implanted region; it is possible that some dislocations change direction at the interface and angle towards the surface, thus "doubling over" themselves from the plane view and appearing shortened although actually they extend to the surface.

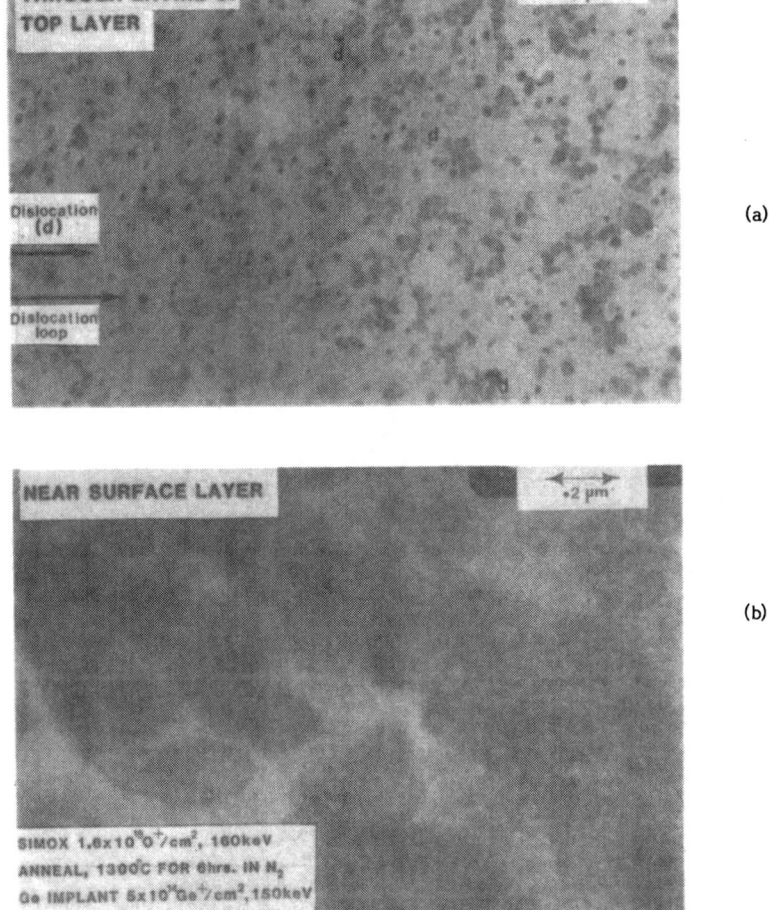

Figure 4. Plane view TEM of Ge-implanted SIMOX. (a) Through entire Si
top layer, where dislocation loops and threading dislocations
were observed; (b) Through only the near-surface region, where
no defects were observed.

CONCLUSIONS

Simultaneous improvement of the top silicon layer and degradation of the silicon at the Si/buried SiO_2 interface has been demonstrated for SIMOX wafers implanted with Ge. Improving the crystalline quality of the surface region is essential for high performance device fabrication. On the other hand, degrading the Si/buried SiO_2 interface region may be advantageous for reducing the back channel leakage current in devices exposed to ionizing radiation. The mechanisms involved are not well understood. Further studies of the basic mechanisms are required and device fabrication and testing in the resulting material is needed for verification.

ACKNOWLEDGEMENTS

This work was supported in part by DOD, RADC at Hanscom AFB, under Contract No. F19628-86-C-0069. We are very grateful to Dr. W. Shedd at RADC for his direction and support of this work. The authors would also like to acknowledge the assistance of Dr. S.N. Bunker, Dr. E.A. Johnson, M.M. Sanfacon, G.S. Beals, L.M. Geoffroy, and J. Breen. We would like to extend our gratitude to Dr. V. Jalan (Electrochem, Inc) and to Dr. P. Ling for their help with TEM work, and to Professor Q. Kessel (U. of Connecticut) for his help with RBS work.

REFERENCES

1. B.Y. Tsaur, Mat. Res. Soc. Symp. Proc. 35, 641 (1985).
2. G.K. Celler, P.L.F. Hemment, K.W. West, and J.M. Gibson, Appl. Phys. Lett. 48, 532 (1986).
3. B.Y. Mao, P.H. Chang, H.W. Lam, B.W. Shen, and J.A. Keenan, Appl. Phys. Lett. 48, 794 (1986).
4. D. Hill, P. Fraundorf, and G. Fraundorf, J. Appl. Phys. 63, 4932 (1988).
5. T.F. Cheek, Jr., and D. Chen, Mat. Res. Soc. Symp. Proc. 107, 53 (1988).
6. J. Margail, J. Stoemenos, C. Jaussaud, M. Bruel, Appl. Phys. Lett. 54, 526 (1989).
7. F. Namavar, E. Cortesi, and P. Sioshansi, in Selected Topics in Electronic Materials, edited by B.R. Appleton et al., (Mat. Res. Soc. Extended Abstracts, Pittsburgh, PA, 1988), 109.
8. F. Namavar, E. Cortesi, and P. Sioshansi, Symposium A, Fall 1988 Mat. Res. Soc. Meeting, accepted for publication.
9. K.S. Jones, S. Prussin, and E.R. Weber, Appl. Phys. A 45, 1 (1988).
10. B.L. Buchanan, D.A. Neamen, and W.M. Shedd, IEEE Trans. Electron Devices ED-25, 959 (1978).
11. B.L. Buchanan, in VLSI Handbook, edited by N.G. Einspruch (Academic Press, Inc., Orlando, FL, 1985), 571.
12. J.R. Srour, O.L. Curtis, Jr., and K.Y. Chiu, IEEE Trans. Nucl. Sci. NS-21, 73 (1974).
13. R.C. Hughes, E.P. EerNisse, and H.J. Stein, IEEE Trans. Nucl. Sci. NS-22, 2227 (1975).
14. S.S. Lau, S. Matteson, J.W. Mayer, P. Revez, J. Gyulai, J. Roth, T.W. Sigmon, and T. Cass, Appl. Phys. Lett. 34, 76 (1979).
15. W.E. Ham, M.S. Abrahams, C.J. Buiocchi, and J. Blanc, J. Electrochem. Soc. 124, 634 (1977).
16. H. Beneking, P. Narozny, N. Emeis, and K.H. Goetz, J. Electronic Materials 15, 247 (1986).

ON THE FORMATION OF THICK AND MULTIPLE LAYER SIMOX STRUCTURES
AND THEIR APPLICATIONS

F. NAMAVAR,* E. CORTESI,* R.A. SOREF** AND P. SIOSHANSI*
*Spire Corporation, Patriots Park, Bedford, MA 01730
**Rome Air Development Center, Hanscom AFB, MA 01731

ABSTRACT

This paper will address the formation of SIMOX structures with thick and multiple buried SiO_2 layers by multiple oxygen implantation and growth of epitaxial Si by chemical vapor deposition (CVD). Our results indicate that SIMOX material can be produced with a buried layer of any thickness or with any number of distinct buried oxide layers and distinct silicon layers. Thick and double buried SiO_2 layer material may be useful for high voltage isolation and electric field shielding.

In addition, we have demonstrated optical waveguide action in SIMOX wafers. This suggests that in a double buried SiO_2 layer system, three dimensional stacked integration of silicon waveguides is possible, including two level optical interconnects.

INTRODUCTION

The Separation by IMplantation of OXygen (SIMOX) process has been widely used in the formation of silicon-on-insulator (SOI) material because of the high quality and large area of the silicon top layer in addition to the speed and radiation hardness advantages of SOI in general. Standard SIMOX material is formed by a single implantation of oxygen into Si with a dose of 1.6 - 1.8 x 10^{18} O^+/cm^2 at 160 - 200 keV, followed by high temperature annealing. This results in a Si top layer about 1200-2500 angstroms thick and a buried SiO_2 layer about 3500-4500 angstroms thick.

SIMOX material with a very thick buried oxide layer would be advantageous because it could provide higher voltage isolation than standard SIMOX. SIMOX with double buried oxide layers [1] may be useful for electric field shielding [2] as well as possibly for three-dimensional, vertically integrated, electro-optical guided-wave devices [3,4].

We have previously studied the formation of SIMOX with thick buried oxide layers by implanting at 180 keV with multiple steps (without growth of epitaxial Si) until the quality of the Si top layer was significantly degraded. Our results [5] indicate that the maximum thickness of the buried oxide layer that can be produced by oxygen implantation using the conventionally and conveniently available beam energy ($\leq$ 200 keV) using high current ion implanters, while still maintaining a high quality silicon top layer, is about 6000-6500 angstroms (with a dose of 2.2 x 10^{18} O^+/cm^2 at 180 keV). The quality of the silicon top layer (about 1000 angstroms thick) in the sample implanted with a total dose of 2.2 x 10^{18} O^+/cm^2 is comparable to standard SIMOX. On the other hand, a sample which was implanted with a total dose of 2.5 x 10^{18} O^+/cm^2 has a heavily damaged silicon top layer (about 700 angstroms thick) with many twins and defects.

FORMATION OF A THICK BURIED OXIDE LAYER BY MULTIPLE IMPLANTATION AND GROWTH OF EPITAXIAL SILICON

The aim of this work was to use a multiple high dose oxygen implantation process with high temperature annealing and growth of epitaxial silicon in order to produce SIMOX material with a thicker buried layer than can be produced without growth of epitaxial silicon. Thicker buried oxide layers can be produced by growing epitaxial silicon on a standard SIMOX wafer to a thickness that allows the subsequent oxgyen implantation to merge with the front of the existing buried SiO_2 layer.

For this study, a 4-inch-diameter p-type silicon (100) wafer was implanted with a dose of 1.7×10^{18} O^+/cm^2 at 160 keV at 500-550°C and then annealed for six hours at 1300°C in N_2 (standard SIMOX process). The silicon top layer was about 1500 angstroms thick and the buried SiO_2 layer was about 3700 angstroms thick. To ensure that the thickness of the silicon top layer was comparable to the range of the implanted oxygen, about 2000 angstroms of epitaxial silicon were grown on the wafer so that the total thickness of the silicon top layer was about 3500 angstroms.

The sample was then reimplanted with a dose of 1.5×10^{18} O^+/cm^2 and annealed for six hours at 1300°C in N_2. Figure 1 is an XTEM of this sample (total dose of 3.2×10^{18} O^+/cm^2). The XTEM shows that we formed a buried oxide with a thickness of about 7500 angstroms with a silicon top layer about 1400 angstroms thick. The quality of the silicon top layer is comparable to standard SIMOX. No silicon islands were observed in the SiO_2 near the silicon surface layer. However, there are silicon islands in the buried layer near the interface of the SiO_2 and the Si substrate.

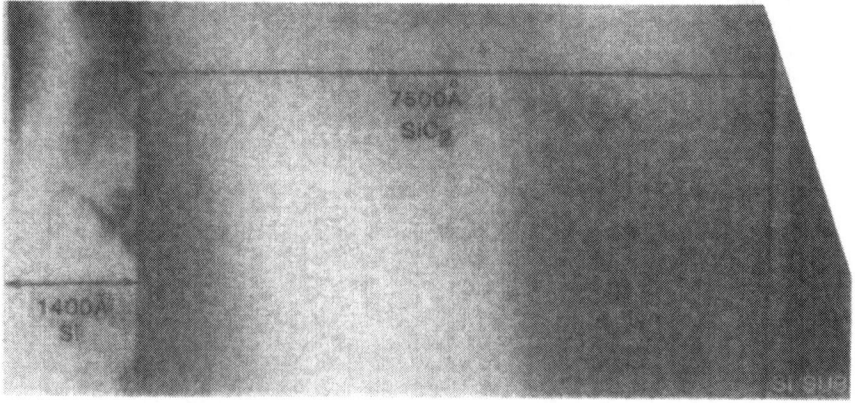

100nm

Figure 1.   XTEM shows fabrication of SIMOX with a thick buried layer.

FORMATION OF TWO BURIED OXIDE LAYERS BY MULTIPLE IMPLANTATION AND GROWTH OF EPITAXIAL SILICON

The process we used for formation of two buried layers is similar to that used to produce SIMOX with very thick buried layers. In this case, however, the thickness of the epitaxial silicon is increased so that the second implantation is not in contact with the already formed buried SiO_2 layer.

For this study, a 3-inch-diameter p-type silicon (100) wafer was implanted with a dose of 1.75×10^{18} O^+/cm^2 and then annealed for six hours at 1300°C in N_2 (standard SIMOX process). Five thousand five hundred angstroms of epitaxial silicon were grown on the wafer. The wafer was then implanted with a dose of 1.7×10^{18} O^+/cm^2 and annealed for six hours at 1300°C in N_2.

Figure 2 is an XTEM of a SIMOX wafer with two buried oxide layers (5500 angstroms epitaxial Si grown). As can be seen, the first buried layer (produced first) is about 4000 angstroms thick, the buried silicon layer is about 3000 angstroms thick, the second buried SiO_2 layer is about 4000 angstroms thick, and the silicon top layer is about 1500 angstroms thick. The buried silicon and the silicon top layer remained single crystal and are of a quality comparable to standard SIMOX. The buried oxide layers have sharp and smooth interfaces.

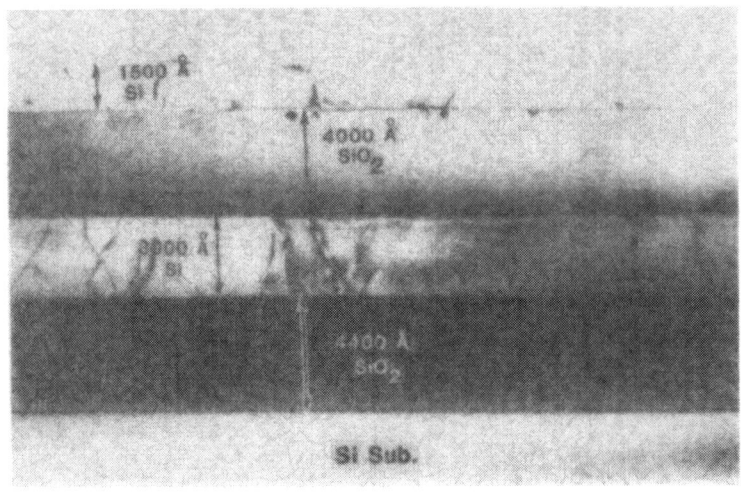

0.5um

Figure 2.   XTEM illustrates fabrication of double buried layer SIMOX.

OPTICAL APPLICATIONS OF SIMOX MATERIAL

Crystalline silicon, surrounded by lower-index claddings, is a low-loss waveguide for infrared light in the 1.3 to 10 micron wavelength range. At 1.3 microns, the refractive index of SiO_2 is 1.45, compared to the 3.50 index of Si. Thus, Si-on-SiO_2 (SOI) provides tight confinement of light in Si; greater confinement than in III-V heterostructure waveguides. Also, curved SOI channel waveguides can be made with smaller radii of curvature than those of III-V guides.

Figure 3 demonstrates optical waveguiding at a wavelength of 1.3 microns in a SIMOX sample with a silicon top layer about 2 microns thick on a buried SiO_2 layer about 4000 angstroms thick which was formed by the standard process. The single crystal Si top layer was thickened by growth of epitaxial Si by CVD. The photo shows near-field pattern of cleaved output end of planar waveguide. The waveguide length is 1.1 cm. Light was focused into the cleaved input end with a 40x lens. Figure 4 shows the experimental apparatus. SIMOX material with much lower density of defects and better Si crystal quality than that shown in Figures 1 and 2 has been produced at Spire and used for waveguiding (see Figure 3). Defects contribute to optical guided-mode scattering. However, this is not a large effect in the sample shown in Figure 3, and the optical propagation loss due to defects is estimated to be less than 1 dB/cm.

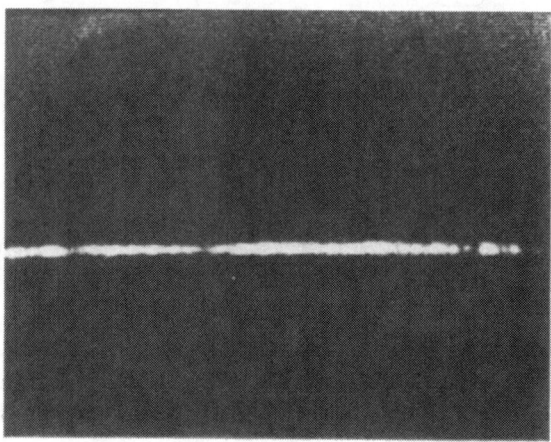

Figure 3.   Optical waveguiding at a wavelength of 1.3 microns in a 2 micron thick layer of silicon on a 4000 angstrom thick buried layer of SiO_2 (SIMOX). Photo shows near-field pattern of cleaved output end of planar waveguide. The waveguide length is 1.1 cm. Light was focused into the cleaved input end with a 40x lens.

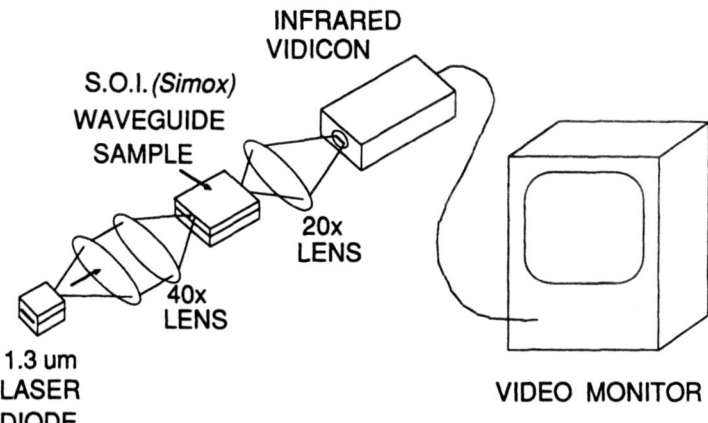

Figure 4.   Arrangement used to observe optical waveguiding in SIMOX.

Silicon guided-wave components and silicon electronics can be integrated
in one SIMOX wafer. When optical sources and optical detectors are added to
the wafer, the electronic ICs can communicate with each other by optical sig-
naling at high data rates. In other words, guided optical interconnects are
possible in SIMOX. (For the 1.3 micron wavelength, Si LEDs and Si Schottky-
barrier photodiodes are being developed [3]).

We speculate that optical waveguiding can be obtained in both silicon layers of a double buried-layer SIMOX wafer (in Si level #1 and Si level #2).
When the SiO_2 layer between 1 and 2 has a thickness of one micron, the
guided optical signals in 1 will be independent of those in 2 because the
guides are not coupled. Alternatively, optical signals can be transfered
from one level to another at certain locations. For example, if the thickness of the 1-2 SiO_2 layer is reduced locally to 1000 angstroms, waveguide
theory predicts that the waveguides in 1 will couple strongly to waveguides
in 2 at that location. As reference 6 suggests, it should be possible to
construct complex networks of Si channel waveguides in both levels by appropriate photomasking and etching during the SIMOX processing.

The free-carrier plasma effect in SOI [7] provides a means for electro-
optic modulation and switching in SOI waveguides. For example, electro-
optical switching between levels 1 and 2 of the double SIMOX should be feasible. For this purpose, a novel 2 x 2 silicon switch has been proposed [8].
It employs a voltage-controlled phase shift in a 3-D waveguide-coupling region
between levels 1 and 2.

Two-level SIMOX ($Si/SiO_2/Si/SiO_2$ on silicon) offers 3-D opto-electronic
integration. There are two novel and real possibilities for 3-D integration:
(1) optical interconnects in both Si levels, and (2) electronic ICs in the top
level with a "subterranean" optical interconnect in the lower level. In either
case, the waveguides are not altered by the annealing cycles.

CONCLUSION

In this work, we have demonstrated the formation of thick and double-layer buried oxide with a multiple implantation process and growth of epitaxial Si by CVD. The Si/SiO_2 interfaces are abrupt and the density of defects in both the buried Si layer and the top Si layer is comparable to standard SIMOX.

Optical waveguiding at the 1.3 micron wavelength has been clearly demonstrated in our SIMOX wafers [9]. These results, combined with recent progress made in the field of silicon electro-optics, suggest the possibility of guided-wave optical interconnects and 3-D opto-electronic integration in double buried-layer SIMOX.

ACKNOWLEDGEMENT

This work was supported in part by the Department of Defense (DOD), Rome Air Development Center (RADC) at Hanscom Air Force Base, MA, under Contract No. F19628-86-C-0069 and by a SBIR from DOD, Defense Nuclear Agency/Strategic Defense Initiative Organization. We are very grateful to Dr. Walter Shedd at RADC for his direction and support in this work. The authors would also like to acknowledge the assistance of Dr. S.N. Bunker, O. DeSilvestre, R. Bricault, M. Sanfacon, G. Beals, N. Martell, L. Geoffroy, and J. Breen. We would like to extend our gratitude to B. Buchanan for fruitful discussions; and to Dr. V. Jalan and to Professor Q. Kessel for their help with TEM and RBS/channeling experiments respectively. We would also like to thank J.P. Lorenzo of RADC for helpful discussions of waveguiding.

REFERENCES

1. P.L.F. Hemment, K.J. Reeson, J.A. Kilner, R.J. Chater, C. Marsh, G.R. Booker, J.R. Davis, and G.K. Celler, Nucl. Instr. and Meth. B21, 129 (1987).

2. S. Nakashima, Y. Maeda, and M. Akiya, IEEE Trans. Elec. Dev. ED-33, 126 (1986).

3. R.A. Soref and J.P. Lorenzo, Solid State Technology, Nov, 95 (1988).

4. S. Cristoloveanu, private communication; and S. Cristoloveanu, Proc. of Int. School on Silicon Technol., in press (1988).

5. F. Namavar and E. Cortesi, to be published.

6. J.P. Lorenzo and R.A. Soref, "Method for Fabricating Low-Loss Crystalline Silicon Waveguides by Dielectric Implantation," U.S. patent 4,789,642, issued December 6, 1988.

7. R.A. Soref and J.P. Lorenzo, paper MEE-1, presented at the Integrated and Guided-Wave Optics Conference (Optical Society of America), Houston, TX, February 6, 1989.

8. J.P. Lorenzo and R.A. Soref. "Electro-Optical Silicon Devices," U.S. Patent 4,787,691, issued November 29, 1988.

9. Waveguiding in Silicon-on-Insulator has also been observed experimentally by Professor Dennis G. Hall of the University of Rochester, and by Professor Bernard L. Weiss of the University of Surrey, U.K. (private communication, February 1989).

OXYGEN DOPED SILICON SURFACE LAYERS BY ION IMPLANTATION

K.Srikanth and S. Ashok
Department of Engineering Science & Mechanics and Center for Electronic Materials and Processing, The Pennsylvania State University, University Park, PA 16802

ABSTRACT

Oxygen doping of silicon with ion implantation has been attempted with the aim of modifying the surface electrical properties of the substrate. The change in current-voltage (I-V) characteristics after 1200 °C anneal has been ascribed to the formation of Oxygen Doped Silicon. The independence of the I-V characteristics on the choice of top metal contact and the reversal in direction of rectification strongly suggest the formation of an alloy layer in Si by the implantation. SIMS measurements indicate a sharp oxygen profile after the anneal and FTIR data show that oxygen is in the form of SiO_2 embedded in a silicon matrix.

INTRODUCTION

Oxygen implanted silicon has been widely studied for creating buried SiO_2 layers in silicon-on-insulator (SOI) applications. The implanted/annealed layers, with good stoichiometry have been characterized by various analytical tools such as HREM, SIMS, RBS, and AES [1-4]. However there is still much to be understood with regard to defect formation. The ultimate aim in most experiments is to form a buried insulating layer for device isolation. However, there is a relatively unexplored area of reseach in which the oxygen introduced does not form a buried stoichiometric oxide layer, but instead modifies the electrical propertis of silicon itself. Tabe et al.[5] have done some pioneering work in this area by evaporating Si under an oxygen ambient in a molecular beam epitaxy (MBE) system. They have fabricated a new wide-gap material, an Oxygen Doped Silicon Epitaxial Film (OXSEF), for application as emitter in silicon heterobipolar transistors. In this paper we present our results of a study to "oxygen dope" the silicon by ion implantation. Such an approach shares with MBE the advantage of a clean and well-controlled heterointerface, while offering compatibility with standard Si processing.

The choice of ion implant conditions is critical in our experiment as our intention is *not* to realize stoichiometric SiO_2 as needed in SOI. Implantation was performed over a set of energies to yield a uniform implant profile, and the doses were adjusted to give the same peak concentration, which was approximately 10 % of Si bulk concentration. The beam current was kept below 1 $\mu A/cm^2$ to avoid beam heating of the wafers. Post implant annealing was done at temperatures up to 1200 °C. These conditions are seldom encountered in SOI fabrication and the results presented here are mostly complementary and are expected to give a better insight into the interaction of oxygen in silicon.

EXPERIMENTAL PROCEDURE

Boron-doped Si wafers of (100) orientation and 1-10 ohm-cm resistivity were implanted in a Varian 350D ion implanter at energies of 20, 30, 50 and 70 keV. The peak of each of these implants was maintained constant at 7×10^{21} cm^{-3}, resulting in a uniform layer about 220 nm deep. The cumulative dose was 2.32×10^{17} cm^{-2}. In addition, a few samples were implanted at only 20 and 30 keV and others with just 50 and 70 keV, keeping the peak concentration same as before. The beam current density was kept below 1 uA/cm^2 to limit the temperature rise to below 100 °C.

Annealing of the implanted sample was done in a furnace with flowing Ar at 1000, 1100 and 1200 °C for 1 hour. 1-2 % O_2 was added to the Ar flow in order to avoid pitting the Si surface. Following the implantation and annealing, 1mm dia. metal dots (Al or Ti/Al) were thermally evaporated on the front side and an Al ohmic contact was formed over the entire back side. Electrical transport across the implanted layers was studied by measuring the current-voltage (I-V) characteristics as a function of temperature. SIMS and FTIR measurements were carried out on unmetallized samples. Selected samples were also studied for deep levels created by the implantation process using deep level transient spectroscopy (DLTS).

RESULTS AND DISCUSSION

As a result of the substoichiometric nature of the implants with no substrate heating during the implantation and anneal temperatures not exceeding 1200 °C, there is electrical conduction across the oxygen implanted layers. Hence our electrical characterization dwells primarily on the current-voltage-temperature measurements.

Fig. 1 shows the *log*I-V plots of the cascaded implant sample with 1000 ,1100 and 1200 °C annealing. These rectifying I-V characteristics are the result of an electrical barrier arising at the metal/implanted layer or the implanted layer/p-Si substrate interface. The as-implanted sample shows rectifying behavior in contrast to the "ohmic" behavior of the unimplanted control sample, indicating an increase in the electrical barrier. This enhanced barrier in the Al/p-Si structure is due to donor-like defects created by ion damage and has been universally observed [6,7]. If the ion damage alone were to be responsible for the change in I-V characateristics, one would expect reversion of the plots to those of the control after the anneals. With Ar implants it has been found that thermal anneal at temperatures up to 1000 °C is needed to regain the control I-V characteristics[7]. In the present case however, there is a simultaneous formation of an implant-generated alloy layer containing oxygen and silicon, whose stoichiometry is as yet undetermined. After a 1000 °C anneal of the oxygen implanted sample, the I-V characteristics do not coincide with those of the control. Further, the characteristics are no longer Schottky diode-like. The deviation from Schottky diode behavior is result of both the implanted oxygen and any remaining implant damage. The characteristics of the samples with 1100 °C anneal are similar to those of the 1000 °C samples but exhibit increasing dominance of the implanted layer. At 1200 °C, there is a *reversal* of the direction of rectification. Now the carrier transport across the structure can be limited by the Al/Si-O layer interface or by a more complex Si-O layer/substrate Si heterojunction interface.

It has been reported that the MBE-synthesized OXSEF/p-Si heterojunction has a valence band discontinuity [5]. It is possible that after the 1200 °C anneal, a similar effect of the band gap discontinuity leads to the observed I-V characteristics of Fig. 1. One other possibility is the generation of donors by oxygen, and the characteristics exhibited may be arising from the interface between the Al contact and the surface

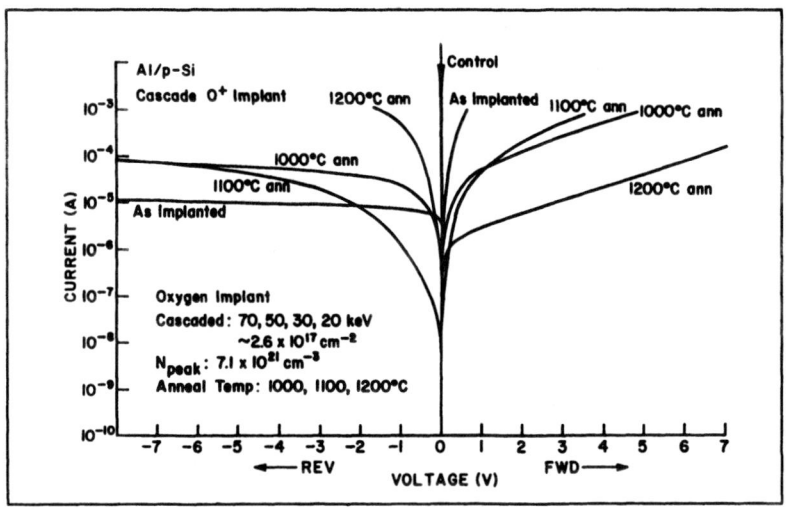

Figure 1   *log*I-V plots of cascade O⁺ implanted Al/p-Si devices as a function of post-implantation anneal temperature. The plots for unimplanted (control) and as-implanted (no anneal) samples are also shown.

layer doped by donors (A metal/n-type semiconductor Schottky barrier, consistent with the observed polarity of rectification after the 1200 °C anneal). Oxygen-related thermal donors have been observed in oxygen-implanted SOI material giving rise to high carrier density in the silicon layer close to the surface [8]. In order to gauge the carrier concentration profiles, spreading resistance measurements were done on these samples. The 1100 °C sample showed low surface carrier concentration, indicating partial compensation, but after the 1200 °C anneal, the compensation is greatly reduced and hence one may rule out the possibility of thermal donor compensation of the boron dopant atoms as the cause for the observed I-V reversal. As a further check a different contact metallization was used on the front side. Ti is known to form a high barrier on p-type Si and a correspondingly low barrier ("ohmic contact") on n-type Si. Despite the change in metallization to a low work function metal, Ti, the characteristics continued to show a very similar trend. The I-V characteristics hence do not originate from the metal/substrate contact but instead is attributable to the heterojunction between the oxygen-implanted layer and Si. More detailed examination of the electrical characteristics is necessary to determine the band diagram of the heterointerface causing the reversed I-V behavior after the 1200 °C anneal.

Secondary ion mass spectrometry (SIMS) data on the as-implanted and 1200 °C-annealed sample are shown in Fig. 2. Fig. 2(a) shows the oxygen profile resulting from the 20,30,50 and 70 keV cascade implant. Comparing this with Fig. 2(b) for the annealed sample one can see a distinct redistribution of the implanted oxygen with no discernable saturation, suggesting a level well below the stoichiometric limit. Another observation is the steep drop in oxygen concentration from 10^{22} to 3×10^{19} cm^{-3} over a distance of about 30 nm, which compares well with the oxygen profile of MBE-grown OXSEF [9].

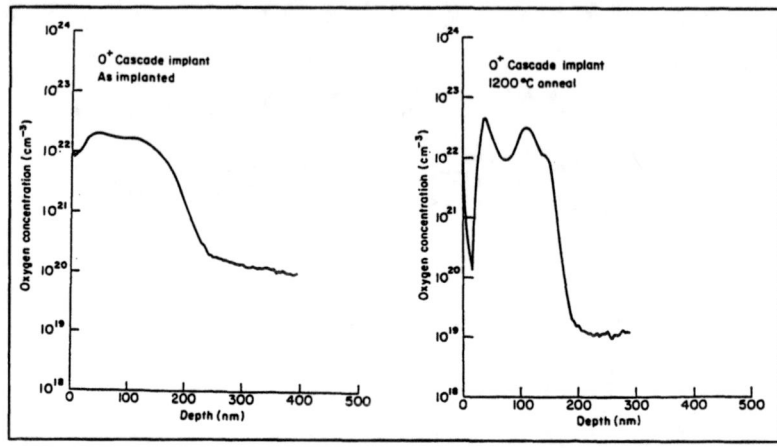

Figure 2   SIMS profile of oxygen concentration for (a) the as-implanted sample, and (b) sample annealed at 1200°C for 1 hour.

In an effort of identify the Si-O phase resulting from the implantation/anneal, Fourier Transform Infrared (FTIR) spectroscopy was done on selected samples. The spectrum for the as-implanted sample, shown in Fig. 3(a) has an absorption peak near a wavenumber of 970 cm^{-1}. This peak shifts to a higher wavenumber of 1087 cm^{-1} for 1000 °C anneal and stays the same for the 1200 °C case (Figs. 3(b) & 3(c)). The peaks after post-implant annealing are very close to the SiO_2 peak of 1100 cm^{-1}. The sharpness and intensity of the peaks increase with annealing, indicating increased incorporation of oxygen and restoration of substrate crystallinity.

The SIMS data of Fig. 2 shows clearly that we are not reaching the stoichiometric limit, but the FTIR spectrum exhibits a strong SiO_2 signal strength. This leads us to believe that the oxygen is predominantly in the form of SiO_2 embedded in a Silicon matrix. This is being further confirmed by spectroscopic ellipsometry.

The nature of the defects created by high dose implantation has recently been a subject of intense study. In our investigation, we have modified the property of the substrate silicon with a high-dose oxygen implant, yet have significant carrier transport across the implanted layer. This gives us the opportunity to perform DLTS measurments on these samples. A preliminary investigation reveals a trap with an activation energy of 455 meV for the 1200 deg. C anneal sample. Studies comparing the differently annealed samples will be discussed in a future publication.

CONCLUSIONS

We have successfully modified the surface of silicon by implanting oxygen at about 10 % of the silicon concentration. This implanted Si-O layer significantly affects carrier transport across the interface. The modified layer appears to contain oxygen in the form of imbedded SiO_2 in crystalline Si, even though the implant dose is well below the stoichiometric limit.

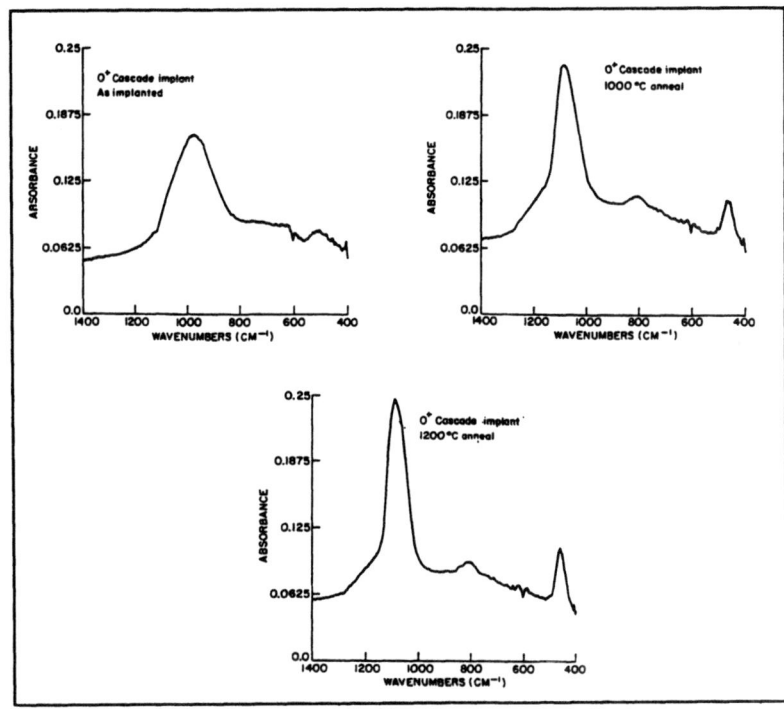

Figure 3   FTIR spectra of cascade O⁺ implanted samples: (a) as-implanted, (b) after 1000 °C - 1 hour anneal, and (c) after 1200 °C - 1 hour anneal.

ACKNOWLEDGMENTS

This work was initiated with a seed grant furnished by the General Electric Foundation to the Penn State Center for Electronic Materials and Processing. We are also grateful to Cheryl Houser for performing the SIMS and FTIR measurements.

REFERENCES

1.   P. L. F. Hemment in Mat. Res. Soc. Symp. Proc. **53**, 207 (1986).
2.   I. H. Wilson, Nucl. Instr. Methods in Phys. Res. **B1**, 331 (1984).
3.   A. E. White, K. T. Short, J. L. Batstone, D. C. Jacobson, J. M. Poate and K. W. West, Appl. Phy. Lett., **50**, 19 (1987).
4.   J. Stoemenos, J. Margail, C. Jaussaud, M. Dupuy, and M. Bruel, Appl. Phy. Lett. **48**, 1470 (1986).

5. M. Tabe, M. Takahashi and Y. Sakakibara, Jap. J of Appl. Phys., **26**, 1830 (1987).
6. S. Ashok, K. Giewont and H. P. Vyas, Phy. Stat. Sol. (a) **98**, K-99 (1986).
7. K. Giewont, S. Ashok and A. Mogro-Campero, Thin Solid Films **142**, 13 (1986).
8. A. Golanski, A. Perio, J. J. Grob, R. Stuk, S. Maillet and E. Clavelier, Appl. Phys. Lett., **49**, 1423 (1986).
9. M. Takahashi, M. Tabe and Y. Sakakibara, IEEE Electron Dev. Lett., **EDL-8**, 475 (1987).

STUDY OF STRESS AND MORPHOLOGY OF SILICON—ON—INSULATOR
BY MEANS OF ULTRAVIOLET REFLECTANCE SPECTROSCOPY

BEA M LACQUET, PIETER L SWART
Sensors Sources and Signal Processing Research Group, Faculty of Engineering,
Rand Afrikaans University, P O Box 524, Johannesburg, 2000, South Africa.

ABSTRACT

SOI material was prepared by implanting a high dose of nitrogen into crystalline silicon at an energy of 160 keV to form a buried layer of silicon nitride. Ultraviolet reflectance spectroscopy was employed to characterize the material. In the current presentation the isothermal annealing behaviour of the surface material was investigated with regard to amorphization and stress. Measured UV—reflectance of the samples are compared to simulated reflectance data which were obtained by presenting the implanted material as a layered structure with the surface layer consisting of a mixed layer of amorphous and crystalline silicon. From these simulations the percentage amorphization at each stage during the annealing cycle is quantitatively determined. Stress is qualitatively determined by considering the shift in the position of the frequencies at which reflectance maxima associated with Van Hove singularities are observed in the ultraviolet range. The annealing cycle used in this work proved to be adequate for returning the material back to single crystal status and relieving the initial stress in this layer.

INTRODUCTION

Material technologies for high speed integrated circuits are becoming increasingly important with the demand for real time digital signal processing. Silicon—on—insulator formed by ion implantation is one of the materials currently under investigation. With the availability of high current ion implanters and the relatively simple operation involved in both the implantation and the post—implant annealing step, ion implanted SOI material could have the edge on contenders such as silicon—on—sapphire, recrystallized polysilicon[1], oxidized porous silicon[2], epitaxial lateral overgrowth[3], graphoepitaxy, laser photochemical deposition, molecular beam epitaxy and silicon epitaxy on cubic zirconia[4,5]. In all of these technologies characterization of the material during the experimental stages as well as during production could be performed in various manners. We investigated optical characterization as a non—destructive evaluation technique which could be used in both environments. By scanning the energy range from 0.025 eV to 6.2 eV, a vast amount of information regarding the properties of the insulator as well as the silicon surfaces can be determined. In this paper we are presenting results on the reflectance measurements in the 3.1 eV to 6.2 eV range. The energy of the incident light and the absorption of the material are such that only a layer stretching approximately 100 nm into the substrate, is probed.

SAMPLE PREPARATION AND REFLECTANCE MEASUREMENTS

The SOI sample used was prepared on a p—type substrate, 76.2 mm in diameter and (100) orientation. The resistivity was 8.6 ± 1.4 ohm—cm and the thickness 381 ± 20 μm. The substrate dopant was boron. The silicon wafer was implanted to a dose of 1.8 x 10^{18} cm^{-2} at an energy of 160 keV. The substrate temperature during the implant was held at 550 °C. The beam current was approximately 2 mA. Before the annealing step, the wafer, now cut into a number of pieces, was cleaned, dipped in a 10% HF solution and covered by a 100 nm layer of SiO_2 by RF sputtering. During the sputtering the substrate temperature rose to 80 °C. The subsequent anneal was carried out at a temperature of 1200 °C in a nitrogen atmosphere, for various lengths of time between 15 and 240 minutes. After annealing the SiO_2 cap layer was removed and the samples characterized. Apart from the SOI material, reference samples of the substrate and a thin layer of amorphous silicon prepared by RF sputtering were also analyzed.

A Hitachi model UV–3400 UV–VIS spectrophotometer with a 5° reflectance attachment was used for the characterization in the mentioned wavelength range. The angle of incidence approximates normal incidence[6]. The ultraviolet reflectance spectra are depicted in figures 1 a, b, c and d. These graphs are of the as–implanted (#P000), 30 minute (#P030), 60 minute (#P060) and the 120 minute annealed (#P120) samples, respectively. Each reflectance curve is compared to that of a substrate reference wafer, #PREF, and an amorphous silicon thin film. The implanted material shows in general a higher reflectance than the amorphous film. After the implant, there is a large difference in the spectrum of sample (#P000) in comparison with the single crystal reference. The main features are reduced intensity of the reflected light and reflectance maxima which appear to be somewhat rounded and shifted towards lower energies. The position of the peaks associated with the Van Hove singularities at approximately 3.36 eV and 4.48 eV shifted by 0.11 eV and 0.09 eV towards lower energies. The intensities of these maxima decreased by 5.1 and 9.8 percentage points, respectively. After annealing of all the samples at 1200 °C, there appears to be negligible shift in the peak position at 4.48 eV. The maximum at 3.36 eV, however, is shifted by between −0.02 eV and 0.01 eV. A very small difference in the reflectance (<1 percentage point) is observed after a 60 minute anneal (#P060).

DISCUSSION

Determination of the physical properties and characteristics of material in the ultraviolet are extremely valuable in elucidating structure and morphology of surface layers or thin films. The presence of lattice damage, surface roughness and strain in the surface layer of the semiconductor has a significant effect on the optical properties of the material [e.g. 7]. In the absence of very thin surface layers such as a native oxide, no detectable interference fringes are usually observed in the ultraviolet energy range of the spectrum. The optical constants of the type of material found in SOI structures such as silicon, silicon nitride, silicon dioxide and amorphous silicon, change especially rapidly as a function of energy in the ultraviolet. [8,9] This energy dependence should be kept in mind when the reflectance curves are analyzed.

In the set of samples under discussion, a reduction in ultraviolet reflectance and a shift in the peak positions with respect to the reference unimplanted silicon wafer are observed. These two observations will be discussed separately.

Damage of surface layer

Reduction in reflectance with respect to a reference can possibly be attributed to surface roughness, a surface overlayer, bulk and surface plasmons, and lattice damage. When analyzing reflectance curves in the ultraviolet, it is not always possible to distinguish between the effects that surface roughness, a very thin native oxide, unintentional mechanical damage[10] or damage caused by ion implantation[11] have on the measured reflectance curve. In these samples the possibility of the presence of a very rough surface is discarded, since the frequency dependence of the decrease in reflectance after the implant with respect to the reference curve is not as expected from a rough surface[12]. The plausible causes could be (i) a very thin native oxide and (ii) amorphization of the surface layer caused by the high dose implant.

The effect of the mentioned native oxide and an amorphous surface layer on the reflectance of the implanted silicon wafer has been investigated by

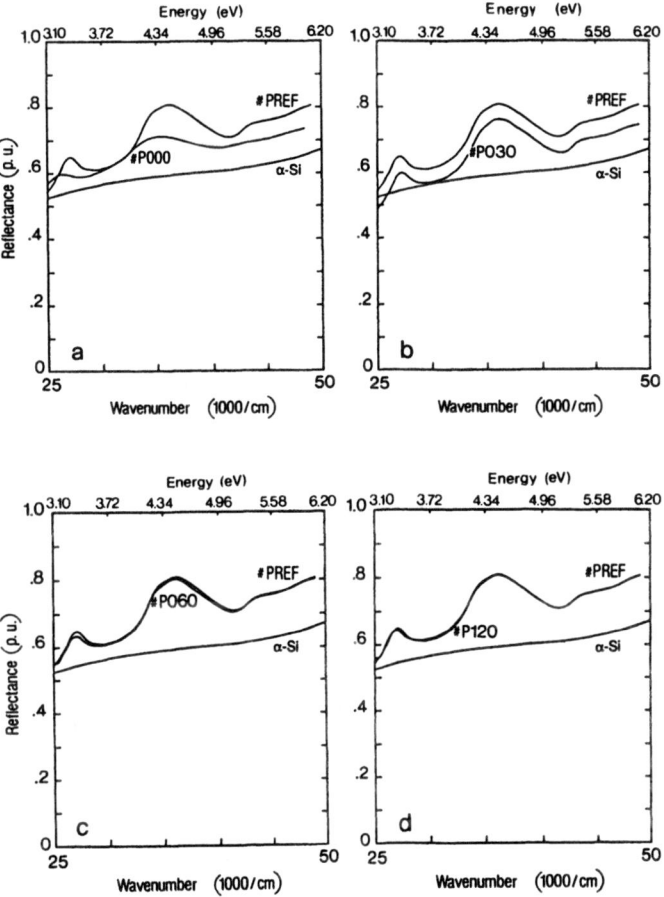

Figure 1.   Measured reflectance in the ultraviolet of samples #P000, #P030, #P060, #P120, #PREF and that of an amorphous silicon thin film.

simulating the possible results and comparing them to the measured reflectance curves. Figure 2 depicts the reduction in reflectance of crystalline silicon due to the presence of a thin native oxide. The thickness was varied between 4 and 20 nm.

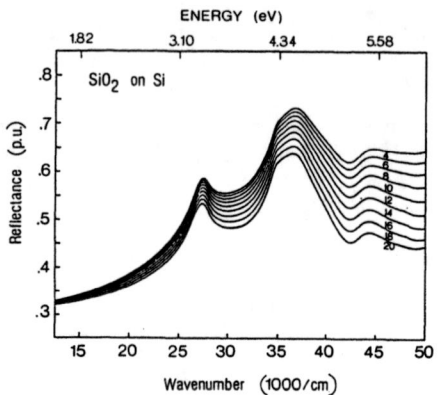

Figure 2.   Calculated normal incidence reflectance of single crystalline silicon with an oxide surface layer varying between 4 and 20 nm.

The effect of having amorphized material as a result of the ion implantation damage is modelled by using an effective—medium approximation [10,13] to calculate the optical properties of the material, employing a mixture of crystalline and amorphous silicon. Figure 3 shows the result of the calculated reflectance curves at normal incidence of a silicon wafer in which the volume fraction of the amorphous silicon in the surface layer is varied between 0 and 80%. The optical constants of the amorphous silicon and crystalline silicon were taken from Palik[9]. It is evident that an increasing fraction of amorphous silicon results in a reduction in reflectance at large energies.

It is noticed in Figure 3 that an increase in the volume fraction amorphous silicon reduces the amplitude of the reflectance peak, both in percentage reflectance and relative to that of the unmixed crystalline silicon. At lower energies the reflectance increases above that of the crystalline silicon. In the case of the presence of a thin surface oxide, the absolute reflectance is reduced over the whole energy range depicted, and a slight increase in peak height at 4.48 eV with respect to the uncovered crystalline silicon surface are observed. The deposited oxide layer was removed prior to the measurements. In the case of the as—implanted p—type wafer (#P000), it is therefore suggested with reference to figure 1a that the reduction in reflectance is probably caused by an amorphous silicon/crystalline silicon mixed layer reaching to the surface of the material as a result of the high dose implant. A comparison between the calculated and measured reflectance curves, indicates that the volume fractions of amorphous and crystalline silicon are approximately 70% and 30% respectively. After annealing at 1200 °C the 30 minute annealed sample, #P030, still shows reduced reflectance with respect to the substrate reference sample #PREF. In this instance. however, the volume fractions have changed to 30% amorphous silicon and 70% crystalline silicon according to Figure 3. Samples #P060 and #P120 show reflectance curves which approximates that of the reference wafer quite well. Any differences in the reflectance can be attributed to the presence of a thin native oxide layer.

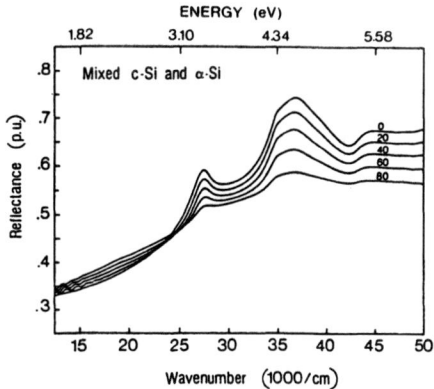

Figure 3.   Calculated normal incidence reflectance of a mixed layer of crystalline and amorphous silicon with the volume fraction amorphous material varying between 0 and 80%.

Strain in the surface layer

A second observation in connection with the reflectance spectra in the ultraviolet is the shift in the position of the energies of the maxima associated with the Van Hove singularities. The position of the maxima shifted towards lower energies after the high dose nitrogen implants. After annealing the direction of the shift is reversed and the peaks almost returned to the values associated with crystalline silicon. In the case of these samples #P, shifts of approximately 0.11 eV and 0.09 eV towards lower energies were recorded after the implant for the maxima at 3.36 eV and 4.48 eV, respectively. After annealing at 1200 °C for 30 minutes, the initial shift of the maximum at 4.48 eV is reduced to zero. Longer high temperature processing, samples #P060 and #P120, does not change the position of this peak. The maximum occurring at 3.36 eV, however, shows a different behaviour as a function of annealing time. After an initial shift, the 30 minute anneal restores the peak to an energy of 3.378 eV, which is higher than the value of the reference sample #PREF. Further annealing for 60 and 120 minutes at 1200 °C leads to this peak shifting to lower energies of 3.355 and 3.336 eV, respectively.

The shift in the position of the maxima is explained by a change in the band structure of the material[14,7]. This can be caused by stress in the surface layer as a result of the forced incorporation into the lattice of a large number of atoms. The large dose of implanted nitrogen ions causes swelling (warping) of the crystal. Since the surface is free, the implanted wafer is then modelled by a thin membrane on bulk material with anisotropic stress and strain in the plane of the wafer.

The available literature on the effects of stress on the properties of silicon pertains to hydrostatic and uniaxial stress. Even though the implanted wafer is in fact a two–dimensional problem, results obtained for uniaxial stress will be used as a basis in the discussion of the observations. It is observed in the work by Gerhardt[15] on the piezoreflectance of silicon in the ultraviolet that a stress applied along the [100] axis yields a much smaller stress–optic shift of the reflectance peaks near 4.5 eV, compared to a stress applied to the [111] axis. For stress along the [111] axis, yielding a tensile strain component of 0.5%, the shift in the peak position is approximately 0.0006 eV towards lower energies. For approximately the same tensile strain, the peak at 3.4 eV shifts by approximately 0.0025 eV in the same direction. The shift in the position of this particular peak at 3.4 eV is much larger than in the case of stress applied to the crystal [100] axis.

Consider tensile stress applied to the [111] axis of samples #P. Should the observed shift in the peak at 3.4 eV be a linear function of the applied stress, it would correspond to a tensile strain component of 2.9%. However, the shift in the position of the maximum in reflectance at 4.5 eV, is extraordinarily high and cannot be explained. One will possibly have to consider the whole two-dimensional problem.

CONCLUSION

Ultraviolet reflectance measurements in the 3.1 eV to 6.2 eV energy range on SOI material gave information as to the possible composition of the surface layer, the presence of surface oxide overlayers and stress of the material. The information could then be used to determine annealing sequences after high dose implant to obtain a crystalline surface layer before further processing can be performed.

ACKNOWLEDGEMENTS

The authors wish to express their gratitude towards the CSIR for the use of their spectrophotometer. Financial assistance by the Rand Afrikaans University and the Foundation for Research Development is gratefully acknowledged.

REFERENCES

1. V.Y. Chen, G.R. Stegen; J. Geophys. Res., 79 (1974) 3019.
2. K. Imai, H. Unno; IEEE Trans. Elect. Dev., ED-31 (1984) 297.
3. J.H. Douglas; High Technol. (1983) 55.
4. K.E. Bean; J. Electrochem. Soc., 124 (1977) 50.
5. S.L. Partridge; Proc. IEEE. 133 (1986) 106.
6. T.S. Moss, G.J. Burrell, B. Ellis; Semiconductor Opto-Electronics, Butterworths, London, 1973.
7. D.L. Greenaway, G. Harbeke; Optical Properties and Band Structure of Semiconductors, Pergamon Press, Oxford, 1968.
8. E. Schmidt; Appl. Optics, 8 (1969) 1905.
9. E.D. Palik; Handbook of Optical Constants of Solids, Academic Press, Florida, 1985.
10. D.E. Aspnes, J.B. Theeten, F. Hottier; Phys. Rev. 20 (1979) 3292.
11. R.V. Collins, J. Cavese; J. Vac. Sci. Technol., A5 (1987) 2797.
12. B.M. Lacquet; Optical Modelling and Characterization of silicon-on-insulator layers and related structures, D.Eng. Thesis, RAU 1988.
13. D.A.G. Bruggeman; Annal. Phys. (Leipzig), 24 (1935) 633.
14. V. Paul; Proc Int School of Phys. "Enrico Fermi," Course 34 (1966) 256.
15. U. Gerhardt; Phys. Rev. Lett., 15 (1965) 401.

PART VI

Implantation in III-V Materials and Multilayers

ION IMPLANTATION PROCESSING OF GaAs AND RELATED COMPOUNDS

S. J. Pearton, W. S. Hobson and C. R. Abernathy
AT&T Bell Laboratories, Murray Hill, NJ 07974

ABSTRACT

The formation of doped or semi-insulating layers by ion implantation in both Ga- and In- based semiconductors is reviewed. The Ga-based materials (GaAs, AlGaAs, GaP, GaSb) tend to show similar characteristics in terms of producing relatively low ($n \leq 3 \times 10^{18}$ cm^{-3}) maximum carrier densities for donor implanted layers, and much higher values for acceptor implants ($p \leq 5 \times 10^{19}$ cm^{-3}). Ion-induced damage is widely used for device isolation in these materials, with midgap levels associated with the damage trapping free carriers and leading to semi-insulating behaviour. By contrast, the In-based materials (InP, InAs, InSb and InGaAs) show higher maximum carrier densities for acceptor implants than for donor implants, and the use of ion damage for isolation purposes is much less effective than in GaAs. All of these materials display singularly poor regrowth characteristics, requiring in some cases the use of elevated temperature implantation to prevent amorphization.

INTRODUCTION

Ion implantation is now firmly established in III-V device technology as the technique of choice for selectively changing the resistivity of the semiconductor[1,2]. This may either be to create doped regions, or to produce high resistivity layers for device isolation purposes. In III-V materials the binary nature of the lattice for GaAs and InP makes the damage removal and dopant activation steps more complex than in Si. Such differences are exacerbated for ternary (InGaAs, InAlAs, AlGaAs) and quaternary (InGaAsP) semiconductors. There have been a number of previous reviews concentrating predominantly on aspects of implantation in GaAs, such as amorphization and recrystallization[3], annealing using conventional furnaces or transient heating[4-6], electrical activation[7,8], range statistics[9] or practical considerations in applying the technology[10]. One of the key differences between implantation in III-V materials relative to Si is that the best electrical activation in the former is achieved when amorphization is avoided. For device fabrication, light ions such as Si and Be are typically used, because good activation is attained for room temperature implantation. By contrast, the use of S, Se, Te or Zn usually requires elevated temperature (80-200°C) implantation to avoid amorphization if doses above ~10^{14} cm^{-2} are used. Apart from poor electrical activation of implanted dopants amorphous layers in III-V materials show poor recrystallization characteristics, and exhibit high degrees of residual disorder in the form of microtwins, stacking faults and point defects. This is in sharp contrast to the regrowth of amorphous Si, which leads to defect-free layers in a single solid phase epitaxial step around 550°C. In III-V's, implantation can also lead to the creation of regions with local deviations from stoichiometry resulting from the different displacement properties of the lattice constituents, which have unequal masses. Incongruent evaporation of the group V element from the material upon high temperature annealing is also characteristic of compound semiconductors.

In this paper we review the current state of understanding of damage introduction and removal steps, dopant activation, chemical and damage-related compensation and some future directions for implantation in III-V materials. Emphasis will be placed on GaAs where most information is available, but InP, AlGaAs and other commonly used semiconductors will be covered.

Mat. Res. Soc. Symp. Proc. Vol. 147. ©1989 Materials Research Society

Damage Introduction and Removal

In III-V's damage accumulation and possible eventual amorphization are modelled using either a heterogeneous mechanism, in which individual damage clusters are considered to be amorphous and overlapping of these regions results in complete amorphization (heavy ions), or a homogeneous mechanism in which the crystal becomes unstable and collapses to an amorphous state when the defect concentration reaches a critical value (light ions). Depending on the ion, the dose and the implant temperature the implant damage can consist of either amorphous layers or extended crystalline defects (dislocations and stacking faults). In GaAs, amorphous layers recrystallize epitaxially during annealing at 150-200°C, but the recrystallized layer is invariably highly defective, consisting of twins, stacking faults and other defects. These defects anneal out to leave only a high density of dislocation loops in the range 400-500°C. These loops grow and annihilate above about 700°C, and the remaining point defect clusters begin to anneal out above about 750°C. Dopant atoms appear to make a short range diffusion to lattice sites around 600°C, but optimum electrical activity is not obtained until ≥750°C for acceptors and ≥850°C for donors. N-type dopants are more difficult to activate and give lower electrically active concentrations than p-type dopants in GaAs and AlGaAs, while the reverse is true for InP. Above annealing temperatures of ~600°C (500°C for InP) the GaAs or AlGaAs surface must be protected against dissociation by As loss, and at elevated temperatures some dopants also display excessive diffusion (S and all of the Ga-site acceptors in GaAs and AlGaAs and the acceptors in InP.

In III-V's the lattice elements are distinguishable and because they recoil unequally due to their different masses, local perturbations in stoichiometry are created. The lighter element recoils further, leading to an excess of the heavier element near the surface (shallower than R_p) and an excess of the lighter element at greater depths (between R_p and $R_p + \Delta R_p$). Repair of the lattice during subsequent annealing requires displaced atoms diffusing back to appropriate sites, and in III-V's the diffusion lengths are not great enough to accomplish complete regrowth. The displaced lattice elements are unable to move quickly enough to keep up with the growth front, leading to highly twinned material and eventually to a complete stop of the regrowth if the initial amorphous layer is thicker than ~2000Å. The electrical activation in regrown III-V's is significantly worse because of remanant disorder than if amorphization is avoided. Implantation at elevated temperatures prevents the formation of an amorphous layer because of the increased mobility of point defects which are able to recombine and annihilate each other, otherwise known as dynamic annealing. For room temperature implantation amorphization of InP (and GaP) occurs at much lower ion doses than for GaAs (and InAs). Table 1 shows the critical dose for amorphization (ϕ_{AM}) and corresponding nuclear stopping energy density (G_{AM}) for various ion-substrate combinations. These energy densities correspond to ~10eV per molecule for GaAs which should be compared to a requirement of ~100eV per molecule for Si. In GaAs for LN_2 temperature implantation the defect concentration is determined by the nuclear stopping energy deposition and amorphization occurs by the heterogeneous mechanism. With increasing implantation temperature homogeneous nucleation becomes dominant. By contrast in InP it appears that even at room temperature, heterogeneous defect nucleation plays a role and less energy deposition is required to damage the material.

In crystalline semiconductors unwanted channelling of implanted ions can be reduced or eliminated by pre-amorphizing the near-surface region. This is not feasible in III-V's, and in practice to minimize both axial and planar channelling, GaAs wafers are oriented with an appropriate azimuthal or twist direction (the angle between the wafer flat and the direction of beam tilt) in addition to being tilted with respect to the beam direction. This is especially necessary for low dose implants where control of the ion profile is critical for reproducible device performance. Figure 1 shows measured carrier profiles for ^{29}Si ions implanted at 100keV energy and a dose of 5×10^{12} cm^{-2} and annealed at 850°C for a variety of tilt and rotation angles during the implant step[12]. Channelling is minimized for tilt angles of $\geq 7°$ and rotation angles of $\geq 30°$. Increasing the ion dose creates damage in the uppermost GaAs layer which randomizes the beam and reduces subsequent channelling. Increasing the implant temperatures can either increase or decrease channelling through two mechanisms-dynamic annealing which reduces scattering out of channels and increased lattice vibrations which increases randomisation of the beam. Implantation through a thin layer deposited on the sample surface can also reduce channelling.

Table 1. Critical values of ion flux (ϕ_{AM}) and nuclear stopping energy density (G_{AM}) for III-V materials at room temperature[11].

Substrate	Ion	Energy (keV)	ϕ_{AM} (10^{13} cm^{-2})	G_{AM} (10^{20}keV · cm^{-3})
GaAs	Ar$^+$	200	25	1
GaP	Ar$^+$	200	10	5
InAs	Ar$^+$	180	100	52
GaAs	Se$^+$	300	7	8
InP	Se$^+$	300	2	3
InAs	Se$^+$	300	32	39

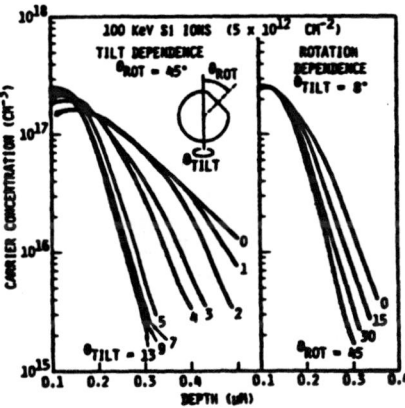

Figure 1. Carrier profiles in ^{29}Si implanted (5×10^{12} cm^{-2}, 100 keV) GaAs annealed at 850°C, 15 min, for a variety of tilt and rotation angles during the implantation[12].

Annealing and Dopant Activation

A summary of the characteristics of the most common implant species in GaAs is given in Table 2. The comments given there are also relevant for AlGaAs and InP. For most device applications n-type regions are formed by Si implantation. The profile shape is given by a Pearson IV distribution up to ~700keV and no significant diffusion occurs up to annealing temperatures of 920°C for GaAs and AlGaAs, and 850°C for InP. In GaAs the projected range of Si is given by $R_p = 2.5E^{0.85}$ (in nm), where E is in keV. The activation for low doses ($<10^{13}$ cm^{-2}) is $\leq 80\%$ for either rapid or furnace annealing, whereas at high doses a marked saturation in activation occurs as the amphoteric nature of Si becomes evident and the ions begin to occupy Ga and As sites in equal numbers. The practical limit to the carrier concentration achievable by high doses is $\sim 2 \times 10^{18}$ cm^{-3} which is far below that required to fabricate non-alloyed ohmic contacts in GaAs ($\sim 10^{20}$ cm^{-3}). For this reason either alloyed (usually AuGeNi eutectic) contacts are utilized, or a thin contact layer of doped InGaAs or InAs is grown on top of the structure.

Table 2.   Implanted Species Characteristics in GaAs.

Ion	Comments
	(a) Donors
Si	Good activation for RT implant, versatile range, amphoteric but mainly donor, $D < 10^{-15}$ cm$^2 \cdot$ S^{-1} at 850°C.
Se	Good activation for low dose, RT implant, limited range, $D = 5 \times 10^{-15}$ cm$^2 \cdot$ S^{-1} at 850°C.
S	Diffuses rapidly during anneal.
Te	Poor activation for RT implant.
Ge	Amphoteric species, poor activation.
Sn	Amphoteric species, diffuses during anneal, metastable solubility.
	(b) Acceptors
Be	Good activation at low temperatures, versatile range, toxic source, $D = 6 \times 10^{-9}\ e^{-0.7/kT}$, getters to surface during RTA.
Mg	Good activation, versatile range, difficult implant source, getters to surface during RTA, $D = 2 \times 10^{-16}\ e^{-0.6/kT}$.
Zn	Reasonable activation. Like other acceptors will diffuse during anneal if present at high concentration.
Cd	Reasonable activation, diffuses during implantation.
C	Amphoteric. Requires Ga co-implant to achieve $p \sim 10^{19}$ cm^{-3}. $D < 10^{-15}$ cm$^2 \cdot$ S^{-1} at 850°C.

Selenium is the other major n-type dopant. The ion distribution can be described by a joined half-Gaussian profile due to a small amount of radiation-enhanced diffusion[10]. However there is no significant diffusion during subsequent annealing. The projected range in nm is given by $2.8E^{0.67}$ up to 400keV[9]. Optimum activation resembles that of Si, but only if the implant is done with substrate held at ≥ 150°C. Se is not an amphoteric species but for high dose implants at elevated temperatures the maximum doping concentration is $\leq 10^{19}$ cm^{-3}. This is true also for the other donor species S and Te. Figure 2 shows Te to be highly soluble (90% substitutional) after either rapid or furnace annealing, or even as-implanted for elevated temperature implantation[13]. The soluble fraction is approximately two orders of magnitude higher than the electrically active fractions which range from 0.8-3% depending on the experimental conditions. In other words, substitutionality is a necessary, but not sufficient, condition for electrical activity.

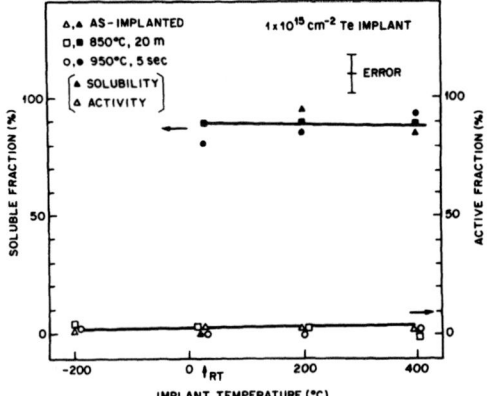

Figure 2. Summary of Te solubility data for 1×10^{15} cm^{-2}, 100keV implants performed at various temperatures. The electrically active fractions are shown at the bottom of the figure.

Using Extended X-Ray Absorption Fine Structure (EXAFS) it has been shown the the limitation in doping is related to the formation of donor-vacancy complexes (D-V_{As}) which are acceptors. Figure 3 shows the situation for S - it occupies two substitutional sites of equal concentration: (i) an unperturbed As site (S_{As}), characterized by a S-Ga distance of 2.42Å, and (ii) a relaxed configuration with S on an As site (S_{As}) and an As-vacancy (V_{As}) in the second neighbor shell, characterized by a S-Ga distance of 2.31Å. These two populations lead to electrical self-compensation, with the net difference in concentration being the measured activity. It is possible to increase the activation somewhat by co-implanting As to reduce the number of As vacancies but this has only limited effectiveness because of the extra damage created by the additional implant.

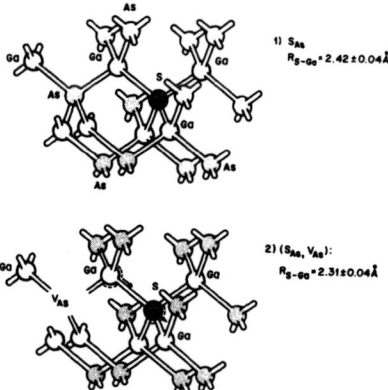

Figure 3. Schematic of GaAs lattice showing two configurations of S in implanted material.

The formation of p-type layers by implantation in GaAs is more straightforward than for n-type layers. All of the acceptor species show high activation to a much higher dose level than do the donors. Because of its light mass, Be can be activated at temperatures as low as 500°C, although the optimum electrical properties are obtained for 800-850°C anneals. Be exhibits a Pearson IV-type distribution with a range of $9.9E^{0.78}$ up to 400keV[9]. There is marked redistribution of all of the acceptors during furnace annealing and loss of the dopant to the surface if the wafer is uncapped during the anneal, even for RTA (Figure 4). Peak carrier concentrations above 10^{19} cm^{-3} are easily achievable by Be, Mg, or Zn implants, with the maximum reported values near 3×10^{20} cm^{-3}. A comparison of the sheet activation versus dose for donor and acceptor implants in GaAs is shown in Figure 5.

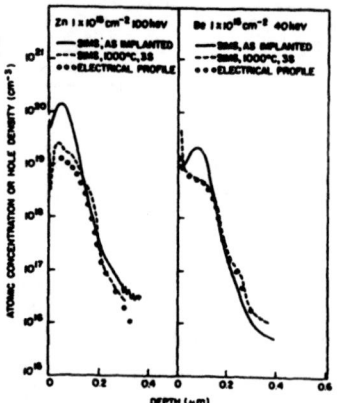

Figure 4.   Atomic profiles and carrier distributions for Zn and Be implanted at a dose of 10^{15} cm^{-2} into GaAs. During an anneal at 1000°C for 3 sec, a substantial amount of both acceptors is lost to the surface.

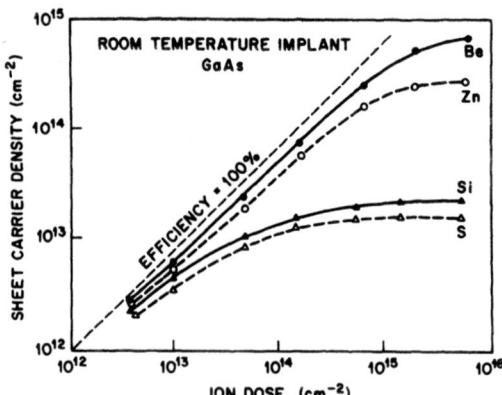

Figure 5.   Comparison of sheet activation versus dose for donors (annealed at 900°C, 10 sec) and acceptors (annealed at 800°C for 3 sec) in GaAs.

In contrast to the situation for GaAs, high electron concentrations are readily produced by implantation in InP, but hole concentrations above 2×10^{18} cm^{-3} are difficult to obtain. There is also a clear advantage to implanting all of the species, both donor and acceptor (except Be) at temperatures $\geq 150°C$.

Some form of surface protection must be provided for III-V materials during implant activation anneals. In principle annealing GaAs (InP) in an AsH$_3$ (PH$_3$) or As–H$_2$ ambient is an ideal solution, but in practice there are problems related to safety issues and gas-purity when conventional furnace annealing is used. Rapid capless annealing under AsH$_3$ is a must for many heterostructure-based devices, and commercially available annealing systems are just beginning to appear. Capless proximity annealing, in which the wafer is placed face-to-face with another uncapped GaAs wafer, is still the most commonly used method for RTA, but it is difficult to maintain pristine surfaces over the whole wafer area. Encapsulation of the GaAs with a dielectric film can also be used, but this is far from ideal because of the introduction of considerable near-surface strain due to the quite different thermal expansion coefficients. This can lead to a significant enhancement in the diffusivity of some implanted dopants. Plasma-deposited SiN$_x$ is the most commonly used encapsulant but these films are often subject to cracking and peeling. SiO$_2$ allows preferential outdiffusion of Ga from the surface, while two promising encapsulants are AlN and PSG (phospho-silicate glass) both of which have similar expansion coefficients to GaAs. PSG is particularly useful for InP encapsulation.

One of the key features about implant activation in compound semiconductors is shown in Figure 6. Carrier activation does not occur until most of the implant damage is removed, and thereafter the gradual removal of point defects with higher annealing temperatures leads to the optimum activation[15]. The reasons for this are dealt with in the next section. The activation of commonly used donor and acceptor ions in Al$_x$Ga$_{1-x}$As as a function of AlAs mole fraction are shown in Figure 7. The activation efficiency decreases with increasing Al composition. It appears that a good deal of the energy required to move an implanted donor ion onto a substitutional site in GaAs and AlGaAs actually comes from the need to dissociate defect complexes in which the donor is incorporated[16]. This is in agreement with the EXAFS data discussed earlier.

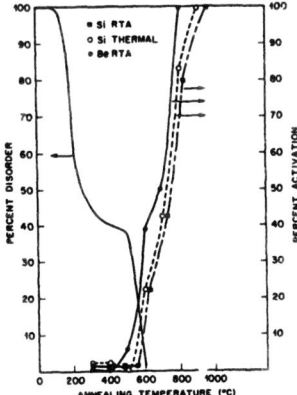

Figure 6.   Carrier activation (plotted as % of the final value) in Si and Be implanted GaAs (10^{15} cm^{-2} dose) after rapid or furnace annealing, contrasted with the disorder remaining at each temperature.

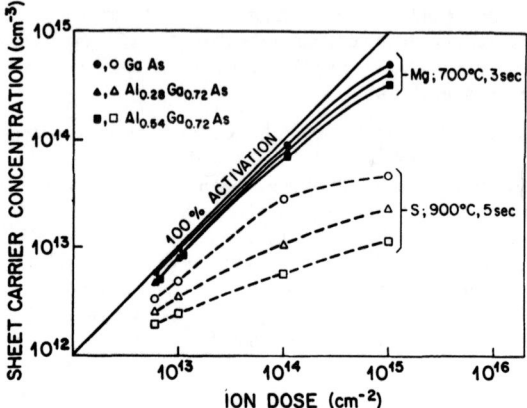

Figure 7.   Sheet carrier densities in Mg or Si implanted $Al_xGa_{1-x}As$ as a function of ion dose and Al composition.

Chemical and Damage-Induced Isolation

There are two different mechanisms for the production of high-resistivity regions in compound semiconductors using ion implantation. The first relies on the implantation of a species which either by itself, or in combination with impurities or dopants already present in the material, creates a chemical deep level state. The second method which is more commonly used, uses ion bombardment by neutral species like H, B, or O to create damage-related deep levels in the material. In either case compensation results from the trapping of free carriers by the deep level centers, and these levels are not thermally ionized at device operating temperatures. The two methods are somewhat complementary in that bombardment-induced isolation is effective to a temperature at which the damage anneals out, typically $\leq 600°C$ in GaAs, whereas chemically-induced isolation requires substitutionality of the implanted species, which occurs at $\geq 600°C$.

In n-type GaAs, chemical compensation can be achieved by implantation of Cr or Fe. These have rather limited ion ranges however, and are not generally useful in device applications. In p-type GaAs there are in general no deep donor elements that can be implanted in order to create thermally stable high-resistivity material. For the specific case of Be-doped GaAs however, oxygen implantation at doses such that the oxygen concentration is above the Be concentration causes formation of a deep Be-O donor complex with an energy level near $E_V + 0.59eV$. This complex is not found with any other acceptor or donor dopant.

In n-type InP chemical compensation can be achieved with Fe implantation and annealing[17], although once again the layer thickness over which this can be achieved is limited by the relatively large mass of the Fe. There is little available information on chemical compensation of p-type InP.

In n-type AlGaAs, oxygen implantation at concentrations above the donor density creates high resistivity material by formation of a deep acceptor level at $E_c - 0.49eV$ which compensates the electrons from the donor dopants[18]. Figure 8 shows the sheet resistivity of an AlGaAs (2000Å, $n = 1.5 \times 10^{18}$ cm^{-3}) - GaAs (5000Å, $p = 10^{14}$ cm^{-3}) structure on a semi-insulating GaAs substrate as function of oxygen or nitrogen implant dose (40,200 and 400keV implant energy) and subsequent annealing temperature. For low doses ($<10^{13}$ cm^{-2}

at 40keV) the evolution of the resistivity follows the usual damage-related compensation result. For doses above 10^{14} cm^{-2}, the oxygen concentration is well above the donor concentration and this correlates with a considerably greater thermal stability of the now high-resistivity AlGaAs. The fact that this is a chemical effect is obvious from the result with N implantation, where the variation of the resistivity of the structure is exactly as expected for damage-only compensation with a complete return of the initial conductivity for temperatures above 700°C. The microstructure of the oxygen-related complex is at present unknown[18]. This effect does not occur for p-type AlGaAs.

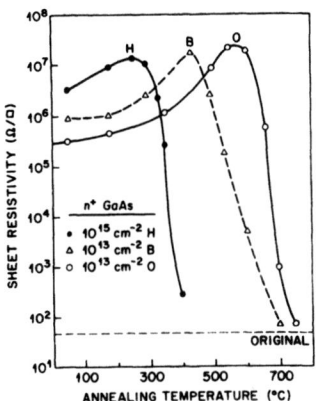

Figure 8.   Sheet resistivity of AlGaAs-GaAs structure implanted with O or N at various doses as a function of post-implant annealing temperature (60s anneals). In each case the dose for 200 and 400keV ions was 2 and 3 times, respectively, the dose for the 40keV ions.

Damage-induced compensation is widely used in device isolation schemes. In general for GaAs and AlGaAs heavier ions are observed to have higher carrier removal rates and higher thermal stability of the compensation effect than for protons, for example. This is shown in Figure 9 for ion bombardment of n-type GaAs with H, B, or O. A 200keV O$^+$ ion creates enough damage to trap or compensate 10-30 electrons, and for doses in the range $10^{13} - 5 \times 10^{13}$ cm^{-2} the maximum resistivity is achieved after annealing at ~500°C. Below this temperature hopping conduction between closely spaced damage sites leads to a low mobility conductivity in the material. Annealing up to 500°C reduces this hopping conduction and the resistivity increases to a maximum. Above 500°C the damage site density falls below the electron density, and the resistivity returns to its initial value. Implant damage in GaAs and AlGaAs creates electron and hole traps so that both n- and p-type material can be made semi-insulating.

Ion bombardment is not as effective in creating high resistivity material in InP[7]. In p-type material high resistances can be achieved with strict control of the dose, while limiting values of the resistivity of $10^3 - 10^4$ Ωcm are obtained in n-type material. In$_{0.53}$Ga$_{0.47}$As has a small bandgap (~0.7eV), and only very low resistivities are achievable by ion bombardment, while InAlAs can be made highly resistive[19,20].

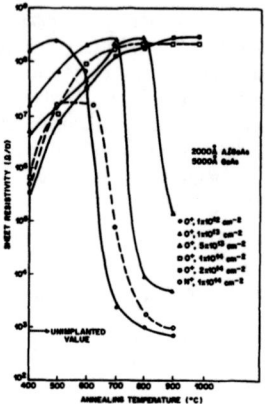

Figure 9. Evolution of sheet resistance in H, B, or O bombarded n-type GaAs as a function of post-implant annealing temperature.

Future Directions

There are a number of areas in which emphasis is likely to be placed in the future:

(a) Schottky barrier enhancement - we have recently observed very large increases in the Schottky barrier height upon high dose As^+ ion implantation into GaAs, followed by annealing at 800°C. At As^+ doses of 10^{17} cm^{-2}, a barrier height in excess of the GaAs bandgap was obtained for TiPtAu contacts. This appears to be due to the formation of a thin, near-surface layer with properties such that extremely high barrier heights are possible.

(b) Co-implantation with acceptor species - while co-implantation has been studied over a long period[21,22], the results for donor species are generally somewhat variable. However for acceptor species, particularly Be, co-implants of As (in GaAs and AlGaAs) or P (in InP) lead to higher carrier concentrations and much reduced diffusion upon annealing because of the greater degree of substitutionality of the acceptor[23-25]. As an example, Figure 10 shows the effect of P co-implantation with Be into InP, after annealing at 850 and 900°C[25], and the advantage over a single Be implant is obvious. The formation of highly doped, thin p$^+$ regions are required for a number of electronic (HBTs) and photonic (APDs) devices. It is likely that carbon implants will play a role here because of the very low diffusivity of this dopant ($\sim 10^{-16}$ cm$^2 \cdot$ S^{-1} at 800°C in GaAs). Co-implantation of Ga is necessary to achieve high doping concentrations with C.

(c) MeV implantation - buried doped layers have been formed in GaAs by 1-6 MeV implants of Si and S without the creation of much damage to the near-surface region because of the relatively small amount of nuclear stopping for high energy ions[26,27]. Normal energy implants can then be used to form doped layers from the surface to the buried MeV implant region. The creation of thick high resistivity regions by MeV ion bombardment may be even more useful.

(d) Ternary and quaternary compounds - the increasing use of InGaAs, InAlAs, InGaAsP and related materials in photonic devices requires a greater understanding of the activation efficiencies and diffusivities of implanted dopants than currently exists.

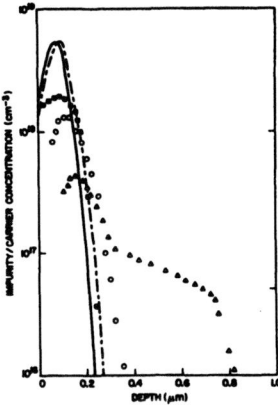

Figure 10.   Carrier profiles for P + Be (100 + 20keV at 6×10^{13} cm^{-2}) implanted InP
after annealing at 850°C, 15 sec (solid squares) or 900°C, 5s (open circles).
The profile for a Be implant only (open triangles) after annealing at 850°C,
15 sec, the LSS profile for Be (continuous line) and P (broken line) are also
included - from ref. 25.

SUMMARY

Table 3 summarizes the most commonly used implant species, the activation
temperatures and efficiencies for GaAs, AlGaAs and InP. In the future the limitations to
down-scaling of devices imposed by lateral straggling of implanted ions under masks are
likely to become more evident, and fundamental issues related to the limitations of carrier
concentration in n-GaAs and p-InP need to be understood.

Table 3.   Summary of Implantation Characteristics in GaAs, AlGaAs and InP - after
Donnelly[7].

	GaAs	AlGaAs	InP
Common Implant Species	Si(RT)	Si(RT)	Si(RT)
	Be(RT)	Be(RT)	Be(RT)
	Se(200°C)		Se,Zn(200°C)
Activation temperature			
- Donor	850°C	900°C	750°C
- Acceptor	750°C	800°C	700°C
Activation Percentage			
Donors - low dose	80%	70%	· 80%
- high dose	$n_{max} = 2 \times 10^{18}$ cm^3	$n_{max} = 1 \times 10^{18}$ cm^{-3}	$n_{max} = 2 \times 10^{19}$ cm^3
Acceptors - low dose	100% (Be)	90%	60%
- high dose	$p_{max} = 2 \times 10^{19}$ cm^3	$p_{max} = 8 \times 10^{18}$ cm^{-3}	$p_{max} = 2 \times 10^{18}$ cm^{-3}

REFERENCES

1. D. V. Morgan and F. H. Eisen, Galluim Arsenide, ed M. J. Howes and D. V. Morgan (Wiley & Sons, NY, 1985), Chapter 5.

2. S. J. Pearton, Solid State Phenomena, *1+2* 247 (1988).

3. D. K. Sadana, Nucl. Instr. Meth. in Phys. Res. B*7/8* 375 (1985).

4. J. S. Williams, Laser Annealing of Semiconductors, ed. J. M. Poate (Academic Press, Sydney 1984), Chapter 10.

5. D. E. Davies, Nucl. Instr. Meth in Phys. Res. B*7/8* 387 (1985).

6. S. J. Pearton, J. M. Gibson, D. C. Jacobson, J. M. Poate, J. S. Williams and D. O. Boerma, Mat. Res. Soc. Symp. Proc. *51* 351 (1986).

7. J. P. Donnelly, Nucl. Instr. Meth. *182/183* 553 (1981).

8. K. G. Stephens, Nucl. Instr. Meth. *209/210* 589 (1983).

9. R. Anholt, P. Balasingam, S. Y. Chou, T. W. Sigmon and M. Deal, J. Appl. Phys. *64* 3429 (1988).

10. R. T. Blunt, Solid State Devices, ed. P. Balk and O. G. Folberth (Elselvier, The Netherlands 1986) pp. 133-148.

11. W. Wesch, E. Wendler, G. Gotz and N. D. Kekelidse, J. Appl. Phys. *65* 519 (1989).

12. D. H. Rosenblatt, W. R. Hitchens, R. E. Anholdt and T. W. Sigmon, IEEE Electron Dev. Lett. *9* 139 (1988).

13. S. J. Pearton, J. S. Williams, K. T. Short, S. T. Johnson, D. C. Jacobson, J. M. Poate, J. M. Gibson and D. O. Boerma, J. Appl. Phys. *65* 1089 (1989).

14. F. Sette, S. J. Pearton, J. M. Poate and J. E. Rowe, Phys. Rev. Lett. *56* 2637 (1986).

15. K. D. Cummings, S. J. Pearton and G. P. Vella-Coleiro, J. Appl. Phys. *60* 163 (1986).

16. R. Bensalem and B. J. Sealy, Vacuum *11* 921 (1986).

17. J. P. Donnelly and C. E. Hurwitz, Solid State Electron *21* 475 (1978).

18. S. J. Pearton, M. P. Ianuzzi, C. L. Reynolds, Jr. and L. Peticolas, Appl. Phys. Lett. *52* 395 (1988).

19. M. V. Rao, R. S. Babu, H. B. Deitrich and P. E. Thompson, J. Appl. Phys. *64* 4755 (1988).

20. B. Tell, T. Y. Chang, K. F. Brown-Goebeler, J. M. Kuo and N. J. Sauer, J. Appl. Phys. *64* 3290 (1988).

21. R. Heckingbottom and T. Ambridge, Rad. Eff. *17* 31 (1973).

22. E. Stoneham, G. Patterson and J. Gladstone, J. Electron. Mater. *9* 371 (1980).

23. J. Kasahara, K. Taira, Y. Kato, M. Arai and N. Watanabe, Japan J. Appl. Phys. *32* L373 (1983).

24. S. Adachi and S. Yamahata, J. Appl. Phys. *64* 3312 (1988).

25. K. W. Wang, Appl. Phys. Lett. *51* 2127 (1987).

26. P. E. Thompson and H. B. Dietrich, J. Electrochem. Soc. *135* 1240 (1988).

27. S. Tong Lee, G. Braunstein and S. Chen, Mat. Res. Soc. Symp. Proc. *126* 183 (1988).

COMPENSATION IN GaAs/GaAlAs HETEROSTRUCTURES BY ION IMPLANTATION :
COMPARISON OF OXYGEN AND BORON

B. DESCOUTS, J. TASSELLI*
Centre National d'Etudes des Télécommunications
Laboratoire de Bagneux
196 avenue Henri Ravera - 92220 BAGNEUX - FRANCE

* LAAS - 7 avenue du Colonel Roche - 31055 TOULOUSE

ABSTRACT

Ion implantation has been used to form an insulating layer in GaAs/GaAlAs heterostructures for bipolar transistor applications with the aim of reducing the base-collector capacitance. Two ions have been compared : boron and oxygen. In both cases magnesium has been implanted to contact the base layer and rapid thermal annealing has been used to activate this dopant. We show that the base-collector capacitance can be lowered by a factor of ~ 2 with oxygen, but high oxygen doses ($\geqslant 10^{14}$ ions/cm^2) are necessary to obtain reproducible results. The capacitance is lowered by a factor of ~ 4 with optimized boron dose. With high boron doses ($\geqslant 10^{13}$/cm^2) we have decreased the capacitance by a factor of ~ 6 but the defects created during the implantation affect the properties of the emitter and base layer.

INTRODUCTION

The performances of heterojunction bipolar transistors are limited by extrinsic parasitic elements and more particularly by the capacitance of the base-collector region. Among the different solutions oxygen or hydrogen ion implantation in the collector region has been reported by several authors [1-6]. Particularly, P.M. Asbeck et al [1] have lowered the base-collector capacitance by a factor of 2.5 (at 0V) with oxygen implantation and Nakajima et al [2] have reported a lowering factor of 2.6 (at 0V) for hydrogen doses $\geqslant 5 \times 10^{12}$ ions/cm^2. The advantage of the compensation by oxygen is its high thermal stability (up to 900°C [1,5]) which makes it adequate for a planar technology, as it avoids a technological step. As a matter of fact, Mg and 0 implantations can be performed using the same implantation mask, followed by an annealing step at high temperature. However, oxygen has the disadvantage of requiring high doses ($\geqslant 10^{14}$ ions/cm^2) to achieve a significant decrease of the capacitance. So, in the case of a mesa technology where no high temperature anneal is necessary boron or hydrogen ions are more suitable than oxygen. Boron has been proved to be efficient for compensation in GaAs devices at low doses ($\sim 10^{12}$ B$^+$/cm^2) [6,7] and it has a higher thermal stability than hydrogen, up to 500°C which is around the usual temperature used for contact alloying.

We have therefore compared the results obtained with oxygen and boron implantation in GaAs/GaAlAs heterostructures. In order to work in similar conditions both for boron and oxygen and to be able to compare the same characterizations of the compensation we have used only a planar technology. Mg ion implantation has been used to contact the base layer.

In addition to capacitance measurements, we have carried out electrochemical profiling thus obtaining the carrier profile in each layer of the heterostructure. We have also performed SIMS measurements to study the behaviour of oxygen and boron during the annealing step.

EXPERIMENTAL PROCEDURES

Mg^{24}, O^{16} and B^{11} were implanted at room temperature using a High Voltage Engineering 400 kV accelerator. The samples were 7° off the incidence direction to minimize channeling effects. In consideration to the thickness of the different layers of the heterostructures (table I) we have chosen the following conditions for Mg implantations : 30, 60, 200 keV at doses of $8x10^{13}$, $2x10^{14}$, $8x10^{14}$ Mg^+/cm^2 respectively. To reach the collector layer it has been necessary to perform the oxygen implantation with an energy of 550 keV, we have used doubly charged oxygen ions. We varied the oxygen dose from $2.5x10^{12}$ to $2x10^{14} O^+/cm^2$.

Boron has been implanted with an energy of 320 keV and doses from $1x10^{11}$ to $2x10^{13}$ B^+/cm^2. Mg has been activated by a rapid thermal anneal, a peak temperature at 900°C (unless when notified) in a commercial halogen lamps furnace |8| . The method to measure the temperature and the configuration used during RTA have been described elsewhere |9| . In the case of oxygen, the anneal has been performed after the two implantations (Mg and O) but in the case of boron the Mg activation has been performed before boron implantation. In order to simulate the heat treatment used for contact alloying and to verify the thermal stability of the compensation, the boron implanted samples have been annealed at 500°C for 1 to 15 mn.

The compensation of the collector layer has been verified by two methods : we have measured the carrier concentration profiles in the heterostructures with a Polaron semiconductor profile plotter, and we have determined the capacitance of the extrinsic base-collector region. Mesa etching has been used to prepare individual diodes of area ($120x120\mu m^2$).

SIMS analysis were performed before and after anneal with a cameca IMS 3F ion microanalyser equipped with a cesium source.

TABLE I : EPITAXIAL LAYER STRUCTURES USED IN THIS WORK

LAYER	Al FRACTION	TYPE	DOPANT	CONCENTRATION	THICKNESS
CONTACT LAYER	0	n^+	Si	$1-3 \times 10^{18} cm^{-3}$	0.50–0.58 μm
COLLECTOR	0	n^-	Si	$1-3 \times 10^{16} cm^{-3}$	0.39–0.52 μm
BASE	0	p^+	Be	$4-5 \times 10^{18} cm^{-3}$	0.19–0.23 μm
EMITTER	0.3	n	Si	$3-6 \times 10^{17} cm^{-3}$	0.20–0.33 μm
CONTACT LAYER	0	n^+	Si	$3-5 \times 10^{18} cm^{-3}$	0.10–0.15 μm

RESULTS AND DISCUSSION

a - Compensation by oxygen

Figure 1 shows the carrier concentration profile for a sample implanted with Mg only (curve a) and for a sample implanted with Mg and O with a dose of $5x10^{13} O^+/cm^2$ (curve b). On curve a, in the p-type region the hole profile corresponds to Mg up to about 0.4μm and to Be (dopant of the base) from 0.4μm up to 0.8μm. We have previously reported |5| the "stairs-like" shape of the Mg^+ profile and the broad Be^+ profile ; and attributed this to an interaction between Be and Mg leading to an anomalous diffusion of these two impurities. In the n-type region the lower doping level (n∼ $2x10^{16} cm^{-3}$) corresponds to the collector layer and the higher doping level (n ∼ $4x10^{18} cm^{-3}$) corresponds to the collector contact layer. After the oxygen implantations (curve b), the hole profile is identical but we do not have a p-n junction. We measure a very low hole concentration with a corresponding large depletion region. We can measure the carrier profile in the collector contact layer, but it is necessary to etch all the collector layer which is highly compensated.

From this figure it is clear that we can form an insulated layer by oxygen implantation. However the minimum oxygen dose to obtain such a compensation is not reproducible from an heterostructure to another (with the same doping level in the collector layer), and can vary in a wide range (from 5×10^{12} to 1×10^{14} ions/cm^2). We found that it is necessary to implant a high oxygen dose ($> 1 \times 10^{14}$ ions/cm^2) to obtain reproducible results. It seems that the compensation is strongly dependant on the material quality.

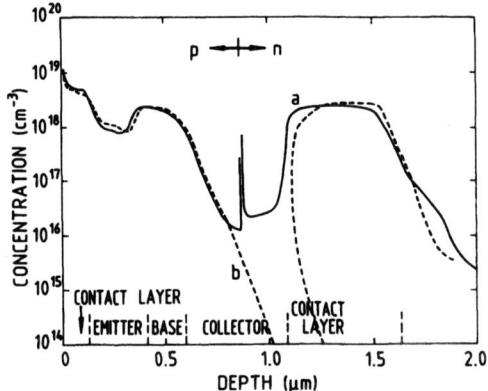

Fig. 1 : Carrier concentrations as a function of depth after anneal at 900°C, (a) for a sample implanted with Mg, (b) for a sample implanted with Mg and O (550 keV, 5×10^{13} O$^+$/cm^2).

b - Compensation by boron

In this case, the experiments have been carried out as follows : 1) Mg ion implantation, 2) annealing at 900°C (peak temperature), 3) B ion implantation, 4) annealing at 500°C (to simulate contact alloying).

For Boron doses lower than 1×10^{12} ions/cm^2 the carrier concentration profile is identical to the profile reported as curve a in figure 1, it is to say that we have not an insulated layer. With a boron dose equal to 1×10^{12} ions/cm^2 the annealing step following boron implantation is quite critical. After annealing at 500°C for 1 mn we have an insulated layer (carrier concentration profile identical to curbe b of figure 1). But if we increase the annealing duration up to 5 mn the collector layer is no longer compensated.

We have therefore implanted higher boron doses. With $5x10^{12}B^{+}/cm^{2}$ the compensation is stable in a wider range of time (up to 15 mn). With a boron doses of $1x10^{13}$ ions/cm^{2}, the current voltage characteristics of the Schottky diode (electrolyte-semiconductor) show a strongly compensated material with an n-type from the middle of the emitter layer up to the collector layer. This can be attributed to an enhanced hopping conduction due to the defects created by the high dose boron implantation. But it is clear that the emitter and the base layers are strongly pertubated, which is, of course, unacceptable for HBT's process.

It seems then, from these carrier profile measurements, that the optimum boron dose is equal to $5x10^{12}$ ions/cm^{2}.

c - SIMS measurements

In order to lower the detection limit for oxygen by SIMS we have implanted the isotope 18 of oxygen. The atomic profiles of O^{18} and Si^{28} are shown in figure 2.a. Two facts must be pointed out : first, a peak of oxygen at about 1μm appears after anneal ; it can be attributed to an oxygen accumulation at the interface collector contact-collector contact layer during the anneal. Secondly, in the collector contact layer oxygen diffuses toward the interface collector contact layer - S.I. substrate. This underlines the importance of the quality of the interfaces and can explain the non reproducibility of the results obtained with low oxygen doses.

The atomic profiles of boron before and after anneal are shown in figure 2.b. There are all identical and do not show any accumulation or diffusion of boron during the anneal.

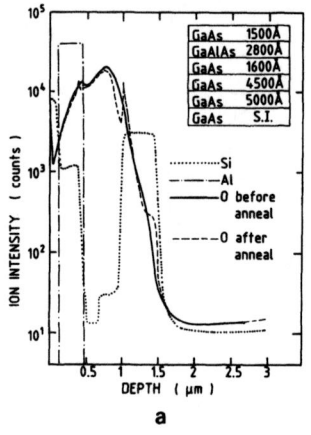 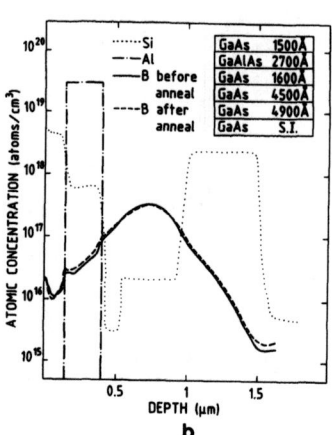

a b

Fig. 2 : SIMS atomic concentration as a function of depth before and after anneal
a.- for a sample implanted with Mg^{24} and O^{18} (550 keV, 5x10^{13} O^{+}/cm^{2}) after RTA at 750°C
b.- for a sample implanted with Mg^{24} and B^{11} (320 keV, 1x10^{13} B^{+}/cm^{2}) after RTA at 750°C or classical furnace anneal at 500°C for 1 mn (the two anneals give the same concentration profile).

d - Capacitance measurements

We have performed oxygen and boron implantations at various doses in the same heterostructure whose collector layer was 4200 A thick with a Si concentration of 4×10^{16} cm^{-3}. Figure 3 shows the behaviour of capacitance versus voltage for a sample implanted with Mg only (curve a) and for samples implanted with 0 (curves b and c) and boron (curves d, e and f).

The capacitance decreases with increasing both oxygen and boron doses. For the oxygen dose of 2×10^{14} 0^{+}/cm^{2} the capacitance is almost independant of bias voltage, indicating that we have formed an insulated layer. At zero bias, the capacitance with oxygen is equal to 3.5 pF, i.e. 2.3 times lower than the value obtained without oxygen (8 pF).

With boron implantation, the capacitance is totaly independant of bias voltage for the two higher doses (5×10^{12} and 2×10^{13} ions/cm^{2}) and is much lower as compared to the samples implanted with oxygen. The capacitance at zero bias is decreased by a factor of 4.2 with a dose of 5×10^{12} B^{+}/cm^{2} and a factor of 5.8 with a dose of 2×10^{13} B^{+}/cm^{2} as compared to the sample implanted with Mg only. So, eventhough the emitter and base layers are strongly pertubated with boron doses equal to or higher than 10^{13} ions/cm^{2} (as we have shown in paragraph b) the capacitance measurements indicate a satisfactory result. This underlines the importance and usefulness of the carrier profile measurements.

We also see on figure 3 that the higher boron dose (2×10^{13} ions/cm^{2}) is not very efficient as compared to 5×10^{12} B^{+}/cm^{2}. As a matter of fact the capacitance decreases from 1.9 pF to 1.38 pF eventhough the dose increases by a factor of 40. We end up again with an optimum boron dose around 5×10^{12} ions/cm^{2}.

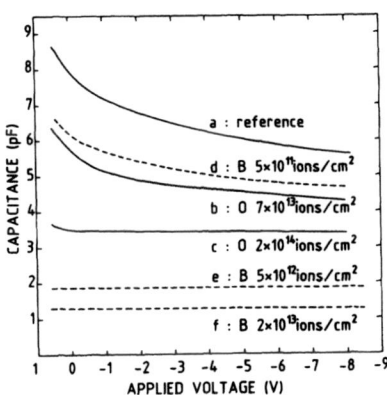

Fig. 3 : Capacitance versus voltage for diodes structures fabricated with a sample implanted with Mg only (a) ; for samples implanted with Mg and 0 (b,c) and for samples implanted with Mg and B (d, e,f).

278

CONCLUSION

We have compared in this work the compensation of the base-collector region of HBT's structures obtained with boron and oxygen ion implantation.

We have shown that both ion allows to obtain high resistivity layers in GaAs/GaAlAs heterostructures. The compensation obtained with oxygen is stable at high temperature (up to 900°C) but it is very dependant on the quality of the starting material and to obtain reproducible results it is necessary to implant high oxygen doses ($\geq 10^{14}$ ions/cm^2).

On the other hand a low boron dose of 5×10^{12} ions/cm^2 decreases the capacitance by a factor of 4.2 (as compared to 2.3 with oxygen). This dose seems to be the optimum value.

The carrier profile measurements have given important informations on each separate layer of the heterostructures. They indicate a strong pertubation of the emitter and base layers for high boron doses ($\geq 10^{13}$ ions/cm^2).

The authors acknowledge F. BAMOUENI for his help in the technological steps, P.M. GAUNEAU and S. GODEFROY for SIMS measurements. They would like to thank Dr F. ALEXANDRE and J. RIOU who furnished the heterostructures.

References

1 P.M. ASBECK, D.L. MILLER, R.J. ANDERSON, F.H. EISEN, IEEE Electron. Device Lett. EDL-5, 310 (1984).
2 O. NAKAJIMA, K. NAGATA, Y. YAMAUCHI, H. ITO, T. ISHIBASHI, IEEE Transactions on Electron. Devices, ED-34 (12), 2393 (1987).
3 M.F. CHANG, P.M. ASBECK, K.C. WANG, G.J. SULLIVAN, D.L. MILLER, Electron. Lett., 22(22), 1173 (1986).
4 Y. YAMAUCHI, K. NAGATA, O. NAKAJIMA, H. ITO, T. NITTONO, T. ISHIBASHI, Electron. Lett., 23 (17), 881 (1987).
5 B. DESCOUTS, N. DUHAMEL, Y. GAO, Journal de Physique, Colloque C4, 49, 437 (1988).
6 F. CLAUWAERT, P. VAN DAELE, R. BAETS, P. LAGASSE, J. Electrochem. Soc. : Solid State Science and Technology, 134(3), 711 (1987).
7 E.V.K. RAO, N. DUHAMEL, J. Appl. Phys. 49 (7), 3898 (1978).
8 Halogen lamps furnace from Atelier d'Electrothermie : AET Grenoble (France).
9 B. DESCOUTS, N. DUHAMEL, K. DAOUD-KETATA, P. KRAUZ, S. GODEFROY, J. Appl. Phys. 60, 450 (1986).

Correlation Between Defect Characteristics and Layer Intermixing In Si Implanted GaAs/AlGaAs Superlattices

SAMUEL CHEN, S.-TONG LEE, G. BRAUNSTEIN, G. RAJESWARAN and P. FELLINGER
Corporate Research Laboratories, Eastman Kodak Company, Rochester, New York 14650

ABSTRACT

Defects induced by ion implantation and subsequent annealing are found to either promote or suppress layer intermixing in III-V compound semiconductor superlattices (SLs). We have studied this intriguing relationship by examining how implantation and annealing conditions affect defect creation and their relevance to intermixing. Layer intermixing has been induced in SLs implanted with 220 keV Si^+ at doses $> 1 \times 10^{14}$ ions/cm^2 and annealed at 850°C for 3 hrs or 1050°C for 10 s. Upon furnace annealing, significant Si in-diffusion is observed over the entire intermixed region, but with rapid thermal annealing layer intermixing is accompanied by negligible Si movement. TEM showed that the totally intermixed layers are centered around a buried band of secondary defects and below the Si peak position. In the near-surface region layer intermixing is suppressed and is only partially completed at $\geq 1 \times 10^{15}$ Si/cm^2. This inhibition is correlated to a loss of the mobile implantation-induced defects, which are responsible for intermixing.

INTRODUCTION

The presence of ion-implanted impurities can, after annealing, induce layer intermixing in GaAs/AlGaAs superlattice (SL) structures [1, 2, 3]. The potential use of this technique to modify the opto-electronic properties of III-V materials is of much interest for fabricating devices such as heterostructure lasers [4]. By carefully selecting parameters such as the implantation energy (keV or MeV), the ion dose (ranging from 1×10^{14} to 1×10^{17} /cm^2), and annealing conditions (furnace annealing (FA) or rapid thermal annealing (RTA)), layer intermixing can be induced at select depths below the surface and with varying widths [5, 6, 7]. However, the layer intermixed region does not necessarily increase with an increase in impurity concentration or annealing times, and is especially true for the near-surface region where the enhanced Al-Ga intermixing process appears inhibited [8, 9, 10]. Microstructural investigation of the intermixed region has shown the presence of dislocation loops in the vicinity of the intermixed layers but these defects bear no clear relationship to the mixed region [11, 12]. While the enhanced Al-Ga intermixing has been attributed to impurity diffusion [13], radiation damage [8, 14] and Fermi level effect [15], their relative importance and the nature and concentration of defects on the enhancement as well as the inhibition of the intermixing phenomenon remain unclear. We have studied the morphology of the implantation damage, and in this paper we report the interplay and relevance of the intermixed zone with the implanted impurities as well as with the lattice defects. In particular we have investigated the microstructure in the near-surface region to probe for possible features that can cause mixing inhibition.

EXPERIMENTAL

A 1.6 μm thick SL, with the composition GaAs (200Å) / $Al_{0.5}Ga_{0.5}As$ (200Å), grown by molecular beam epitaxy on undoped (100) GaAs wafer, was implanted 7° off-normal axis at room temperature using 220 keV Si^+. The dose ranged from 3×10^{14} to 3×10^{15} /cm^2. In order to reduce beam heating to the sample, the ion current was decreased to about 0.5 μA/cm^2 and the back of the sample was heat-sink glued to the holder. We have determined, under similar implantation conditions, that the temperature of the sample does not rise more than 20-30°C during irradiation. Subsequently, the samples were annealed either by RTA at 1050°C, 20 s (N_2 ambient) or by FA at 850°C, 3 h (H_2/Ar ambient) with the SL surface placed face down on another GaAs substrate. The Si impurity depth profile and the extent of the

Al-Ga interlayer diffusion was monitored by secondary ion mass spectrometry (SIMS) using a 5.5 keV O_2^+ beam for sputtering, while the microstructural morphology of the dislocation loops and defects in the near-surface region were imaged by transmission electron microscopy (TEM) operated at 200 kV. Samples suitable for TEM observation were prepared by mechanically polishing and dimpling the cross-section until it was about 20 μm thick and then Ar ion thinning the sample while it was cooled by liquid nitrogen to reduce ion milling damage.

RESULTS

The cross-section impurity concentration and defect morphology profiles of the Si implanted SLs have been examined, and significant compositional and structural differences, associated with layer intermixing, are found to exist between RTA and FA processed samples.

With RTA, in the SLs containing a dose of 3 x 10^{14} /cm^2 (Fig. 1(a)), the near-surface region (layers 1 to 3 beneath the surface) was free of any secondary defects and contained individual GaAs and AlGaAs layers with sharp interlayer boundaries. In the intermediate region (layers 4 to 12) the SL layers are well resolved even though this region had a dense network of interstitial dislocation loops. Beneath this region (layers 13 and deeper) the SL appeared unaffected by implantation damage. At the higher dose of 1 x 10^{15} /cm^2 (Fig 1(b)), the dislocation-free near-surface region was wider (layers 1 to 6) and contained distinct individual III-V layers although their layer boundaries appeared slightly diffuse. Dispersed in the GaAs layers were small near-spherical defects (20-200 Å in diameter) and they are found as deep as the fifth and sixth SL layers. In the intermediate region (layers 7 to 13), the GaAs and AlGaAs layers were still distinct although their boundaries have become more diffuse. Dislocation loops (with sizes ranging from 150 to 450 Å) are observed between layers 2 and 13. Further down, the layers appeared to be undamaged, and the sharpness of the layer boundaries was similar to that found in the pristine SL. At the highest dose of 3 x 10^{15} /cm^2, the (200) dark field micrograph showed a near-surface region (layers 1 to 6) that contained diffuse layer boundaries between the GaAs (dark contrast) and the AlGaAs (light contrast) layers (Fig. 1(c)). Near-spherical defects (~200 Å in diameter), primarily in GaAs layers, were also visible down to the sixth SL layer. The intermediate region (layers 7 to 13) had a uniform contrast, indicating a totally intermixed region. Interstitial dislocation loops, lying in the (111) lattice planes, are observed between the surface and the fourteenth layer. Most of these dislocation loops were large (~500 Å in diameter) and showed signs of having coalesced with nearby dislocation loops, presumably to reduce their surface strain energy with the lattice. With further increase in depth, the SL layer boundaries became increasingly sharp, and beyond the sixteenth layer, the SL layers appeared undamaged.

Layer intermixing can be directly correlated with the oscillation of the Al signal in SIMS profile. The Al and Si signals for the SL implanted at 3 x 10^{15} Si/cm^2 are shown in Fig. 1(d), which has the same depth scale as the TEM micrographs in Fig. 1(c). From the surface down to the seventh layer, the Al signal oscillation rapidly dampened, indicating that the Al-Ga interlayer diffusion is increasing with depth. Between the eighth and the twelfth layer, the signal was flat, showing a totally mixed region, and beyond the thirteenth layer, the Al signal quickly recovered to that seen in unimplanted SLs. The center of the totally mixed region is at a depth of 0.43 μm beneath the surface, corresponding approximately to the tenth layer. The Si concentration profile, upon RTA processing, showed no detectable impurity movement with respect to the as-implanted SL [8]. Since the R_p position is located at 0.28 μm, the completely intermixed zone is found below the Si peak position.

The extent of layer intermixing in the furnace-annealed SL, containing a dose of 3 x 10^{15} Si/cm^2, is seen in Fig. 2. The (200) dark field image showed that the near-surface region (layers 1 to 6) is significantly intermixed with very diffuse layer boundaries. Distinctly seen are the small near-spherical defects (~200 Å in diameter), located predominantly in the GaAs layers. Underneath, a totally intermixed region (with uniform contrast) is found from the eighth to the twenty-fourth layer. In this sample, the interstitial dislocation loops, lying in the (111) planes, are observed from the surface to a depth of 0.7 μm (or approximately down to the seventeenth layer). Dislocation loops were large (~ 600 Å) and showed signs of having coalesced with each other. None was found near the lower boundary

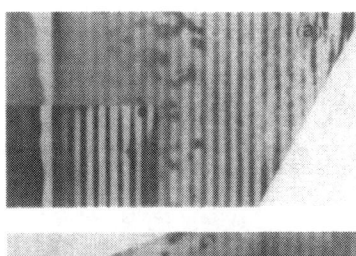

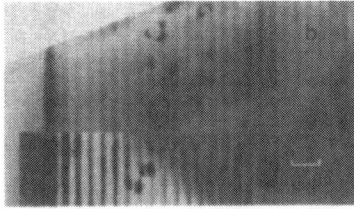

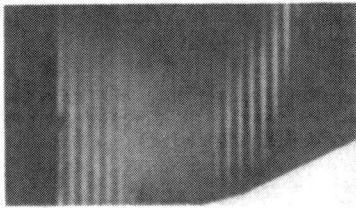

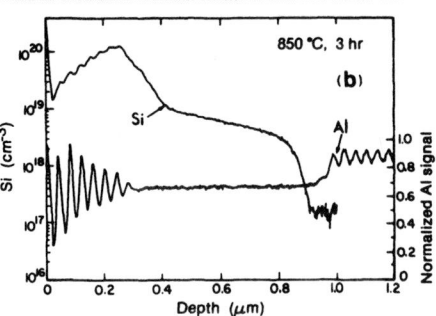

Fig. 2 Si implanted and furnace annealed (850°C, 3 h) SLs.

(a) (200) dark field image of the SL implanted with 3×10^{15} Si/cm^2, showing the position of the totally layer intermixed region relative to the spherical defects and dislocation loops.

(b) SIMS Si and Al profiles of the SL implanted with 3×10^{15} Si/cm^2.

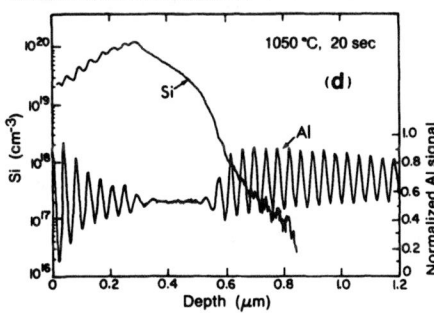

Fig. 1 Si implanted and rapid thermally annealed (1050°C, 20 s) SLs.

(a) TEM micrograph of the SL containing a dose of 3×10^{14} /cm^2.

(b) Cross-section image of the SL containing a dose of 1×10^{15} /cm^2 showing the distribution of dislocation loops and small spherical defects. In (a) and (b), superimposed on the bottom left corners are (200) dark field images of the near-surface region.

(c) (200) dark field image of the SL with a dose of 3×10^{15} /cm^2, showing the position of the totally layer intermixed region relative to the spherical defects and dislocation loops.

(d) SIMS Si and Al profiles of the SL implanted with 3×10^{15} Si/cm^2.

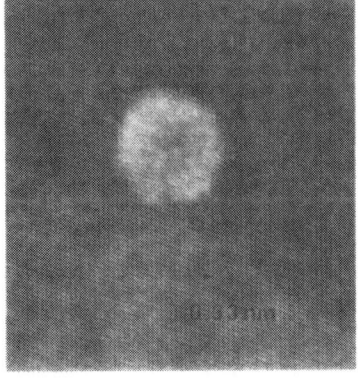

Fig. 3 High resolution image of a spherical defect (in a GaAs layer) in the near-surface region, showing its non-crystalline central region.

of the completely intermixed region. Beneath the mixed region (twenty-fifth layer and deeper), the sharpness of the layer boundaries quickly recovered to the level found in the pristine SL. In the corresponding SIMS profiles, Fig. 2(b), the oscillation in the Al signal quickly dampened from the surface down to the seventh layer; became flat from the eighth to the twenty-fourth layer; and beyond the twenty-fifth layer, the Al signal rapidly recovered to that seen in unimplanted SL. The center of the totally mixed region is at a depth of 0.6 μm beneath the surface. The impurity concentration profile showed little Si diffusion from the surface to the R_p position, but with increasing depth extensive Si in-diffusion took place (Fig. 2(b)).

The small near-spherical defects, observed in the near-surface region for the SLs implanted at doses $\geq 1 \times 10^{15}$ Si/cm^2, were further characterized by TEM. Weak-beam imaging using either the (111) or (200) diffracted beam revealed no fringes inside these defects, and high resolution lattice imaging showed that these defects have a non-crystalline central region (Fig. 3). Even though this image was formed from a SL region with very little crystalline III-V lattice above and below the spherical defect, the presence of amorphous surface contaminants makes it unclear whether these defects actually contain any amorphous III-V material or are small cavities in the lattice. Unlike dislocation loops, these defects did not induce significant strain on the surrounding III-V lattice and become observable only in thin cross-section SL regions. (Based on thickness contour contrast, this value is estimated to be less than 2000 Å.)

DISCUSSION

The presence of small near-spherical defects containing a non-crystalline core region seen in the implanted SL has not been reported, to the best of the authors' knowledge, in implanted bulk GaAs. Our own observations in implanted and annealed GaAs (220 keV Si at a dose of 3 $\times$ 10^{15} /cm^2, processed at 850°C, 3 h), however, did show the presence of such defects [16]. While the cause of these defects is as yet unclear, several features have been noted.

We have attempted to analyze the composition of these non-crystalline defects by energy dispersive spectroscopy, using a 100 Å diameter focussed electron probe. But significant beam spreading to the surrounding GaAs matrix prevented a conclusive determination of any difference in composition between the defect and its surrounding. We have, however, found that these defects are not necessarily Si-rich in composition. In the first place, our TEM observations (Fig. 2(a)) showed that while these defects are formed at implantation doses $\geq 1 \times 10^{15}$ Si/cm^2, yet once formed, their distribution through the first six GaAs layers beneath the surface did not follow the increasing Si concentration profile. Secondly, these defects are distributed approximately equally in the near-surface GaAs layers which contained Si concentrations ranging between 0.6 to 10 $\times$ 10^{19} Si/cm^3 (first to sixth GaAs layers in SLs implanted at doses of 1 $\times$ 10^{15} and 3 $\times$ 10^{15} /cm^2). Lastly, in the absence of Si, we have also observed these spherical defects in a SL implanted with 192 keV Al to a dose of 3 $\times$ 10^{15} /cm^2 [17]. These results suggest the possibility that the spherical defects may be due to annealing of the lattice damage formed in the near-surface region.

Aside from the lack of correlation to a specific dopant species, we have noted that these lattice imperfections are non-crystalline, even after annealing at 1050°C (Fig. 3). This points to their thermal stability and suggests that their local composition is not the stoichiometric 1Ga:1As, since either epitaxial or twinned re-growth would have occurred, as has been observed in the regrowth of implantation-amorphized GaAs [12, 20]. Aside from implanted SLs, Mei et al [21] recently reported spherically shaped defects, resembling those observed here, in the near-surface region of a GaAs/AlGaAs SL containing a grown-in Si concentration of 1 $\times$ 10^{20} /cm^3. They suggested that these defects may be Si-rich. While further characterizations are needed to determine the composition of these defects, we have noted that they are found near the surface in highly implanted or possibly in highly Si-doped SLs [21]. This may point to the important role of the surface in terms of its contribution to the annealing of defects in GaAs materials located close to the surface.

Layer intermixing inhibition can be seen in Figs. 1(c,d) and 2(a,b), primarily in the first seven SL layers of the implanted and annealed GaAs/AlGaAs superlattices. In this region the Si concentration profile showed little annealing-induced diffusion, and the dampening of the oscillating Al signal did not follow the Si implantation profile. Inhibition is further demonstrated by comparing the width and position of the totally intermixed region in the RTA and FA processed SLs. At a dose of 3×10^{15} Si/cm^2, rapid thermal annealing induced total layer intermixing over a region approximately 0.2 μm wide (seventh to twelfth layers, Fig 1(c,d)), while under FA conditions mixing extended about 0.6 μm (eighth to twenty-fourth layers, Fig 2(a,b)). This difference in width occurred primarily due to changes to the lower boundary of the totally mixed region. Were it not due to surface mixing inhibition, the upper boundary of the totally mixed region also would have been different between RTA and FA processed SLs.

The near-spherical defects and intermixing inhibition have several characteristics in common in the room-temperature implanted and annealed SLs. First, both are observed in the near-surface region (extending approximately 0.3 μm from the surface, Figs. 1(d) and 2(b)). Second, both are seen at similarly high concentrations of impurity (at implant doses ≥ 1×10^{15} Si/cm^2). Furthermore, it is of interest to note that layer intermixing and Si diffusion near the surface was found to be more extensive in SLs implanted at a dose of 3×10^{14} Si/cm^2, and annealed with SiO$_2$ cap, than in SLs implanted at higher doses [6]. On the other hand, in the absence of impurities or damages, layer intermixing, as observed in variable depth single quantum well heterostructures [19], showed decreasing magnitude with increasing distance away from the surface. Thirdly, mixing inhibition and the appearance of defects are observed under both RTA and FA conditions (Figs. 1(c,d) and 2 (a,b)), as discussed above. These correlations point to the possibility that the formation of these near-spherical defects during annealing can affect the concentration of available Ga vacancies that are believed to contribute to the observed enhanced Al-Ga interlayer diffusion [15]. If these defects are actually cavities in the lattice, formed during annealing, then they are effectively similar to open surfaces which can act as sinks for fast diffusing point defects so that in the vicinity of the spherical defects, the concentration of vacancies (whether formed at the surface or from the implantation process) becomes low resulting in the observed mixing inhibition in the near-surface region. Further studies to correlate the presence of such defects with mixing inhibition in SLs implanted at different temperatures are currently in progress.

The position of the observed dislocation loops relative to the depth and width of the layer-intermixed region, as a function of annealing conditions, suggests that these crystalline lattice defects do not have a direct relevance to the enhanced Al-Ga diffusion. In the RTA processed SLs containing a dose of 3×10^{15} /cm^2, dislocation loops are found from the surface down to a depth of about 0.6 μm while the totally mixed region is seen between 0.3 and 0.6 μm (Fig. 1(c)). On the other hand, upon furnace annealing, dislocation loops are not found near the lower boundary of the totally intermixed zone, which has extended down to about 0.9 μm (Fig. 2(a)). We have recently found that upon implantation and rapid thermal annealing, the depth distribution of dislocation loops, lying in the (111) lattice planes, can be correlated to the lattice damage caused by the momentum transfer of the implantation process [22]. The lattice damage, manifested as dislocation loops after annealing, follows the nuclear energy loss profile so that the defect peak density coincides approximately with the R_p position. Such a distribution of these defects is, in fact, clearly observed in the SL containing a dose of 3×10^{14} /cm^2, Fig. 1(a), where the band of dislocation loops are seen between the fourth and twelfth SL layers, and is centered around R_p. Previously, we have shown that in MeV Si-implanted SLs the dislocation loop bandwidth increases with an increase in ion dose as the near-surface region becomes increasingly dislocation loop free [18]. By comparison, the implantation damage at 220 keV is closer to the surface so that the annealing-induced annihilation of the dislocation loops by the surface becomes more significant. This is consistent with our observation that with an increase in implant dose (≥ 1×10^{15} /cm^2) the defect bandwidth increased; the dislocation sizes also increased, together with a decrease in defect density (due to the coalescing of dislocation loops, Fig. 1 (b,c)).

At a dose of 3×10^{15} Si/cm^2, our data also revealed that dislocation loops are found deeper in FA than in RTA processed SLs, and this suggests that such defects can be formed in regions of extensive dopant diffusion. This is consistent with the observations made in Si doped SL [21].

In summary, we have shown evidence that two types of lattice defects can form in ion-implanted and annealed GaAs/AlGaAs SLs. In the near-surface region, small near-spherical defects, possibly cavities in the lattice, are dispersed primarily in GaAs layers at ion doses ≥ 1×10^{15} /cm^2. These defects appear to correlate well with the observed inhibition of the enhanced Al-Ga diffusion process in layer intermixing. On the other hand, the formation of dislocation loops, caused by implantation damage or impurity diffusion, have no direct relevance to layer intermixing.

ACKNOWLEDGEMENTS

The authors would like to thank Professor T. Y. Tan and Dr. A. W. West for valuable discussions. we would also like to also acknowledge J. Madathil for technical assistance in ion implantation and J. J. DeJohn for TEM sample preparation.

REFERENCES

[1] J. J. Coleman, P. D. Dapkus, C. G. Kirkpatrick, M. D. Camras and N. Holonyak, Jr., Appl. Phys. Lett. **40**, 904 (1982).

[2] K. Kash, B. Tell, P. Grabbe, E. A. Dobisz, H. G. Craighead and M. C. Tamargo, J. Appl. Phys. **63**, 190 (1988).

[3] Samuel Chen, S.-Tong Lee, G. Braunstein and G. Rajeswaran, Mater. Res. Symp. Proc., Dec, 1988.

[4] D. F. Welch, D. R. Scifres, P. S. Cross and W. Streifer, Appl. Phys. Lett. **51**, 1401 (1987).

[5] T. Venkatesan, S. A. Schwarz, D. M. Hwang, R. Bhat, M. Koza, H. W. Yoon, P. Mei, Y. Arakawa and Y. Yariv, Appl. Phys. Lett. **49**, 701 (1986).

[6] K. Matsui, J. Kobayashi, T. Fukunaga, K. Ishida and H. Nakashima, Jpn. J. Appl. Phys. **26**, L1122 (1987).

[7] R. P. Bryan, M. E. Givens, J. L. Klatt, R. S. Averback and J. J. Coleman, J. Electron. Mater. **18**, 39 (1989).

[8] S.-Tong Lee, G. Braunstein, P. Fellinger, K. B. Kahen, and G. Rajeswaran, Appl. Phys. Lett. **53**, 2531 (1988).

[9] S. A. Schwarz, T. Venkatesan, D. M. Hwang, H. W. Yoon, R. Bhat and Y. Arakawa, Appl. Phys. Lett. **50**, 281 (1987).

[10] S. A. Schwarz, T. Venkatesan, R. Bhat, M. Koza, H. W. Yoon, Mater. Res. Soc. Symp. Proc. **56**, 321 (1986).

[11] Y. Arakawa, J. S. Smith, A. Yariv, N. Otsuka, C. Choi, B. P. Gu and T. Venkatesan, Appl. Phy. Lett. **50**, 92 (1987).

[12] J. Ralston, G. W. Wicks. L. F. Eastman, B. C. De Cooman and C. B. Carter, J. Appl. Phys. **59**, 120 (1986).

[13] D. P. Deppe and N. Holonyak, Jr., J. Appl. Phys. **64**, R93 (1988)

[14] J. Cibert, P. M. Petroff, D. J. Werder, S. J. Pearton, A. C. Gossard and J. H. English, Appl. Phys. Lett. **49**, 223 (1986).

[15] T. Y. Tan and U. Gosele, Appl. Phys. Lett. **52**, 1240 (1988).

[16] Samuel Chen and S.-Tong Lee, unpublished data.

[17] Samuel Chen and S.-Tong Lee, to be submitted to Appl. Phys. Lett.

[18] S.-Tong Lee, Samuel Chen, G. Rajeswaran, G. Braunstein, P. Fellinger and J. Madathil, Appl. Phys. Lett. **54**, 1145 (1989).

[19] L. J. Guido, N. Holonyak, Jr., K. C. Hsieh and J. E. Baker, Appl. Phys. Lett. **54**, 262 (1989).

[20] D. K. Sadana, T. Sands and J. Washburn, Appl. Phys. Lett. **44**, 523 (1984).

[21] P. Mei, S. A. Schwarz, T. Venkatesan, C. L. Schwarz, J. P. Harbison, L. Florez, N. Theodore and C. B. Carter, Appl. Phys. Lett. **53**, 2650 (1988).

[22] Samuel Chen, G. Braunstein and S.-Tong Lee, Mater. Res. Soc. Symp. Proc., Dec, 1988.

IMPLANT DAMAGE IN AlGaAs BASED SUPERLATTICES AND ALLOYS AT 77K

E. A. Dobisz[*], H. Dietrich[*], A. W. McCormick[**], J. P. Harbison[***]
[*]Naval Research Laboratory, Washington, D. C. 20375
[**]Universal Energy Systems, 4401 Dayton-Xenia Rd., Dayton, Ohio 45432
[***]Bellcore, Red Bank, NJ 07701

ABSTRACT

Previously, it was shown that superlattices implanted with Si at 77K, exhibited more extensive damage and uniform compositional mixing upon subsequent annealing than samples implanted at room temperature.[1,2] The current work focuses on the damage in samples implanted with Si at 77K. The study shows that for a given dose, the amount of damage depends upon the layer thickness and the composition. Specimens of bulk GaAs, $Al_{.3}Ga_{.7}As$, 7.5 nm GaAs - 10 nm $Al_{.3}Ga_{.7}As$ superlattice (SL1), 5.5 nm GaAs -3.5 nm AlAs superlattice (SL2), and 8.0 nm GaAs - 8.0 nm AlAs superlattice (SL3) were implanted at 77K with 100 KeV Si, with doses ranging from 3 X 10^{13} cm^{-2} to 1 X 10^{15} cm^{-2}. The samples were examined by ion channelling and cross sectional transmission electron microscopy (TEM). At 77K and a dose of 1 X 10^{14} cm^{-2}, the GaAs and SL1 showed an amorphous layer, while no damage peak was observed in SL2. The 77K amorphization thresholds of the $Al_{.3}Ga_{.7}As$ alloy, SL2, and SL3 were 2.5 X 10^{14} cm^{-2}, 4 X 10^{14} cm^{-2}, and 1 X 10^{15} cm^{-2} respectively. The sharpness of the amorphization threshold varied with the material.

INTRODUCTION

Impurity enhanced compositional disordering [1 - 6] is important as a means of patterning a heterostructure without interrupting the single crystal structure with an interface, as caused by etching. Due to the control in depth and lateral space [1,3,7], ion implantation is an important method to introduce the impurities. Several authors have suggested an interaction between the ion implant damage and the interdiffusion [1,2,5,6], while others have used the damage itself to induce the interdiffusion [7,8]. It is important to understand the damage introduced by the implantation.

Previously we reported that superlattices implanted with Si at 77K showed more extensive and uniform compositional disordering than those implanted at room temperature.[1,2] Because there is little dynamic annealing at 77K, Williams and Austin [9] have shown that GaAs samples develop amorphous layers at lower implant doses than at room. The low dose, amorphous 77K implanted samples required lower annealing temperatures to remove the implant damage than the room temperature equivalent samples [9].

The present work focuses on lattice damage in the AlGaAs based superlattices resulting from ion implantation at 77K. K. Matsui et. al [10] have shown that room temperature implanted superlattices exhibited less damage than GaAs and the amount of damage was correlated to average alloy composition. B. C. DeCooman et. al [6] reported that the implant damage centered about the GaAs layers in GaAs -AlGaAs superlattices implanted with Si at room temperature. In this work, we examine the evolution of damage with implant dose in GaAs, an $Al_{.3}Ga_{.7}As$ alloy, and superlattices of 7.5 nm GaAs -10 nm $Al_{.3}Ga_{.7}As$, 5.5 nm GaAs -3.5 nm AlAs, and 8.0 nm GaAs -8.0 nm AlAs. The present work shows that layer thickness and barrier composition are also important in the damage behavior.

EXPERIMENTAL

Mat. Res. Soc. Symp. Proc. Vol. 147. ©1989 Materials Research Society

The samples examined consisted of GaAs, $Al_{.3}Ga_{.7}As$, 10 nm $Al_{.3}Ga_{.7}As$ -7.5 nm GaAs superlattice (SL1), 5.5 nm GaAs - 3.5 nm AlAs superlattice (SL2), and a 8 nm GaAs - 8 nm AlAs superlattice(SL3). All samples except the bulk GaAs were grown by MBE. The samples were implanted with 100 KeV Si^{28}, 0.7 $\mu A/cm^{-2}$, with doses ranging from 3 X 10^{13} $/cm^{-2}$ to 1 X $10^{15}/$ cm^{-2} at 77K. The sample holder consisted of a hollow Cu block that was filled with liquid nitrogen, to which samples were attached with MUNG II heat sinking compound. A summary of the samples, implant doses and RBS results, discussed below, are given in Table I.

The implant damage profiles were measured with backscattered ion beam channelling, about the (100) axis, using 2 MeV He^{++}, at a 168° scattering angle, with detector resolution of 4 keV. Cross sectional samples for transmission electron microscopy were made by mechanically thinning, followed by Ar ion milling at 77K.

RESULTS

TABLE I.   SUMMARY OF RBS CHANNELLING RESULTS

Sample	Implant Dose cm^{-2}	X Damage Peak	Minimum X
GaAs	Unimplanted	-----	0.04
	3 X 10^{13}	-----	0.09
	7 X 10^{13}	0.86	0.38
	1 X 10^{14}	0.99	0.48
	2.5 X 10^{14}	1.00	0.52
	4 X 10^{14}	1.00	0.52
	1 X 10^{15}	1.00	0.53
AlGaAs	Unimplanted	----	0.04
	3 X 10^{13}	----	0.08
	7 X 10^{13}	0.33	0.15
	1 X 10^{14}	0.86	0.41
	2.5 X 10^{14}	1.00	0.49
SL1	Unimplanted	----	0.03
	3 X 10^{13}	----	0.09
	7 X 10^{13}	0.86	0.38
	1 X 10^{14}	1.00	0.45
	2.5 X 10^{14}	1.00	0.50
	1 X 10^{15}	1.00	0.53
SL2	Unimplanted	-----	0.03
	3 X 10^{13}	-----	0.08
	7 X 10^{13}	-----	0.09
	1 X 10^{14}	-----	0.10
	2.5 X 10^{14}	0.87	0.39
	4 X 10^{14}	1.00	0.47
	1 X 10^{15}	1.00	0.47
SL3	Unimplanted	-----	0.03
	3 X 10^{13}	-----	0.07
	7 X 10^{13}	0.31	0.11
	1 X 10^{14}	0.59	0.26
	2.5 X 10^{14}	0.85	0.41
	4 X 10^{14}	0.92	0.46
	1 X 10^{15}	1.00	0.48

The RBS channelled and random orientation yield spectra of GaAs implanted with different doses are shown in Figure 1. The GaAs is exemplary of all the samples in that: the unimplanted sample exhibits a minimum X (ratio of the channelled yield to the random orientation yield) of 0.04 and the spectrum of the sample implanted with a dose of 3×10^{13} cm^{-2} shows no damage peak. The minimum X of GaAs implanted with a dose of 3×10^{13} cm^{-2} was $\approx 9\%$. GaAs exhibits a damage peak, with a X of 0.86, at a dose of 7×10^{13} cm^{-2}. The height of the damage peak, in GaAs reaches the random yield at a dose of $\approx 1 \times 10^{14}$ cm^{-2}, indicating an amorphous region. The width of the damage peak increases with higher doses of 2.5×10^{14} cm^{-2} and 1×10^{15} cm^{-2}.

There are significant differences in the damage in the materials. As an example, the channelled and random yield spectra of SL2 are shown in Figure 2. No damage is observed for doses as high as 1×10^{14} cm^{-2}. A large damage peak with a X of 0.87 appears at a dose of 2.5×10^{14} cm^{-2}. A dark field (200) TEM micrograph, shown in Figure 3, shows a highly damaged region near the surface, which has retained sufficient crystallinity for AlAs to diffract electrons. The damage peak reaches the random yield at a dose of 4×10^{14} cm^{-2} and broadens as the dose increases to 1×10^{15} cm^{-2}.

Summarized in Table I, is the X value for the damage peak and the minimum X value for the region immediately beneath the damage region for each sample and implant dose. The channelled spectra of SL1 is identical, within experimental error, to that of GaAs for all doses. A dark field (200) TEM micrograph SL1, implanted at 2.5×10^{14} cm^{-2}, shown in Figure 4, shows an amorhous layer, characterized by the total absence of the (200) reflection, of 150 nm width, at the surface of the sample, consistent with the RBS. In the two GaAs layers beneath the amorphous region there is evidence of some damage.

The behavior of the damage peak in the different materials is shown in Figure 5, where the X of the damage peak is plotted vs. implant dose for the different samples. At a given dose prior to amorphization, the height of the damage peak of the Al$_{.3}$Ga$_{.7}$As alloy, SL2, and SL3 is less than that of GaAs. The Al$_{.3}$Ga$_{.7}$As alloy exhibits a lower amount of damage than GaAs, but develops an amorphous layer at a dose of 2.5×10^{14} cm^{-2}. SL2

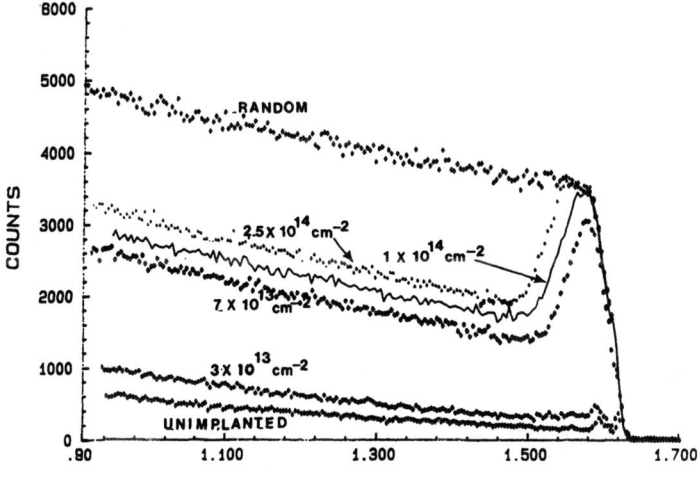

Figure 1. RBS spectra of GaAs samples implanted with 100 KeV Si at 77K. The channelled spectra about the (100) axis are shown for different implant doses.

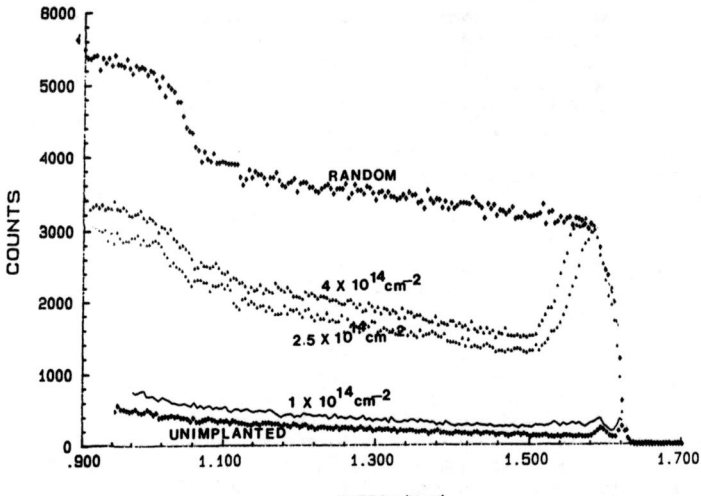

Figure 2. RBS spectra of 3.5 nm GaAs - 5.5 nm AlAs superlattice implanted with 100 KeV Si at 77K. The channelled spectra about the (100) axis are shown for different implant doses.

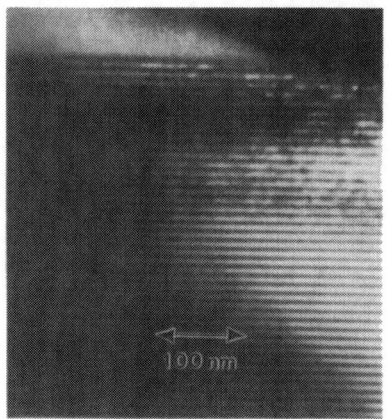

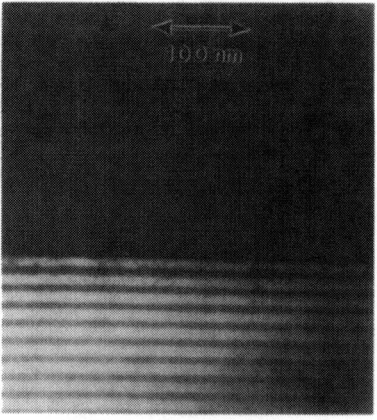

Figure 3. Dark field (200) TEM micrograph of 5.5 nm GaAs - 3.5 nm AlAs superlattice implanted with 100 KeV Si at 77K with a dose of 2.5×10^{14} cm^{-2}.

Figure 4. Dark field (200) TEM micrograph of 7.5 nm GaAs - 10 nm Al$_{.3}$Ga$_{.7}$As superlattice implanted with 100 KeV Si at 77K with dose of 2.5×10^{14} cm^{-2}.

shows only the sudden appearance of a damage peak at a dose of 2.5×10^{14} cm^{-2} and the development of an amorphous layer at a dose of 4×10^{14} cm^{-2}.

At low doses the damage peak of SL3 follows that of the Al$_{.3}$Ga$_{.7}$As alloy, but at higher doses SL3 dose not develop an amorphous region until a dose of 1×10^{15} cm^{-2} is reached. Shown in Figure 6 is a dark field (200) TEM micrograph of SL3 implanted with a dose of 2.5×10^{14} cm^{-2}. Although, there is evidence of damage, there is still strong (200) reflection from

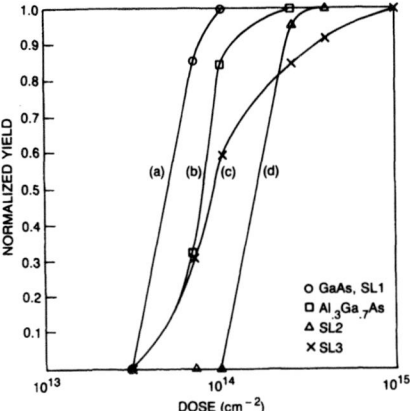 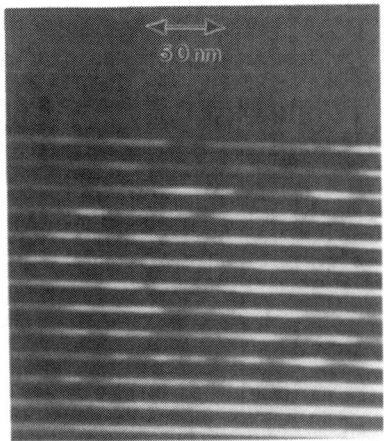

Figure 5. Height of damage peak in the channelled RBS spectra, normalized to the random yield vs. dose for 100 KeV Si at 77K. The curves connecting the data points are for visual aid.

Figure 6. Dark field (200) TEM micrograph of 5.5 nm GaAs - 3.5 nm AlAs superlattice implanted with 100 KeV Si at 77K with a dose of 2.5×10^{14} cm^{-2}.

the AlAs layers, indicating a large degree of crystallinity. The GaAs layers in the implanted region are darker (less (200) reflection) than those beneath the implanted region. The width of the damage peak of SL3 is always greater than that of SL2, even after SL2 has become amorhous.

DISCUSSION

Since the scattering cross section of Al is less than that of Ga, one would expect samples with a higher Al content to exhibit less damage. A TRIM [11] simulation predicts only a 30 % variation in the average number of displaced atoms per ion between GaAs and AlAs targets. The effects will be much larger if secondary scattering cross sections (i. e. Ga - Ga vs. Ga - Al or Al - Ga) are considered. However, the difference in elemental scattering cross sections cannot entirely explain the data, since the effects are not monotonic with average composition.

The TEM micrographs show some evidence that the GaAs layers in the superlattices damage more readily than the AlAs layers. Again, this does not explain the evolution of damage in all of the superlattices. One would expect the damage peak to reflect the amorphization thresholds of the two different layers. In this case, the superlattices would exhibit a damage peak at the same doses that produced a damage peak in GaAs. The damage peak would be expected to grow, in proportion to the fraction of the SL period, comprised of GaAs. However, the damage peak of the superlattices would not be expected to reach the random yield until the amorphization threshold of the $Al_xGa_{1-x}As$ layer. Such an explanation may explain the behavior of SL3, but not SL1 and SL2.

Since the differences in the materials for SL2 and SL3 are only layer thickness, it appears that interfacial strain may also play an important role, particularly at 77K. Hamdi et. al [12] have reported that lattice strain in $Al_xGa_{1-x}As$ stabilized implant strain at room temperature. In SL2, the damage of the GaAs layers is repressed as evidenced by the lack of a damage peak, unless implanted at high doses.

CONCLUSIONS

We have shown GaAs to have an amorphization threshold of 1×10^{14} cm^{-2} with 100 KeV Si at 77K. The 77K amorphization thresholds of the Al$_{.3}$Ga$_{.7}$As alloy, SL1, SL2, and SL3 were 2.5×10^{14} cm^{-2}, 1×10^{14} cm^{-2}, 4×10^{14} cm^{-2}, and 1×10^{15} cm^{-2} respectively. The sharpness of the amorphization threshold varied with the material. No dynamic annealing has been observed at 77K. The effects may in part be due to a larger scattering cross section of Ga than Al. There is evidence that the GaAs layers damage more readily than the Al$_x$Ga$_{1-x}$As layers. However, it neither of these explanations can describe the total behavior of all of the samples. Layer thickness and barrier composition also play an important role, suggesting interfacial strain.

REFERENCES

1. E. A. Dobisz, B. Tell, H. G. Craighead, S. A. Schwarz, M. C. Tamargo, and J. P. Harbison, Mat. Res. Soc. Symp. Proc., 77, 423 (1987).

2. E. A. Dobisz, B. Tell, H. G. Craighead, and M. C. Tamargo, J. Appl. Phys., 60, 4150 (1986).

3. E. A. Dobisz, H. G. Craighead, S. A. Schwarz, P. S. D. Lin, K. Kash, L. M. Schiavone, A. Scherer, J. P. Harbison, SPIE, 797, 194 (1987).

4. W. D. Laidig, N. Holonyak, M. D. Camras, K. Hess, J. J. Coleman, P. D. Dapkus, J. Bardeen, Appl. Phys. Lett., 38, 776 (1981).

5. T. Venkatesan, S. A. Schwarz, D. M. Hwang, R. Bhat, M. Koza, H. W. Yoon, P. Mei, Y. Arakawa, A. Yariv, Appl. Phys. Lett., 49, 701 (1986).

6. B. C. De Cooman, C. B. Carter, J. Ralston, G. W. Wicks, L. F. Eastman, Mat. Res. Soc. Symp. Proc., 77, (1987).

7. J. Cibert, P. M. Petroff, G. J. Dolan, S. J. Pearton, A. C. Gossard, J. H. English, Appl. Phys. Lett., 49, 1275 (1986).

8. K. Kash, B. Tell, P. Grabbe, E. A. Dobisz, H. G. Craighead, J. Appl. Phys., 63, 192 (1988).

9. J. S. Williams and M. W. Austin, Appl. Phys. Lett., 36, 994 (1980).

10. K. Matsui, S. Takatani, T. Fukunaga, T. Narusawa, Y. Bamba, and H. Nakashima, Jpn. J. Appl. Phys., 25, L391 (1986).

11. J. F. Ziegler, J. P. Biersack, U. Littmark, The Stopping and Range of Ions in Solids, (Pergamon Press, Inc. New York, 1985).

12. A. H. Hamdi, J. L. Tandom, M. -A. Nicolet, Nucl. Instrum. & Methods, B10 - 11, 588 (1985).

ENHANCED INTERDIFFUSION OF GaAs-AlGaAs INTERFACES FOLLOWING
ION IMPLANTATION AND RAPID THERMAL ANNEALING

K. B. KAHEN, G. RAJESWARAN, D. L. PETERSON, L. R. ZHENG, AND M. L. OTT
Corporate Research Labs, Eastman Kodak Company, Rochester NY 14650-2011

ABSTRACT

The interdiffusion of GaAs-AlGaAs interfaces has been shown to be enhanced following ion implantation and rapid thermal annealing at approximately 950°C. A model is presented which explains this phenomenon. It is based
on the solution of coupled diffusion equations involving the excess vacancy
and Al distributions following ion implantation. Both initial distributions
are obtained from the solution of a three-dimensional Monte Carlo simulation
of ion implantation into a heterostructure sample. The model is found to be
in excellent agreement with several sets of experimental data. More specifi-
cally, the model is shown to be valid for ions which do not diffuse apprecia-
bly in the time frame of the rapid thermal annealing and for as-implanted
vacancy concentrations below ~6 x 10^{19} cm^{-3}. Above that concentration, some
vacancies are hypothesized to coalesce, thus being unavailable to assist in
the enhanced interdiffusion process.

I. INTRODUCTION

Integrated optoelectronics based on III-V technology is currently an
active area of research. As the complexity of optoelectronic devices grows,
it is important that the individual elements be simple to manufacture and
integrate. With the development of impurity-induced disordering of lasers
[1], it is possible to produce planar, laterally guided devices. The impurities are introduced either by diffusion from the sample surface or by ion
implantation. The intermixing is energized by multi-hour furnace annealing.
Recently, a number of groups [2-5] have shown experimentally that ion implantation followed by rapid thermal annealing (RTA) also causes intermixing of
GaAs-AlGaAs interfaces. Previously, a model [6] was introduced to explain
this phenomenon and it was shown to be in agreeement with the experimental
data [2,5]. In this paper, the model is compared with additional intermixing
data and used to show that the intermixing phenomenon weakens above an ion
dose which corresponds to a calculated vacancy concentration of approximately
6 x 10^{19} cm^{-3}. For larger ion doses some vacancies are proposed to agglo-
merate in, for example, dislocation loops and, therefore, be unavailable to
assist in the Al-Ga interdiffusion.

II. INTERMIXING MODEL

The model describing intermixing via ion implantation followed by RTA
consists of two parts. A Monte Carlo simulation is used to obtain the distributions of interstitials, vacancies, and occupied lattice sites following
ion implantation into a heterostructure sample. The RTA processing is modeled
by simultaneously solving the diffusion equations for the distributions of
lattice-site Al and vacancies.
The Monte Carlo program is a modified version [7] of TRIM85 [8], which
is capable of tracking the ions and the defects in three dimensions, handling
arbitrarily complicated sample structures, and following collision cascades.
Since this Monte Carlo simulation only produces the distributions of vacancies,
ions, and displaced target atoms, an additional Monte Carlo scheme was devised
to decide whether the displaced target atoms and ions remain on interstitial

sites or whether they recombine with the vacancies to become substitutional atoms. Assuming negligible thermal diffusion during implantation, each vacancy was allowed to recombine only with interstitials which were within 3 Å of its lattice position. Both the type of interstitial and the type of empty lattice site were chosen using a Monte Carlo scheme. After applying this algorithm to each vacancy, each defect atom (Ga, Al, As, and the ion) was assigned either to a neighboring lattice site or remained as an interstitial in its original position.

In Fig. 1, the above algorithms were used to calculate the atomic distributions resulting from the implantation of 220 keV Si^+ ions at a dose of 3×10^{15} cm^{-2} into 50 periods of a 200 Å GaAs-200 Å $Al_{0.5}Ga_{0.5}As$ superlattice. Figure 1a shows a plot of the distributions of Si atoms; vacancies; interstitial Al; and the sum of the three (Al, Ga, and As) interstitial concentrations. The figure also shows a plot of the experimental ion distribution [5] which was measured using secondary ion mass spectrometry (SIMS). In Fig. 1b the distributions of Ga, Al, and As on their respective lattice sites are shown. The experimental and theoretical ion distributions are in good agreement for depths less than the projected range, R_p, but are somewhat different deeper into the sample probably owing to the neglect of channeling effects in the ion implantation program. However, the model will be compared quantitatively with the experimental data [2,3] only for depths less than R_p. Figure 1 also shows that both substitutional and interstitial Al atoms are present in the GaAs layers, which demonstrates that the ion implantation process alone has caused some intermixing of the layers.

The distributions from the Monte Carlo simulations are used as inputs for solving the coupled diffusion equations involving the Al lattice atoms and the five point defects, namely, the Ga, Al, and As interstitials, the vacancies, and the ions. Assuming that the diffusion of the Al is representative of the intermixing process, the diffusion and interaction of the Ga and As lattice atoms are ignored. The equations involving the interstitial atoms and the ions are eliminated based on the following approximations. It is assumed that Ga, Al, As, and many implanted ions, for example, Si, diffuse small distances via interstitial sites in the time frame of RTA, i.e., < 30 seconds. Evidence for this assumption for the case of the implanted ion being

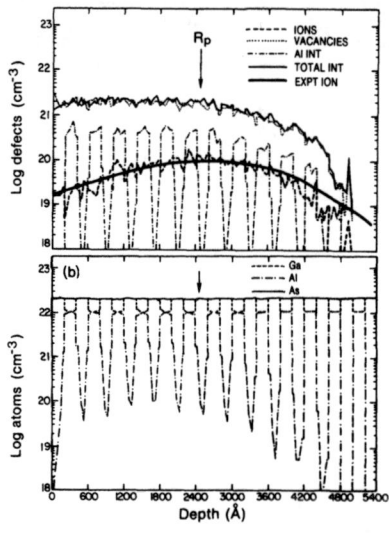

Fig. 1. Calculated results for the atomic density distributions following 220 keV ion implantation of Si at a dose of 3×10^{15} cm^{-2} into a superlattice having 50 periods of 200 Å of GaAs and 200 Å of $Al_{0.5}Ga_{0.5}As$. (a) Point defect distributions. (b) Distributions of atomic Ga, Al, and As on their respective lattice sites.

Si has been presented elsewhere [6]. Consequently, these types of atoms influence the Al interdiffusion process solely by recombining with the excess vacancies. The interaction of the interstitial atoms was included in the diffusion equation for the vacancies by adding to it a recombination term of the form, $(C_v - C_{v,eq}) / \tau$, where C_v is the transient vacancy distribution, $C_{v,eq}$ is the thermal equilibrium vacancy concentration at the annealing temperature, and τ is a phenomenological time constant that describes the decay of the transient vacancy distribution. $C_{v,eq}$ has been determined both experimentally [9] and theoretically [10], while τ was fit to one of the experimental RTA results [5]. The general form of the recombination term was used to include the possibility of the decay of C_v by other mechanisms, such as gettering.

Using the above approximations, the initial set of eight coupled equations reduces to a coupled set between the vacancies and the lattice site Al. This equation set can be writtten as

$$\frac{\partial C_v}{\partial t} = \frac{\partial}{\partial x}\left(D_v \frac{\partial C_v}{\partial x}\right) - \left(C_v - C_{v,eq}\right) / \tau \tag{1a}$$

$$\frac{\partial C_{Al}}{\partial t} = \frac{\partial}{\partial x}\left(D_{Al} \frac{\partial C_{Al}}{\partial x}\right) \tag{1b}$$

where D_v is the vacancy thermal equilibrium diffusion coefficient, C_{Al} is the Al lattice site distribution, and the Al transient diffusion coefficient is given by

$$D_{Al} = D_{Al,eq} \left(C_v / C_{v,eq}\right), \tag{2}$$

where $D_{Al,eq}$ is the thermal equilibrium Al diffusion coefficient. Values for D_v, $C_{v,eq}$, and $D_{Al,eq}$ can be found in the literature [11,9,12]. The two important assumptions made in eqs. (1-2) are that the transient enhancement of the Al diffusion constant is given by eq. (2) and that thermal equilibrium constants remain valid during the transient interdiffusion process. The latter assumption has been discussed in detail, elsewhere [13]. Note that eq. (2) implies that the Al interdiffusion is proportional to the total vacancy concentration, and not just to the concentration of vacancies on type III lattice sites. In this way the As vacancies can assist in the Al-Ga interdiffusion by opening an unobstructed path between the Al and Ga sites. The boundary conditions for eqs. (1-2) are that at the sample surface there is zero net flux for both the time-dependent vacancy and Al concentrations, while deep within the sample, both concentrations remain constant in time [13].

The important point to note is that there is only one parameter, τ, the vacancy decay constant, that is not known from available experimental data. As discussed above, τ is fit to one experimental RTA result and then held constant for all other comparisons with experimental data, provided that the annealing conditions are analogous.

III. RESULTS AND DISCUSSION

The value of τ was determined by comparing the results of the model with the experimental intermixing data of Lee et al. [5]. The best fit to the data was obtained with a value of τ of 4 sec, with an uncertainty of 1 sec. Keeping τ fixed at 4s, the results of the model will be compared with a number of different experimental results. The first comparison will be with the

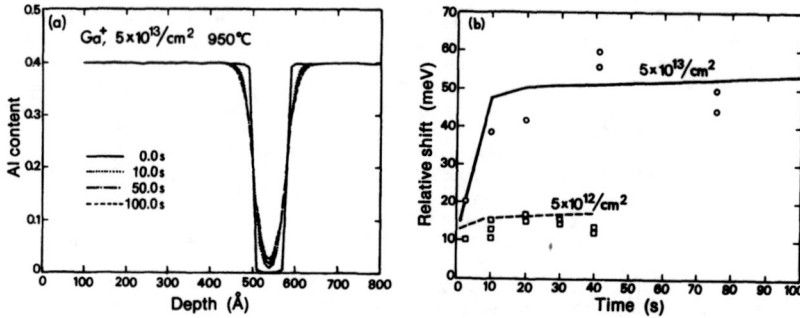

Fig. 2. (a) Time sequence of the calculated change in the Al composition surrounding the quantum well. (b) Energy shift as a function of time between the first electron-heavy hole peak of the implanted and non-implanted areas following RTA. The circles and squares are the experimental data while the solid and dashed lines are the corresponding theoretical results.

experimental data of Cibert et al. [2]. In their experiment they used an 80 Å GaAs quantum well sandwiched between two $Al_{0.4}Ga_{0.6}As$ barrier layers. The sample was implanted with 210 kev Ga^+ ions at doses of 5×10^{13} and 5×10^{12} cm^{-2}. This was followed by RTA for up to 130 s at 950°C. They used cathodoluminescence (< 10 K) to measure the change in the first electron-heavy hole peak following RTA. Instead of reporting the absolute peak values, they plotted the peak shifts relative to an annealed but nonimplanted sample area.

Figure 2a shows the calculated time evolution of the Al composition for a region which surrounds the 80 Å GaAs quantum well. The figure shows that interdiffusion begins rapidly, and then after ~20 s decays quickly to its equilibrium value. This phenomenon is also borne out in Fig. 2b which compares the calculated shifts for both Ga doses with the experimental data. In order to obtain electron-heavy hole peaks from the results given in Fig. 2a, the Numerov process [14] is used to solve Schrodinger's equation numerically for an arbitrary potential profile. The important feature about Fig. 2b is that it shows that the model is in excellent agreement with the experimental data for $\tau=4s$. (For $\tau=3s$, the calculated shifts would be 10% smaller.) This agreement also gives some justification for the simple Monte Carlo interstitial-vacancy annihilation scheme discussed above, the reason being that if nearby vacancies and interstitials did not recombine, then the model would predict a much larger cathodoluminescence shift and, therefore, would not be consistent with the experimental data.

Additional support for the model is presented in Fig. 3 which compares the calculated results, squares, with the experimental data, circles, of Kash et al. [3]. Their sample was a 60 Å GaAs quantum well embedded between two $Al_{0.4}Ga_{0.6}As$ barrier layers. Figure 3 shows the peak shifts (relative to the peaks in the as-grown sample) for the quantum well after RTA annealing for 20 s at 925°C following the implantation of 75 keV Al^+ ions at doses of 2×10^{13}, 1×10^{14}, and 5×10^{14} cm^{-2}. For the two lower doses the model is in very good agreement with the experimental results, while for the highest dose the theoretical value is high by ~20 %. At the dose of 5×10^{14} cm^{-2}, the as-implanted calculated vacancy concentration is ~3 $\times 10^{20}$ cm^{-3}, while at the dose of 1×10^{14} cm^{-2}, the calculated vacancy concentration is ~6 $\times 10^{19}$ cm^{-3}. We postulate that at a vacancy concentration between these two values, some vacancies begin to coalesce and might not be free to assist in the Al inter-

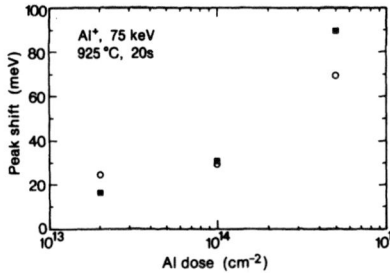

Fig. 3. Energy shift as a function of dose between the first electron-heavy hole peak of the implanted area following RTA and the peak of the as-grown sample. The theoretical and experimental results are squares and circles, respectively.

diffusion. Therefore, at this and larger vacancy concentrations, the model overestimates the number of freely diffusing vacancies and, thus, would predict a greater degree of intermixing.

In order to verify this postulate of a transitional vacancy concentration, an experiment similar to that of Kash et al. [3] was performed. A heterostructure sample was implanted with various doses of 390 keV Ar. The sample was composed of three GaAs quantum wells having thicknesses of 37, 55 and 111 Å (QW1, QW2, and QW3, respectively), surrounded by barrier layers of $Al_{0.36}Ga_{0.64}As$. The sample was capped with 250 Å of GaAs, and QW1, QW2, and QW3 were 750, 1687, and 2642 Å from the surface, respectively. Following implantation the samples were encapsulated with 1000 Å of SiN and subjected to RTA at 950°C for 30 s. Peak shifts were measured analogously to the method of Kash et al. [3]. Figure 4 compares the theoretical shifts with the experimental values for Ar doses of 2×10^{13}, 7×10^{13}, and 1.4×10^{14} cm^{-2}. Results are not given for QW3 because of the large uncertainties associated with small shifts. In Fig. 4 the model was applied with $r = 3$ s. This value is within the uncertainty for r stated above. In the Ar dose experiment, a SiN cap layer was used to prevent As out-diffusion instead of proximity annealing on a fresh GaAs surface [2-5]. This resulted in weaker thermal shifts of the quantum wells, under no-implant conditions, compared with reported experimental values [12]. This difference is postulated to occur as a result of some out-diffusion of As into the SiN cap layer and would produce a lower Ga vacancy concentration in the near-surface region of the sample, resulting in smaller $D_{Al,eq}$ values. Instead of using a different 950°C value for $D_{Al,eq}$ in eq. (2), r was fit to the data for the dose of 2×10^{13} cm^{-2}. The good agreement between the theoretical and experimental results (Fig. 4) for the dose of 7×10^{13} cm^{-2} shows that this procedure appears to be reasonable. The figure also shows that the model begins to overestimate the extent of intermixing for the highest Ar dose, which corresponds to a calculated as-implanted vacancy concentration of $\sim 10^{20}$ cm^{-3}. Therefore, Fig. 4 verifies

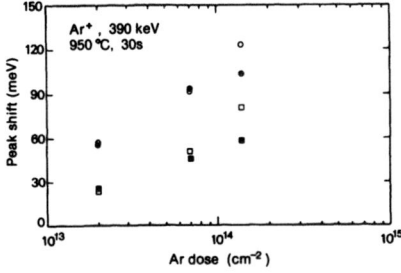

Fig. 4. Energy shifts for QW1 (circles) and QW2 (squares) as a function of dose. The theoretical and experimental results are open and solid symbols, respectively.

the results of Fig. 3 and narrows the proposed upper limit for free vacancy movement to between 6×10^{19} and 1×10^{20} cm^{-3}. Figure 4 also shows that since the model overestimates the degree of intermixing for both QW1 and QW2 at the highest dose, the inhibition of intermixing is not dominated by surface effects.

IV. CONCLUSIONS

In summary, a model for the transient-enhanced interdiffusion of Al at GaAs-AlGaAs interfaces has been presented. The model is found to be in excellent agreement with three independent sets of interdiffusion data. Using the results of the model, the vacancies are postulated to begin to coalesce above an as-implanted concentration of 6×10^{19} cm^{-3} and, therefore some are unavailable to assist in the interdiffusion process.

ACKNOWLEDGEMENTS

We would like to thank M. Demay, C. Derby, B. Higberg, and S. Wilson for technical support.

REFERENCES

1. K. Meehan, J.M. Brown, N. Holonyak, Jr., R.D. Burnham, T.L. Paoli, W. Streifer, Appl. Phys. Lett. 44, 700 (1984).
2. J. Cibert, P.M. Petroff, D.J. Werber, S.J. Pearton, A.C. Gossard, J.H. English, Appl. Phys. Lett. 49, 223 (1986).
3. K. Kash, B. Tell, P. Grabbe, E.A. Dobisz, H.G. Craighead, M.C. Tamargo, J. Appl. Phys. 63, 190 (1988).
4. J. Kobayashi, T. Fukunaga, K. Ishida, H. Nakashima, J.D. Flood, G. Bahir, J.L. Merz, Appl. Phys. Lett. 50, 519 (1987).
5. S.T. Lee, G. Braunstein, P. Fellinger, K.B. Kahen, G. Rajeswaran, Appl. Phys. Lett. 53, 2531 (1988).
6. K.B. Kahen, G. Rajeswaran, S.T. Lee, Appl. Phys. Lett. 53, 1635 (1988).
7. J.P. Biersack, Nucl. Instrum. Methods B19, 32 (1987).
8. J.F. Zeigler, J.P. Biersack, U. Littmark, The Stopping and Range of Ions in Solids, Vol. 1 (Pergamon, New York, 1986).
9. H.R. Potts and G.L. Pearson, J. Appl. Phys. 37, 2098 (1966).
10. R.M. Logan and D.T.J. Hurle, J. Phys. Chem. Solids 32, 1739 (1971).
11. K.B. Kahen, D.L. Peterson, G. Rajeswaran, D.J. Lawrence, submitted to Appl. Phys. Lett.
12. T.E. Schlesinger and T. Kuech, Appl. Phys. Lett. 49, 521 (1986).
13. K.B. Kahen and G. Rajeswaran, to be published in J. Appl. Phys.
14. P.C. Chow, Amer. J. Phys. 40, 730 (1972).

ION BEAM MIXING OF GaAs/AlGaAs SUPERLATTICE AND ITS RELATIONSHIP TO AMORPHIZATION

P.P. Pronko, A.W. McCormick, D.B. Patrizio, and A.K. Rai
Universal Energy Systems, Inc., 4401 Dayton-Xenia Road, Dayton, OH 45432.

R.M. Kolbas and B.S. Frank
Department of Electrical and Computer Engineering, North Carolina State
University (NCSU), Raleigh, NC 27695-7911

ABSTRACT

This work is part of a study to understand the process by which energetic ion bombardment can be used to mix the chemical components of AlGaAs and GaAs superlattice (S/L) layers of nominal 35 to 50Å thickness. Data reported here involve the retention and build-up of collision cascade damage in the S/L and its relationship to amorphization and chemical mixing in these systems.

INTRODUCTION

As part of an investigation to use ion beam mixing as a process to compositionally disorder superlattice layers in GaAs/AlGaAs, we have studied the response of GaAs and $Al_{0.35}Ga_{0.65}As$ individually and as a superlattice, to ion implantation (cascade) damage at room temperature. Gallium ions at 890 and 960 keV were used in the fluence range of 10^{13} to 10^{15} ions/cm^2. These energies were chosen to correspond to implant damage and mixing in 3000Å and 4000Å superlattice stacks where the mean projected ranges (R_p) are 1500Å and 2000Å, respectively. The periodicity of the superlattice was typically in the range of 35 to 50Å.

We observed in these superlattices that the GaAs layers accumulate damage more rapidly than the AlGaAs layers, a result consistent with literature reports of others [1]. There is a need, therefore, to clarify the damage production and retention processes occurring in these systems and the extent to which it is a property of the superlattice or an intrinsic property of the individual materials that make up the superlattice. We demonstrate below that the preferential build up of damage in GaAs is an intrinsic property of the GaAs compared to $Al_{0.35}Ga_{0.65}As$, there being a distinct and radical difference between the room temperature annealing characteristics of these two materials and, therefore, a difference in the amount and the nature of the heavy ion collision damage retained by each.

The intermixing of superlattices as a controllable and spatially selectable process has received considerable interest in the literature [2]. This has been an outgrowth of the discovery that chemical intermixing of superlattices can be achieved by the in-diffusion of certain chemical species [3,4]. In the case of Al in GaAs/AlGaAs, it has been found that Zn [3] or Si [4] act as disordering species. The chemical disordering achieved in this way has also been produced by ion-implanting these same diffusional species and then subjecting the system to a thermal annealing cycle in order to re-distribute them [5]. In so doing, a chemical intermixing of the superlattice occurs. Implantation damage has been found to play some role in the compositional disordering during these annealing cycles [6].

In general, the superlattices of interest will have periodicities of <100Å. In such cases, it is reasonable to expect that cascade intermixing of the AlGaAs with the GaAs layers can be achieved. This should occur as a result of dynamic collision cascade events where net transport of Al occurs thru dense sub-cascade atomic intermixing which is known to occur across dimensions of <100Å. This is especially true in highly miscible systems such as GaAs/AlGaAs. Our interest in using this proposed mixing mechanism resulted in the research being presented in this report.

EXPERIMENTAL

The superlattice (S/L) samples prepared for these experiments were fabricated at North Carolina State University (Electrical Engineering Department) using a Varian MBE system. The structure of the samples was based on having an active GaAs quantum well (QW) approximately 300Å thick positioned above an AlGaAs buffer layer with a superlattice stack above the QW and a top 500Å capping layer of AlGaAs above the S/L. This design was chosen for its relevance to a possible laser device that is being considered for development. Figure 1 is a cross-sectional TEM micrograph of such a sample showing the relative position of the various component layers.

The experiments being reported here involved the use of Ga^+ implantation at relatively high energies (890 and 960 keV) in order to have a heavy ion, capable of producing dense sub-cascades along its damage track, and also sufficiently energetic to penetrate a S/L stack of 3000Å and 4000Å, respectively. The choice of Ga as bombarding ions would also be helpful in explaining the observed S/L intermixing as entirely due to implantation mixing and not due to chemical effects. Implant fluence levels were chosen to produce damage that ranged from lightly dispersed (10^{13} cm^{-2}) to very dense and ultimately to saturated damage (10^{15} cm^{-2}). The dynamic progression of the system to an amorphous phase in the high damage regime was of particular interest. Ga^+ implantation under the aforementioned conditions were also performed in pure GaAs and $Al_{0.35}Ga_{0.65}As$ substrates.

Samples were analyzed by various techniques: RBS channeling (RBS-C), cross-sectional transmission electron microscopy (X-TEM), and sputter Auger analysis.

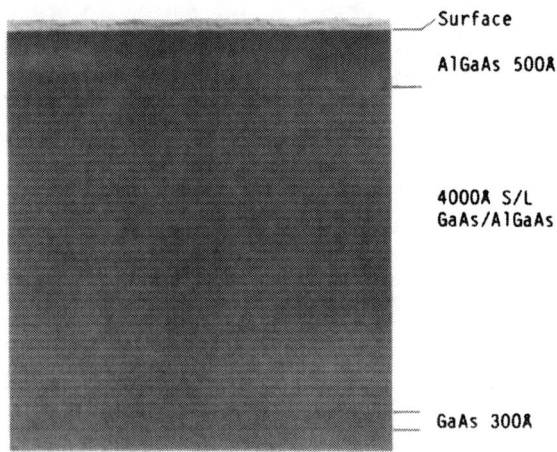

Fig. 1.
X-TEM Bright Field
Micrograph of a
Multi-Quantum
Well/Superlattice
(MQW/SL) Used in
Mixing Experiments

Surface

AlGaAs 500Å

4000Å S/L
GaAs/AlGaAs

GaAs 300Å

RESULTS

Prior to evaluating compositional mixing from the collision cascade process, it is instructive and sensible to determine the extent of damage production in binary GaAs and the fluence levels needed, at these energies, to produce multiple DPA (displacements per atom) events that will, in addition to mixing, ultimately lead to amorphization of the crystalline material. The fluence levels normally used for ion beam mixing of such layers (10¹⁵ to 10¹⁶ cm⁻²) should produce amorphous conditions in GaAs under room temperature bombardment. Figure 2 is an RBS channeling spectrum for bulk GaAs material which was subjected to 960 keV Ga⁺ bombardment to fluence levels of 4x10¹³, 7x10¹³, and 1x10¹⁴ cm⁻². It is clear from these results that a fluence level of approximately 2x10¹⁴ cm⁻² will result in GaAs channeling spectrum with a yield equivalent to the random yield representing an amorphous layer with maximum damage and mixing in the range of 2000 to 4000Å.

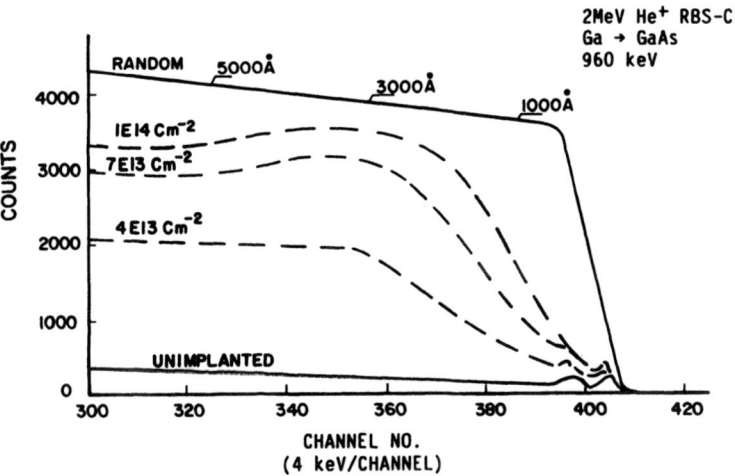

Fig. 2. RBS-Channeling Spectra. 960 keV Implanted GaAs

Comparison of these channeling results for GaAs with that of our S/L (Fig. 3) shows that in the S/L a fluence of 3.88x10¹⁴ cm⁻² results in less than 50% total damage. Further comparison of Fig. 3 with the S/L X-TEM results of Fig. 4 shows that the GaAs quantum well sustains more damage than the AlGaAs portion of the S/L and that the AlGaAs capping layer has retained very little observable damage. This is further supported by the near surface crystal string channeling oscillations in the AlGaAs capping layer.

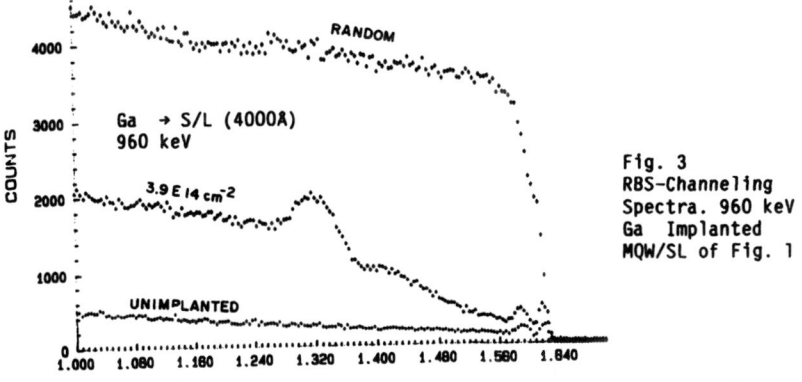

Fig. 3
RBS-Channeling
Spectra. 960 keV
Ga Implanted
MQW/SL of Fig. 1

Additionally, examination of the X-TEM micrograph of Fig. 4 reveals heavy black spot damage clusters in the GaAs quantum wells and the GaAs portions of the S/L but very little clustered damage in the AlGaAs capping layer and the AlGaAs portions of the S/L. In order to further understand these effects, a bulk specimen of AlGaAs was implanted with Ga at 890 keV with fluences ranging from 1×10^{13} cm^{-2} to 1×10^{16} cm^{-2}. Figure 5 shows the RBS channeling spectra obtained from that sample. It is clear from these results that the AlGaAs is building its damage level at a

Fig. 4.
X-TEM MQW/SL
of Fig. 1
After Implant.
960 keV Ga
1E14 cm^{-2}

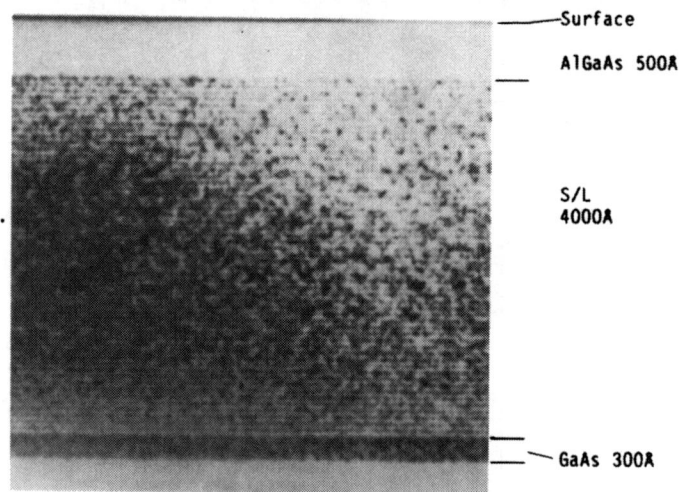

————Surface

AlGaAs 500Å

S/L
4000Å

GaAs 300Å

much lower rate than the GaAs. The spectra in Fig. 5 have the characteristic appearance of dechanneling spectra. These are typical results normally seen for heavy ion damage in ionic crystals where amorphous damage zones are not retained, but in their place is observed small clusters of defects such as Frank loops embedded between the crystal planes of the host lattice. This result argues in favor of a very rapid room temperature annealing process that allows the material to retain its crystallinity with the displacement damage accommodated in clustered interstital or vacancy loops. This response to heavy ion damage is known to be very temperature dependent for ionic crystals or for compound semiconductor material that exhibits a high degree of ionic bonding. In fact, it is usually possible to achieve some degree of amorphization by reducing the bombarding temperature to the liquid nitrogen temperature range. The binary compound AlAs has been shown to become amorphous under ion bombardment at 77°K [7]. However, for room temperature bombardment, it is clear from Fig. 5 (as well as from TEM analysis) that the crystallinity of the AlGaAs material is retained. Notwithstanding this effect, it is also clear that under heavy ion bombardment, there will be a large DPA rate in effect and the aluminum atoms in the AlGaAs will in fact be displaced many times and transported distances on the order of 25, 50 or 100 Å by the dynamic mixing effects of the collision subcascades. In view of this, continued bombardment of the S/L should result in eventual transport of aluminum out of the AlGaAs and into the GaAs. Since the GaAs will become amorphous prior to extensive mixing of the Al into it, a continued amorphous condition will be sustained during and after Al intermixing into the original GaAs material. However, as the Al is depleted from the AlGaAs, the high degree of ionic bonding will be converted to the more co-valent bonding of the GaAs type material and at some minimum critical concentration of Al, it too becomes amorphous under RT bombardment.

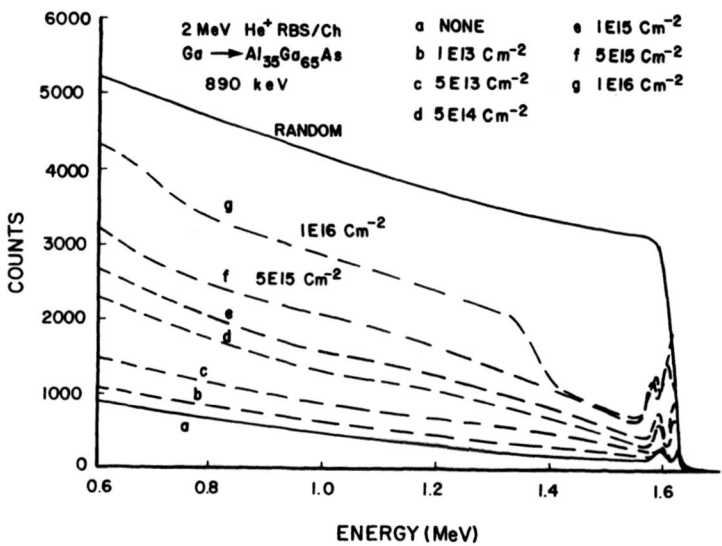

Fig. 5 RBS Channeling Spectra. 890 keV Ga Implanted AlGaAs

In order to confirm that the S/L could be driven into the amorphous state, we continued with bombardment of the S/L up to 5×10^{14} cm^{-2} Ga ions. Figure 6 shows the RBS channeling spectra for this case and confirms the above argument that a sufficient re-distribution of Al occurs across the 35 to 50 Å superlattice layer to achieve a uniform and compositionally average aluminum concentration necessary for amorphization.

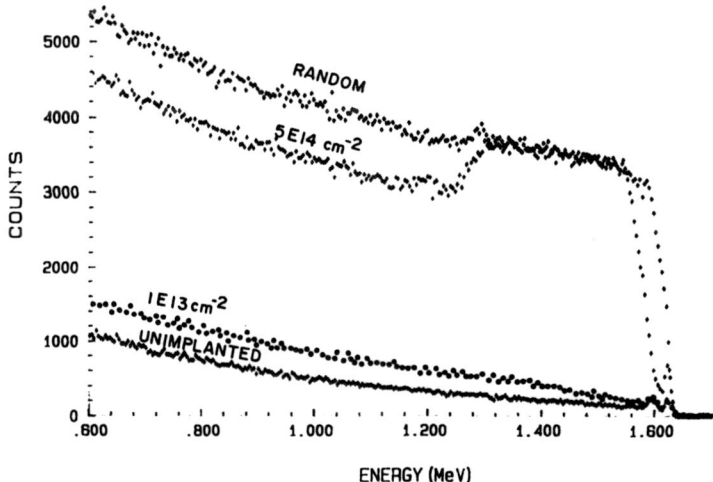

Fig. 6. RBS-Channeling Spectra. 890 keV Ga Implanted MQW/S1 of Fig. 1.

The sample in Fig. 6 has in fact been shown to be compositionally mixed at this level of bombardment by examining the 64 eV Al Auger line before and after bombardment. The initial well-defined periodicity of the Al signal is found to be completely homogeneous after the 5×10^{14} cm^{-2} Ga implant of Fig. 6. Thus, achieving amorphization in this system is a sufficient condition for mixing.

CONCLUSION

From this investigation, it was seen that for 960 keV Ga ions a fluence level of approximately 1.5×10^{14} cm^{-2} produced an amorphous layer in GaAs. At a fluence level of twice that necessary to produce an amorphous layer in GaAs, the S/L achieves less than 50% total damage, with the GaAs quantum well sustaining more damage than the S/L region containing AlGaAs.

Even though bulk AlGaAs does not become amorphous at a fluence of 1×10^{16} cm^{-2}, the aluminum atoms in the AlGaAs will be displaced several times at high fluences. This displacement of Al atoms will allow mixing of the GaAs/AlGaAs superlattices as was shown by bombarding a S/L with a fluence of 5×10^{14} cm^{-2} at 890 keV. After implantation of this S/L, Auger analysis showed the Al concentration to be homogeneous and the X-TEM analysis indicated the amorphization of the superlattice [8]. It is determined that the production of bombardment induced amorphicity in GaAs/AlGaAs is a sufficient condition for compositional mixing at room temperature.

REFERENCES

1. J. Ralston, G.W. Wicks, L.F. Eastman, B.C. DeCooman and C.B. Carter, J. Appl. Phys. 59, 120 (1986).

2. D.G. Deppe and N. Holonyak, Jr., J. Appl. Phys. 64, R93 (1988) and references therein.

3. W.D. Laidig, N. Holonyak, Jr., M.D. Camras, K. Hess. J.J. Coleman, P.D. Dapkus and J. Bardeen, Appl. Phys. Lett. 38, 776, (1981).

4. K. Meehan, K.C. Hsieh, G. Costrini, R.W. Kaliski, N. Holonyak, Jr., and J.J. Coleman, Appl. Phys. Lett. 48, 861 (1986).

5. E.A. Dobisz, H.G. Craighead, S.A. Schwarz, P.S.D. Lin, K. Kash, L.M. Schiuvone, A. Scherer and J.P. Harbison, SPIE, Advanced Processing of Semiconductor Devices, 797, 194 (1987).

6. S. Tong Lee, G. Braustein, P. Fellinger, K.B. Kahen and G. Rajeswaran, Appl. Phys. Lett. 53, 2531 (1988).

7. E.A. Dobisz, H. Dietrich and A.W. McCormick. These proceedings.

8. P.P. Pronko et al. To be published.

Work supported by U.S. Naval Air Systems Command under contract No. N00019-87-C-0267 with Universal Energy Systems.

EFFECTS OF HELIUM ION IMPLANTATION ON THE OPTICAL AND CRYSTAL PROPERTIES OF GaAs

R. C. Bowman, Jr., P. M. Adams, J.F.Knudsen, S. C. Moss,
P. A. Dafesh, D. D. Smith, M. H. Herman*, and I. D. Ward*

The Aerospace Corporation, P. O. Box 92957, Los Angeles, CA 90009
*Charles Evans & Associates, 301 Chesapeake Drive, Redwood City, CA 94063

ABSTRACT

The damage to GaAs crystals caused by helium ion implants has been monitored by changes in the Raman scattering phonon modes, double-crystal x-ray diffraction rocking curves, photoreflectance (PR), and electron beam electroreflectance (EBER) band edge transitions. As the implanted helium ion dose was increased, the various techniques revealed threshold damage behavior at very different levels. Although PR and EBER were the most sensitive to the defects created at the lowest ion doses, all techniques indicated substantial disorder for implants greater than 10^{14} ions/cm^2.

INTRODUCTION

Moderate and high dosage multiple energy ^{4}He$^+$ ion implants will reduce the carrier lifetime in GaAs and thus alter [1] the temporal response as well as the dark- and photo-current-voltage characteristics of picosecond photoconductive switches fabricated from this semiconductor. In order to understand the performance of these picosecond switches, the density and distribution of the defects created by the ^{4}He$^+$ implants, and perhaps their specific nature, needs to be determined. Although the general behavior of ion implantation for GaAs has been extensively studied [2, 3], few results explicitly correspond to the conditions used to prepare the picosecond switches [1]. Consequently, several optical methods (e.g., Raman scattering, photoreflectance [PR], and electron beam electroreflectance [EBER]) and double-crystal x-ray diffraction have been used to examine the GaAs samples after the ^{4}He$^+$ ion implants. Although some effects were noted in the PR and EBER spectra for total ^{4}He$^+$ doses below 10^{13} ions/cm^2, changes were most pronounced for the larger doses. The ^{4}He$^+$ implants lead to lattice strains, whose magnitudes were directly proportional to the ion dose in a given volume, as well as the eventual formation of amorphous regions. The present observations are generally consistent with the behavior reported for other implanted ion species [2-6] in GaAs when the influences of mass, implant energy, and dose are considered.

SAMPLE DESCRIPTION AND IMPLANT CONDITIONS

Undoped and semi-insulating (100)-GaAs wafers from LEC grown crystals were purchased from Sumitomo for these studies. The polished (100)-faces were oriented 2° off-axis towards a (110)-direction. ^{4}He$^+$ ions were implanted into the samples with the beam currents sufficiently low (beam fluence < 0.22μA/cm^2) to avoid extraneous heating above room temperature. As summarized in Table I, four ion energies were used for

Table I. Summary of ^{4}He$^+$ Ion Implant Conditions

Sample	Implant Energy Order from First to Last	Dosages (cm^{-2}) at each Energy	Dose (cm^{-2}) Total
A	Unimplanted	None	None
B	200/100/50/20 keV	1E12/2E12/2E12/2E12	7E12
C	200/100/50/20 keV	1E13/2E13/2E13/2E13	7E13
D	200/100/50/20 keV	1E14/2E14/2E14/2E14	7E14
E	200/100/50/20 keV	1E15/2E15/2E15/2E15	7E15

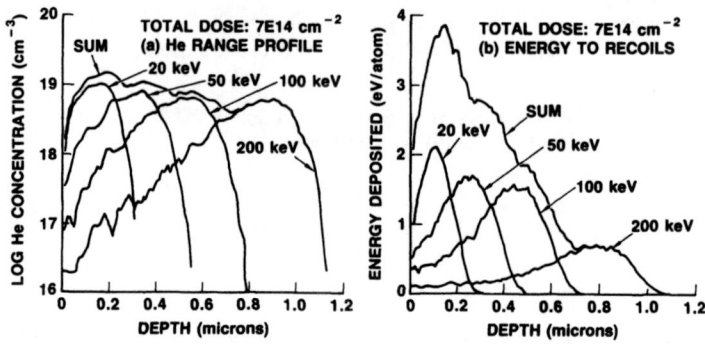

Figure 1. (a) ^{4}He$^+$ concentration profile and (b) energy deposition profile from a TRIM Monte Carlo simulation for the four energy (i.e., 20 keV, 50 keV, 100 keV, and 200 keV) implants with a total dose of 7×10^{14} ions/cm^2.

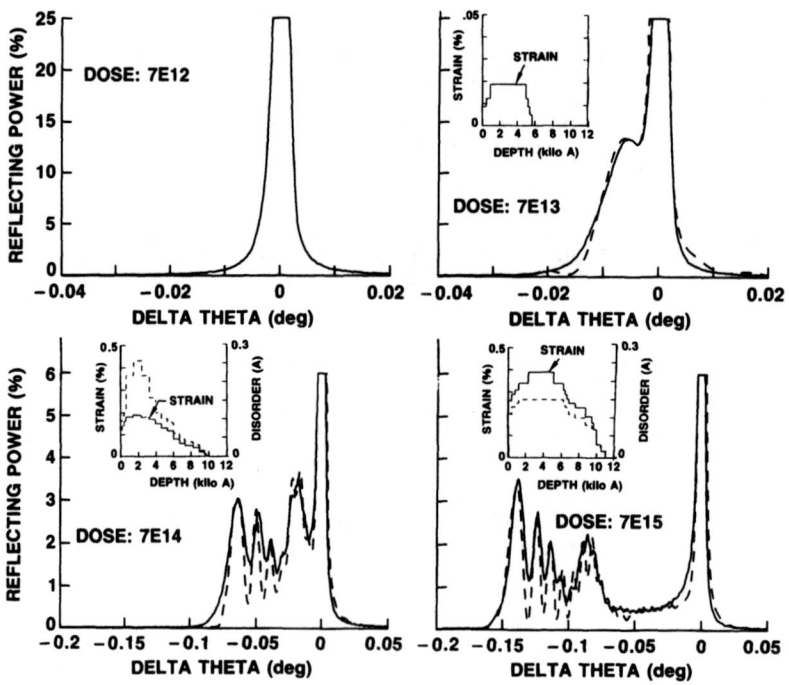

Figure 2. X-ray rocking curves measured (i.e., solid lines) for the (400)-reflection from ^{4}He$^+$ implanted (100)-GaAs with four different total doses. The inserts show the strain and disorder profiles used to obtain the fits given by the dashed rocking curves.

each sample over the total range from 7 x 10^{12} ions/ cm^2 to 7 x 10^{15} ions/cm^2. This combination of ion energies leads to an asymmetric deposition with a peak in the implanted ^{4}He$^+$ content toward the surface. Figure 1(a) gives the predicted ^{4}He$^+$ distribution for the total ion dose of 7 x 10^{14} cm^{-2} from a Monte Carlo computer simulation [7] using the TRIM-88 code. Identical profiles (albeit with different concentrations) are obtained for the other implant conditions. As shown in Figure 1(b), the maximum crystal damage (i.e., energy dissipated by the recoiling ions) occurs between 100 nm and 200 nm, but substantial disorder will occur to depths approaching a micron. While, maximum energy deposited for the 7x10^{14} ^{4}He$^+$/cm^2 dose is below the empirical threshold [2] of $\approx$5.5 eV/atom for amorphization from room temperature implants, the maximum energy deposited for the 7x10^{15} ^{4}He$^+$/cm^2 dose is above this threshold to a depth of greater than 600 nm.

RESULTS AND DISCUSSION

Among the many techniques that are used to determine structural changes induced by ion implantation, double-crystal x-ray diffraction can provide information [4] on the magnitudes and profiles of the strain arising from atomic displacements. The rocking curves from the GaAs samples covering the range of ^{4}He$^+$ implants in Table I were recorded with Cu K$_{\alpha_1}$ x-rays about the (400) Bragg reflection and are presented in Figure 2. Depth profiles of strain perpendicular to the sample surface and lattice disorder parameter [4] were found by fitting the rocking curves with a kinematical model [4]. These profiles and calculated curves are also shown in Figure 2. However, no strain component was discernable from the rocking curve for the 7 x 10^{12} cm^{-2} total dose implant (which was identical to the rocking curve for an unimplanted sample). The maximum strain increased substantially with the ion dose up to 0.39% for the 7x10^{15} cm^{-2} total dose. Because multi-energy implants were used, the extracted profile shapes were rather complicated, but they do correspond rather closely to the TRIM predictions for the recoil energy distribution.

The effects of the ^{4}He$^+$ implants on the Raman scattering from the GaAs phonon modes are shown in Figure 3. The measurements were performed at room temperature in the conventional backscattering manner [6] with excitation by the 514.5 nm line of an argon ion laser. The Raman spectra obtained with this source will probe unimplanted GaAs to a depth of about 110 nm but much less deeply when the crystal is damaged [5, 6]. Virtually no difference was found between the Raman spectra in unimplanted GaAs and the lowest two ^{4}He$^+$ implants. The sharp dominant peak arises from the allowed zone-center longitudinal optic (LO) phonon at 292.0 cm^{-1} while the symmetry forbidden transverse optic (TO) mode may be caused by the slight misorientation of the crystal axis from the (100)-direction. However, the two larger ^{4}He$^+$ implant fluences produced several well known [5, 6, 8] changes in the GaAs Raman spectra: (1) Intensity of the LO-phonon is decreased; (2) the LO peak broadened significantly and it's peak position shifted by -1.0 cm^{-1} and -3.5 cm^{-1} for the 7x10^{14} cm^{-2} and 7x10^{15} cm^{-2} total doses, respectively; and (3) a very broad and weak peak centered around 260 cm^{-1} that has been associated with an amorphous phase [5, 8] became apparent. Although Tiong, et al. [8] have attributed the shifts and broadening in the LO phonon mode to confinement effects of relatively undamaged crystalline regions surrounded by amorphous/disordered volumes, Burns, et al. [5] proposed an entirely different mechanism for these changes after ion implantation. Namely, local strain from isolated point defects (i.e., vacancies and/or antisites) are primarily responsible for the shift in the LO phonon Raman line. The correlation between the maximum strains from the x-ray rocking curves and the negative shifts of the LO phonon frequencies are consistent with this latter viewpoint. Although Raman indicates that the surface region of the heavily ^{4}He$^+$ implanted GaAs does contain either very highly disordered or amorphous volumes, crystalline regions still remain. From the TRIM simulations, it is unlikely that the 7x10^{14} ^{4}He$^+$/cm^2 implant would convert a layer of GaAs completely into the amorphous phase even at the peak of the projected range (i.e., about 100 nm - 300 nm below the surface). Nevertheless, at least partial conversion is expected for the 7x10^{15} ^{4}He$^+$/cm^2 implant. Consequently, contributions to the observed LO phonon shifts and broadening due to confinement effects [5] cannot be separated from the effects of large quantities of point defects with their associated strain contributions.

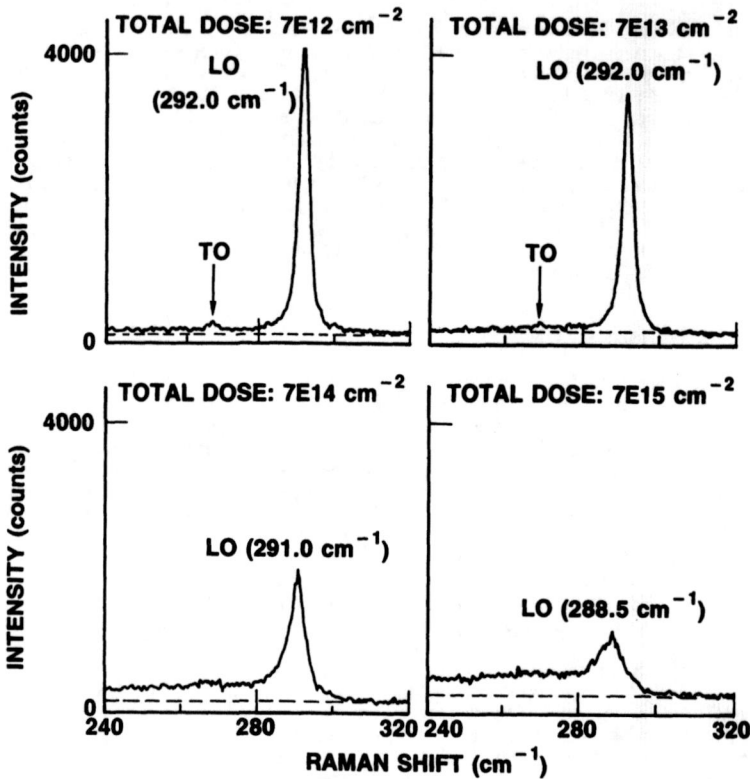

Figure 3. Raman spectra for ^{4}He$^+$ implanted (100)-GaAs with four different total doses. Peak energies of the LO phonons shift downwards with increasing dose.

Modulation techniques such as electroreflectance [9] and photoreflectance [6] have provided useful insights on the effects of ion implants. A previously described [10] EBER spectrometer system was used to observe optical transitions in the vicinity of the direct bandgap E_o. Room temperature spectra are presented in Figure 4. The spectrum for the unimplanted GaAs crystal exhibits several peaks which include an unspecified impurity transition [11] at 1.389 eV, the exciton transition at 1.426 eV and the interband transition at 1.443 eV with a linewidth of 15.2 meV. The 7×10^{12} cm^{-2} ^{4}He$^+$ implants caused the exciton transition to vanish and width of bulk interband transition at 1.43 eV to substantially increase to about 30 meV. These changes must arise from the point defects already created at this low dose which had no influence on the x-ray and Raman measurements. Furthermore, the E_o peak became barely detectable after the 7×10^{13} cm^{-2} total dose and the EBER measurements did not reveal any observable E_o transitions for the larger ^{4}He$^+$ ion fluences. Hence, EBER is extremely sensitive to the disorder produced by the ^{4}He$^+$ implants. As illustrated in Figure 5 for the unimplanted and lowest ^{4}He$^+$ ion dose samples, very similar behavior is also apparent in the PR spectra but with some differences in the relative phases for the various peaks. However, the signal-to-noise ratios from the PR spectra are somewhat lower than were obtained in the EBER measurements, which are mainly due to the higher electric field modulation available in EBER. An explanation for the disappearance of the exciton component to the EBER and PR transitions can be based upon an extension of the mechanism for the annihilation of excitons by free carriers [12]. Both free and impurity-bound excitons in

GaAs have binding energies of 5-8 meV which correspond to Bohr radii between 7-12 nm and densities of 1.2-7×10^{17} cm^{-3}. Consequently, implants which generate defects in excess of 10^{17} cm^{-3} (i.e., fluences $> 10^{12}$ cm^{-2} according to the TRIM simulations) should annihilate free excitons in GaAs. In addition, lower defect densities would reduce the lifetimes of the excitonic states which produces well-known [13] line broadening effects. Hence, the EBER and PR spectra exhibit about the same sensitivity to implant disorder as would be obtained by photoluminescence. Unfortunately, little direct information is available from reflectance measurements on the microscopic identities of the responsible defects.

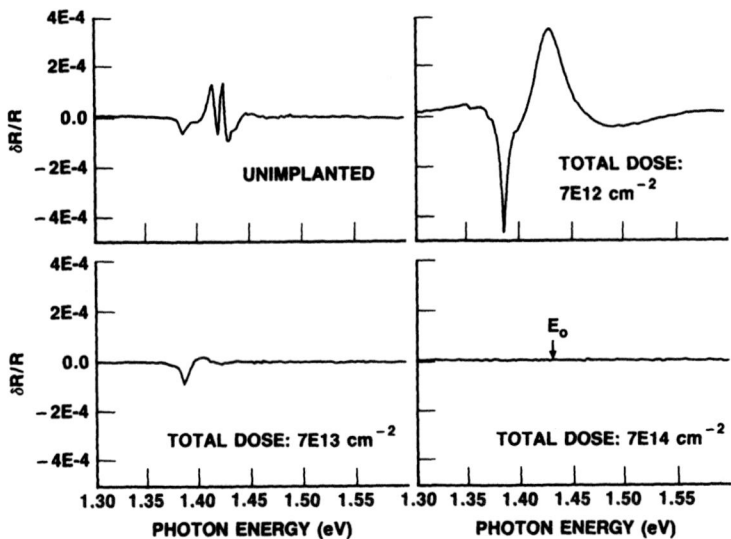

Figure 4. Room temperature electron beam electroreflectance (EBER) spectra for unimplanted GaAs crystal and three lowest ^{4}He$^+$ implant doses. The transition at 1.389 eV is attributed to an impurity while the change in the peaks above 1.40 eV are described in the text.

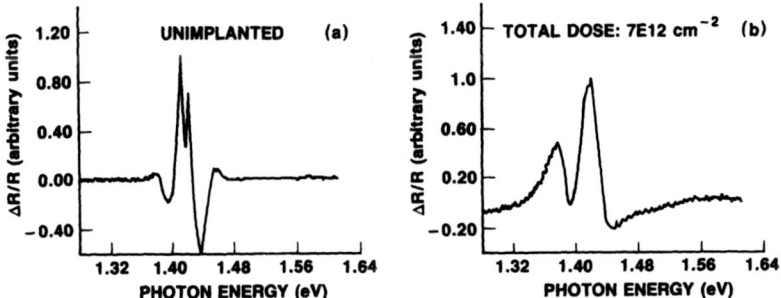

Figure 5. Room temperature photoreflectance (PR) spectra for unimplanted GaAs crystal and after ^{4}He$^+$ implants to a total dose of 7×10^{12} ions/cm^2.

SUMMARY AND CONCLUSIONS

Even though ^{4}He$^+$ ions have low mass, substantial disorder and lattice strain are found in (100)-GaAs after room temperature implants. The strain profiles deduced from the x-ray rocking curves correspond with the predicted damage distributions from Monte Carlo calculations. Furthermore, the optical EBER and PR measurements are strongly influenced by implant damage of ion fluences that give rise to little or no effect on the x-ray and Raman studies. These techniques allow a more complete understanding of the density and distribution of defects created in GaAs by these ^{4}He$^+$ ion implants, but yield little information concerning their specific nature. Further measurements, e.g., XTEM, DLTS, or EPR, may help ascertain the identities of these defects.

ACKNOWLEDGEMENTS

The work at The Aerospace Corporation has been supported by the United States Air Force Space Systems Division under Contract No. F04701-85-C-0086. The EBER measurements were supported in part by the Defense Advanced Research Projects Agency under contract DAA H01-88-C-0873 and by the Office of Naval Research under contract N00014-88-C-0221.

REFERENCES

1. D. D. Smith, J. F. Knudsen, R. C. Bowman, Jr., M. H. Herman, and S. C. Moss, in Proc. IEEE Lasers and Electro-Optics Society Meeting, Santa Clara, CA, November 2-4, 1988.
2. D. K. Sadana, Nucl. Instru. Meth. Phys. Res. B 7/8, 375 (1985).
3. S. J. Pearton, J. M. Poate, F. Sette, J. M. Gibson, D. C. Jacobson, and J. S. Williams, Nucl. Instru. Meth. Phys. Res. B 19/20, 369 (1987).
4. B. M. Paine, N. N. Hurvitz, and V. S. Speriosu, J. Appl. Phys. 61, 1335 (1987); B. M. Paine and V. S. Speriosu, ibid., 62, 1704 (1987).
5. G. Burns, F. H. Dacol, C. R. Wie, E. Burstein, and M. Cardona, Solid State Commun. 62, 449 (1987).
6. R. C. Bowman, Jr., D. N. Jamieson, P. M. Adams, and R. L. Alt, Symp. Proc. Soc. Photo-Optical Instrum. Eng. 946, 65 (1988).
7. J. P. Biersack and L. G. Haggmark, Nucl. Instru. Meth. 174, 257 (1980).
8. K. K. Tiong, P. M. Amirtharaj, F. A. Pollak, and D. E. Aspnes, Appl. Phys. Lett. 44, 122 (1984).
9. R. L. Brown, L. Schoonveld, L. L. Abels, S. Sundaram, and P. M. Raccah, J. Appl. Phys. 52, 2950 (1981).
10. P. M. Raccah, J. W. Garland, S. E. Butrill, Jr., L. Francke, and J. Jackson, Appl. Phys. Lett. 52, 1584 (1988).
11. A. N. Pikhtin, V. M. Airaksinen, H. Lipsanen, and T. Tuomi, J. Appl. Phys. 65, 2556 (1989).
12. W. A. Albers, Phys. Rev. Lett. 23, 410 (1969).
13. D. E. Aspnes, Surf. Sci. 37, 418 (1973).

NEW FEATURES OF DARK AND PHOTOCONDUCTIVITY RESPONSE OF LOW ENERGY Ar+ ION BOMBARDED GaAs

A. VASEASHTA and L. C. BURTON
Bradley Department of Electrical Engineering
Virginia Polytechnic Institute and State University
Blacksburg, VA 24061

ABSTRACT

Photoconductivity measurements have been made on three SI GaAs sample types: Ar+ ion etched at 1eV, 3keV, and unetched. Measurements were made versus time, wavelength, and temperature. Photoquenching was done at 1.19 and 2.0eV energy.

Some distinct changes in photoconductive properties are caused by the ion etch: 1) Increase in dark resistance;   2) presence of persistent photoconductivity; 3) increased photosensitivity; 4) increase in quenching rate under 2eV illumination; 5) destruction of the 0.47eV thermally activated photoconductivity.

All samples exhibit Shockley-Read recombination controlled photoconductivity below a temperature of 125K, with the same apparent trap location, at 0.26eV above E_F.

INTRODUCTION

Semi-insulating (SI) GaAs substrates are used for a variety of GaAs applications, and are often processed in a low energy ion ambient. It is desired that such substrates not be damaged during low energy ion processing steps. However, near-surface damage does occur, and has been extensively reported, both for GaAs [1-6] and other substrate materials [7-9].   Measurement techniques reported in these studies include I-V, C-V, DLTS, Auger, XPS, LEED, photoluminescence, Rutherford backscattering, photovoltaic response, thermally stimulated current, Raman spectroscopy, optical absorption, ellipsometry and Kelvin probe.   These techniques generally find substantial changes within a few hundred Angstroms of the surface.

We would like to report on the use of photoconductivity, and its temperature, spectral, and time dependence, as a way to probe near-surface damage resulting from low energy ion processing.   Photoconductivity measurements on SI GaAs have been reported, related to intentional and unwanted doping effects, and the nature of other defects, such as EL2 [10-13]. We have, however, found no reports of the use of photoconductance as a probe of low energy ion damage.

EXPERIMENTAL

SI GaAs pieces of (100) orientation were etched in 8:1:1 $H_2SO_4:H_2O_2:H_2O$ and rinsed prior to Ar+ ion bombardment in a Perkin Elmer 5300 XPS system, at 1 and 3 keV energies and flux of $10^{16}cm^{-2}$.   Four Au-Ge-Ni stripe contacts were evaporated onto each sample, followed by anneal (440°C, 2 min, forming gas) to provide ohmic contacts. The pieces were then bonded to TO-8 headers and leads attached for measurement.

The optical characterization system used for the measurements has been described in detail elsewhere [14].   For these measurements, photons were provided by a Jarrell Ash Monospec 50 monochromator (for spectral response), and a Gemini arc lamp with filters for quenching. Photocurrents (at constant voltage) were measured by a Keithley 617 electrometer. Sample temperatures were controlled between 77 and 400K in a liquid nitrogen cryostat. All measurements were automated using ASYST software and an IBM-AT PC [14].

RESULTS AND DISCUSSION

For all samples measured, photoconductive response depends strongly on time, wavelength and temperature, as expected - these all being dependent on the degree of ion etch (virgin, 1keV, 3keV). Time dependent characteristics at 77K are shown in Figure 1; spectral responses at 77K in Figures 2 and 3; and inverse temperature behavior in Figure 4.

The time dependent curves (Fig. 1) exhibit several interesting characteristics: a) The rather slow rise in photocurrent at 1040nm (1.19eV excitation), and the much more rapid increase for above band gap (600nm, 2.06eV) illumination, are strikingly similar to results reported by Desnica and Santic [10]. The very sharp rise for the 600nm curves is attributed to the rapid equilibrium of electron - hole pair generation. The much slower time response of the 1020nm curves can be attributed to the slower generation of electrons and holes from EL2 levels, with subsequent electron trapping and dominance of holes [10,15,16]. b) There is considerable persistent photoconductance (PPC) evident for the ion etched samples, with none evident for the virgin. (PPC was reported by Kiminez et al. [13] for SI GaAs, but not by Mita [12]. The samples in those reports would presumably correspond to our virgin case. One possible mechanism for the PPC is the existence of longer lifetime electron traps in the damaged and partly-amorphous near-surface region, with the EL2-generated hole current dominating for a much longer time after the photons are turned off. These could be the same traps which give rise to the low frequency capacitive dispersion reported elsewhere for ion-etched Schottky diodes [6]. (Analagous to the persistent photoconductance, no significant dispersion was seen for diodes made on un-etched material). A second mechanism could be the photogeneration of metastable states which have smaller capture cross sections for electrons [13]. For these samples, PPC is a clear signature of the ion-etch, with a memory existing for a long period after illumination. c) The degree of photosensitivity is increased by the ion etch. This could be due to the fact that the ion etch causes an increase in the dark resistivity by several orders of magnitude, which is to a large extent reduced at photon energies above 1.2eV.

Spectral response curves are shown in Figures 2 and 3. Several features deserve note: a) The un-quenched spectra are similar for ion-etched and virgin samples, showing a rather narrow peak at 0.75eV, a broad peak from about 0.9 to 1.5eV, and a fairly flat region beyond 1.6eV. b) The broad central peak of the ion-etched samples can be totally quenched by flooding with 2eV photons for 30 min, whereas quenching of the central peak of the virgin sample required more than 60 min. (Fig. 3). Since the central peak can be restored by heating in the dark to above 140K, it is attributed to the EL2 center. This more rapid quenching of the central peak is another feature that results from ion etching.

Dark and illuminated (1040nm, 1.19eV) temperature curves are shown in Figure 4. The most prominent features of these curves are: a) The photoconductivity increases exponentially with $1/T$ for all samples at temperatures below about 125K, with the same activation energy. This indicates, according to Bube et al. [11, 16], that Shockley-Read recombination is occurring at a center located about 0.26eV above the Fermi level. It is interesting that this level is not influenced significantly by the ion etch. b) There is evidence for three distinct segments on the virgin curve as temperature decreases, with activation energies of 0.78, 0.47, and 0.26eV.

The 0.47eV segment has been substantially reduced by the 1keV ion etch, and totally eliminated by 3keV.

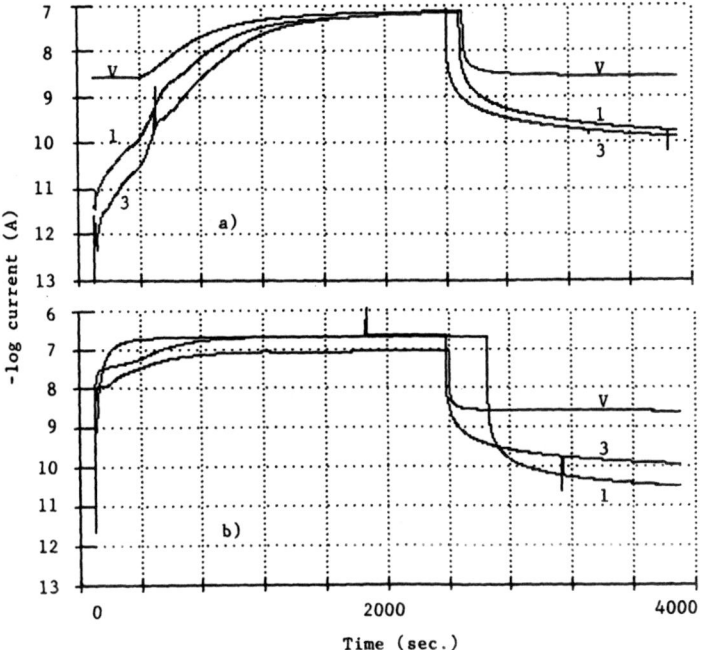

Figure 1. Time dependence of photocurrent for virgin, 1keV etched and 3keV etched samples (labelled V, 1, 3 respectively) at 77K for a) 1.19eV excitation, b) 2.06eV excitation.

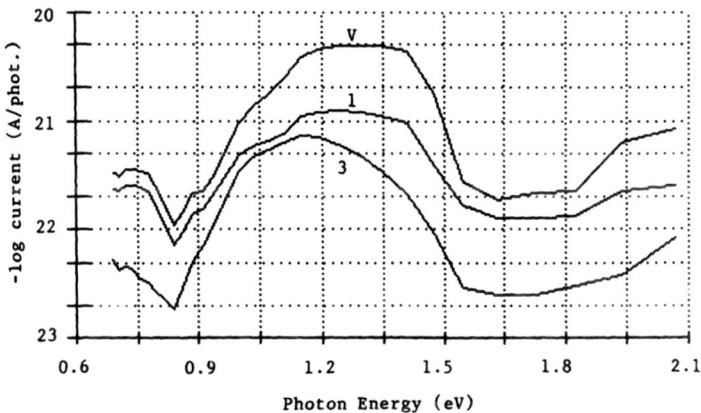

Figure 2. Spectral response (current per incident photon) for virgin, 1keV etched and 3keV etched samples at 77K (30 sec. soak time at each wavelength.)

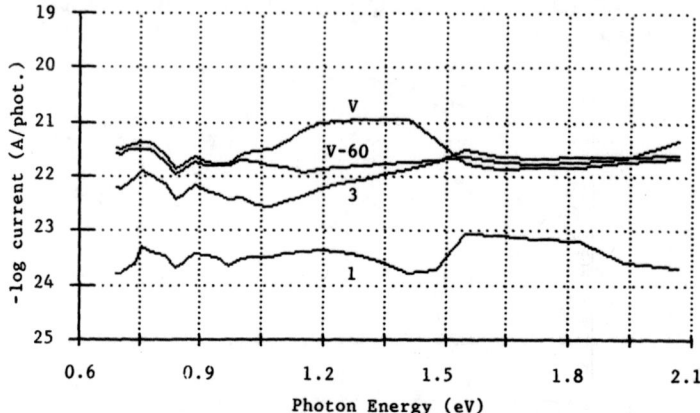

Figure 3. Spectral response (current per incident photon) for virgin, 1keV etched and 3keV etched samples at 77K after 30 min. white light soak (except for V-60, which received 60 min. soak.)

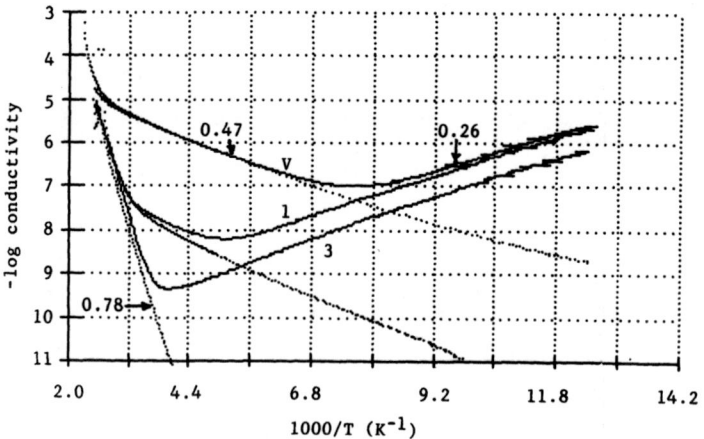

Figure 4. Inverse temperature conductivity plots for virgin, 1keV etched and 3keV etched samples in the dark (lower curves) and under 1.19eV illumination (upper curves).

SUMMARY AND CONCLUSIONS

Ar$^+$ ion etching in the 1-3keV energy range causes significant changes in the photoconductive behavior of SI GaAs. Some of the salient features are:

1. A persistent photoconductivity that is not present for the virgin (un-etched) sample;

2. The degree of photosensitivity is increased;

3. Ion etching increases the quenching rate of the large 0.9-1.5eV photoconductivity peak under 2eV illumination;

4. The 0.47eV thermal activation energy mechanism of conductivity - temperature is not present after 3keV ion etching.

All samples indicate the presence of similar Shockley-Read recombination at temperatures below 125K. Additional studies to illucidate the nature of ion etch caused changes are in progress, and will be reported.

ACKNOWLEDGEMENTS

We would like to acknowledge Texas Instruments for their partial support of this work. The technical assistance of Uma Kommineni, June Epp, N. Muthukrishnan and Eric Ellis is greatly appreciated.

REFERENCES

1. S. W. Pang et al., J. Vac. Sci. Technol. **B1** (4), 1334 (1983).

2. S. K. Ghandi et al., IEEE Elec. Dev. Lett. **EDL-3**, 48 (1982).

3. D. A. Vandenbroucke et al., Semiconduct. Sci. Technol. **2**, 293 (1987).

4. M. Kawabe et al., Appl. Optics **17**, 2556 (1978).

5. T. Hara et al., J. Appl. Phys. **62**, 4109 (1987).

6. E. D. Cole, S. Sen and L. C. Burton, J. Elec. Mater. (in press).

7. S. J. Fonash, S. Ashok and R. Singh, Thin Sol. Films **90**, 231 (1982).

8. C. Peng and H-H. Sun, Semiconduct. Sci. Technol. **2**, 779 (1987).

9. F. Y. Chen, Lin I and C. H. Lin, Semicond. Sci. Technol. **2**, 533 (1987).

10. U. V. Desnica and B. Santic, Appl. Phys. Lett. **54**, 810 (1989).

11. A. L. Lin and R. H. Bube, J. Appl. Phys. **47**, 1859 (1976).

12. Y. Mita, J. Appl, Phys. **61**, 5325 (1987).

13. J. Jimenez et al., Sol. St. Commun. **55**, 459 (1985).

14. A. Vaseashta, P. Johnson and L. C. Burton, Measurement and Control (in press).

15. B. Dischler and V. Kaufmann, Revue Phys. Appl. **23**, 779 (1988).

16. A. L. Lin, E. Omelianovski and R. H. Bube, J. Appl. Phys. **47**, 1852 (1976).

SHALLOW DOPING OF GALLIUM ARSENIDE BY RECOIL IMPLANTATION

D.K. Sadana, J.P. de Souza*, R.F. Rutz, F. Cardone and M.H. Norcott
IBM T.J. Watson Research Center, Yorktown Heights, NY 10598.
*Permanent address: Instituto de F*sica, UFRGS, Porto Alegre, R.S.,
Brasil.

ABSTRACT

Si atoms were recoil-implanted into GaAs by bombarding neutral (As+) or dopant (Si+) ions through a thin Si cap. The bombarded samples were subsequently rapid thermally or furnace annealed at 815-1000°C in Ar or arsine ambient. The presence of the recoiled Si in GaAs and resulting n+-doping was confirmed by secondary ion mass spectrometry and Hall measurements. It was found that sheet resistances of < 150 $\Omega/\square$ can be achieved by this method. Capless furnace annealing in arsine ambient generally yielded better electrical results (especially for shallow implants, i.e., < 100 nm deep) compared to those obtained by RTA in an inert ambient with a Si cap. In the latter case, electrical activation deteriorated above 900°C due to high As loss and the deterioration was pronounced for shallow implants. Significant Si redistribution occurred during arsine annealing whenever the Si concentration (from recoil or direct implant) in GaAs exceeded $1 \times 10^{19} cm^3$ and the annealing temperature was > 850°C. Our present electrical data show that the recoil implant method is a viable alternative to direct shallow implant for n+ doping of GaAs.

INTRODUCTION

Shallow dopant profiles (<0.2 μm) are essential for high performance GaAs digital IC fabrication. Conventionally, such dopant profiles are obtained by direct low energy ion implantation into the substrate or through a thin dielectric film deposited onto the substrate. In this work we studied the application of recoil implantation (1-4) to achieve very shallow n+ dopant profiles in GaAs. Recently, the same approach was used to counter dope the surface region of the n-GaAs with acceptors to increase its Schottky barrier height (5). To achieve ultra shallow doping by recoils alone, neutral ions (preferentially IIIrd or Vth group element) are implanted through a dopant cap deposited on the III-V substrate. The doping efficiency can be enhanced when bombarding neutral ions are replaced by dopant ions. There are two components of doping in this case : (i) doping via recoiled Si and (ii) doping via partially introduced direct ^{29}Si. The recoil implant method has the following advantages over conventional direct implant : (i) the surface of the substrate is prevented from direct sputtering during high dose, low energy (10 -20 keV) ion implant, (ii) the deposited layer acts as a mask for the contaminants present in the implant chamber and (iii) very shallow profiles are readily attained using implantation energies (> 20 keV) for which conventional implanters are designed. The recoil method is also advantageous over the through dielectric method because it avoids the introduction of undesired chemical species from the dielectric cap during implantation by the latter method.

EXPERIMENTAL

For n+ recoil-doping Si layers of 41-90 nm were sputter deposited onto undoped GaAs wafers at room temperature. To achieve doping via recoils only, an As+ beam was either implanted through a 41 nm Si layer at 55 keV to a dose of

$1.0 \times 10^{16} \text{cm}^{-2}$. To achieve a combination of recoil and direct bombardment doping, $^{29}\text{Si}^+$ was implanted with 30, 50 or 60 keV to doses of $1 - 10 \times 10^{14} \text{cm}^{-2}$ through a 50 or 70 nm Si cap. Subsequent annealing of the recoil- implanted samples was performed with the deposited Si at 900 - 1000°C in argon (RTA) or without the Si in arsine ambient (2% AsH$_3$, 90% Ar, 8% H$_2$)(furnace anneal) at 815 - 900 °C. For capless studies the Si was removed by reactive ion etching (RIE) in CF4:02 (9:1) plasma. The secondary ion mass spectrometry analysis (SIMS) of Si from a control unimplanted sample that was subsequently annealed at 850°C/20s showed Si concentrations of $\simeq 10^{17} \text{cm}^{-3}$ at the surface indicating a near-complete removal of Si by the etching. Cross-sectional transmission electron microscopic examination of the above samples (not included) showed a sharp Si/GaAs interface indicating the absence of any significant reaction between Si and GaAs during the annealing. However, the implanted samples into which recoil doping was performed, occasionally showed Si residues at the surface of the GaAs after the RIE. In such cases, the SIMS analysis was repeated until a reasonable agreement between the the Si distributions obtained by SIMS and those obtained by computer simulations using a TRIM program (6) was achieved in the surface region (0-50 nm). Sheet carrier concentration and resistance from the recoiled samples were obtained by van der Pauw/Hall measurements.

RECOIL DOPING : CAPLESS ARSINE ANNEALING

Figure 1 shows the recoil phenomenon in sample I. The SIMS profiles of the recoiled Si prior to and after capless annealing in arsine ambient at 815°C for 10 min are included in the figure. The peak of the As$^+$ distribution as calculated by TRIM simulations was located within the deposited layer, such that only the tail end of the As$^+$ profile penetrated into the GaAs substrate. This minimized the creation of damage in the near surface region of the GaAs. It is evident from Fig.1 that the introduction of Si recoils occurs in the depth range 0-200 nm with an exponential distribution which is in agreement with the theory (6,7). Only a minor redistribution of the recoiled Si occurred in the annealed sample. The Hall measurements showed n$^+$-type doping in the implanted region with sheet resistance of 341 $\Omega/\square$ and carrier mobility of 1600 cm^2/V.s. The sheet resistance increased to 514 $\Omega/\square$ with a carrier mobility comparable to the above when the annealing was performed at 900°C for nominal 0 min (rise time of 60 s from 450°C and fall time of 40 s to 700°C). Such degradation in electrical activation with increasing annealing temperature may be related to the As loss from the GaAs substrate during the annealing and is discussed later in the text.

In order to further enhance surface doping of GaAs, 30-60 keV $^{29}\text{Si}^+$ implants were conducted through the Si(50, 70 or 90 nm)/GaAs to doses of $1.0 \times 10^{14} \text{cm}^{-2}$ and $1.0 \times 10^{15} \text{cm}^{-2}$. There were two components of doping in this case : (i) doping via recoiled ^{28}Si and (ii) doping via partially introduced direct ^{29}Si implant. This is illustrated in Fig. 2 which shows ^{29}Si and ^{28}Si distributions from the GaAs sample into which $1.0 \times 10^{15} \text{cm}^{-2}$ $^{29}\text{Si}^+$ was implanted through a 70 nm Si cap. A significant redistribution occurred when the sample was subsequently annealed in AsH$_3$ at 850°C for 20 min and is shown in Fig.3. The Hall measurements again showed n-type carriers in the implanted region with sheet resistance of 125 $\Omega/\square$.

Both Figs 1 and 3 showed anomalous diffusion of Si during arsine annealing, however, the diffusion was more pronounced in Fig. 3 because of higher temperature/time employed here compared to that in Fig. 2. Comparison of Figs 2 and 3 reveals that the Si (both 29 and 28 isotopes) in 0-50 nm depth range remains almost pinned during the annealing whereas fast diffusion occurs beyond 50 nm into GaAs. This is believed to occur by a combination of the following two events : (i) enhanced Ga vacancies formation in the near-surface region and (ii) the presence of trace amounts of oxygen in the

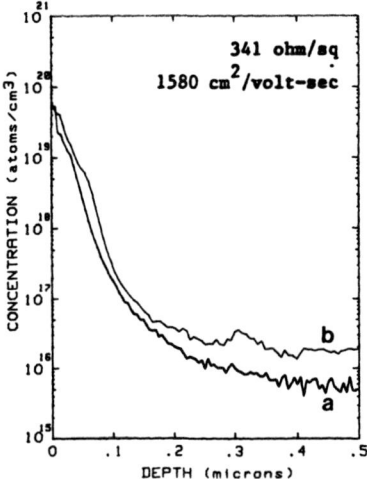

Fig. 1 SIMS profile of ²⁸Si recoiled during implantation of As⁺ through 41 nm Si layer into GaAs; (a) unannealed and (b) after capless furnace annealing at 815°C for 10 min in AsH₃. The Hall data from the sample is inset. The Si layer was removed by RIE prior to annealing.

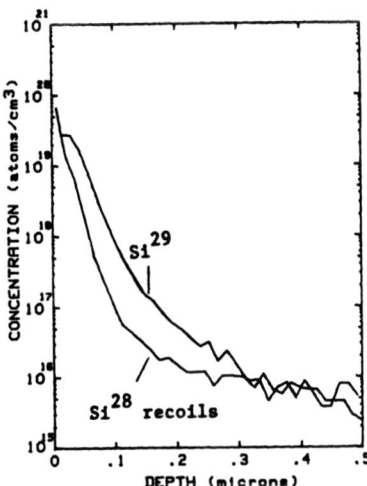

Fig. 2 SIMS profiles of implanted ²⁹Si and recoiled ²⁸Si in the unannealed GaAs sample that received a 50 keV ²⁹Si⁺ implantation (1.0×10¹⁵cm⁻², room temperature) through a 70 nm thick Si layer deposited over it.

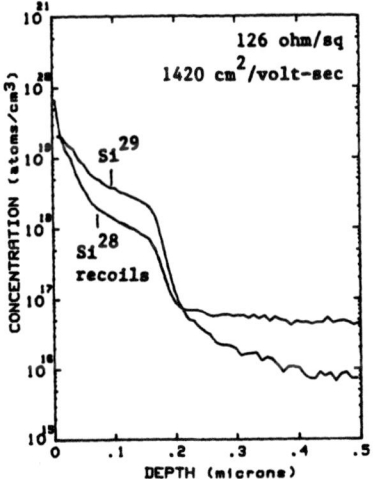

Fig. 3 SIMS profiles of the same sample as in Fig.3, but after the GaAs sample was capless annealed in AsH₃ at 850°C for 20 minutes. The Hall data from the sample is inset. The deposited Si was removed prior to annealing.

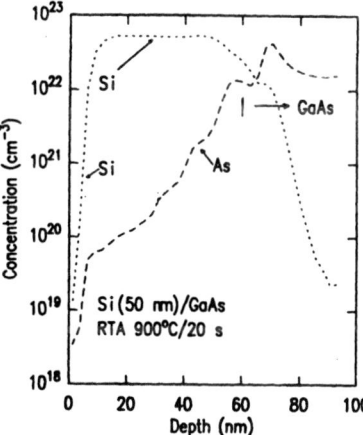

Fig. 4 SIMS measured outdiffusion of As into the Si cap (50 nm) from a GaAs substrate during RTA at 900°Cfor 20 s.

annealing ambient. The oxygen in the ambient interacts with high concentrations of Si in the surface region to create Si-O complexes which presumably retard the Si diffusion near the surface. In the deeper regions, the Si diffusion is probably aided by Ga vacancies being injected from the surface during the arsine annealing.

RTA WITH A SILICON CAP

The electrical activation behavior of recoil implanted samples was also studied after rapid thermal annealing (RTA) in Ar ambient at 900°C/20s, 950°C/10s or 1000°C/3s. The annealing time was reduced at higher annealing temperatures to reduce the As loss during annealing. The electrical activation was found to deteriorate rapidly with increasing time for RTA conducted at temperatures > 900°C. The Hall/sheet resistance data from 30 and 60 keV $^{29}Si^+$ implants through 50nm Si/GaAs (doses of 1.0×10^{14} or $1.0\times10^{15}cm^{-2}$ at each energy) is listed in Table 1 below. In all these cases, the Si layer used for recoil implant was also used as a cap during the subsequent RTA. The lowest sheet resistance was always obtained after an 900°C/20s anneal and it deteriorated with increasing RTA temperature. For example, for a 30 keV, $1.0\times10^{14}cm^{-2}$ $^{29}Si^+$ implant, sheet resistances were 2250 $\Omega/\square$, 20,000 $\Omega/\square$ and indeterminate for the 900, 950 and 1000°C anneals, respectively. The magnitude of the increase in the sheet resistance with RTA temperature was higher for shallow implants irrespective of the their dose. For example, the sheet resistance of a 30 keV, $1.0\times10^{15}cm^{-2}$ implant deteriorated more rapidly than the 60 keV, $1.0\times10^{14}cm^{-2}$ implant (Table 1) (the through-cap dose of the former is almost an order of magnitude higher than the latter). This indicates that As loss during the anneal may be governing the electrical properties of the ultra shallow implants. Indeed, the SIMS analysis of the Si cap from the annealed samples showed the As loss in excess of $1.0\times10^{16}cm^{-2}$ even after an 900°/20s anneal(Fig.4). The Ga loss into the Si cap was $<10^{15}cm^{-2}$ for the 900°C (not shown in Fig.4). The reduced activation of Si by

TABLE 1: Van der Pauw/Hall results of samples implanted with $^{29}Si^+$ after RTA in Ar atmosphere. The Si cap (50 nm thickness) was removed by RIE after the annealing.

ANNEAL (°C/s)	DOSE ($\times10^{14}cm^{-2}$)	ENERGY (keV)	R_s ($\Omega/\square$)	μ (cm^2/V.s)	N_s ($\times10^{13}cm^{-2}$)
900/20	1	30	2251	1710	0.16
900/20	1	60	259	1875	1.28
900/20	10	30	330	1970	0.96
900/20	10	60	169	1747	2.12
950/10	1	30	>2×10⁴	425	0.05
950/10	1	60	457	1637	0.83
950/10	10	30	1027	1350	0.45
950/10	10	60	214	1775	1.64
1000/3	1	30	****	****	****
1000/3	1	60	541	1540	0.75
1000/3	10	30	1545	1215	0.33
1000/3	10	60	249	1518	1.65

**** indeterminate

As loss can be explained either by considering the creation of high concentration p-type point defects/complexes (As vacancies?) in the near-surface region of GaAs and/or assuming increased occupancy of Si on As sites in the presence of high concentration of As vacancies. Si on an As site is expected to behave as an acceptor.

To reduce the As loss into the Si cap during subsequent annealing, the overlying Si layer was saturated with As before the anneal by implanting As^+ to a dose of $1.0 \times 10^{16} cm^{-2}$ through 50 nm Si cap into GaAs at 80 keV. The peak of the As distribution was thus in the vicinity of Si/GaAs interface. Subsequent RTA even at $950^{\circ}C/10s$ in Ar (with the Si layer) gave sheet carrier concentration of $6 \times 10^{13} cm^{-2}$ and sheet resistance of 136 $\Omega/\square$ with the doping depth of < 0.25 μm (SIMS data not included here). This is despite the extensive damage created during the As implant, a shallow Si profile and a high temperature RTA treatment. Presumably the presence of high concentration of As in the Si layer reduced the As loss from the GaAs during the RTA in this case.

SUMMARY

In conclusion the major findings of the present investigation are as follows:

1. The Si atoms recoiled by As^+ implants through a Si cap indeed provide shallow n^+-layers (0.2- 0.3 μm) in GaAs the sheet resistance of which can be adjusted by varying the As^+ dose. The lowest sheet resistance by As^+ induced Si recoils was 136 $\Omega/\square$.

2. The As loss during subsequent annealing apparently degrades the electrical activation of Si in GaAs. With the increasing As loss the Si activation deteriorates and becomes quite pronounced for shallow/low dose implants.

3. Capless furnace annealing in arsine ambient at $850^{\circ}C$ typically yielded better electrical activation compared to RTA in Ar ambient with a Si cap at $900^{\circ}C$ or above.

ACKNOWLEDGMENTS

The authors would like to acknowledge technical contributions of Thermon Mckoy for implantation, Joe DeGelormo for Si layer deposition and Harry Hovel for arsine annealing and useful discussions.

REFERENCES

1. A. Grob, J.J. Grob, I. Mesli, D. Sales and P. Siffert, Nucl. Instrum. Methods, 182/183, 93 (1981).

2. M. Bruel, M. Flocari and J.D. Gailliard, Nucl. Instrum. Methods, 182/183, 93 (1981).

3. R. Erichsen Jr., I.J.R. Baumvol and J.P. de Souza, Nucl. Instrum. Methods, B7/8, 316 (1985).

4. Y. Yamamoto, S. Fujima, H. Takada, Y. Segawa and K. Ishibashi, Nucl. Instrum. Methods, B19/20, 392 (1987).

5. M. Eizenberg, A.C. Callegari, D.K. Sadana, H.J. Hovel and T.N. Jackson, Appl. Phys. Lett., 54, 1696 (1989).

6. J.F. Ziegler, J.P. Biersack and U. Littmark, "The Stopping and Range of Ion in Solids", Pergamon Press, 1985.

7. L.A. Christel, J.F. Gibbons and S. Mylroie, Nucl. Instrum. Methods, 182/183, 187 (1981).

MOLECULAR ION S_2^+ AND SiF_n^+ IMPLANTATIONS INTO GaAs

W.D. FAN AND W.Y. WANG
Shanghai Institute of Metallurgy, Academia Sinica
Shanghai 200050, China

ABSTRACT

Molecular ion S_2^+ and SiF_n^+ implantations into GaAs have been investigated to form very thin active layers. After implantation, the transient annealing (TA) and furnace annealing (FA) were used. The measurements of activation efficiency, mobility, carrier concentration profiles and PL spectra were carried out. The experiments show that after TA, the activation efficiency, mobility and carrier distribution are almost the same between samples implanted with S^+ at an energy of 50KeV to a dose of $3 \times 10^{13} cm^{-2}$ and S_2^+ at 100KeV to $1.5 \times 10^{13} cm^{-2}$. It shows that the damage of S_2-implanted samples can be removed by TA, and a very thin active layer can be formed by the implantation of S_2^+ at 50KeV. For SiF_n-implanted samples, the activation efficiency and mobility decrease with increase of the implanted ion mass. As^+ co-implantation into SiF-implanted samples has been used to improve both activation efficiency and mobility. After comparison with the properties of the SiF_n^+ implantation, S_2^+ implantation is more acceptable to form thin active layers.

INTRODUCTION

In order to meet the demand for fabricating GaAs ICs with high quality, it is important to form very thin active layers with high carrier concentration by means of low-energy ion implantation. But the low-energy ion implantation is not only limited by the apparatus, but also it has the ion channeling effect which broadens the doping profile [1], brings about the doping profile tail and changes the sheet carrier concentration [2]. Molecular ion implantation can be used to reduce the implantation energy of dopant ions. Recently, GaAs MESFETs have been fabricated by SiF^+ [3] or SiF_2^+ [4] implantation. Properties of SiF_n- and SF_n-implanted samples were reported [5]. However, the detail characteristics of molecular ion implantation are still not clear.

In this paper, S_2^+ and SiF_n^+ implantations into GaAs have been investigated by measurements of activation efficiency, mobility, carrier concentration profiles and 20K PL spectra, and the effect of F atoms on active layer properties is discussed.

EXPERIMENTAL PROCEDURES

Samples used in this work were undoped LEC SI [100] GaAs single crystal wafers. Prior to implantation, the wafers were chemically cleaned and etched in $3H_2SO_4:1H_2O_2:1H_2O$ solution at $50°C$ for 5 min to remove polishing damage. Ions were implanted at room temperature at $7°$ off the [100] crystal axis. S^+ and S_2^+ were implanted at an energy of 50KeV or 100KeV to a dose of 3×10^{13} or $1.5 \times 10^{13} cm^{-2}$, respectively. The implanted energy of Si^+, SiF^+, SiF_2^+ and SiF_3^+ was 43, 72, 100 and 128KeV respectively. The dose was the same, $2 \times 10^{13} cm^{-2}$. The energy was chosen to give the same S^+ or Si^+ range from theoretical consideration. In order to improve electrical properties, As^+ and SiF^+ dual implantations were used. After SiF^+ implantation, As ions were implanted at 25KeV to $5 \times 10^{12} cm^{-2}$. F^+ and Si^+ dual implantations were also carried out to examine the effect of F atoms. For this purpose, F ions were implanted at 180KeV to a various dose, and then Si ions were implanted at 43KeV to $1.5 \times 10^{13} cm^{-2}$. After implantation, the close-contact capless furnace annealing (FA) or transient annealing (TA) [3] was performed.

The Hall effect and sheet resistivity measurements were made using the Van der Pauw method to examine the activation efficiency and mobility of the implanted wafers. Carrier concentration profiles were obtained by Polaron automatic concentration profile meter. Also, 20K PL measurements were carried out.

RESULTS

S₂-implanted GaAs layer characteristics

Fig.1 and 2 show the dependence of the activation efficiency and mobility on TA temperature and TA time respectively. The optimum condition

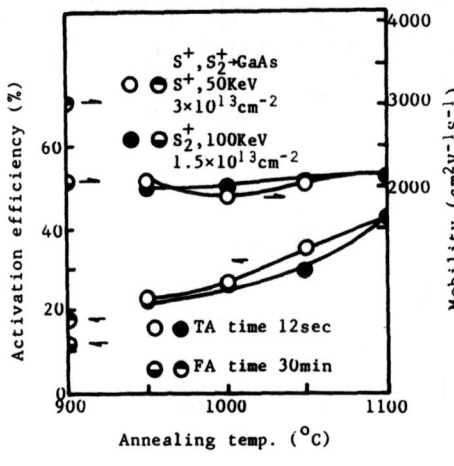

Fig.1 Dependence of activation efficiency and mobility on annealing temperature

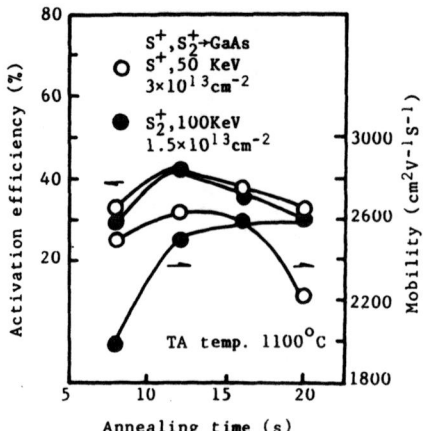

Fig.2 Dependence of activation efficiency and mobility on TA time

for annealing is at 1100°C for 12s. The properties of both samples after FA at 900°C for 30 min are also shown in Fig.1. The activation efficiency and mobility of the S-implanted samples are almost the same as those of the S_2-implanted samples after TA at 1100°C for 12 or 15s, but they are different from each other after FA. Those of the S_2-implanted sample after TA at 1100°C are 43% and 2520cm^{2}V^{-1}S^{-1} respectively, better than those of that after FA.

The carrier concentration profiles of S- and S_2-implanted samples after TA are shown in Fig.3. It can be seen that the profiles are similar between the samples implanted with the same effective energy and dose, and the S redistribution occurs during annealing after comparison with the LSS theory. By reducing the implantation energy of S_2^+ to 50KeV, a very thin active layer is formed.

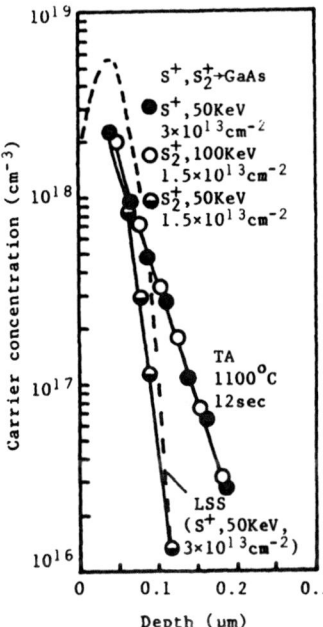

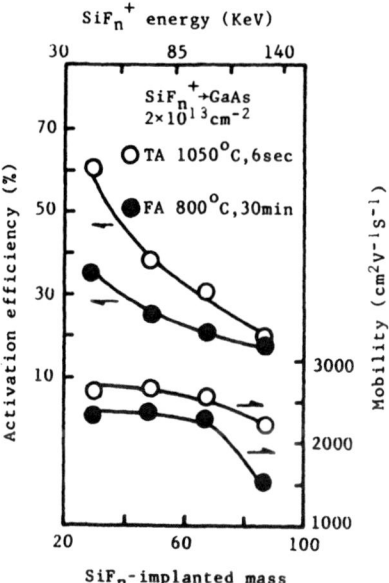

Fig.3 Carrier concentration profiles

Fig.4 Relation between activation efficiency or mobility and SiF_n-implanted mass or its energy

SiF_n-implanted GaAs layer characteristics

After TA, the relation between the electrical properties and SiF_n-implanted ion mass or its energy is shown in Fig.4. Also shown in Fig.4 is the relation after FA at 800°C for 30 min. It indicates that even at the same Si-implanted dose, the electrical parameters decrease with increase of the implanted ion mass. For the Si- or SiF-implanted sample, the activation efficiency and mobility are 60 or 38% and 2630 or 2590cm^{2}V^{-1}S^{-1} respectively after TA at 1050°C for 6s, which is more better than that after FA at 800°C for 30 min. The activation efficiency and mobility of the sample dually

implanted with As$^+$ and SiF$^+$ and annealed at 1050°C for 6s are 44% and
2620cm^{2}V^{-1}S^{-1} respectively, which are better than those of the singly
SiF-implanted sample.

Fig.5 shows the carrier concentration profiles of Si- and SiF$_n$-
implanted samples after TA. All the samples have almost the same depth of
active layers, but the peak carrier concentration decreases with increase
of implanted ion mass or its implanted energy.

After TA for samples dually implanted with F$^+$ and Si$^+$, the dependence
of the activation efficiency, mobility and normalized PL intensity
I(1.49eV)/I$_0$(1.51eV) on the dose of F$^+$ is shown in Fig.6. With increase
of the F-implanted dose, the electrical properties decrease obviously and
the normalized intensity increases. The normalized intensity is 0.074 and
1.33 for the sample implanted with single Si$^+$ (43KeV, 1.5×10^{13}cm^{-2}) and
dual F$^+$ (180KeV, 5×10^{12}cm^{-2}) +Si$^+$ (43KeV, 1.5×10^{13}cm^{-2}), respectively. But
it is 0.11 for the sample implanted with SiF$^+$ (72KeV, 2×10^{13}cm^{-2}, the data
is not shown in Fig.6).

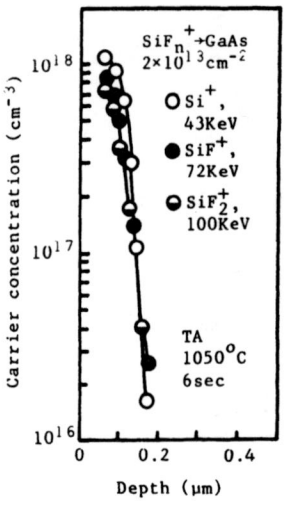

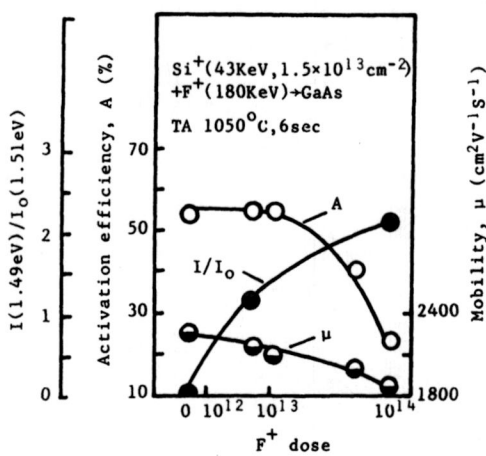

Fig.5 Carrier concentra-
tion profiles

Fig.6 Dependence of activation efficiency
and mobility on F$^+$ dose

DISCUSSIONS

The range distribution of molecular ion implantation is a complex
problem due to the multiple-body interaction. The SIMS results of SiF$_n^+$
implantation into GaAs by Tamura et al [6] indicate that if the implanted
energy increases with the implanted mass, the Si or F ions have the same
distribution. They suggested that the implanted molecular ions (SiF$_n^+$) are
completely decomposed into Si and F atoms at the substrate surface. In our
experiments, after annealing for samples implanted with S$_2^+$ and S$^+$, or SiF$_n^+$
and Si$^+$, at the same effective energy, the carrier concentration profiles
are similar (Fig.3 and 5), and their active layer depths are almost equal.
Even so, it is hardly said that the molecular ions are completely decomposed
at the surface. Considering implantation of SiF$_n^+$ in our experiments, for
example, if the molecular ions are completely decomposed, the projected
ranges of Si$^+$ and F$^+$ are 0.037 and 0.039μm by LSS theory. It can be imagined that no matter where the molecular ions are decomposed, the pro-

jected range of Si^+ is from 0.037 to 0.039μm. For such a variation in Si distribution, it is difficult to detect by SIMS or carrier concentration profiles.

The molecular ion (SiF_n^+) implantation produces heavier damage [7]. The results of P_2^+ and P^+ implantations into silicon by Fang et al [8] indicate the conclusion, too. In our experiments, the properties of S_2-implanted GaAs samples are inferior to those of S-implanted samples after FA (Fig.1). The properties of SiF_n-implanted samples are inferior to those of Si-implanted samples after FA or TA (Fig.4). Due to the fact that the molecular ion implantation produces heavior damage than the single ion implantation even at the same effective energy and dose, we have to consider that the implanted molecular ions are not completely decomposed at the surface. But the questions how and when the decomposition happens need more studying in the future.

The implanted F atoms affect the PL spectra of active layers (Fig.6). But the normalized PL intensity $I(1.49eV)/I_0(1.51eV)$ of the sample implanted with F^+ (180KeV, $5\times10^{12}cm^{-2}$) and Si^+ (43KeV, $1.5\times10^{13}cm^{-2}$) is much higher than that of the sample singly implanted with SiF^+ (72KeV, $2\times10^{13}cm^{-2}$). We attribute it to the difference of the F^+ ranges and the F out-diffusion. The deeper F atoms may out-diffuse more difficultly. Therefore, we suggest that part of F atoms in the SiF-implanted sample are out-diffused, and the other F atoms make the properties of the active layer inferior. The peak at 1.49eV indicates that F^+ implantation produces the acceptor level.

CONCLUSION

The results on the activation efficiency, mobility, carrier concentration profiles and 20K PL spectra for GaAs wafers implanted with S^+ and S_2^+ or Si^+ and SiF_n^+ after TA or FA show that:
1. The properties for samples implanted with molecular ions after TA are superior to those after FA.
2. A very thin active layer can be formed by low energy S_2^+ implantation.
3. The F atoms in SiF_n-implanted samples affect the activation efficiency and mobility of the active layers after TA or FA.
4. Comparison with SiF_n^+ implantation, the S_2^+ implantation may be more acceptable.

ACKNOWLEDGEMENT

The authors would like to acknowledge their indebtedness to Z.S. Lin, R.D. Wu and GaAs MESFET group of our institute for their efficient cooperation in ion implantation, PL measurements and technological process.

REFERENCES

1. S. Sugitani, K. Yamasaki, and H. Yamazaki, Appl. Phys. Lett. 51, 806 (1987).

2. K.T. Short, and S.J. Pearton, J. Appl. Phys. 64, 1206 (1988).

3. H. Jaeckel, V. Graf, B.J. Van, Zeghbroeck, P. Vettiger, and P. Wolf, in Gallium Arsenide and Related Compounds 1986, edited by W.T. Lindley (IOP Publishing Ltd, Bristol, 1987), pp.471-476; W.D. Fan, X.Y. Jiang, G.Q. Xia and W.Y. Wang, ibid., pp.277-282.

4. M. Kuzuhara, Y. Ogawa, S. Asai, T. Furutsuka, and T. Nozaki, in
   1986 International Electron Device Meeting Technical Digest,
   sponsored by Electron Devices Society of IEEE (The Institute of
   Electrical and Electronics Engineering, Inc., 1986), p.763.

5. A. Tamura and T. Onuma, J. Appl. Phys. 64, 2044 (1988).

6. A. Tamura, K. Inoue, T. Onuma and M. Sato, Appl. Phys. Lett. 51, 1503
   (1987).

7. A.E. Geissberger, R.A. Sadler, M. Holtz and R. Zallen, in Semi-
   insulating Materials, edited by H. Kukimoto and S. Miyazawa
   (Ohmsha Ltd, Tokyo, 1986), p.249.

8. Z.W. Fang, C.L. Lin and S.C. Zou, Acta Physica Sinica, 37, 1426
   (1988).

DUAL ION IMPLANTATIONS OF Si+As AND S+As INTO GaAs

W.D. FAN, W.Y. WANG and B.L. ZHOU[*]
Shanghai Institute of Metallurgy, Academia Sinica, Shanghai 200050, China
*Xsirius Superconductivity Inc., Scottsdale, Arizona 85260, USA

ABSTRACT

As$^+$ implantation has been used to compensate the As loss during annealing for Si- or S-implanted GaAs. The As$^+$ implantation was at an energy range of 25-100KeV to a dose range of 10^{12}-10^{15}cm^{-2}. The Si$^+$ or S$^+$ implantation was at 150 or 50KeV respectively to the same dose, 2×10^{13}cm^{-2}. By using the measurements of the Hall effect and sheet resistivity, the activation efficiency and mobility of the dually implanted sample with Si$^+$ and As$^+$ (25KeV, 1×10^{13}cm^{-2}) after TA at 1050°C for 6s increase by 21 and 18% respectively over those of the singly Si-implanted sample after TA at 1100°C for 6s. Those of the dually implanted sample with Si$^+$ and As$^+$ (100KeV, 1×10^{13}cm^{-2}) increase by 63 and 30% respectively over those of the singly Si-implanted sample after FA at 800°C for 30min. Also, the density of Si$_{As}$ and V$_{Ga}$ determined by 20K PL spectra decreases of dually Si- and As-implanted samples. The activation efficiency and mobility of the dually implanted sample with S$^+$ and As$^+$ (40KeV, 1×10^{13}cm^{-2}) after TA at 1100°C for 12s increase by 17 and 16% respectively over those of the singly S-implanted sample. The annealing behaviour after As$^+$ implantation is discussed.

INTRODUCTION

The GaAs MESFET performances depend strongly on the electrical properties of active layers. The layers with high quality are usually fabricated by Si$^+$ implantation into semi-insulating (SI) GaAs substrates. In order to increase the activation efficiency and uniformity, and control As evaporation during annealing, transient annealing (TA) [1,2] and annealing at trimethylarsenic overpressure [3] were used. Recently, the improvements of the activation efficiency and the vacancy concentration ratio of As to Ga were reported on Si-implanted GaAs by P$^+$ co-implantation [4]. In the present paper, As$^+$ implantation has been used to compensate the As loss during annealing for Si- or S-implanted GaAs in order to obtain active layers with high quality. The activation efficiency, mobility, carrier concentration profiles and PL spectra of the active layers were measured, and the effects of As$^+$ implantation are discussed.

EXPERIMENTAL PROCEDURES

Samples used in this work were undoped LEC SI [100] GaAs single crystal wafers. Prior to implantation, the wafers were chemically cleaned and etched in $3H_2SO_4$:$1H_2O_2$:$1H_2O$ solution at 50°C for 5 min to remove polishing damage. Ions were implanted at room temperature at 7° off the [100] crystal axis. Si$^+$ and S$^+$ were implanted at an energy of 150 and 50KeV respectively to the same dose of 2×10^{13}cm^{-2}, and then As ions were implanted at an energy of 25-100KeV to a dose of 10^{12}-10^{15}cm^{-2}. The close-contact capless furnace annealing (FA) or TA was performed [1].

After annealing, the Hall effect and sheet resistivity measurements were made using the Van der Pauw method to examine the activation efficiency and mobility of implanted wafers. Carrier concentration profiles were obtained by Polaron automatic concentration profile meter. Also, 20K PL

328

measurements were carried out.

RESULTS

<u>Layer characteristics of GaAs implanted with Si$^+$ and As$^+$</u>

Fig.1 shows that the activation efficiency and mobility for dually
implanted samples depend on the As-implanted dose and TA temperature. The
data of the dually implanted samples with As$^+$ dose of 1×10^{12} or 1×10^{15}cm^{-2}
are not shown in Fig.1. Generally speaking, both activation efficiency and
mobility of the dually implanted samples are higher than those of the singly
implanted samples after TA at any temperature, and the former increases with
increase of the temperature. Samples implanted with As$^+$ below the dose of
1×10^{13}cm^{-2} and annealed at 1050°C have high activation efficiency and
mobility, about 60% and 3000cm^{2}V^{-1}S^{-1} respectively. When the temperature
further increases to 1100°C, the activation efficiency increases a little,
but the mobility decreases. With increase of the As-implanted dose from
1×10^{13} to 1×10^{15}cm^{-2}, the activation efficiency and mobility of the dually
implanted samples decrease after TA at 1050°C or higher.

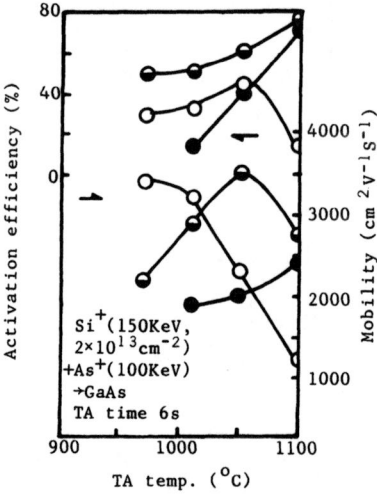

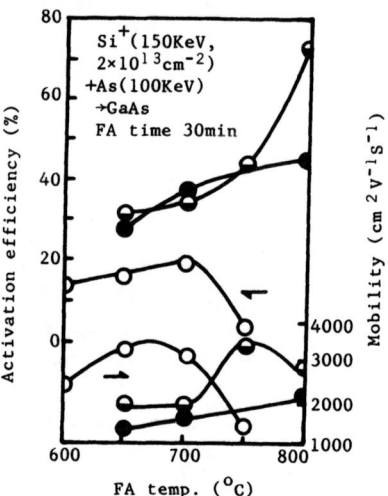

Fig.1 Dependence of activation
efficiency and mobility on TA temperature
   As$^+$ dose: (cm^{-2})
   ◐ 1×10^{13}
   ○ 1×10^{14}
   ● 0

Fig.2 Dependence of activation
efficiency and mobility on FA temperature
   As$^+$ dose: (cm^{-2})
   ◐ 1×10^{13}
   ○ 1×10^{14}
   ● 0

The dependence of activation efficiency and mobility on the FA temperature for dually and singly implanted samples is shown in Fig.2. The data of
the dually implanted samples with As$^+$ dose of 1×10^{12}cm^{-2} are not shown in
Fig.2. When the As-implanted dose is below 1×10^{13}cm^{-2} and annealing temperature is 800°C, the activation efficiency and mobility of dually implanted samples, about 60% and 2600cm^{2}V^{-1}S^{-1} respectively, are higher than

those of singly implanted samples. The increases of those for the dually implanted sample with As^+ at 100KeV to $1\times10^{13}cm^{-2}$ are 63 and 30% respectively. For the sample co-implanted with As^+ to $1\times10^{14}cm^{-2}$, both activation efficiency and mobility are very low.

The activation efficiency and mobility of dually implanted samples are shown in Fig.3 as a function of the As-implanted energy after TA or FA. In general, with increase of the As-implanted energy, the activation efficiency decreases, but the mobility increases. The activation efficiency and mobility for the dually implanted sample with As^+ at 25KeV to $1\times10^{13}cm^{-2}$ after TA at $1050^{\circ}C$ for 6s (Fig.3) increase by 21 and 18% resepctively over those for the singly implanted sample after TA at $1100^{\circ}C$ for 6s (Fig.1).

The carrier concentration profiles after annealing are shown in Fig.4. After TA or FA, the peak concentrations and active layers of the dually implanted samples are higher and thinner respectively than those of the singly implanted samples.

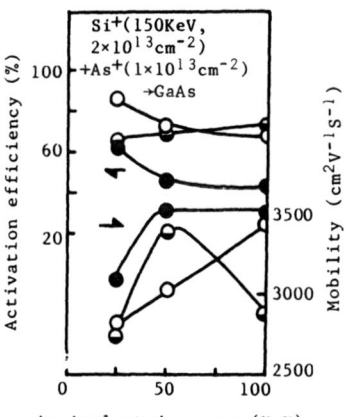

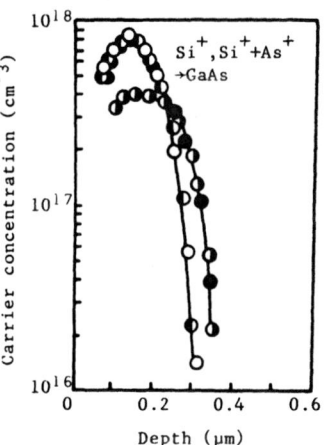

Fig.3 Activation efficiency and mobility as a function of As-implanted energy
○ TA $1050^{\circ}C$, 6s
● FA $750^{\circ}C$, 30min
◑ FA $800^{\circ}C$, 30min

Fig.4 Carrier concentration profiles
○ ◐ $Si^+(150KeV,2\times10^{13}cm^{-2})$ $+As^+(100KeV,1\times10^{13}cm^{-2})$
● ◐ $Si^+(150KeV,2\times10^{13}cm^{-2})$
○ ● TA $1100^{\circ}C$, 6s
◐ ◑ FA $800^{\circ}C$, 30min

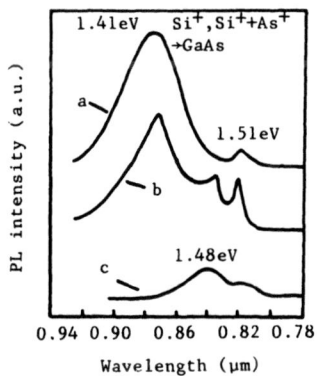

Fig.5 20K PL spectra
a $Si^+(150KeV,2\times10^{13}cm^{-2})$, FA $800^{\circ}C$, 30min
b $Si^+(150KeV,2\times10^{13}cm^{-2})$ $+As^+(100KeV,1\times10^{12}cm^{-2})$, FA $800^{\circ}C$, 30min
c $Si^+(150KeV,2\times10^{13}cm^{-2})$ $+As^+(100KeV,1\times10^{14}cm^{-2})$, FA $750^{\circ}C$, 30min

Fig.5 shows the 20K PL spectra for dually and singly implanted samples after FA. For the dually implanted samples, a peak at 1.48eV appears, and a peak at 1.41eV is weaker than that for the singly implanted sample. The peaks at 1.48 and 1.51eV are very weak for the sample co-implanted with As$^+$ to 10^{14}cm^{-2}. The peaks at 1.41, 1.48 and 1.51eV are attributed to complexes (Si$_{As}$-V$_{As}$) composed of Si atoms on As sites (Si$_{As}$) and As vacancies (V$_{As}$) [5], Si$_{As}$ [6] and band-to-band emission respectively.

Layer characteristics of implanted GaAs with S$^+$ and As$^+$

The dependence of activation efficiency and mobility on the TA temperature for samples implanted with dual S$^+$ and As$^+$ or single S$^+$ is shown in Fig.6. It can be seen that both properties of the dually implanted samples are better than those of the singly implanted samples after TA at any temperature. After TA at 1100°C for 12s, the activation efficiency and mobility for the dually S- and As-implanted sample increase by 17 and 16% over those for the singly implanted sample respectively.

Shown in Fig.7 are electrical parameters as a function of the As-implanted energy after TA. The As-implanted energy should be chosen between 40-60KeV in order to obtain the sample with high activation efficiency and mobility.

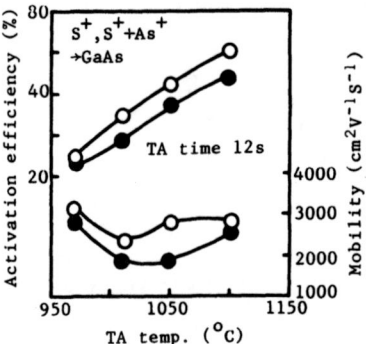
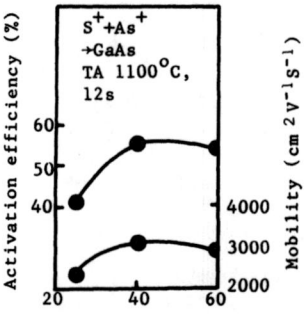

Fig.6 Dependence of activation efficiency and mobility on TA temperature
○ S$^+$(50KeV, 2×10^{13}cm^{-2})
+As$^+$(40KeV, 1×10^{13}cm^{-2})
● S$^+$(50KeV, 2×10^{13}cm^{-2})

Fig.7 Activation efficiency and mobility as a function of As-implanted energy
● S$^+$(50KeV, 2×10^{13}cm^{-2})
+As$^+$(1×10^{12}cm^{-2})

DISCUSSION

It is well-known that electrical parameters of GaAs implanted with ions are associated with the As loss at the surface during annealing. Ten years before, Stolte [7] used the As$^+$ co-implantation to the dose of 1×10^{13}cm^{-2} to compensate the As loss for Si-implanted samples, and found that the activation efficiency of dually Si- and As-implanted samples is lower than that of singly Si-implanted samples. The difference from our results is thought to be the different annealing method. Recently, Neida et al [1] carried out such dual implantations into GaAs with Si$^+$ at 100KeV and As$^+$ at 180KeV. Their results indicated that the activation efficiency increases after As^+ implantation only at the dose of $1 \times 10^{13} cm^{-2}$, which is identical with ours. They attributed the decrease of the activation efficiency to the increase of the damage related compensation by As^+ implantation to the dose over $1 \times 10^{13} cm^{-2}$. Here, we propose an additional interpretation.

It can be imagine that for dually Si- and As-implanted samples, the high activation efficiency (Fig.1 and 2) is obtained for the decrease of the V_{As} density after As co-implantation since that is proportional to the Si_{Ga}-V_{As} density. But when As dose is over $1 \times 10^{14} cm^{-2}$, the following reaction may occur during annealing:

$$As_i + Si_{Ga} = As_{Ga} + Si_i$$

where As_i and Si_{Ga} represents the As interstitial sites and Si atoms on Ga sites respectively. Therefore, the increase of As implants brings about the decrease of the Si_{Ga} density, and low activation efficiency are obtained (Fig.1 and 2). For As implants are partly compensated the As loss during annealing, the mobility can be improved (Fig.1 and 2).

Similarly, the dually S- and As-implanted samples with higher activation efficiency and mobility are obtained due to the decrease of the complex $(S_{As}$-$V_{Ga})$ [8].

From the above discussion, the density of V_{As} of dually Si- and As-implanted samples is lower than that of singly Si-implanted samples. Since the lower the vacancy density, the less is the diffusion of Si atoms [9], the active layers are thinner for the dually Si- and As-implanted samples (Fig.4).

CONCLUSION

The results on the activation efficiency, mobility, carrier concentration profiles and PL spectra for GaAs wafers dually implanted with Si^+ and As^+ or S^+ and As^+ after TA or FA show that:

1. When a proper As-implanted energy and dose are chosen, the dually Si- and As- or S- and As-implanted samples with higher activation efficiency and mobility, and thinner active layers can be obtained than singly Si- or S-implanted samples after TA or FA.

2. The higher activation efficiency and mobility and thinner active layers of dually implanted samples can be explained by the increase of As_i and the recrystallization at an early stage of annealing.

ACKNOWLEDGEMENTS

The authors would like to acknowledge their indebtedness to Z.S. Lin, R.D. Wu and GaAs MESFET group of our institute for their efficient cooperation in ion implantation, PL measurements and technological process.

REFERENCES

1. W.D. Fan, X.Y. Jiang, G.Q. Xia and W.Y. Wang, in Gallium Arsenide and Related Compounds 1986, edited by W.T. Lindley (IOP Publishing Ltd, Bristol, 1987), pp.277-282; A.R. Von Neida, S.J. Pearton, M. Stavola and R. Caruso, ibid., pp.57-62.

2. A. Tamura, T. Uenoyama, K. Nishii, K. Inoue and T. Onuma, J. Appl. Phys. 62, 1102 (1987).

3. S. Reynolds, D.W. Vook, W.G. Opyd and J.F. Gibbons, Appl. Phys. Lett. 51, 916 (1987).

4. F. Hyuga, H. Yamazaki, K. Watanabe and J. Osaka, Appl. Phys. Lett. 50, 1592 (1987).

5. T. Itoh and M. Takeuchi, Jpn. J. Appl. Phys. 16, 227 (1977).

6. W.Y. Lun and H.H. Wieder, J. Appl. Phys. 49, 6187 (1978).

7. C.A. Stolte, in Ion Implantation in Semiconductors 1976, edited by F. Chernow, J.A. Borders and D.K. Brice (Plenum Publishing Corporation, New York, 1977), p.149.

8. R.S. Bhattacharya, A.K. Rai, Y.K. Yeo, P.P. Pronko and Y.S. Park, Nuclear Instruments and Methods, 209/210, 637 (1983).

9. M.E. Greiner and J.F. Gibbons, J. Appl. Phys. 57, 5181 (1985).

ION IMPLANTATION INTO InP/InGaAs HETEROSTRUCTURES GROWN BY MOVPE

WULF HÄUSSLER, J.W. WALTER, J. MÜLLER
Siemens Research Laboratories, Otto-Hahn-Ring 6, 8000 München 83
Fed. Rep. Germany

ABSTRACT

Annealing of Beryllium implantations into epitaxial InP/InGaAs hetero-structures was investigated. Different ion energies were used to position the profile maximum in the InP cap, at the hetero-interface, or in the underlying InGaAs layer. By measuring the Be atom and carrier profiles, it is shown that annealing conditions necessary for optimum Be activation in the InP cap layer are compatible with good profile control for Be in InGaAs. There is a slight transfer of Be across the interface during annealing, which is important, if the pn-junction is intended to be close to the hetero-interface.

INTRODUCTION

Advanced electronic and optoelectronic devices based upon III-V-com-pounds contain several epitaxial layers of different composition. The behavior of dopants in such a multi-layered structure is complicated by differences with respect to diffusion lengths and electrical activation. Disordering of superlattices by diffusion or ion implantation is currently investigated intensively. However, few reports concern doping of hetero-structures consisting of thick (>0.1 μm) layers. Ion implantation into GaAs/AlGaAs heterostructures has been studied recently [1,2,3]. Few results have been published on diffusion [4] and ion implantation [5] into InP/InGaAs heterostructures.

This paper discusses the annealing of Beryllium ion implants in the InP/InGaAs system.

EXPERIMENTAL BACKGROUND

The epitaxial layers were grown on InP:S or InP:Fe substrates by metalorganic vapor phase epitaxy (MOVPE) and consisted of an InP buffer layer, a lattice-matched InGaAs layer and an InP cap layer. Two basic structures were investigated with the InP cap having a thickness of 0.3 μm or 0.95 μm. The InGaAs layer was about 2 μm thick. All epitaxial layers were not intentionally doped having donor concentrations below $5*10^{15}$ cm^{-3}.

Beryllium ions were implanted with energies of 20, 300 and 600 keV to a maximum Be concentration exceeding $5*10^{18}$ cm^{-3}. The low-energy implant was accompanied by an additional P co-implant and confined the Be profile to the top InP layer. The high-energy implants had their profile maximum either right at the InP/InGaAs interface or deep in the InGaAs layer.

Annealing was performed in a rapid thermal annealing system in a PH_3/H_2 atmosphere [6]. The annealing temperature of 850 °C, held up to 30 seconds, was based on the optimum results obtained for Be-implanted InP [7,8].

Secondary ion mass spectrometry (SIMS) was used to measure the Be atom profiles and the behavior of residual dopants, e.g. S and Fe. The SIMS data were calibrated with unannealed ion-implanted InP and InGaAs samples to give atomic concentrations. Measurements on different standards showed the calibration for Be in InP to be accurate to within 10 % and that for Be in InGaAs to be correct within 25 %.

Carrier profiles were obtained by the electro-chemical capacitance-voltage (ECV) technique [9]. Since profiling through the InP/InGaAs interface leads to erroneous results for n-type material [10], unimplanted samples were measured twice, removing the InP cap layer chemically to determine the doping level in the InGaAs layer. We estimate the ECV technique to be accurate to within 20 %.

RESULTS

1. Redistribution of implanted Beryllium

1a. Implantation maximum in the top InP layer

Fig. 1a shows the Be atom profile of a shallow (20 keV) implant into the InP cap layer before annealing. Since it has been established that Be out-diffusion occurs during annealing and can be avoided by co-implantation of heavy ions [7,8], a P implant (60 keV) of equal dose was added. Fig. 1b shows the result after annealing at a temperature of 850 °C held for 6 seconds. The SIMS profile in Fig. 1b shows that there is an injection of Be from the InP layer into the InGaAs layer during annealing leading to a discontinuity in the Be profile. The Be concentration on the InGaAs side of the interface is about five times higher than that on the InP side. The same behavior was observed for a 50 keV Be implant into a thicker InP cap that needed no additional P implant. Fig. 1b shows the hole concentration profile to follow the atomic profile. The pn-junction was determined to lie 0.2 µm below the interface where $p = n = 2*10^{16}$ cm^{-3} (see section 2). This result shows that it will be difficult to position the pn-junction in InP right in front of the InGaAs interface.

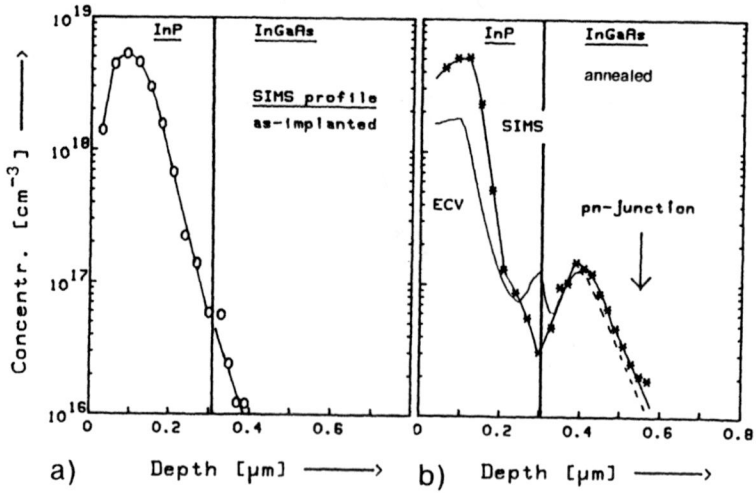

Fig. 1: Results of a 20 keV (+ 60 keV P) implant into InP/InGaAs.
    a) SIMS profile of Be before annealing (dose: $7*10^{13}$ cm^{-2})
    b) Be atom and hole concentration profiles after annealing at 850 °C
       for 6 seconds showing injection of Be into InGaAs.

1b. Implantation maximum at the InP/InGaAs interface

Fig. 2 illustrates the redistribution of Be during annealing when the profile maximum is positioned right at the interface. The SIMS profiles show that upon annealing, the Be profile is split in two parts with the interface sealing off the Be-containing InGaAs layer. The Be level in InP is reduced to about $2*10^{18}$ cm^{-3} by rapid out-diffusion, whereas Be in InGaAs shows a well-behaved profile broadening due to diffusion even at 850 °C. The latter is in contrast to Be redistribution in InGaAs annealed without an InP cap, where substantial Be redistribution is observed for high doses at temperatures as low as 600 °C [11,12]. The carrier profile closely follows the atomic profile giving values of 50 % and 100 % for the electrical activation of Be in InP and InGaAs, respectively.

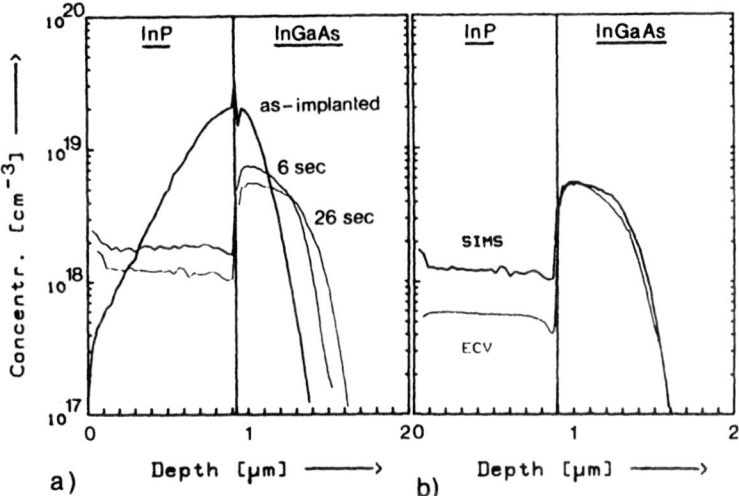

Fig. 2: Redistribution of a 300 keV Be implant (dose: $1*10^{15}$ cm^{-2}) into InP/InGaAs occuring during annealing at 850 °C.
  a) Be profiles of unimplanted sample and samples annealed for two hold times (measured by SIMS).
  b) Atomic and hole profile for a hold time of 26 seconds.

1c. Implantation maximum buried in the InGaAs layer

Fig. 3 shows the Be redistribution during annealing for the case where most of the Be is buried in the InGaAs layer, exhibiting a behavior similar to that observed in Fig. 2. Again, the Be profile in the InP cap is flattened during annealing. A step-like profile results, with a Be level of $1*10^{18}$ cm^{-3} in InP and of $1*10^{19}$ cm^{-3} in InGaAs. The Be diffusion front in InGaAs is very sharp. Again, electrical activation of Be is 100 % in InGaAs.

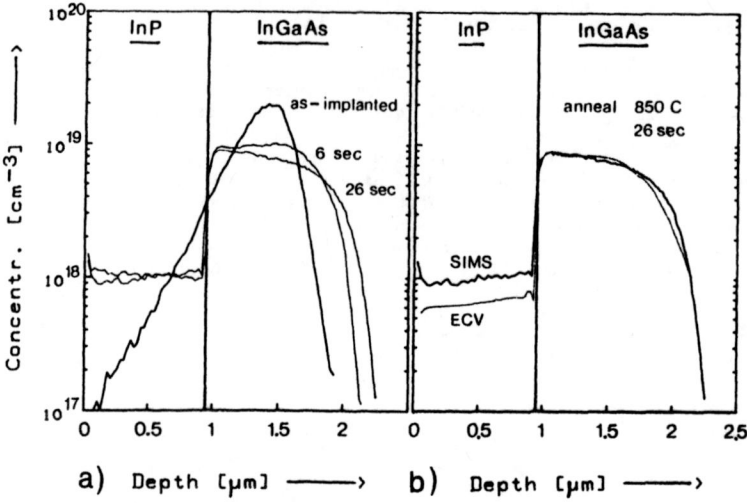

Fig. 3: Redistribution of a 600 keV Be implant (dose: $1*10^{15}$ cm^{-2}) into InP/InGaAs occuring during annealing at 850 °C.
a) and b) same as in Fig. 2.

2. Annealing of unimplanted InP/InGaAs heterostructures

There are some doubts as to the stability of the InP/InGaAs system with respect to high temperature processing. In fact, the present study seems to be the first subjecting InP/InGaAs heterostructures (grown at about 650 °C) to temperatures of 850 °C. No detrimental influence of our anneals on the surface quality of the samples is observed. X-ray rocking measurements on annealed heterostructures showed that the lattice matching of the epitaxial layers is not modified by the high temperature process [13]. Occasionally, however, we observe a change in the background doping of the InGaAs layer to a level of about $2*10^{16}$ cm^{-3}, as is illustrated in Fig. 4. The increase in doping level in the InGaAs layer has been observed for both Fe-doped and S-doped substrates. Attempts to identify a residual impurity to account for this increase have yet been unsuccessful.

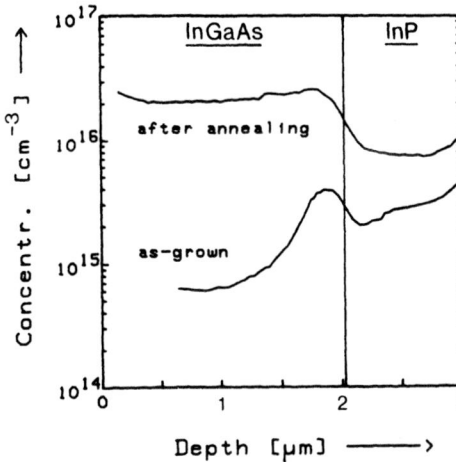

Fig. 4: Effect of annealing of an InP/InGaAs heterostructure at 850 °C. The doping level of the InGaAs layer increases from $n < 1*10^{15}$ cm^{-3} to $n = 2*10^{16}$ cm^{-3}. The InP cap layer has been removed for ECV profiling.

3. Out-diffusion of Fe and S from the InP substrate

The SIMS investigation of implanted and annealed heterostructures grown on InP:Fe shows no signs of Fe out-diffusion. The Fe level in the InGaAs and InP layers stays below our SIMS detection limit of about $1*10^{15}$ cm^{-3} and rises sharply at the substrate interface (Fig. 5). The peak at the surface seen in Fig. 5 is most likely a SIMS artifact 14] caused by mass interference with a ^{56}(BeOP) molecule.

Heterostructures grown on S-doped InP exhibit no detectable ($< 5*10^{15}$ cm^{-3}) increase in S concentration in the epitaxial layers upon annealing.

We thus conclude that the dominant substrate impurities do not diffuse into the epitaxial layers for our anneals at 850 °C.

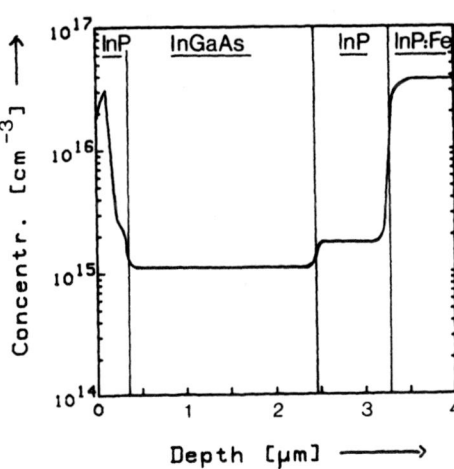

Fig. 5: SIMS profile of Fe in an InP/InGaAs/InP heterostructure grown on InP:Fe after Be implantation and annealing for 6 seconds at 850 °C (for sample details see Fig. 1).

DISCUSSION AND CONCLUSION

Annealing of Be implantations into InP/InGaAs has been studied at a temperature of 850 °C. The annealing behavior of Be in InP is identical to that observed for Be implantation into InP substrates exhibiting a strong tendency for out-diffusion. On the other hand, Be in InGaAs shows a rather good-natured behavior with respect to redistribution. Although there is some diffusion of Be in InGaAs, control of the depth of the pn-junction is satisfactory. The hetero-interface shows a good blocking behavior with respect to Be diffusion, if the Be concentration in the InGaAs is equal to or higher than that in the InP layer. However, some Be diffusion across the interface does occur, if there is more Be in InP than in InGaAs. This makes it difficult to position a pn-junction in InP close to the interface. No gettering of the acceptor at the InP/InGaAs interface is seen in contrast to results reported for Zn diffusion [4].

Electrical activation of Be in InP is the same as that observed for implantations into InP substrates. In InGaAs, electrical activation is approximately 100 %.

No diffusion of Fe and S from the substrate into the epitaxial layers is observed.

In some samples, there is an increase in the doping level of the InGaAs layer during annealing. The phenomenon is under investigation.

In conclusion, implantation of deeply penetrating Be ions appears to be promising method for p-type doping of InP/InGaAs heterostructures encountered in optoelectronic devices.

ACKNOWLEDGEMENTS

We would like to thank E. Frenzel of GeMeTec, Munich, for SIMS measurements and G. Ebbinghaus and T. Scherg for the MOVPE samples.

REFERENCES

1. B. Descouts, N. Duhamel, K. Daoud-Ketata, P. Krauz and S. Godefroy J. Appl. Phys. 60, 450 (1986)
2. B. Descouts, N. Duhamel and Y. Gao, Proc. ESSDERC'88, J. de Physique 49, C4-437 (1988)
3. T. Humer-Hager, R. Treichler, P. Wurzinger, H. Tews, and P. Zwicknagl to be published in J. Appl. Phys., June (1989)
4. M. Geva and T.E. Seidel, J. Appl. Phys., 59(7), 2408 (1986)
5. Y. Akahori, S. Hata, M. Ikeda, M. Yuda, Y. Kawaguchi, and S. Uehara, Electronics Lett. 25, 37 (1989)
6. W. Häussler, D. Römer, M. Plihal, Siemens Res. & Dev. Rep. 17(4), 177 (1988)
7. W. Häussler, Proc. ESSDERC'87, Bologna, Italy, 121 (1987)
8. K.W.Wang, Appl. Phys. Lett. 51, 2127 (1987)
9. T. Ambridge and M.M. Faktor, Inst. Phys. Conf. Ser. 24, 320 (1975)
10. W. Kuebart, Proc. E-MRS '86, Strasbourg, France, 245 (1986)
11. L. Vescan, J. Selders, M. Maier, H. Kräutle, H. Beneking, J. Cryst. Growth 67, 353 (1984)
12. W. Häussler, unpublished work (1987)
13. R. Strzoda, (private communication)
14. M. Gauneau, R. Chaplain, A. Rupert, E.V.K. Rao, N. Duhamel, J. Appl. Phys. 57(4), 1029 (1985)

OBSERVATION OF THE WURTZITE PHASE IN OMVPE GROWN ZnSe/GaAs:
EFFECT ON IMPLANTATION AND RAPID THERMAL ANNEALING

K.S. Jones, J. Yu, P.D. Lowen and D.Kisker*

Department of Materials Science and Engineering, University of Florida, Gainesville, FL 32611
*AT&T Bell Laboratories, Crawfords Corner Road, Holmdel, NJ 07733

ABSTRACT

Transmission electron diffraction patterns of cross-sectional TEM samples of OMVPE ZnSe on GaAs indicate the existence of the hexagonal wurtzite phase in the epitaxial layers. The orientation relationship is $(0002) // (111)$; $(11\bar{2}0) // (220)$. Etching studies indicate the phase is internal not ion milling induced. The average wurtzite particle size is 80Å-120Å. Because of interplanar spacing matches it is easily overlooked. Electrical property measurements show a high resistivity ($10^{10}\Omega$/square) which drops by four orders of magnitude upon rapid thermal annealing between 700°C and 900°C for 3 sec. Implantation of Li and N have little effect on the electrical transport properties. The Li is shown to have a high diffusivity, a solid solubility of $\sim 10^{16}$/cm^3 at 800°C and getters to the ZnSe/GaAs interface.

INTRODUCTION

Interest in ZnSe stems from its 2.7eV direct bandgap which makes it possible to fabricate a blue electroluminescent device.The problem with realizing this potential is the inability to produce stable p-type material. Ion implantation offers many potential advantages such as an extremely pure dopant source and low temperature processing. Several groups have tried to achieve p-type doping via implantation into ZnSe.[1,2,3] However, as with doping during growth, attempts to produce p-type ZnSe generally result in high resistivity films.

Before implantation or any other method of doping can be realized it appears the problem of high resistivity arising from impurities or self compensation[4] must be solved. Much work has been done over the past 8-10 years to develop high purity ZnSe layers grown by OMVPE and MBE. The resistivity can now be varied by adjusting the growth conditions. For example by adjusting the growth temperature one can drop the resistivity from 10^7 Ω-cm to 1 Ω-cm resulting in n-type material.[5] The high resistivity of ZnSe has been attributed to the so-called self-activated deep centers which compensate the films. These S-A peaks are believed to be associated with impurity-V_{zn} complexes.[1]

ZnSe has both a possible wurtzite and zincblende crystal structure and thus an inversion twin can occur. Inversion twins as well as other stacking defects in bulk ZnSe crystals grown by solid-solid reaction have been observed in high resolution TEM by Shiojiri et al.[6] Because of the nearly perfect match with the zincblende structure formation of the hexagonal phase is believed to have a very low free energy of formation. It has also been proposed that stacking defects (e.g. microtwins) can be electrically active either intrinsically (n-type, p-type or neutral)[7] or extrinsically based on aliovalent impurities lowering the electrostatic bonding energy of the interface.[8] The results of Shirojiri et al. indicate that the host ions themselves may behave as aliovalent impurities and could thus possibly create atomic interstitials or vacancies. Thus it is possible that planar defects such as the interface between wurtzite and zincblende regions may give rise to V_{zn} and could contribute to the high resistivity observed.

Transmission electron diffraction results indicate, for the first time, the hexagonal wurtzite phase can form during OMVPE growth of ZnSe/GaAs. Additional I-V resistance measurements indicates the ZnSe has a high resistivity ($>10^7\Omega$-cm) until between 700° C and 900°C rapid thermal annealing.

EXPERIMENTAL

ZnSe epitaxial layers on GaAs were grown by OMVPE at AT&T Bell Laboratories using either thermal decomposition or plasma assisted decomposition of the organometallic precursors.

The substrates were n-type (Si doped) for the thermal decomposition growth process and undoped for the plasma assisted growth process. Diethylselenium and diethylzinc were used as organometallic precursors. The growth temperature was 500°C for the thermal growth and 310°C for the plasma assisted growth. The Se/Zn ratio was varied between 2 and 4.

Four crystal high resolution X-ray rocking curve analysis of the films was done prior to implantation. The full width half maximum (004) was about 250 arc-seconds for both growth techniques. There was a slight (<1/3°) tilt between the substrate and the film, which is consistent with previously reported tilt values between ZnSe and Ge.[9] Photoluminescence results using either an argon ion laser (100mW/cm^2) or a pulsed nitrogen dye laser showed good room temperature band edge luminescence. Low temperature PL measurements are currently in progress. Cross-sectional TEM sample preparation was conventional except it was concluded by using reactive iodine ion milling. The milling conditions were 2.5 kV and 10-12° for 10-30 minutes to finish cleaning the surface. This dramatically reduced the ion milling damage in both the ZnSe and the GaAs substrates.

RESULTS

Figure 1a shows a cross-sectional TEM micrograph of a ZnSe film after optimized iodine ion milling. Bright field 2-beam (g_{220}) imaging conditions were used. It will be discussed why this reflection shows so many defects. Figure 1b shows the same sample after etching in Br$_2$/methanol to remove 1000-2000Å of ZnSe. The sample showed a marked improvement in the larger damage but the smaller points of contrast remained. Selected area diffraction of the etched sample (Figure 2a) at the (111) pole showed the existence of an extra set of reflections which have been indexed as the {1100} type reflections and the interplanar spacing matches exactly with the wurtzite ZnSe phase. The orientation relationship between the zincblende (cubic) and the wurtzite (hexagonal) phase is $(0002)_{hex} // (111)_{cubic}$ and $[1120]//[220]$. In addition, the interplanar spacing d_{0002}=3.25Å and d_{111}=3.27Å therefore g_{0002}=g_{111}. Also d_{1120}=1.997Å and d_{220}=2.003 therefore g_{1120}=g_{220}. Thus, the {0002} and {111} type reflections are superimposed as are the {1120} and {220} reflections. Thus only the {1100} reflections are seen. They appear as a faint spot surrounded by 3 spots. This may arise from multiple variants of the hexagonal phase. This is being further investigated.

Figure 2b shows the (111) SADP from the GaAs substrate. No extra spots were observed. Even upon heavily damaging the GaAs by Ar ion milling to the point where in the image there is a large amount of ion milling damage in the image, no hexagonal phase is observed. Figure 3a is the centered dark field from the (1100) reflection and shows a high concentration of the hexagonal phase (even after etching). The average domain size is 80-120Å. In Figure 3b & 3c weak beam dark field results also confirmed the orientation relationship described. For reflections g_{220} and g_{111} bright areas were observed as predicted because of the hexagonal superposition whereas for g_{040} there is no corresponding hexagonal reflection and nothing is observed.

The hexagonal phase was observed in both the thermally grown and plasma-assisted grown films. This phase would not appear in normal x-ray rocking curves as the hexagonal plane closest

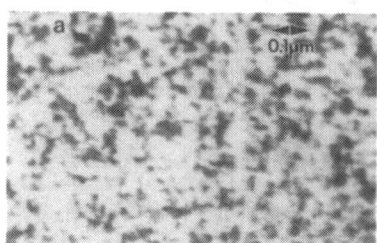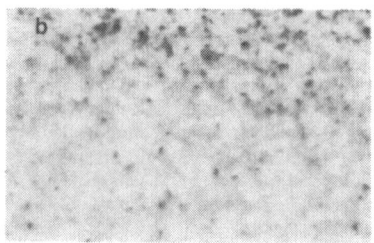

Figure 1. Bright field XTEM micrographs (g_{220}) of a ZnSe film after a) iodine ion milling and b) etching in Br$_2$/methanol.

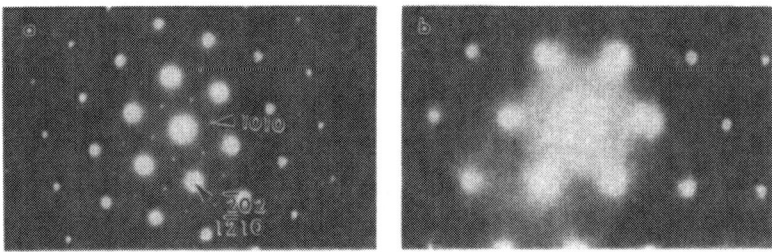

Figure 2. Selected area diffraction patterns at the (111) pole of the etched a) ZnSe thin film and the b) GaAs substrate.

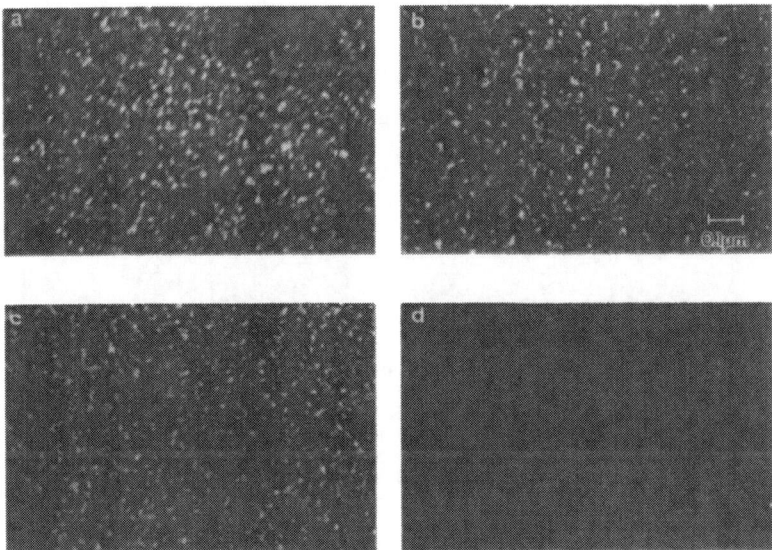

Figure 3. Dark field (DF) XTEM micrographs of a ZnSe thin film a) centered DF $g_{10\bar{1}0}$ b) weak-beam DF g_{220} c) weak beam DF g_{111} d) weak-beam DF g_{040}.

to being parallel with the surface ($\{20\bar{2}3\}$) is $\approx$3 degrees from the (004) reflection. The wurtzite phase also would not be observed in the (100) or ($\bar{1}10$) TED patterns because of the superposition. A Laue back reflection experiment is currently in progress to further study this phase.

IMPLANTATION RESULTS

Three sets of implantations were done. $^7Li^+$ was implanted at 30 keV to a dose of 1 x $10^{14}/cm^2$ into all 3 samples (A, B, C). No further implantations were done to sample A. Sample B was subsequently implanted with $^{14}N^+$ at 70 keV and a dose of 5 x $10^{13}/cm^2$ and sample C was implanted with $^{14}N^+$ at 70 keV with a dose of 5 x $10^{14}/cm^2$. Figure 4 shows cross-sectional TEM micrographs of the implant damage and the corresponding TRIM '88 Monte Carlo simulations for the ion concentration and distribution. For these ions (Li and N) at low doses,

342

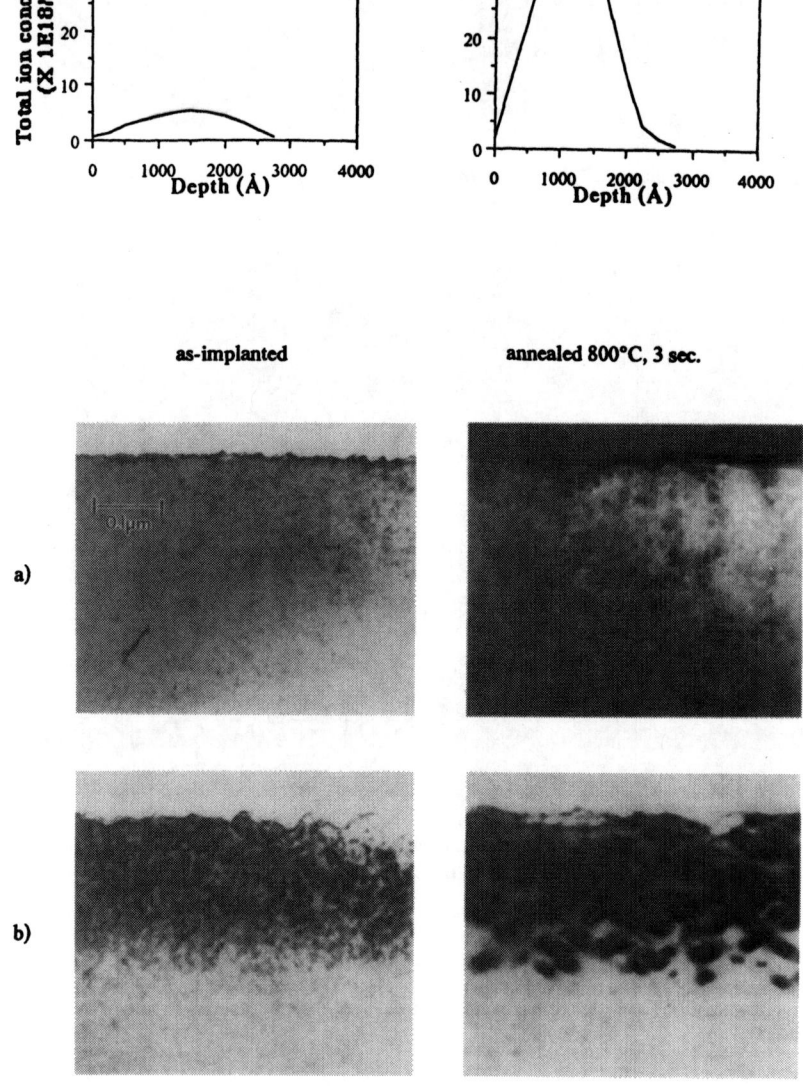

Figure 4. Bright field XTEM micrographs of ZnSe films implanted with a) 30 keV Li[+] to a dose of 1×10^{14}/cm2 and b) an additional 70 keV N[+] to a 5×10^{14}/cm2 dose. Also shown are the TRIM simulations of the impant profiles.

amorphization of the ZnSe was not observed and thus the defects formed are category I or sub threshold defects.[10] In Si the depth of these defects follows the projected ion range (not the damage density distribution). In ZnSe the defects appear to follow the damage density distribution better as they are shallower than the ion distribution (or possibly the TRIM simulation for ZnSe is inaccurate).

It is obvious that the concentration of implant damage is a strong function of dose (characteristic of category I damage). Figure 4 also shows the defect annealing kinetics. Rapid thermal annealing was used. The atmosphere was Ar as it was found that dry N_2 had some residual oxidizing effects. The sample was enclosed in a graphite boat (proximity capping). The sample was held at 250°C for 30 seconds prior to the higher temperature 3-second anneal in order to reduce thermal shock. The Li implant damage has completely annealed out by 800°C whereas the higher dose N_2 damage has coalesced into layer dislocation loops. This coalescence has been previously observed in ZnSe implanted with a higher dose of N^+.[11] This threshold of category I defect formation between $1 \times 10^{14}/cm^2$ and $5 \times 10^{14}/cm^2$ also matches well with the threshold in implanted Si of $2 \times 10^{14}/cm^2$.

The SIMS results in Figure 5 show the effect of 800°C RTA on the lithium distribution. The lithium has a very high diffusivity and getters quickly to the ZnSe/GaAs interface (presumably to the misfit dislocation network). The solubility of the Li appears to be around $10^{16}/cm^3$ at 800°C which would definitely limit its ability to dope ZnSe. The transport properties of the ZnSe are shown in Figure 6. I-V resistance was measured between two evaporated Au contacts on the cleaned surface. Four point Van der Pauw measurements were used to determine the sheet resistance. A close match was obtained to the I-V curve values as shown in figure 6. The sheet resistance remains very high ($\approx 10^9$ - $10^{10} \Omega$/sq.) (5μm total film thickness) until between 700°C and 900°C where the sample sheet resistance drops over four orders of magnitude. The high resistivity was confirmed with In contacts. This high resistivity for undoped ZnSe may also be associated with material that was grown "off" stoichiometry.[12]

The I-V curve with the Au contacts was very linear at higher temperatures implying the material might be p-type. Subsequent Hall effect measurements indicated the material was inhomogeneous yielding both positive and negative Hall coefficient signs after a 900°C RTA. The freeze-out slope of this inhomogeneous sample was ≈ 20meV or that of a donor. This drop in resistivity is most likely associated with impuritites. The sample 880622-2A was Li^+ implanted and showed no difference in electrical behavior from the unimplanted sample (22-2).

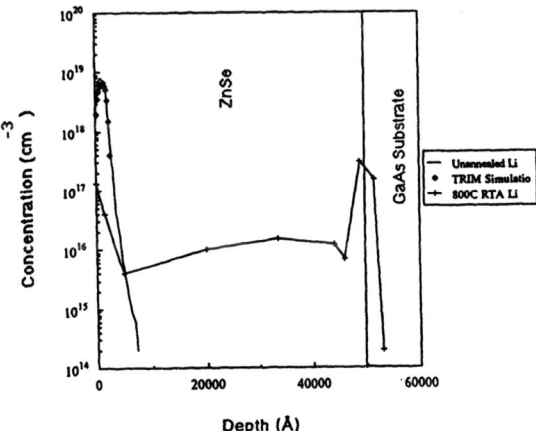

Figure 5. SIMS profiles of Li (30 keV $1 \times 10^{14}/cm^2$) redistribution upon rapid thermal annealing.

344

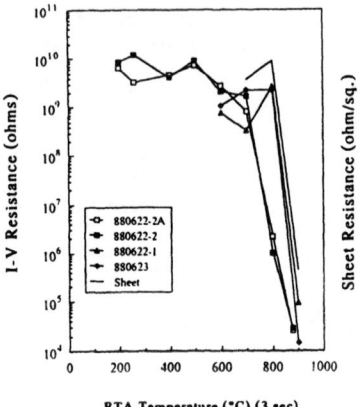

RTA Temperature (°C) (3 sec)

Figure 6. Transport properties of the ZnSe as a function of rapid thermal annealing temperature. Sample 22-2A was implanted with Li; Samples 22-2, 22-1 and 23 were unimplanted.

CONCLUSIONS

The hexagonal wurtzite phase was observed for the first time in OMVPE ZnSe/GaAs. Based on numerous etching results and studies of GaAs ion milling damage observation of this phase does not appear to be associated with electron or ion beam damage. Its existence could help explain the high resistivity of ZnSe. Implantation of ZnSe with Li and Li + N does not appear to have a dramatic effect on the transport properties upon annealing, although further studies using lower resistivity ZnSe are in progress. Li displays a low solubility and high diffusivity. $^{14}N^+$ implants alone would appear more promising as the compensating effects of interstitial Li donors would be removed. Further studies on the wurtzite phase in epitaxial films by high resolution x-ray diffraction are in progress.

ACKNOWLEDGMENTS

This work was supported by the DARPA Florida Initiative on optoelectronics. P. D. Lowen was supported by a U.F. Pittman Fellowship.The authors wish to thank K. Tokuda for assisting in the film growth, J. Eyler and S. Bates for assistance in SIMS and HRXRD analysis, T. Anderson and P. H. Holloway for use of electrical characterization equipment and numerous fruitful discussions and D. Venables and W. S. Rubart for their invaluable technical assistance.

REFERENCES

1. B.J. Skromme, N.G. Stoffel, A.S. Gozdz, M.C. Tamargo, and S.M. Shibli in <u>Advances in Materials Processing and Devices in III-V Compound Semiconductors,</u> edited by D.K. Sadana, L. Eastman and R. Dupuis (Mater. Res. Soc. Proc. <u>144,</u> Pittsburg, PA 1988) .
2. T. Yodo and K.Yamashita, Appl. Phys. Lett. <u>53,</u> 2403 (1988).
3. Z.L. Wu, J.L. Merz, C.J. Werkhoven, B.J. Fitzpatrick, and R.N. Bhargava, Appl. Phys. Lett. <u>40,</u> 345 (1982).
4. K. Kosai, B.J. Fitzpatrick, H.G. Grimmeiss, R.N. Bhargava, and G.F. Neumark, Appl. Phys. Lett. <u>35,</u> 194 (1979).
5. T. Yao, M. Ogura, S. Matsuoka, T. Morishita, Jap. J. Appl. Phys. <u>22,</u> L144 (1983).
6. M. Shiojiri, C. Kaito, S. Sekimoto, and N. Nakamura, Phil. Mag. A <u>46,</u> 495 (1982).
7. D.B. Holt, J. Mater. Sci. <u>19,</u> 439 (1984).
8. S.B. Austuman and W.G. Gehman, J. Mater. Sci. <u>1,</u> 249 (1966).
9. R.M. Park, J. Keiman and H.H. Mar, SPIE Conf. on Advances in Semiconductors and Semiconductor Structure, Baypoint, FL, March 22, 1987.
10. K.S. Jones, S. Prussin, and E.R. Weber, Appl. Phys. A <u>45,</u> 1 (1988).
11. J.S. Vermaak and J. Petruzzello, J. Appl. Phys. <u>55,</u> 1215 (1984).
12. R. Gunshor private communications

Si ION IMPLANTATION IN InAlAs/InGaAs HETEROSTRUCTURES

B. DESCOUTS, N. DUHAMEL*, E.V.K. RAO, Y. GAO, J.P. PRASEUTH
Centre National d'Etudes des Télécommunications
Laboratoire de Bagneux
196 avenue Henri Ravera - 92220 BAGNEUX - FRANCE
* Centre National d'Etudes des Télécommunications
route de Trégastel - 22301 LANNION Cedex

ABSTRACT

Si[29] implants have been performed in InGaAs and InAlAs single layers as well as in InGaAs/InAlAs heterostructures. Nearly 100% activation has been obtained in InGaAs after conventional furnace annealing or rapid thermal annealing. On the other hand, a low activation efficiency (30%) has been observed in InAlAs. A preliminary photoluminescence measurements study shows that an appreciable fraction of Si exists in the form of complex centers. Hall effect and specific contact resistivity maps carried out in the heterostructures indicate a very good homogeneity of the electrical parameters over a 4 cm² sample and give a satisfactory value of the specific contact resistivity (10^{-7} Ω.cm²).

INTRODUCTION

The high electron mobility of InGaAs and the large bandgap of InAlAs make the InGaAs-InAlAs system lattice matched to InP substrates very attractive for high-speed microwave devices as well as long wavelength photonic components. A number of transistors and lasers based on this heterostructure have already been realized [1,6]. Using mesa type structure on InAlAs/InGaAs/InP heterostructures, excellent device performances (a transconductance of 200 ms/mm on devices with 1 μm gate length and an unilateral gain cut-off frequency of 40 GHz) have been reported from our laboratory [1]. A further improvement in the performances can be expected from a planar technology where the access resistances for source and drain are particularly low. In this case, ion implantation is the most suitable technique to realize well controlled selectively doped regions.

Eventhough some experimental results have been reported on n-type implants in InGaAs [7,8], to our knowledge no results have been published on n-type implants in InAlAs. This work is a part of a program towards the realization of planar type InAlAs/InGaAs heterojunction field effect transistors (HFET) and concerns the optimization of Si implants and anneals to achieve low resistivity source and drain ohmic contacts.

In a first step, using both conventional furnace annealing (CFA) and rapid thermal annealing (RTA) at different temperatures, we have investigated the Si implants and anneals conditions in InAlAs and InGaAs single layers to achieve a high n-type doping level. Carrier profiles have been compared with Si atomic distributions. In order to understand the low electrical efficiency obtained in InAlAs, a preliminary study of photoluminescence (PL) measurements has been carried out.

In a second step, we have investigated the Si implants in InAlAs/InGaAs heterostructures. The electrical parameters (sheet resistance, sheet mobility, specific contact resistivity) of these Si implanted heterostructures have been characterized by Hall effect and specific contact resistivity maps.

EXPERIMENTAL PROCEDURES

The single layers and the heterostructures have been grown on Fe doped S.I. InP substrates by molecular beam epitaxy (MBE) in a Riber 2300 system with computer control [9,10]. The implants have been carried out at room temperature using a High Voltage Engineering 400 kV accelerator. The samples were 7° off the incidence direction to minimize channeling effects. To be compatible with the thickness of the different layers in the heterostructures (see figure 5) an implantation energy of 90 keV has been chosen. The implant dose was $1x10^{14}Si^{29}/cm^2$, which gave low specific contact resistivity as reported in Si implanted InP [11]. In accordance with LSS theory these conditions lead to a peak concentration of about $1x10^{19}$ cm^{-3} at 0.1 μm. Two annealing methods have been utilized : classical furnace anneals at 750°C for 15 min and rapid thermal anneals at 800° or 850°C for 10 s in an halogen lamps furnace [12] under flowing (Ar + 10% H_2) gas with the implanted surface in close contact with a silicon substrate.

The carrier concentration profiles were recorded using an electrochemical C-V profiler (Polaron PN4200). Atomic distributions of Si before and after anneal were obtained with a Cameca IMS 3F ion microanalyser equipped with a cesium source. The PL measurements were performed at 77 k using an argon-ion laser as an excitation source and a liquid nitrogen cooled Ge photodetector. Hall effect and specific contact resistivity maps have been obtained using a commercial instrument equipped with probe cards. The elementary test pattern containing a clover-leaf (200 μm x 200 μm) and a line for transient line method (50 μmx10 μm) was repeated each 2 mm in the X and Y directions. The sensitivity of the specific contact resistivity measurements was better than $1x10^{-7}$ $\Omega.cm^2$.

RESULTS AND DISCUSSION

Si implantation in InGaAs single layers

The thickness of the nominally undoped InGaAs layers used in this work ranged from 0.8 to 2 μm. Figure 1 shows the Si atomic distributions before and after anneals for a sample implanted with $1x10^{14}$ Si^{14}/cm^2 at 90 keV. For comparison, the carrier profiles obtained after anneal are also shown.

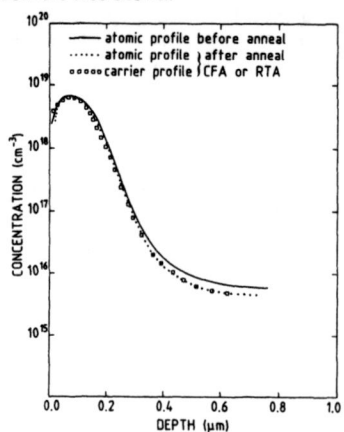

The excellent agreement between Si atomic distributions before and after anneal indicates no migration of Si (indiffusion or out diffusion). Besides, the carrier distributions which are identical for the two annealing processes and for the different annealing temperatures of RTA (800 or 850°C), follow very closely the Si atomic distribution indicating a total activation of implanted Si.

This result has been confirmed by Hall effect measurement on a sample annealed at 750°C for 15 min, which gave 95% activation with a sheet mobility of 2500 $cm^2/V.s$. This value of the mobility is in reasonable agreement with those reported for Si implanted material [8] as well as Si doped material by MBE [13].

Fig. 1 : Si atomic distributions before and after anneal, carrier distributions after anneal obtained in Si- implanted InGaAs ($1x10^{14}$ Si/cm², 90 keV) CFA : 750°C -15 min. ; RTA : 850 or 800°C -10s.

Si implantation in InAlAs single layers

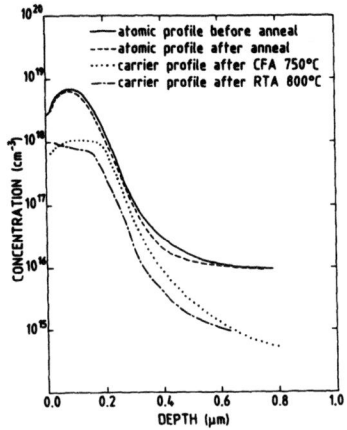

Fig. 2 : Si atomic distributions before and after anneal, carrier distributions after anneal obtained in Si⁺ implanted InAlAs (1x10¹⁴ Si/cm², 90 keV) CFA : 750°C -15 min. ; RTA : 800°C -10s.

The thickness of the layers used for this study ranged from 0.5 to 2 μm. The same Si implants as those in InGaAs have been performed. Figure 2 shows Si atomic concentration profiles before and after annealings, together with their carrier associated profiles after anneals.

Like in the case of InGaAs, no noticeable Si diffusion is detected after anneals. On the other hand, a comparison of carrier and atomic distributions clearly indicates a lower activation of Si implants in InAlAs. Furthermore, as is seen from carrier distribution CFA yields a better activation than RTA with a maximum concentration of $1x10^{18}$ cm⁻³, eventhough the later is carried out at a higher temperature. Hall effect measurements further confirmed these observations, typically 30% activation on CFA samples. This is not surprising considering the low Si activation reported in GaAlAs [14].

It is clear from the above data that identical Si implants and anneals lead to total activation in InGaAs but only to a partial activation in InAlAs. The non availability of any published data on donor implants in InAlAs lead us to suspect one of the following reasons to understand the low Si activation. Firstly a possible increase of the donor activation energy in the presence of Al like in the case of the much studied GaAlAs, for example the formation of deep donors. Secondly, because of its amphoteric nature, Si might take up arsenic sites (Si_{As}) and act as compensating acceptor. Lastly and most importantly, Si either on group III site (Si_{In},Si_{Al}) or group V site (Si_{As}) might interact with residual implant induced defects to form complex defect centers. Indeed, the latter hypothesis is often proposed to explain the low activation of high Si dose implants in InP [15].

To verify either of the above hypothesis, low temperature PL measurements have been carried out on Si implanted InAlAs samples.

PL measurements : Figure 3 shows the 77 k PL spectra recorded on virgin, Si implanted and furnace annealed samples. Let us remind here that the 2 μm thick InAlAs layer employed in this experiment is intentionally doped with $3x10^{16}$ Si/cm³ and has a lattice mismatch Δa/a of -4x10⁻³ suggesting a small excess in the Al content compared to the lattice matched condition. This is reflected in the high energy emission at 1.549 eV in the spectrum of the virgin sample. This emission contains unresolved band to band and donor to valence band transitions. The next high energy emission at 1.504 eV which shows an increased intensity after anneal possibly involves acceptor centers, for example voluntarily introduced Si (Si_{As}). This assignment is not inconsistent with SIMS data where higher Si atomic concentrations compared to carrier concentrations have been measured on two samples with different MBE doping level. Nevertheless it appears from curve c of figure 3 that the important consequence of the Si implant and anneal is the emergence of a new broad emission at 1.34 eV with an overall increase in the PL intensity.

A further confirmation of the involvement of Si in the 1.34 eV band is obtained from a comparative study of similar P and Si implants and anneals as can be seen from figure 4. Let us remind that the atomic masses of P and Si are similar and consequently they lead to the same density of implant induced defects. It is evident from figure 4 that Si implants alone ($\emptyset = 1\times10^{13}$ ions.cm⁻², $\emptyset = 1\times10^{14}$ ions.cm⁻²) result in the emergence of the 1.34 eV band whereas no major spectral evolution is observed for P implants. Besides, one can further notice a dependance of the intensity of this emission on silicon dose.

In light of the above data on PL measurements the following can be ascertained : firstly the possibility of incorporation of Si as acceptors and secondly an unambiguous participation of Si in the 1.34 eV band. Also from the broad nature of this PL signal we think that the center responsible for this emission involves implant induced defects in addition to Si atoms. From this we presume that the existence of this center renders Si inactive and explains at least partially low electrical activation of Si implants in InAlAs.

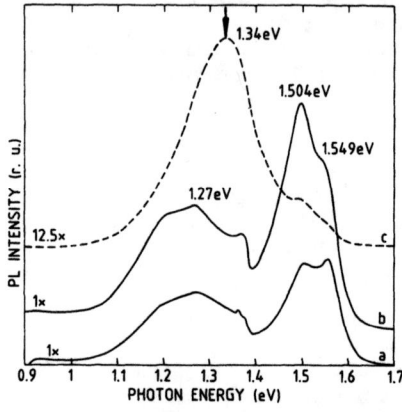

Fig. 3 : 77 K PL spectra recorded on original (a), furnace annealed (b), and Si⁺ implanted (1x10¹⁴ Si/cm², 90 keV) and furnace annealed (750°C - 15 min.) (c) samples. "12.5X" signifies that the PL intensity of curve c is 12.5 times greater than curves b and a. The detection of 1.34 eV band in sample c should be noted.

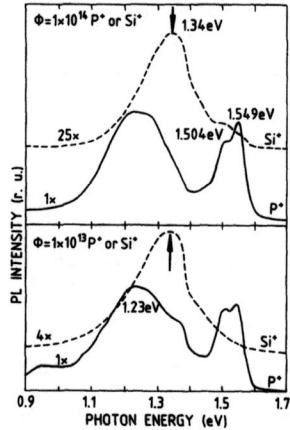

Fig. 4 : Comparison of 77 K PL spectra recorded on samples implanted with Si⁺ and P⁺ and subjected to identical annealings. Note the emergence of 1.34 eV band only in Si⁺ implanted samples.

Si implants in InAlAs/InGaAs heterostructures

As is mentioned earlier in the introduction part, the principal aim of Si implants in InAlAs/InGaAs is to contact the active InGaAs layer in HFET devices. Si implants similar to those described previously (E = 90 keV, $\emptyset = 1\times10^{14}$ Si/cm²) have been performed in several heterostructures. A typical example of implant schedule is shown in figure 5 representing the detail of an heterostructure together with the SIMS measured Si atomic distribution prior to anneal.

It can be seen here that the peak of Si concentration is close to the upper heterointerface of the active layer. This signifies that a large fraction of Si is present also in the active layer. Such structures after anneal have been characterized by performing sheet Hall effect measurements. Our typical results after CFA are a sheet resistance (R_s) of 41.2 $\Omega/\quad$, sheet concentration (N_s) of 4.1×10^{13} cm⁻² and sheet mobility (μs) of 3600 cm²/V.s ; and after RTA R_s = 46.7 $\Omega/\quad$, N_s = 3.8×10^{13} cm⁻² and μ_s = 3850 cm²/V.s. Let us emphasize here that these values are characteristic of the composite of InAlAs and InGaAs layers since Si distribution is present in both layers. The homogeneity of these electrical parameters have been characterized and specific contact resistivity maps using TLM have been performed over a surface of 4 cm². Our best results gave an average sheet resistivity (figure 6) of 41.75 $\Omega/$ with a standart deviation of 2.10 $\Omega/\quad$, i.e a variation of 5%. The average sheet mobility was 2800 cm²/V.s, and the sheet concentration 5×10^{13} cm⁻² Figure 7 shows that 98% of the measured specific contact resistivities are lower or equal to 1×10^{-7} Ωcm².

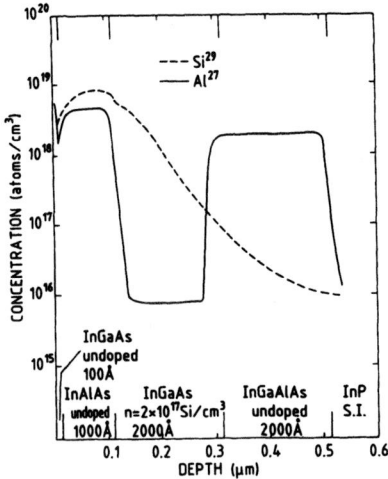

*Fig. 5 : Si atomic distribution in a Si⁺
implanted InAlAs/InGaAs heterostructure
(1x10¹⁴ Si/cm², 90 keV) prior to annealing.*

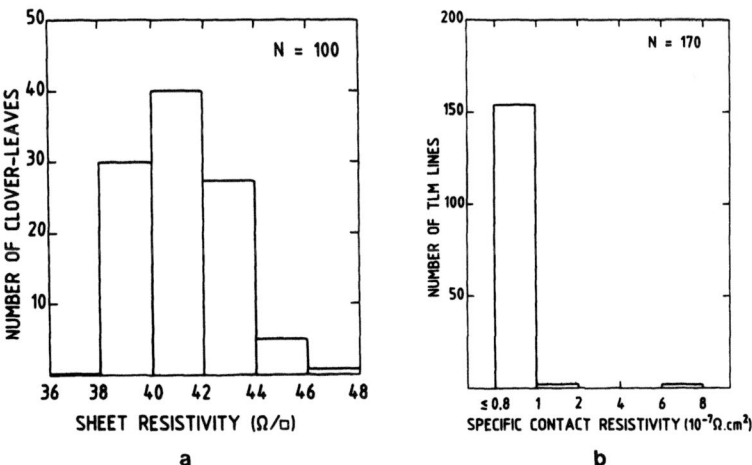

Fig. 6 : Histograms obtained from a Si⁺ implanted (1x10¹⁴ Si/cm², 90 keV) and furnace annealed (750°C - 15 min.) InAlAs/InGaAs heterostructure. The multilayer structure is schematized in figure 5. (a) sheet resistance, (b) specific contact resistivity.

The contact resistance R_c measured is typically 0.1 Ω over a distance of 50 μm which corresponds to an access resistance of 5×10^3 Ωmm. For a typical HFET with an InGaAs active layer of about 0.1 μm doped to 2×10^{17} Si/cm³ and a drain to source spacing of 1.5 μm the estimated resistance of the layer is about 1 Ωmm. This shows that the Si implants described in this paper can be satisfactorily employed to contact the active layer of InAlAs/InGaAs devices.

CONCLUSION

We have shown in this work that Si implants (E = 90 keV, $\varnothing = 1 \times 10^{14}$ ions/cm²) lead to nearly total activation in InGaAs single layers, but only to 30% activation in InAlAs.

By performing low temperature PL measurements we have unambiguously identified that an appreciable fraction of Si implanted in InAlAs exists in the form of complex centers even though its incorporation on As site cannot be totally neglected.

Finally, we have demonstrated that Si implants can be successfully employed for contacting the InGaAs active layer in InAlAs/InGaAs HFET.

ACKNOWLEDGEMENTS

The authors gratefully acknowledge P. Krauz for his help in implantations, L. Nguyen for the technological processes and H. Thibierge for PL measurements.

1. L. Giraudet, N.L. Nguyen, M. Allovon, M. Laporte, A. Scavennec, presented at the 12ª workshop on compound semiconductor devices and integrated circuits, Lugano, Zwitzerland (1988)
2. C.Y. Chen, A.Y. Cho, K. Alavi, P.A. Garbinski, IEEE Electron Device Lett., EDL-3(8), 205 (1982)
3. T.P. Pearsall, R. Hendel, P. O'Connor, K. Alavi, A.Y. Cho, IEEE Electron Device Lett., EDL-4(1), 5 (1983)
4. C.Y. Chen, A.Y. Cho, K.Y. Cheng, J.P. Pearsall, P. O'Connor, P.A. Garbinski, IEEE Electron Device Lett., EDL-3(6), 152 (1982)
5. W. Lee, C.G. Fonstad, IEEE Electron Device Lett., EDL-7(12), 683 (1986)
6. W.T. Tsang, J. Appl. Phys., 52(6), 3861 (1981)
7. A.N.M.M. Choudhury, N.J. Slater, K. Tabatabaie-Alavie, C.G. Fonstad, Appl. Phys. Lett., 40(7), 607 (1982)
8. T. Penna, B. Tell, A.S.H. Liao, J.J. Bridges, G. Burkhardt, J. Appl. Phys., 57(2), 351 (1985)
9. J.P. Praseuth, M.C. Joncour, J.M. Gerard, P. Henoc, M. Quillec, J. Appl. Phy., 63(2), 400 (1988)
10. J.P. Praseuth, L. Goldstein, P. Henoc, J. Primot, G. Danan, J.Appl.Phy., 61(1), 215 (1987)
11. N. Duhamel, B. Descouts, P. Krauz, E.V.K. Rao, J. Dangla, P. Henoc in Rapid Thermal Processing of Electronic Materials, edited by S.R. Wilson, R. Powell, D.E. Davies (Mater. Res. Soc. Proc. 92, Pittsburgh, PA 1987) pp.443-448
12. Halogen lamps furnace from SITESA-ADDAX, MONTBONNOT (FRANCE)
13. K.Y. Cheng, A.Y. Cho, J.Appl.Phys., 53(6), 4411 (1982)
14. C.S. Lam, C.G. Fonstad, J.Appl.Phys., 64(4), 2103 (1988)
15. N. Duhamel, E.V.K. Rao, M. Gauneau, H. Thibierge, A. Mircea, J.Cryst.Growth, 64, 186 (1983)

STUDIES ON PHOSPHORUS-IMPLANTATION AND THE ANNEALING OF CADMIUM TELLURIDE AND COPPER INDIUM DISULFIDE

Y. J. HSU AND H. L. HWANG
Dept. of Electrical Engineering, National Tsing Hua University, Hsinchu, Taiwan 30043, Republic of China
H. Y. UENG
Dept. of Electrical Engineering, National Sun Yat-Sen University, Kaohsiung, Taiwan 80424, Republic of China

ABSTRACT

The doping efficiencies obtained in phosphorus-implanted cadmium telluride and copper indium disulfide annealed by pulse electron beam were higher than those annealed by conventional thermal method. To get insights into this phenomenon, electron paramagnetic resonance measurements were performed for both crystals at various stages during the doping process. The results indicated that the pulse electron beam annealing could effectively eliminate the phosphorus interstitials in the implanted crystals but the thermal annealing could not. This shows the significant effect of melting crystals by pulse electron beam annealing to obtain high doping efficiencies.

INTRODUCTION

To control the electrical conduction by extrinsic impurities is basic in fabricating electronic devices. However, while the n-type conduction is easily controlled in CdTe and $CuInS_2$, the p-type conduction control in these materials has been difficult. In particular, the doping efficiency is not high.[1]

In our previous work, high doping efficiencies were obtained in CdTe and $CuInS_2$ by using phosphorus implantation and pulse electron beam (PEB) annealing.[2,3] Hole concentrations higher than 10^{19} cm^{-3} were reached in both materials, which could not be achieved by conventional thermal annealing. In the present work, electron paramagnetic resonance (EPR) measurements were performed to get more insights into this phenomenon.

EXPERIMENTAL

The $CuInS_2$ single crystals were grown by the travelling heater method (THM)[4] and the chemical vapor transport (CVT) method.[5] The

CdTe single crystals were grown by THM. After polishing and surface etching, phosphorus implantations were carried out at room temperature, at an incident energy of 100 KeV and at various doses. PEB annealing was performed as described in Ref.2. Some of the implanted samples were subjected to thermal annealing at 300-400 °C for the time periods which range from tens of minutes up to about one day and some others were subjected to PEB annealing. In the PEB annealing, the electron beam current ranged between 1 and 60 micro-Amperes and an about 2000 A thick Ta_2O_5 cap was deposited on the sample by electron-gun evaporation to prevent the crystal from dissociation during the annealing. It was found that the capless CdTe showed no significant difference from the capped ones.

The EPR measurement was performed at room temperature using a Bruker 200D 10/12 system with the microwave frequency in the X-band. A sample holder was designed so as to allow the rotation of samples in the cavity. Four kinds of samples(as-grown, P+-implanted, P+-implanted and thermally annealed, as well as P+-implanted and PEB annealed) were subjected to the EPR measurement.

RESULTS

Cadmium Telluride

No EPR signals were detected for the as-grown and the PEB-annealed samples. For the other two kinds of samples, similar spectra were obtained. Fig.1 shows the typical ones. With the magnetic field in the (110) plane, a signal was detected with the g value ranging only slightly around 2.20. The linewidth is about 125 G. The signal is accompanied with the vague satellites.

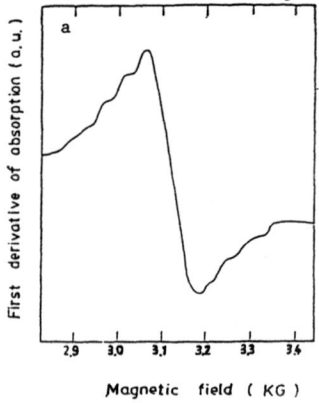 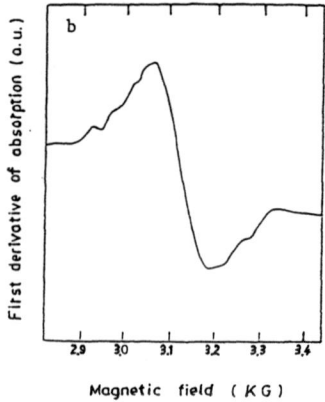

Fig.1 EPR spectra of CdTe single crystals. (a) phosphorus implanted; (b) phosphorus implanted and thermally annealed.

It is not easy to detect EPR signals for undoped CdTe crystals. [6,7,8] Even at 4 K only a very small signal has been reported. [6] This may be due to the fact that most of the defects[9] in undoped CdTe crystals are not paramagnetic and/or to that paramagnetic centers are of low concentrations or short-living.

From Fig.1, the lack of clear hyperfine interaction signals indicates that the unpaired electron is not located near phosphorus(100% natural abundance of I =1/2 atoms) and cadmium(25% natural abundance of I =1/2 isotopes), but rather at tellurium(8% natural abundance of I =1/2 isotopes). The shift of the g value in CdTe has been discussed in Ref.8 and our observed positive g shift coincides with the involvement of Te atoms. The broad linewidth shows the interactions of the electron with the surrounding atoms. By the self-consistent field calculation of the atomic structures, the quantity $< r^{-3} >_p$ for the 5p orbitals of Te atoms was estimated to be about 95×10^{24} cm^{-3} which corresponds to the separation between the symmetric satellites. Along with the vagueness of the satellites, which coincides with the small abundance of I =1/2 Te isotopes, we conclude that the EPR signal come from the electrons near Te atoms.

Copper Indium Disulfide

Both the THM and the CVT samples showed similar results. At the as-grown stage, EPR signals were observed in some samples. From our chemical analysis, [10] the existence of these signals was seen not to depend on the composition of the crystals and thus they were not due to the intrinsic defects. Transition metals, especially Fe, are major contaminants in the as-grown ternary compounds and their EPR signals have been studied for a long time. [11,12] Our signals

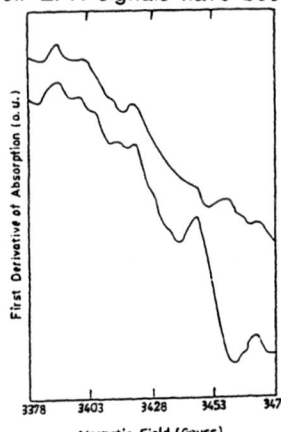

Fig.2 The EPR signal in the as-P$^+$-implanted CuInS$_2$ single crystal (the lower curve). The upper curve is of the as-grown CuInS$_2$.

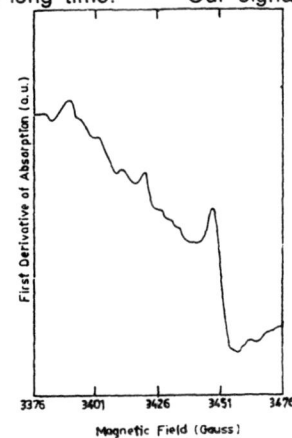

Fig.3 The EPR signal in the P$^+$-implanted and thermally annealed CuInS$_2$ single crystal.

were identified to belong to this category.

At the as-implanted stage, in addition to the signals appearing at the as-grown stage, all the implanted crystals showed a single-line signal as shown in Fig.2. Since it appeared after the implantation, its correlation with the implantation induced defects is obvious. This signal is isotropic and has a g-value equal to 2.0011. Its linewidth is only 15 Gauss wide. The extraordinary narrow linewidth requires exchange effects, therefore high local defect density, which just corresponds to the characteristics of the implants distribution. The lack of any hyperfine structure indicates that the unpaired electrons do not center at copper, phosphorus, and indium atoms, but rather at sulfur atoms, and the isotropy as well as the g-value indicate the s-orbital property. To be the source of this signal, the sulfur 4s-orbital is the most probable candidate.

As to the annealed stages, the g=2.0011 signal remained in the thermally annealed samples (as shown in Fig.3) but disappeared in the PEB annealed ones.

DISCUSSION

During implantation, damages such as vacancies, substitutionals and interstitials are produced in the target crystals. These implantation-induced defects may play important roles in the conduction of carriers. A procedure estimating the types and

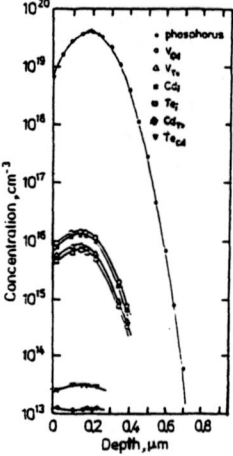

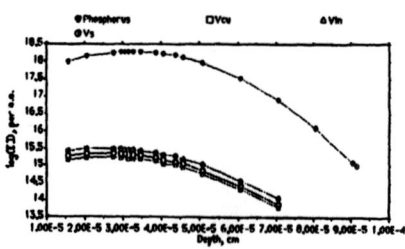

Fig.4 Distributions of the phosphorus and the implantation induced defects in CdTe. (E= 100KeV, dose=10^{15} cm^{-2}).

Fig.5 Distributions of the phosphorus and the implantation induced defects in CuInS2. (E=100KeV, dose=10^{14} cm^{-2}).

quantities of them in CdTe had been developed and the details were described in Ref.13. Figs.4 and 5 show the typical results. It was noted that the number of implanted phosphorus is about three orders higher than those of displaced atoms. therefore, after implantation, only about one out of a few thousands of phosphorus atoms has the chance to occupy the normal lattice site, and the others are located at interstices. To obtain good doping effects, it is necessary to eliminate these phosphorus interstitials.

There are two types of interstices in CdTe single crystals: one is surrounded by Cd atoms and the other by Te atoms. By the electronegativity considerations, most phosphorus interstitials are of the former type and are expected to be more thermally stable than the other type. In this atom arrangement, because of the larger electronegativity of phosphorus atoms, they would break the nearest Cd-Te bonds and construct the bondings between them and the nearest Cd atoms, thus leaving dangling bonds near tellurium atoms which should give rise to EPR signals. Besides, the expected thermal stability of these phosphorus interstitials should lead to the observation of the same EPR signals in the thermally annealed CdTe samples as in the as-implanted ones.

It is similar in the case of $CuInS_2$. After the phosphorus implantation, the majority of the implants stop at the interstitials each of which is surrounded by four nearest sulfur atoms. Because of the large electronegativity of the sulfur atoms, the electrons tend to be attracted from the copper, indium and phosphorus atoms to the sulfur atoms. Part of the sulfur 4s-orbitals would be half filled due to the extra electrons from the phosphorus interstitials and thus should give rise to EPR signals.

All these reasonings are confirmed by our experiments. The fact that the EPR signals at the as-implanted stage still showed up in the thermally annealed samples indicates that the thermal annealing could not effectively remove the phosphorus interstitials. Besides, it is natural that those signals observed in the as-grown $CuInS_2$ existed at the as-implanted stage since the ion implantation only affects the near surface region.

Either laser or electron pulses can be used for transient annealing. The beams couple energies to the targets by different mechanisms but they produce ultimately the same effect, heat, for annealing samples. The samples may or may not be melted to certain depths under different experimental conditions, and it was expected that the melting of the targets plays an important role in eliminating the implantation-induced damages. In our experiment, good doping effects were obtained in the melted samples. The melting depth was about 4 μm which is deep enough to anneal the implanted region. During the PEB annealing, the crystal surface is melted and then recrystalises. Because of the fluidity and the chemical nature, the phosphorus atoms have greater chance to occupy the tellurium sites and the sulfur sites in CdTe and $CuInS_2$ respectively. After the annealing, the amount of the phosphorus interstitials should be small enough to make the EPR signals showing up at the as-implanted stage disappear and the substitutional phosphorus should act as acceptors. All these can also be verified by our experiments as well as the good match between the hole concentration profiles and the phosphorus redistribution profiles. [13]

As a conclusion, EPR experiments have been done on the phosphorus-implanted CdTe and $CuInS_2$ at different stages to reveal the ineffectiveness of the thermal annealing and the significance of the melting effect in the PEB annealing on eliminating the phosphorus interstitials to achieve high p-type doping efficiencies.

ACKNOWLEDGEMENT

The financial support from the National Science Council of the Republic of China is acknowledged.

REFERENCES

1. H. Y. Ueng and H. L. Hwang, Solid State Phenomena 1&2, 343 (1988).
2. C. B. Yang, M. L. Peng, J. T. Lue and H. L. Hwang, IEEE Trans. Electron Devices ED-32, 2293 (1985).
3. J. L. Lin, L. M. Liu, J. T. Lue, M. H. Yang, and H. L. Hwang, J. Appl. Phys. 59, 378 (1986).
4. H. J. Hsu, M. H. Yang, R. S. Tang, T. M. Hsu, and H. L. Hwang, J. Crystal Growth 70,183 (1984).
5. C. Y. Sun, H. L. Hwang, C. Y. Leu, L. M. Liu, and B. H. Tseng, Jpn. J. Appl. Phys. 19, 81 (1980).
6. A. Goltzene and C. Schwab, Rev. Phys. Appl., 12,199(1977).
7. K. Saminadayar, D. Galland and E. Molva, Solid State Commun. 49, 627 (1984).
8. R. M. Bilbe, J. E. Nicholls and J. J. Davies, Phys. Stat. Sol. (b) 121, 339 (1984).
9. F. A. Kroger, Rev. Phys. Appl. 12, 205 (1977).
10. H. L. Hwang, L. M. Liu, M. H. Yang, J. S. Chen, J. R. Chen, and C. Y. Sun, Solar Energy Materials 7, 325 (1982).
11. S. Geschwind, Phys. Rev. 121, 363 (1961).
12. J. M. Tchapkui-Niat, A. Goltzene, and C. Schwab, J. Phys. C 15, 4671 (1982).
13. Y. J. Hsu, H. L. Hwang and C. Y. Sun, J. Crystal Growth 86, 749 (1988).

THE STUDY OF BORON IMPLANTATION IN 1:7 Nd/Fe MULTILAYERS

GUOAN CHENG* AND YINGGUO PENG*
*Institute of Applied Physics, Jiangxi University, Nanchang, The People's Republic of China

ABSTRACT

The structures and magnetic characteristics of multilayered samples with alternatively deposited metal layers (Nd and Fe) were studied for 80 keV boron ion implantation to various doses. Boron ion implantation induced amorphization in Nd/Fe multilayers gradually with increasing irradiation dose. The hysteresis loops of magnetization of the samples showed that a sunken curve appeared in the second and fourth quadrants of the loops for the deposited Nd/Fe films and disappeared when implantation dose was more than $4*10^{16}$ B^+/cm^2 . During ion irradiation, the coercive forces of the films were decreased. The saturation magnetization and the residual magnetization decreased first and then increased with increasing ion dose, the squareness ratio S.R increased first and then decreased. Discussion was carried out in this paper.

INTRODUCTION

Rare earth-transition metal (Re-Tm) films have attracted considerable attention recently. This follows from the novel tapes of random magnetism(1) and phase transition(2) induced by local random anisotropy, and also because certain of these alloys in thin-film form exhibit perpendicular magnetic anisotropy which gives them potential as erasable magneto-optic storage media(3), such as systems of Gd-Co, Gd-Fe, Tb-Fe and Sm-Co, etc.

Many experimental results shows that alloy elements make the structures and the magnetization of Re-Tm film samples change, such as B, Ti and Al, etc. Tsutsumi and Sugahara(4) have recently reported that perpendicularly manetized films of Fe-Nd-Ti could be prepared by RF sputtering. H. Chen has also reported that perpendicular magnetized films of Co-Cr-N alloy could be deposited by magneto-sputtering introducing N_2 gas into the chamber(5).Nd-Fe-B based permanent magnets have been successfully sputtered in thin film form with intrinsic coercive forces, iHc up to 16 kOe(6). From the above results, it is noted that nonmagnetic elements introduced into alloy films can induce changes of magnetic-anisotropy and magnetic-stability.

Ion implantation is a special technique in the investigation of metallurgy. Many amorphous alloys(7), quasicrystalline phases (8), and new magnetic alloys(9) have been formed by ion beam techniques. Amorphous Gd-Co was formed by ion beam mixing, giving a polar kerr rotation of about 0.17^0 which was approximate to that of alloy films formed by sputtering techniques(10). Boron ion implantation in pure Fe and Ni films showed that the magnetization of the films decreased during ion irradiation(11). However, the structures and magnetization of Nd-Fe alloy films with dfferent concentrations of boron have not been reported recently.

The present work was motivated by the above considerations. The structures and magnetic characteristics of Nd/Fe multilayers were studied by boron ion implantation to various doses.

EXPERIMENTAL PROCEDURE

The Nd/Fe multilayers were prepared with a multiple-resisti-
vity evaporating system after the system was pumped to a base
pressure of more than 10^{-5} Torr. A (111) Si wafer, on which SiO_2
film with 250 nm thickness was deposited, acted as the substrate,
and two Nd films were inserted among three Fe films. The total
thickness of the multilayers was about 200 nm.
The composition of the Nd/Fe films was determined by means
of the Electron-Energy-Dispersive Spectrum (EDS). The X-ray diffraction data of the alloy films were obtained using Cu radiation.
The magnetic data were obtained at room temperature using a Vib-
rating Sample Magnetometer (MODEL-9500) in field up 1000 Oe. The
magnetic field was parellel to the film plane. Ion implantation
was carried out to various doses.

RESULTS AND DISCUSSION

The composition of Nd/Fe multilayers measured using the EDS
technique before boron ion implantation is shown in table I. From
table I, it can see that the designed composition of the sample
is near to the observed composition. The composition of irradiated
Nd/Fe multilayers are the same as that of deposited samples.

Table I. Comparison of composition of 1:7
Nd/Fe multilayered samples between designed
and observed by EDS technique.

ELEMENT	Nd(%at)	Fe(%at)
DESIGNED COMPOSITION	12.5	87.5
OBSERVED COMPOSITION	12.2	87.8

The Nd/Fe multilayers were irradiated with 80 keV boron ions
to doses of $1*10^{16}$ B^+/cm^2, $4*10^{16}$ B^+/cm^2 and $8*10^{16}$ B^+/cm^2. The
boron concentrations in the irradiated multilayers were calculated
and are listed in table II. It is noted that different boron concentrations in Nd/Fe multilayers can be introduced by the boron
ion implantation to various doses, and that magnetic Nd-Fe-B can
be realized with an ion dose of $8*10^{16}$ B^+/cm^2. This is to say
that Nd-Fe-B films can be prepared using ion beam techniques.
The X-ray diffraction patterns of the deposited and the irradiated Nd/Fe multilayers are showed in fig. 1. Fig. 1 indicates
that there are crystalline lines in the diffraction pattern of
the deposited multilayers. Checking with the available ASTM cards
and literature, some of the lines could not be indexed to any
known Nd-Fe intermetallic phase structures. This means that a new
Nd-Fe intermetallic phase has been formed at the interface of Nd
layers and Fe layers during evaporation. X-ray diffraction patterns
of the irradiated films showed that there were only diffuse peaks

which were diffraction peaks of amorphous phases. Other lines of

Table II. The composition of 80 keV B^+ ion implantation with various doses in 1:7 Nd/Fe multilayered samples.

DOSE (ions/cm^2)	$1*10^{16}$	$4*10^{16}$	$8*10^{16}$
COMPOSITION (%at)	0.73	2.84	5.33

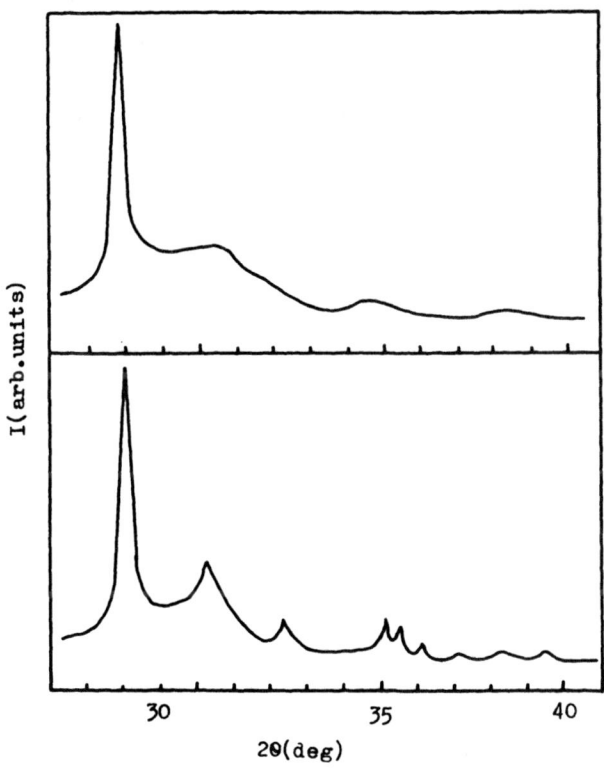

Fig. 1. X-ray diffraction patterns of Nd/Fe multilayered samples. a) Virgin; b) $8*10^{16}$ B^+/cm^2.

the new intermetallic phase disappeared during the ion irradiation.
According to the above results, boron ion implantation has gradually induced amorphization of Nd/Fe multilayers and Nd-Fe-B
amorphous alloy was formed at high dose irradiation.

The hystersis loops of deposited samples and irradiated multilayers to various doses measured parallel to the film plane are
shown in fig. 2. From fig. 2(a), it can be seen that a sunken
curve appeared in the second and the fourth quadrants of the loops
of the deposited Nd/Fe films. This mean that there are two ferromagnetic phases present in the films. This is consistant with the
X-ray diffraction patterns above. The two ferromagnetic phases
with different coercive forces induced the sunken curve in the
hystersis loops of the deposited Nd/Fe multilayers. After ion
processing to various doses, the sunken curve gradually disappeared
in the loops.

At the same time, the saturation magnetization σ_s, the resdual magnetization σ_r and the coercive force H_c of the Nd/Fe multilayers also changed during ion implantation. The saturation
magnetization σ_s decreased from 109.3 EUM/g to 98.2 EUM/g first
and then increased to 128.3 EUM/g with increasing of boron ion
dose. There is a minimum value of saturation magnetization σ_s of
the films at a dose of $1*10^{16}$ B^+/cm^2. The resdual magnetization σ_r
also changed as the saturation magnetization σ_s above during ion
irradiation. On the contrary, the squareness ratio S.R of the
films increased from 0.8165 to 0.8388 first and then decreased
quickly to 0.7686. There is a maximum value of the squareness
ratio at a dose of $1*10^{16}$ B^+/cm^2. During boron ion implantation,
the coercive force H_c quickly decreased from 99.58 Oe to 55.96 Oe.
These results are contrary to that of 1:2 Nd/Fe multilayers(12)
and the pure metal films(1) irradiated with boron ion beam. The
magnetic data of the Nd/Fe multilayers are listed in table III.

Table III. The magnetic data of 1:7 Nd/Fe multilayered
samples irradiated with 80 keV boron ion to various doses

DOSE (ions/cm^2)	σ_s (EUM/g)	σ_r (EUM/g)	S.R	Hc (Oe)
VIRGIN	109.3	89.22	0.8165	99.58
$1*10^{16}$	98.2	82.37	0.8388	79.66
$4*10^{16}$	109.2	87.82	0.8043	60.40
$8*10^{16}$	128.3	98.62	0.7686	55.96

Interaction between ions and atoms in solids can induce the
changes in materials structures and properties. Because energetic
ions gradually deposit energy and are stopped in the solid during
the collisions with atoms in the solid, there is a Gaussian's distribution of the ions in the solid. At the same time, the formation of a mixing layer is induced by collisions among ions and
atoms in the solid. The alloy phase in the mixing layer may be

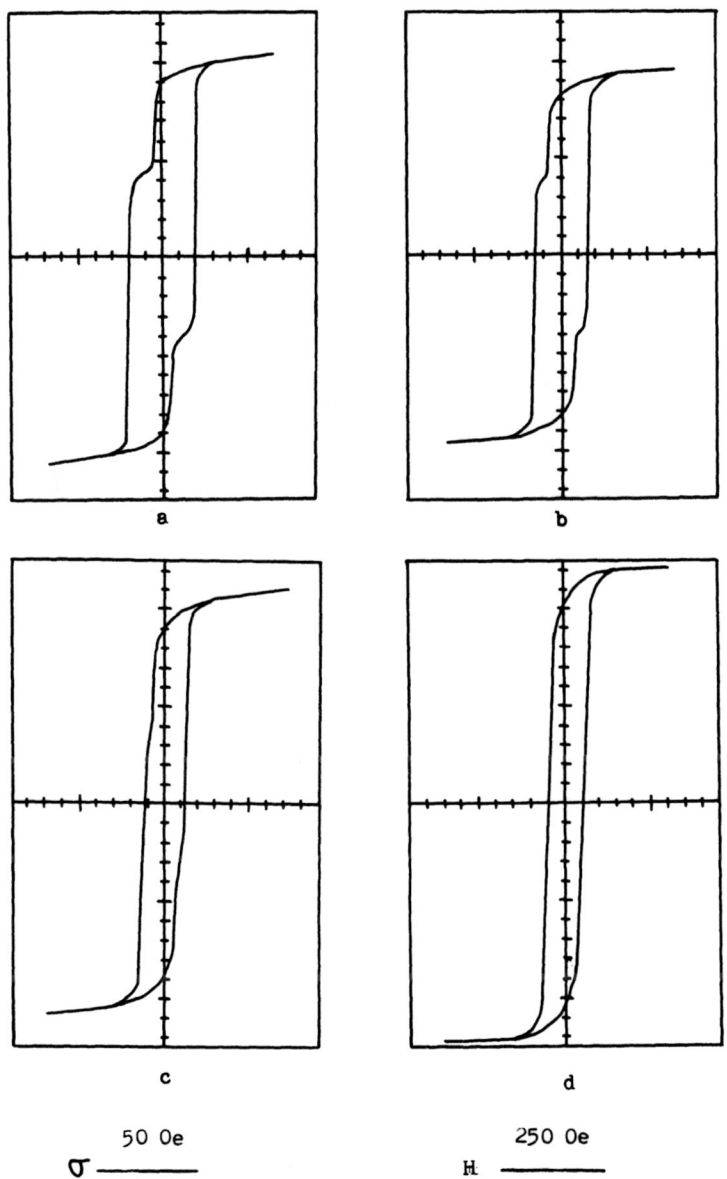

50 Oe   250 Oe

σ ——————   H ——————

Fig. 2. The hystersis loops of Nd/Fe multilayered samples measured by VSM. a) Virgin; b) $1*10^{16}$ B^+/cm^2; c) $4*10^{16}$ B^+/cm^2; d) $8*10^{16}$ B^+/cm^2.

362

amorphous or quasicrystal or new crystal or complex phases. For the present work, fig. 1 indicats that magnetic Nd-Fe-B amorphous was formed during boron ion implantation. This is due to the formation of Nd-Fe mixing layers and the introduction boron which is an element of alloy amorphization. The saturation magnetization σ_s, the residual magnetization σ_r and the coercive force Hc also changed with the changing of structures of Nd/Fe multilayers during ion implantation. Further research is planned for these systems.

CONCLUSIONS

Boron ion implantation induced the formation of magnetic Nd-Fe-B amorphous in the 1:7 Nd/Fe multilayers. During ion irradiation, the coercive force Hc of the films gradually decreased. The saturation magnetization σ_s the residual magnetization σ_r of Nd/Fe films decreased first and then increased. However, the squareness ratio S.R increased first and then decreased with increasing of irradiation dose. There is a minimum value in the saturation and the residual magnetization respectively, and a maximum value for the squareness ratio at an irradiation dose of $1*10^{16}$ B$^+$/cm^2.

ACKNOWLEDGEMENTS

We thank J.G. Zhao for his assistance and helpful discussions, S.Y. Chen for the analysis of X-ray diffraction, and S.G. Wang for the preparing samples. For financial support, we are indebted to the Magnetism Laboratory, Academia Sinica.

REFERECES

1. K.Moorjani and J.M.D.Coey, Magnetic Glasses (Elsevier, new York, 1984).
2. D.J.Sellmyer and S.Nafis, Phy. Rev. Lett. 57, 1173(1986).
3. G.A.N.Connell, J. Magn. Magn. Mater. 54-57, (1986) 1561.
4. K.Tsutsumi and H.Sugahara, Japan J. Appl. Phys. 23 (1984) L169.
5. H.Chen, Thesis, Tsinghua University, 1986.
6. F.J.Cadieu, T.D.Cheung and L.Wickramasekara, J. Magn. Magn. Mater. 54-57 (1986) 535.
7. B.X.Liu and E.Ma, Physics (Chinese Journal) 15 (1986) 91.
8. B.X.Liu and G.A.Cheng, Philosophical Magazine Letters, 55 (1987) 256.
9. T.Yoshiie, C.L.Bauer and M.H.Kryder, J. Appl. Phys. 57 (1985) 2155.
10. Zhihua Yan, Baixi Liu and Hengde Li, Phys. Stat. Sol. (a)94 (1986) 483.
11. Y.L. Chou et al., Proceeding of Fiveth National Conference of Materials and Physics of Amorphous, China, (1988) 264.
12. Guoan Cheng and Yingguo Peng, Thin Film Sciences and Technology (Chinese Journal), to be published.

PART VII

Implantation in Electronic Materials

ULTRA-PURE PROCESSING:
A KEY CHALLENGE FOR ION IMPLANTATION
PROCESSING FOR FABRICATION OF ULSI DEVICES

M.I. CURRENT[*] and L.A. LARSON[**]

[*]Applied Materials, Implant Division, 3050 Bowers Avenue, Santa Clara, CA 95054 USA

[*]National Semiconductor, 2900 Semiconductor Drive, Santa Clara, CA 95052 USA

ABSTRACT

A key issue in modern ion implantation processing is the requirement for dramatic improvements in the purity of the incident ion beam and reductions in the deposition of foreign materials onto the wafer surface. These deposited materials include particles as well as sputtered and vapor deposited metals and dopants. Physical mechanisms which effect the elemental purity of atoms arriving at the surface of ion implanted wafers and progress towards achieving implantation purity levels of below 100 ppm of the implanted dose for sputtered metal and dopant films are discussed.

INTRODUCTION

The design and operation of ion beam processing equipment, and ion implantation systems in particular, has been driven by increases in process capability such as operation over a wider range of energies and higher beam currents for efficient production at high doses. Additional priorities include increases in system automation, reliability and control of yield limiting effects, such as wafer charging [1]. Device yield and reliability requirements for ULSI point to a need for new initiatives for control of the levels and purity of process materials which are deposited on the wafer surface. The systematic approach to reduction of particulate and sputtered film contamination as well as improvements in ion beam purity, known as "ultra-pure processing", is a parallel effort to recent advances in the state of understanding of the need for dramatic improvements in the cleanliness of fabrication areas and in the elemental purity of process materials [2]. The aim of this paper is to review the process requirements, materials characterization issues and recent progress towards those goals.

Mat. Res. Soc. Symp. Proc. Vol. 147. ©1989 Materials Research Society

PARTICLES

Estimates of the total defect density budgets for high-density IC devices indicate a need for levels of less than one critical defect per process layer per 200mm wafer for cost effective manufacturing of 64 M DRAMs [3]. A common working assumption is that, at the present time, the majority of critical device defects are associated with particulate contamination and that other source of defects, such as masking errors, Si materials and other process-induced effects, are relatively minor contributors. This assumption produces an underestimation of the real requirements for particle control by the extent of the impact of these other effects.

Particles are a defect risk for ion implantation as a blocking mask for the incoming ions and as a potential source of elemental contamination from outdiffusion during high temperature processing steps subsequent to the implantation.

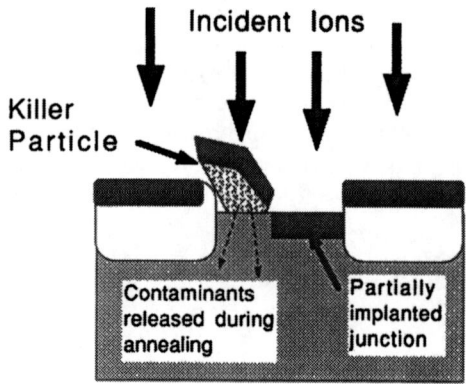

Fig. 1: A schematic diagram of a particle acting as a blocking mask and an elemental contaminant source.

Particle contamination has been linked to direct yield loss conditions such as emitter-base shorts, contact opens and oxide degradation [4,5]. Identified sources of particles include resist abrasion during wafer handling, dirty vent and pump valves, condensation of water vapor and reaction products from ion sources [6,7].

Sensitivity to masking effects extends to particles as small as a few hundred Ångstroms because of the limited range of ions used in a typical IC process. This is illustrated in Figure 2, where the critical particle size is taken to be, D_{crit} = <X>+2<ΔX>, where <X> and <ΔX> are the mean projected range and straggle lengths. Note that particles as small as 0.1 μm are implant masks for As implants at energies typical of modern CMOS source/drain implants (<50 keV).

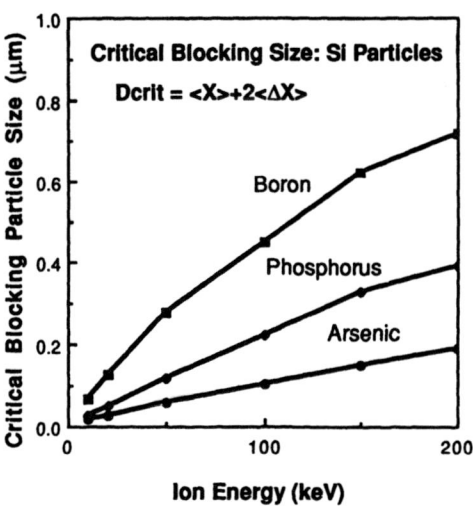

Fig. 2: Critical particle size for 98% blocking of As, P and B implants.

Recent advances in machine technology to meet particulate control requirements are aimed at systematic reductions and elimination of particle generation and transport mechanisms. These advances include increased sophistication in wafer handling and fixturing to eliminate wafer abrasion and front-side contact, introduction of load lock system architectures to isolate the implantation chamber from the clean room atmosphere, controlled pumping and venting rates to avoid turbulent flows and the use of polished process chamber surfaces to enhance the effectiveness of cleaning procedures. Future progress will likely involve the invention of techniques for active detection and removal of particles from the process chamber.

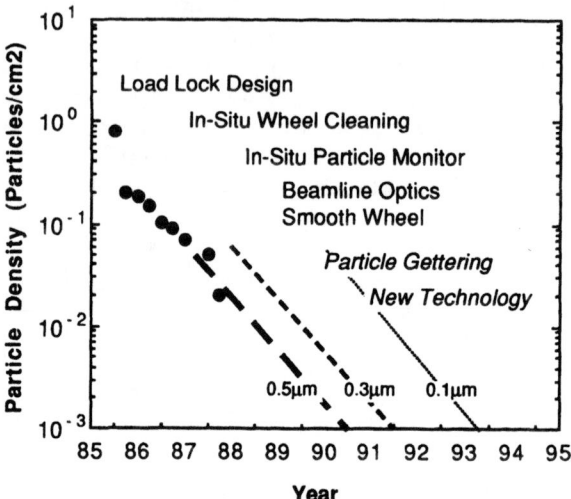

Fig 3 Particle learning curves showing estimated required levels for 0.5, 0.3 and 0.1 μm particles. Progress towards these goals is indicated for the PI9000 implanter along with present and projected machine technology advances.

SPUTTERED FILMS

Concern over the presence of sputtered films on implanted surfaces predates the recent emphasis on particulate studies [8-10]. The active mechanisms include direct sputtering of the wafer fixturing and beamline component and "cross-contamination" by sputtering of previously implanted elements from the parts of the implanter which are exposed to the ion beam [11,12]. In addition, recoil implantation of atoms in surface films and vaporization of volatile elements can be important factors in special cases [13]. Even surfaces which are not directly exposed to the ion beam, the backsides of accelerating electrodes for example, can be sources of contaminant materials following sputtering by atoms back-reflected at high energies from the implanted surface [10].

The general levels of sputtered film contaminants can range as high as several percent of the primary implanted dose for conventional implanter designs [11,12,14]. The principal sources of sputtered contamination are the machine surfaces in the immediate vicinity of the wafer during implantation, such as wafer clamps and implanted surfaces on spinning disks [11,12]. Significantly lower levels of contamination are seen with system designs which eliminate front-side wafer clamps[14] and use "open form" wafer holders which minimize the area of the spinning wheel which is exposed to the ion beam [15,16].

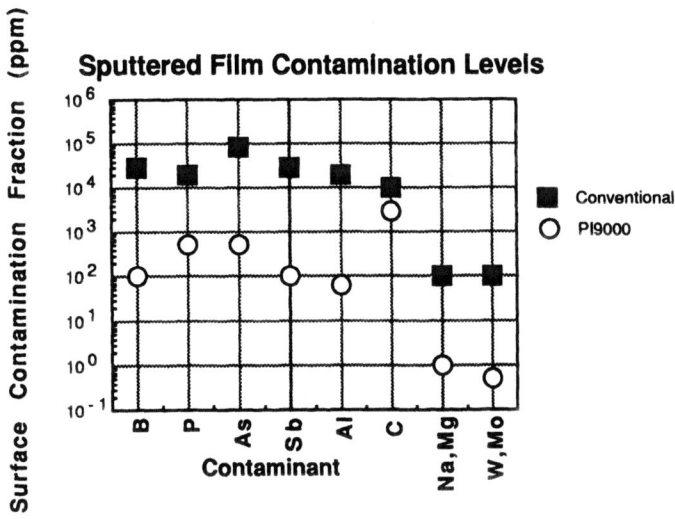

Fig. 4 Measured elemental contamination levels for conventional and new-generation (PI9000) implanters as a fraction of the primary implanted dose.

The relatively high levels of Carbon shown in Figure 4 reflect the significant contribution of hydrocarbon evolution from photoresist coated wafers [5].

Dopants

High levels of dopant cross-contamination, of the order of a few percent of the implanted dose, can result in surface layer junctions (poor metal contacts) and lateral auto-doping of epi layers (inter-transistor collector shorts). For high dose As and Sb junctions (at carrier concentrations higher than $\approx 10^{19}$ n/cm^3), the most significant effect is deep shifts in junction depths related to enhanced diffusion of contaminants such as P [17]. This effect is illustrated in Figure 5 where annealed As and Sb junctions show significant shifts after intentional contamination with 25 keV P at doses as low as 10^{12} ions/cm^2 [11,12]. For the case of the As junction, the onset (taken as a 5% shift) of anomolous junction diffusion occurs at a P contamination level of 100 ppm of the As dose.

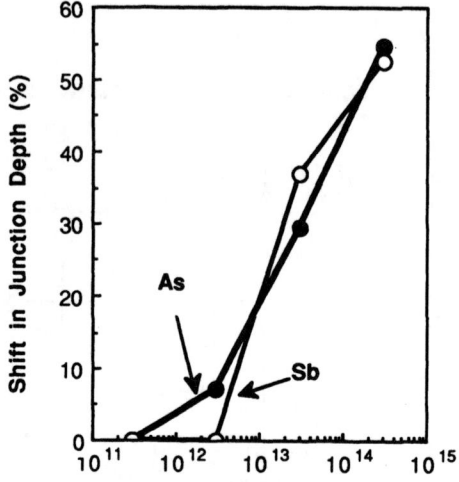

Phosphous Dose (i/cm2)

Fig. 5. Relative junction
depth shifts after
contamination with 25 kev
P implants for high dose
As and Sb junctions.

Advances in machine technology to reduce levels of dopant cross-contamination include elimination of the use of front-side wafer clamping and use of "open structured" wafer scanning components (spoked wheels with wafers held on extended arms rather than imbedded in a solid disk), selectable mass resolving apertures and reduction of the net sputtered flux from all surfaces exposed to the ion beam by careful choice of materials and of beam incidence angles. The status of progress towards attaining dopant cross-contamination levels below 100 ppm of the primary ion dose is shown in Figure 6.

Metals

Detection of sputtered metal contamination in early implantation systems [8] led to the additional of non-metallic beamline liners and the substitution of Al for stainless steel as the material of choice for construction of beamline and wafer transport and clamping components. Significant amounts of recoil implanted Al have been observed in implantation systems which utilize Al front-side clamps to obtain sufficient thermal contact to wafers (see Figure 7). This Al contamination is considered, at present, to be relatively innocuous, however the electrically active components of Al alloys (Mg, Na, and K) have been linked to oxide degradation and stacking fault formation.

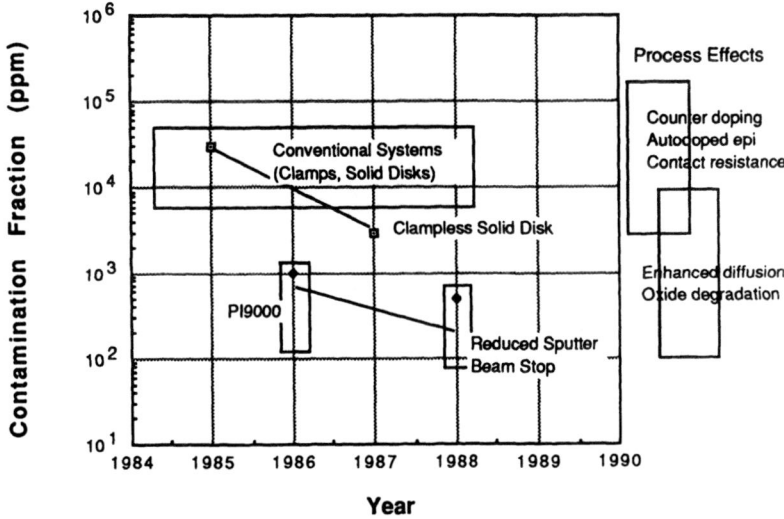

Fig. 6 Progress towards reduction of dopant cross-contamination levels. The range of contamination levels associated with significant process effects are indicated to the right.

The machine technology developments to reduce metals contamination are related to those driven by concern for dopant cross-contamiantion. These include, in addition to the elimination of front-side clamps, reduction in the area of metal surface exposed to the ion beam (by surface cladding with C or Si and use of "open structure" wafer scanning fixtures) and specialized structures in the beamline in the mass analysis region to supress the migration of source metals (such as W from the filament of a Freeman source) by sequential sputtering of the beamliner walls.

BEAM PURITY

Molecular BF_2^+ has been in place as a routine method of obtaining shallow B implants for so long that the two F atoms implanted per B dopant are often not acknowledged as "contaminants". In older process flows, anneal temperatures were high enough to allow outgassing of F from the implanted junction For anneal temperatures below 1000° C, substantial factions (25 to 100%) of the implanted F atoms remain in the Si layer [18-20]. The trapped F contamination has been linked to low B dopant activation [20], microbubbles in the junction area [22], decoration and pinning of lattice defects [18,19], degradation of oxide breakdown characterisitics [23], poor contact resistance and reduced latch-up characteristics [24].

372

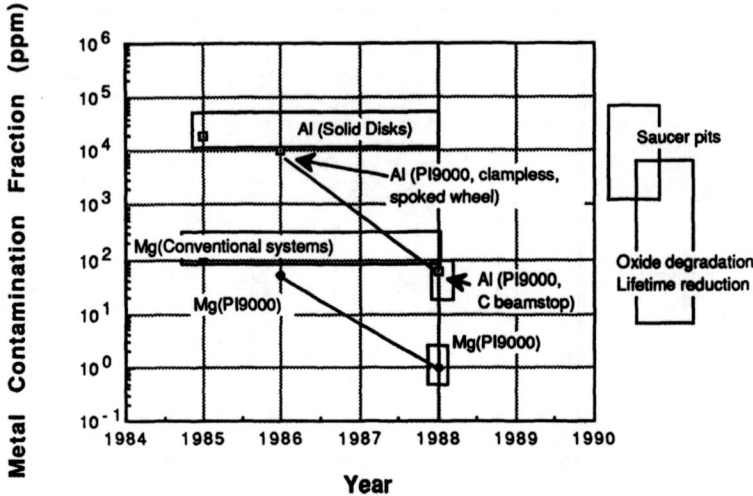

Fig. 7 Progress towards reduction of metallic contamination levels. The range of contamination levels associated with significant process effects are indicated to the right.

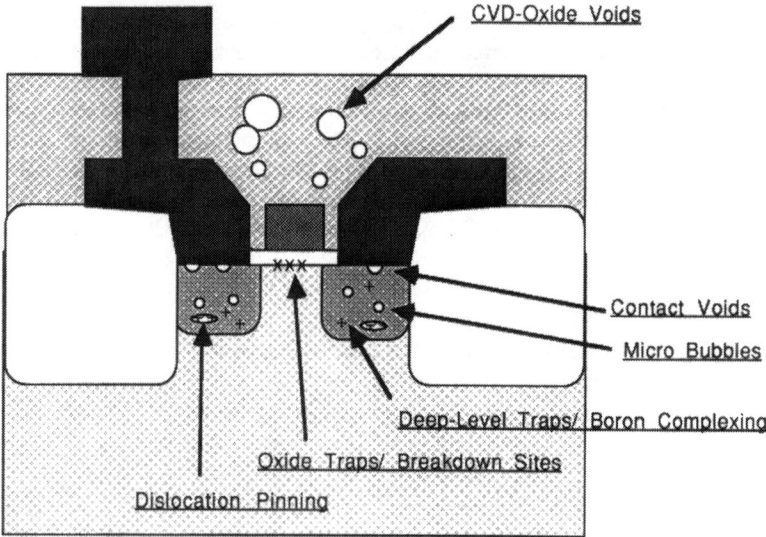

Fig. 8 Schematic summary of device effects associated with trapped F atoms after implantation with BF_2^+.

Additional problems have been reported with metal contact reliability with BF_2^+ implanted junctions and the prevalence of "haze" in p-channel areas (caused by F bubbles outgassed from the p-source/drain channels during PSG reflow and trapped in the upper layers of CVD-oxides). These effects are shown in a schematic fashion in Figure 8.

An additional contaminant effect associated with BF_2^+ implants is the coincident implantation of Mo^{+2} ions (which have the same mass to charge ratio) at twice the kinetic energy of the BF_2^+ ions. The source of the Mo^{+2} contaminant is sputtered surfaces in those source arc chambers which use Mo as a structural material. Deep implanted Mo at peak levels of 10^3 ppm of the BF_2^+ dose is shown in Figure 9 for an implanter which contained a Mo arc chamber.

Fig. 9 SIMS profile of a 50 keV BF_2^+ implant showing recoil implanted Al surface contamination from front-side wafer clamps and 100 keV Mo contamination from the source arc chamber region.

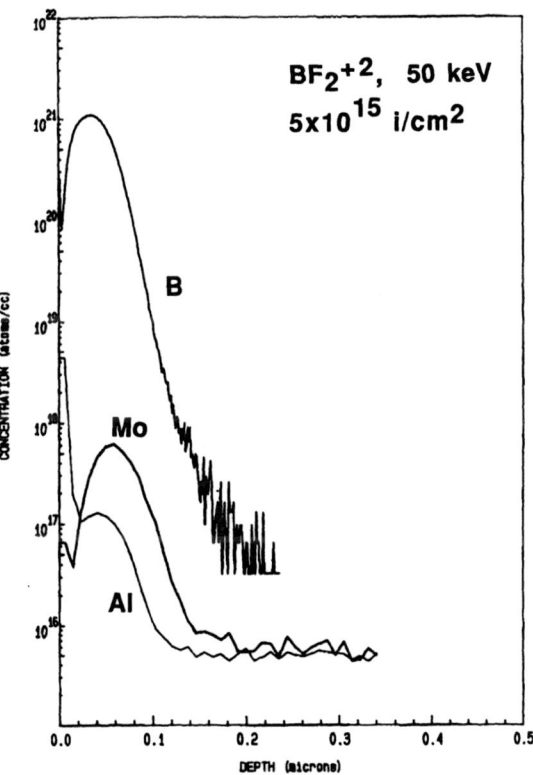

The machine technology advances to eliminate these forms of contamination include use of alternative materials for ion source arc chambers (such as graphite or Ta) and the production of high beam currents at low-energy (10-20 keV) for elemental B^+ as a high-purity alternative to $BF2^+$ for shallow p-type junctions..

CONCLUSION

The process requirements for ULSI device fabrication place stringent limits on the levels of particulate and elemental contamination which can be tolerated during ion implantation processing. The levels of particulate contamination are shifting from coverage of 0.1 particles/cm^2 for particles larger than 1 μm towards levels of 10^{-3} particles/cm^2 for particles as small as 0.1 μm. The levels of sputtered metals and dopants are shifting from several percent of the implanted dose towards process sensitivity level of 100 ppm. In addition, beam purity requirements are challenging many standard implant procedures, such as the use of BF_2^+ for shallow p-type junctions. These broad shifts over orders of magnitude in contamination levels and critical particle size present a key challenge to machine technology and operation as the systematic efforts to attain ultra-pure processing within process tools join the efforts to provide ultra-clean fab environments and process materials.

References

1) For example, see the Proc. 7th Int. Conf. on **Ion Implantation Technology**, Kyoto (June, 1988), Nuc. Inst. Meth. **B37/38**(1989).

2) T. Ohmi, M. Mikoshiba and K. Tsubouchi, "Super Clean Room System- Ultra Clean Technology for Submicron LSI Fabrication", in ULSI Science and Technology/1987, Electrochemical Society Proceedings **87-11** (1987) 761-785.

3) A.S Oberai, "Lithography-Challenges of the Future", Solid State Technology, Vol. **30**, No. 9 (Sept. 1987) 123-128.

4) J.L. Forneris et al., "Implant Processes for Bipolar Product Manufacturing and Their Effects on Device Yield" in **Ion Implantation: Equipment and Techniques**, eds. H .Ryssel and H. Glawischnig, Springer (1983) pp. 407-425.

5) T.C. Smith, "Photoresist Problems and Particle Contamination", in **Ion Implantation Science and Technology**, 2nd Edition, ed. J.F. Ziegler, Academic Press (1988) pp. 345-375.

6) W. Weisenberger, P. Borden and W. Knodle, "Real-time In-situ Particle Monitoring in a High Current Ion Implantation Production Bay", to appear in Nuc. Inst. Meth. **B37/38**(1989).

7) J. Strain, S. Moffatt and M. Current, "Characterization and Reduction of Particle Contamination in Ion Implantation Processing", in **Microcontamination/88** , Santa Clara, CA (1988) 42-54.

8) E.W. Haas, H. Glawischnig, G. Lichti and A. Bleier, "Activation Analysis of Contamination and Cross-Contamination in Ion Implantation", J. of Electronic Materials, **7** (1978) 525-533.

9) B.J. Masters, "RadioTracer Measurements of Sputtered Contamination Incurred During Ion Implantation Processing, IEEE Trans. Nucl. Sci. **NS-20** (1973) 1032-1034.

10) P.L.F. Hemment, "Sample Contamination Caused by Sputtering During Ion Implantation", Vacuum **29** (1979) 439-442.

11) L.A. Larson and M.I. Current, "Metallic Impurities and Dopant Cross-Contamination Effects in Ion Implanted Surfaces", in Materials Research Society Proceedings, **45** (1985) 381-388.

12) L.A. Larson, M.I. Current and C. Healy, "Enhanced Diffusion Effects and Dopant Cross-Contamination in Ion Implanted Surfaces", in **Semiconductor Silicon 1986**, Electrochemical Society Proceedings **86-4** (1986)667-675.

13) D.K. Sadana, N.R. Wu, J. Washburn, M.I. Current, A. Morgan, D. Reed, M. Maenpaa, "The Effect of Recoiled Oxygen on Damage Regrowth and Electrical Properties of Through-Oxide Implanted Silicon" Nuc. Inst. Meth **209/210**(1983)743-750.

14) G.L. Kennedy and L.A. Larson, "An Examination of Implant Cross-Contamination Levels Produced by Clampless Disks" J. Electrochem Soc. **135**(1988)1805-1808.

15) L.S. Steen, J.M. Hall, D.A. Aitken, F.G. Plumb, M.T. Wauk, N. Bright, C. Pruitt, L. Robinson, "The Precision Implant 9000, A New Concept in Ion Implantation Systems", Nuc. Inst. Meth **B21**(1987)328-333.

16) S. Moffatt," A 100 mA Class Ion Implanter for Large Scale Production of Buried Oxide SOI Wafers", Nuc. Inst. Meth **B21**(1987)251-257.

17) R.B. Fair., "Concentration Profiles of Diffused Dopants in Silicon", in **Impurity Doping Processes in Silicon**, ed. F.F.Y. Wang, North-Holland (1981)pp. 315-442.

18) M. Simard-Normandin and C. Slaby, J. Electrochem. Soc. **132** (1985)2218-2223.

19) R.G Wilson, J. Appl. Physics **54** (1983) 6879-6889.

20) O.W. Holland,J.R. Alvis and C. Hance,SPIE Proc.**797** (1987) 14-19.

21) N. Chan Tung, J. Electrochem. Soc. **132** (1985)914-917.

22) C.W. Nieh and L.J. Chen; J. Appl. Physics **60** (1986)3114 and Appl. Phys. Lett. **48** (1986) 1528.

23) K.W. Teng, HH. Tseng, B.Y. Nguyen and F.T. Lou, Electrochem. Soc. Extended Abstracts **87-1** (1987) 328-329.

24) C. Mazure, J. Winnerl and F. Neppl,Solid State Device Research Conf. (Bologna, 1987) pg. 585-588.

ELECTRONIC PROPERTIES OF ION IMPLANTED POLYMER FILMS

R. E. Giedd, J. Shipman, and M. Murphy, Physics and Astronomy Dept., Southwest Missouri State University, 901 South National Ave, Springfield, Mo. 65802.

ABSTRACT

We have implanted a number of insulating polymer (PET) thick films with fluences in the range 1×10^{13} to 5×10^{17} ions/cm^2 in order to induce electronic activity. Finite electrical conductivities were obtained for fluences as low as 3×10^{15} ions/cm^2. The resistivity for these implanted materials increases with decreasing temperature demonstrating highly disordered or semiconducting behavior. Hall effect measurements confirm this result and indicate negatively charged carriers with a carrier density of 1.44×10^{17} cm^{-3} for an implant dose of 1×10^{17} ions/cm^2. The conductivities also seem to be dependent on the implanted ion species. We believe this material will be useful as an accurate temperature sensor near room temperature.

INTRODUCTION

Ion implanted polymers have been the subject of a number of previous investigations[1-6]. These materials are of interest from both a purely scientific and an applications point of view.

Ion implanted polymers have been widely known to display a very large increases in conductivity as a function of ion implant fluence[3]. These ion damaged polymers can increase in conductivity as much as 14 orders of magnitude[3]. Very few physical parameters can vary to this this extent in a single process. This large increase in conductivity seems to suggest a percolative or super critical behavior as a function of fluence. Indeed percolation has been suggested for the conduction mechanism[7]. If these materials are exhibiting percolation behavior at a microscopic scale they may be prime candidates for studying critical behavior in general since the ion implant fluence can be controlled accurately.

Possible applications for ion implanted polymers include:

- Flexible resistor matrix material.

- Temperature sensors from 320K to 100K.

- Infrared sensing material.

- Field effect semiconducting devices.

The implanted polymers can exhibit a large amount of deformation without an affect on the conductivity. We observed that after deformation neither the surface morphology as observed by optical microscopy or the conductivity changed.

Most implanted polymers have a large temperature coefficient of resistance, TCR, at room temperature, making them ideal candidates for temperature sensors.

Infrared detection depends on emmisivity, heat capacity, and thermal isolation from a cooler background. Since these materials are very thin and they are usually black, the requirements of small heat capacity and emmisivity are met. The base polymer is usually a poor thermal conductor making the implanted polymer films good candidates for infrared detectors.

Some previous work on testing the semiconducting behavior of implanted polymer films has been done indicating that they may be good candidates for low cost transistors[8].

Of primary concern in these applications and others is characterization of the materials in the following ways:

1. Obtain maximum electrical conductivity of the implanted polymers by adjusting parameters during processing including fluence and polymer temperature.

2. Further analyze electrical behavior of implanted material including high frequency response, charge carriers, magnetic and thermal effects.

3. Measure TCR and test thermal cycle behavior of materials with an intent to eliminate any hysterisis.

This research in addition to previous work concentrates on items 1 and 2 above with some attention directed to 3.

MATERIALS

The polymers chosen for ion implantation were 5 mill thick amorphous PET films. The PET films were chosen because of their relatively simple structure along the polymer chain and their high temperature performance (T_g = 200°C). There are polymers with much higher temperature performance, e.g. polyimides and epoxies and these will be the subject of future investigations, however we were also interested in the increase in electrical conductivity of PET as the implantation temperature approached the glass temperature.

Two ion species, Ar, and B, were chosen for the implant ion in the PET films. We expected that Ar would remain relatively inert in the structure and that B should increase the electrical activity of the implanted film, revealing the materials sensitivity to chemical character of the implant ion.

EXPERIMENTS

Resistivity

Resistivity measurements were made along the surface of the films by using a modified 4 point probe method. The method was modified to allow us to measure the surface boundary resistance. Four Al strips were evaporated across the implanted side of the PET films. Spring loaded contacts were then compressed against the 4 Al contacts. The boundary resistance could then be measured between the Al contacts and the conducting mechanism in the implanted films. We found that the contribution of contact resistance to the resistance of the implanted films was less than 1%. This is not of consequence in our results, however it may be important in more conductive polymer implanted films.

The measurements were done in a vacuum since the resistance of the films could be as high as 10^{14} Ohms and we wanted to prevent any moisture contribution to the measured resistance. Typical values for the measured resistance were about 10^9 Ohms. Temperature dependence measurements were performed by cooling the outer vacuum chamber with liquid nitrogen. The sample temperature is monitored by a type K thermocouple attached to the rear face (unimplanted side) of the polymer film. Data acquisition is done by a fully automated computer controlled system.

Magnetoresistance

Magnetoresistance measurements were performed on implanted films with higher conductivities. The applied external magnetic field strength was measured by a Hall effect probe in the vicinity of the sample. Magnetoresistance measurements were performed at room temperature in magnetic fields up to 1.6 Tesla.

Contact to the samples was made with evaporated Al contacts similar to that of the resistivity measurement. Graphite paint was used to attach Cu wires to the Al contacts.

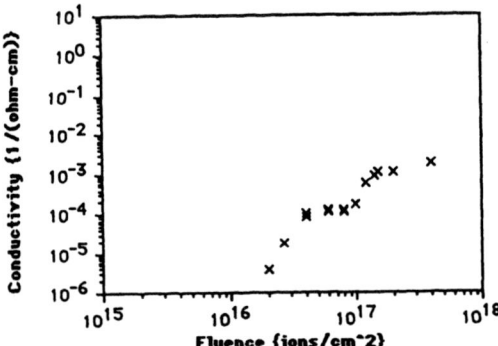

Figure 1: Conductivity of Ar implanted PET thick films with electron flood off.

Hall effect

Hall effect measurements were also made on the implanted films with higher conductivities. A conventional Hall effect mask was evaporated on to the implanted side of the polymer films. Contact to the Al mask was made with graphite paint and Cu wires. Measurements were performed with the same magnetic field strength impinging normal and anti-normal into the implanted film to minimize any geometrical effects due to small errors in the mask. The external magnetic field was provided by an electromagnet capable of generating fields up to 1.6 Tesla.

RESULTS AND DISCUSSION

Room temperature resistivity results.

Previous experiments[1-6] as well as our own observations show that the conductivity of ion implanted polymer films is dramatically higher than the conductivity of the unimplanted material. A typical conductivity verses fluence curve is shown in figure 1. It is not known where in the implanted layer the electrical conduction occurs. Previous investigators calculate the conductivity of the implanted material using R_p as the thickness of the conduction layer. It should be noted that R_p as calculated by TRIM[9] or other ion depth calculations[3] is an accurate representation of the ion penetration depth as measured in previous investigations[8]. However the assumption that the conduction mechanism is uniform from the surface of the film to R_p seems an over simplification. It would seem very plausible to us that the conduction process is not constant within the implanted layer.

For comparison purposes, however we include conductivities calculated using R_p as the thickness of the conduction layer. Table 1 shows the results of these calculations for the energies used and the two implant ions, B and Ar.

The room temperature conductivity of Ar implanted PET film is shown in figure 1. These results show behavior common to most ion implanted polymers. The conductivity increases exponentially with fluence until a plateau is reached when the fluence is between 4×10^{16} and 10^{17}. It has been observed that in the plateau region the conductivity can even decrease as a function of fluence. The suggested mechanism for this plateau is a crossover from a "hopping" type process of charge transfer along a fractal network to a homogeneous region where presumably the "hopping" mechanism is replaced by a continuous conduction mechanism[7].

Table 1: Ion energies and penetration depth as calculated by TRIM[9].

Implant ion	Ion Energy (Kev)	Penetration Depth (nm)
Ar^{+1}	50	90
B^{+1}	50	240

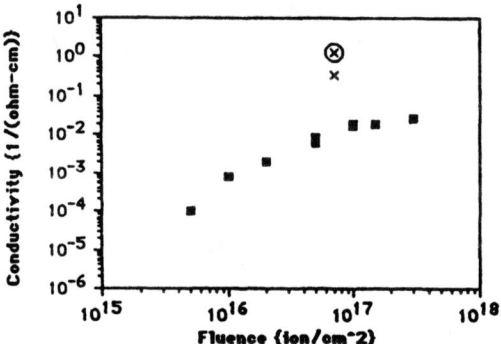

Figure 2: Conductivity of Ar implanted PET at a beam current density of $1.53\,mA/cm^2$ with electron flood (dots). Cross represents an Ar implanted film ran with a beam current density of $5.09\,mA/cm^2$ and no electron flood. Cross in circle represents an Ar implanted film ran at $5.09\,mA/cm^2$ and electron flood on.

Our results show a slightly different trend. The conductivity of the Ar implanted PET materials increases to the plateau occurring at a fluence of 7×10^{16} ions/cm^2 then increases again at 1.1×10^{17} ions/cm^2 to another plateau at about 3×10^{17} ions/cm^2. These results were obtained at an ion beam current density of $1.53\,mA/cm^2$. Upon removing the films at the higher fluences we observed a substantial static charge on the material. Although the materials have finite conductivity at the larger fluences, they build up a rather large static charge due to the small conduction path to ground. Static charges on the surface of insulating ion implanted materials is well known. Some ion implantation systems for Si have a mechanism of neutralization of the surface charge to obtain accurate fluences. We believe this to be even more important in the case of implanted polymers.

We installed an electron flood process to neutralize the static charge built up on the surface of the polymers during the implantation. In this process a filament is installed in the sample chamber and is used to balance the charge of the ion beam so that no net charge is transferred to the sample. The fluence is then calculated from the electron current. Then using an identical beam current density of $1.53\,mA/cm^2$ we repeated the experiment in the above section. The results for the electron experiment are shown in Figure 2. The results indicate a large increase in conductivity over the results obtained without using electron flood. They also indicate an elimination of both plateaus. The electron flood process increases the conductivty of the PET film even more dramatically then the previous results at low fluences. It also appears that at some much larger fluence the two processes will converge as one might expect. The other less intuitive result is the lack of plateaus. We have no explanation for the lack of plateaus except to note that the plateaus may be the result of excessive static charge on the surfaces of the polymers,

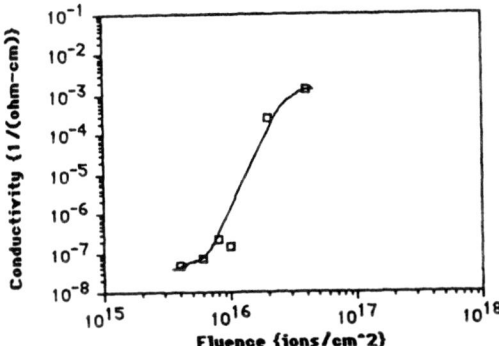

Figure 3: Conductivity of B implanted PET films. Curve is the probable behavior based on the accuracy of the measured values.

disturbing the measurement of the fluence.

We should note that our samples reach very high temperatures during the implant process. This precluded us from using materials other than high temperature polymers. At beam current densities higher than $5.09\,mA/cm^2$ the PET materials reach their glass point temperature, $300\,°C$, and thus do not yield usable materials. This may be part of the explanation as to why our materials with electron flood do not match the results obtained by previous investigators.

Also in Figure 2 are the results for the high beam current density of $5.09\,mA/cm^2$. These materials exhibited the highest conductivities obtained with Ar implanted PET film. At this current density the resulting conductivities of the implanted films are about 3 orders of magnitude over the results obtained for the material processed at the same fluence with a beam current density of $1.53\,mA/cm^2$. As can be seen from figure 2, electron flood on the higher beam current density material makes about a factor of ten difference. This is smaller than the difference between electron flood and no electron flood at $1.53\,mA/cm^2$ presumably a result of the higher overall conductivity of the material processed at higher current densities. Of primary interest to us at this time is the unexpected increase in conductivity with beam current density at the same fluence. Two possible explanations for this are:

1. Increases of the temperature of the sample material to close to glass point temperature increases the efficiency of the mechanism that causes finite electrical conductivity due to density changes.

2. Overall radiation damage to the material is greater for high current/short time than low current/high time exposures, perhaps related to structural relaxation time constants in the material. This may result in some pyrolysis of the material.

Boron results

The results for the room temperature resistivity of the Boron implanted PET films is shown in figure 3. These samples were run with the electron flood process and a beam current density of $1.53\,mA/cm^2$.

It was our hypothesis that Argon may remain inactive in the polymer structure, and that Boron may donate free electrons negatively doping the material. This should result in an enhancement of the conductivity over the Argon implanted material. This is apparently not the

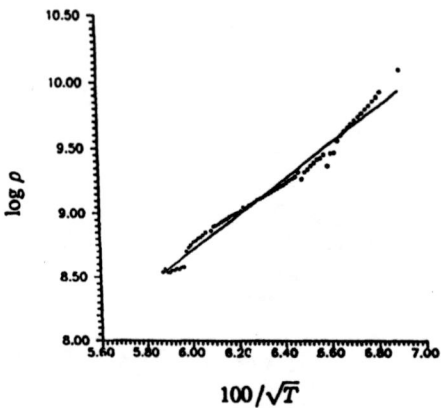

$$100/\sqrt{T}$$

Figure 4: Exponential behavior of resistivity as a function of fluence.

case as at a fluence of 8×10^{16} ions/cm^2 the conductivity of PET implanted with B is about a factor of 5 smaller than the conductivity of PET implanted with Ar. The primary reason for this may be that the boron ions are less massive and penetrate deeper into the film than the argon ions. This spreads the disorder over a larger volume, resulting in a smaller overall conductivity. The conductivity curve as a function of fluence is much different than the argon curve. Boron displays a much larger increase in conductivity in the range of fluences from 7×10^{15} ions/cm^2 to 8×10^{17} ions/cm^2 than the argon implanted PET. It may be that this is the percolation threshold for the structure in boron implanted PET. Presumably the percolation threshold for the argon implanted PET lies at lower fluences than 5×10^{15} ions/cm^2. The implant ion may also contribute to the conduction mechanism since values for the conductivity still differ by factor of 2 for the two materials with the same number of ions per unit volume.

Although we could not quantify the result at this time, the materials implanted with Boron show a remarkably different optical absorption coefficient from the materials implanted with Argon with similar fluence.

Temperature dependence

The resistivity of an argon implanted PET film measured as a function of temperature is shown in figure 4. The resistivity increases exponentially with decreasing temperature. To emphasize this the data is plotted as log ρ verses $100/\sqrt{T}$ where T is the temperature of the sample. The fit to the curve is of the form

$$\rho = Ae^{\sqrt{T_o/T}}$$

where $A = 1.36$ ohm $-$ cm and $T_o = 1.9 \times 10^4$ K. This is a common form for the temperature dependence of ion implanted photoresists and other polymers[3,8]. The deviation from this functional form may be significant and will be the subject of further research. Such a large temperature coefficient of resistance implies a relatively small number of carriers in the material. If we can control this small number of carriers, we should be able to make a useful device.

From these results we can conclude that the number of carriers is very sensitive to the temperature of the implanted region. Small changes in this temperature (perhaps due to infrared radiation) will produce a large change in resistance.

Magnetoresistance

We measured the magnetoresistance of two of the argon implanted PET thick films. The fluences of these materials were 3×10^{17} and 5×10^{16} ions/cm^2. We saw no change in the resistance of the implanted materials up to 1.6 Tesla. This implies that the carriers are not sensitive to magnetic fields and that these materials may be useful in high magnetic fields as temperature sensors. The two implanted films have similar conductivities as can be seen from figure 2. It is possible that below the steep increase in conductivity the materials would be more sensitive to magnetic fields.

Hall effect

Hall voltages were measured for an PET thick film with an argon implant of 1×10^{17} ions/cm^2. The results of these measurements can be summarized as follows.

- The sign of the carriers is negative implying electron conduction rather than hole or ion conduction.

- The number of carriers per unit volume is 1.4×10^{17} cm^{-3}.

Assuming the penetration depth of the argon ions to be 90nm, as in Table 1, the number of ions per unit volume is 1.11×10^{22} cm^{-3}. This is an interesting result in two aspects; first a single ion penetrating the material does not put a carrier into the conduction band through radiation damage, second the ions themselves do not seem to be donating carriers to the material.

The number of carriers per unit volume of 1.4×10^{17} cm^{-3} compares favorably with silicon implanted materials. For instance silicon doped with phosphorus to an amount 4.7×10^{17} phosphorus atoms per cm^{-3} has a carrier density of 4.0×10^{17} cm^{-3} at 25 °C [10].

CONCLUSION

The conductivity of ion implanted PET is a function of ion beam current density. Larger conductivities can be obtained by increasing the ion beam current density to the polymer and using electron flood to prevent static charge buildup on the polymer surface.

The TCR is relatively stable and does indicate possible use of materiel as an infrared detector near room temperature. It also indicates a relatively small number of carriers in the material. Hall effect measurements confirm this observation as well as indicate that electrons are most likely the carriers.

Boundary resistance measurements of evaporated contacts to polymer implant suggest that the conduction layer is at or near the surface.

There is little doubt that at high ion beam currents pyrolysis occurs in the material to some extent. This will be at least one contribution to conduction at room temperature. However, as can be seen from figures 2 and 3, we can control this process precisely with an ion implanter. These materials have promise for use in many applications.

REFERENCES

1. S.A. Jenekhe and S.J. Tibbetts, J.Polym.Sci. B26,201 (1988).

2. I.H. Loh, R.W. Oliver and P. Sioshansi, Nucl.Inst. Methods B34,337 (1988).

3. M.S. Dresselhaus, B.Wasserman and G.E. Wnek, Mat.Res.Soc., Symp.Proc., 27, 413 (1983).

4. T. Venkatesan, R.C. Dynes, B. Wilkens, A.E. White, J.M. Gibson and R. Hamm, Mat.Res.Soc. Symp.Proc. 27,449 (1983).

5. M.L. Kaplan, S.R. Forrest, P.H. Schmidt, and T. Venkatesan, J.Appl.Phys. 55(3), 732 (1984).

6. T. Venkatesan, S.R. Forrest, M.L. Kaplan, P.H. Schmidt, C.A. Murray, W.L. Brown, B.J. Wilkens, R.F. Roberts, L.Rupp Jr. and H. Schonhorn, J.Appl.Phys. 56(10), 2778 (1984).

7. B. Wasserman, Phys.Rev, B34(3), 1926 (1986).

8. B. Wasserman, G. Braunstein, M.S. Dresselhaus, and G.E. Wnek, Mat.Res.Soc., Symp.Proc., 27, 423 (1983).

9. J.F. Ziegler, J.P. Biersack and U. Littmark, "The Stopping and Range of Ions in Solids", Pergamon, Oxford (1985).

10. W. Shockley, "Electrons and Holes in Semiconductors", D. Van Nostrand, Princeton (1950).

SIMS DETERMINATION OF MG$^+$ AND AS$^+$ RANGE PROFILES IN PHOTORESIST
AND POLYIMIDE IMPLANT MASKS.

D. L. Dugger, M. B. Stern and T. M. Rubico,
GTE Laboratories, Incorporated, Waltham, MA   02254

ABSTRACT

The distribution of Mg$^+$ (a p-type dopant for GaAs) and As$^+$ (an n-type dopant for Si) implanted into both photoresist (PR) and polyimide (PI) have been determined experimentally. Range data of Mg ions at 200 keV and 300 keV and As ions at 150 keV have been measured by Secondary Ion Mass Spectroscopy (SIMS). SIMS values for the projected range Rp and the standard deviation ΔRp were compared to range profile data calculated using the Projected Range Algorithm (PRAL) of Biersack [1] as well as the standard LSS theory [2]. While the values for Rp calculated from the PRAL model generally agreed within 10% of the SIMS values, the calculations underestimated Rp for PR but were in good agreement for PI. The LSS calculations underestimated Rp in both materials.

INTRODUCTION

Both positive photoresist and polyimide are used as ion implantation masks for GaAs and Si. For high dose, high energy, and high resolution implant conditions, polyimide has superior thermal resistance enabling it to withstand the temperature rise associated with these implant conditions that are responsible for the degradation observed in photoresist. Recent work has indicated that photoresist films subjected to implantation doses greater than 1×10^{14} ions/cm^2 undergo chemical changes into disordered graphite rendering them extremely difficult to remove from the substrate [3]. In particular, the thermal distortion or change in the photoresist sidewall slope angle when the postbaking temperature exceeds the thermal flow temperature reduces the thickness at the edge of the photoresist feature. This effectively reduces the sharpness of the ion implantation mask as the thin resist edges will only partially mask the doping ions, depending on the ion implant parameters. For high energy, submicron feature size (high resolution implants) this effect becomes more important as the penetration depth of the ions increases. Polyimide has also been shown to outgas less than positive photoresist during ion implantation and to resist degradation and flow at high implant doses.

In device fabrication, prior to implanting wafers, the penetration depth of the ions in the polymer mask is normally precalculated (i.e., the stopping power of a given mask thickness for a set of implant conditions). However, published literature would suggest that the common practice is to correlate data for given implant conditions and species to that of KTFR [4], or use values from Gibbons, Johnson and Mylroie [5]. One such calculation for 150 keV As in AZ 111 negative photoresist yields an Rp value of 0.34 μm, a ΔRp of 0.0575 μm which produces a stopping range of 0.59 μm. These calculations were based on the LSS program. Using these values applied to the polymer film in this study, significant errors would result.

This paper investigates the use of experimental SIMS measurements to determine the actual range data for two polymer films, then use these data to compare the effective use of two similar computer programs to calculate these values.

EXPERIMENTAL

Positive novalac resin photoresist (Shipley's AZ 1450) and polyimide (Ciba-Geigy's probrimide 200 series) with thicknesses of 1-2 μm were spin coated on silicon substrates. Some samples were patterned through steps of prebaking, exposure, development and postbaking. Photoresist samples were cured at a temperature of 120°C for 30 minutes. Polyimide samples were soft cured at a temperatures of 160°C for 30 minutes. The low temperature cured polyimide is removable in organic solvents which do not attack the substrate. An ^{75}As$^+$ or ^{24}Mg$^+$ or ^{24}Mg^{++} ion beam with energy between 150 and 300 keV and a current density of 10 μA/cm^2 was implanted into the polymer film at a dose of 8x10^{14} ions/cm^2 using an Eaton Nova ion implanter. A solid source of Mg was used to generate the singly (energies ≤ 200 keV) or doubly (energies > 200 keV) ionized Mg ions.

The SIMS analyses were performed on a VG SIMSLAB equipped with a duoplasmatron ion gun using oxygen as the primary ion source. An ion current of 200 nA at 10 keV was employed operating in a rastered RAS (reduced area scan) mode at a magnification of 200x. An electron flood gun was used to eliminate surface charging. Secondary ions were collected and channelled into the quadrupole mass spectrometer using an HTO100 ion collection system. The secondary ions were electronically gated at 25% to minimize contributions from the crater walls. In order to optimize the counts, the implant species were counted for 25 seconds/cycle, while the substrate matrix ions (Si) were counted for 5 seconds/cycle. In the two samples that were doubly implanted with both As and Mg, each implant was analyzed in a separate analysis in order to maintain the same counting statistics. All data acquisition was performed using a DEC 11-73 computer with Framestore capability. After calibration of the depth axis using several techniques (Nanospec, Tencor a-step and Sloan Detak IIA), the data were further reduced using a Compaq 286 PC and Enplot plotting software. This included a Spline curve fit and a determination of the FWHM of the pseudo-gaussian peaks used in the calculation of the Rp and ΔRp values. The PRAL program is embodied in a package called SUSPRE [6] which is contained in the VG software package.

RESULTS AND DISCUSSION

The results, both calculated and experimental, are shown in Tables I and II for photoresist and polyimide, respectively.

Table I. PHOTORESIST DATA SUMMARY
(UNITS = Å)

		Rp	**ΔRp**	**4.27*ΔRp+Rp**
Mg 200				
	LSS	4700	980	8900
	PRAL	5300	980	9200
	EXP.	5300	1000	9400
		5400	1100	10000
MG 300				
	LSS	6800	1200	12000
	PRAL	7900	1300	14000
	EXP.	8600	1400	15000
AS 150				
	LSS	1400	350	2900
	PRAL	1400	250	2500
	EXP.	1800	650	4600
		1600	520	3900

Table II. POLYIMIDE DATA SUMMARY
(Units = Å)

		Rp	**ΔRp**	**4.27*ΔRp+Rp**
Mg 200				
	LSS	5100	1100	9600
	PRAL	5700	1000	10000
	EXP.	5500	1200	11000
		5500	1200	11000
MG 300				
	LSS	7400	1300	13000
	PRAL	8400	1300	14000
	EXP.	8300	1400	14000
AS 150				
	LSS	1500	380	3100
	PRAL	1500	270	2600
	EXP.	1400	540	3700
		1700	600	4300

Within experimental error, the Mg 200 keV implants gave range profiles which were very similar to the PRAL calculated values, consistent with duplicate runs and similar in each polymer film. The experimental ΔRp's were slightly larger than calculated, possibly due to cascade mixing by the SIMS primary ion beam, and therefore some broadening of the implants might be possible. The double implant in one sample for each polymer made it possible to provide some information regarding the reproducibility of the technique. The stopping powers for the 200 keV Mg implants were also found to be slightly higher than the calculated values but similar to each other. The higher energy Mg implant revealed a much deeper range profile possibly related to the reported decomposition of the photoresist. This was not observed in the case of the polyimide 300 keV Mg implant. The most significant differences were found with the As implant. Although there was more scatter in the range profile data, it is suggested that the experimentally determined values are at a deeper level than would be expected from the theoretical calculations. This could be due to limitations of the theoretical programs when applied to polymers, since they were primarily designed for simple inorganic systems. Even with the errors involves, the ΔRp's are as much as 65% higher for both materials. This also lead to significant differences in the stopping power. Approximately 50% thicker films are required to effectively prevent As from penetrating into the substrate than would be expected based on calculations. Overall, the data for polyimide appears to be more consistent than the traditional photoresist with regard to both range profiles and stopping power when compared to theoretical calculations. These results would support the theory that the polyimide is more stable under these typical implant conditions [4].

There were several sources of possible errors which are worthy of mention. First, it is difficult to determine with certainty the exact interface between the polymer films and the substrates in SIMS depth profile experiments. For consistency in this work, the onset of the steep upward slope of the silicon signal was used as the reference to calibrate the depth axis. The thicknesses of the films were measured by several techniques including measuring the profile crater of the SIMS experiment, physical removal of the films followed by Detak measurement, and a non-destructive optical technique. There was a 10% reduction of photoresist film thickness after ion implantation and 15% reduction in polyimide. Other errors included uncertainty regarding the actual compositions of the polymers and their corresponding densities. The Mg determination was made difficult because its mass at 24 Daltons was riding on top of a background signal of C_2^+, also at 24 Daltons from the polymer films. Due to the narrow peak and low signal of As, too few data points were available for good curve fitting which might have contributed to the scatter observed in the determinations.

SUMMARY

For an effective stopping distance 4.27ΔRp+Rp (99.999%), the
SIMS results indicated that 200 keV Mg ions were effectively
stopped by 0.97 μm of PR and 1.1 μm of PI, while 300 keV Mg
ions required 1.5 μm of PR and 1.4 μm of PI. As ions at 150
keV were stopped by 0.43 μm of PR and 0.4 μm of PI. A
comparison between the two theoretical calculation programs
revealed that the PRAL program provided results which were
closer to the experimental values than the more often used LSS
program, and much better than the common practice of relating
any polymer film to the value determined for KTFR by LSS.
Except for As, the PRAL program can be used to satisfactorily
calculate the expected range profile and stopping power.

REFERENCES

[1]   J. Biersack, Nucl. Instrum. Methods, 182/183, 199 (1981).

[2]   J. Lindhard, M. Scharff, and H. Schiott, Mat. Fys. Medd.
       Dan. Vid. Selsk, 33, 14 (1963).

[3]   Y. Okuyama, T. Hashimoto, and T. Koguchi, J. Electrochem.
       Soc.: Solid-State Science and Technology, August, 1978.

[4]   T. Herndon, R. Burke, and J. Yasaitis, Solid State
       Technology, November 1984.

[5]   J.F. Gibbons, W.S. Johnson, and S.W. Mylroie, "Projected
       Range Statistics", 2nd ed., Dowden, Hutchinson, and
       Ross, Inc. Stroudsburg, Pa. (1975).

[6]   VG SUSPRE - The VG Ionex and Surrey University Sputter
       Profile Resolution from Energy deposition programme.
       SUSPRE V1.3 Manual, November 1986.

ION BEAM MODIFICATION OF THE Y-Ba-Cu-O SYSTEM WITH THE MEVVA HIGH CURRENT METAL ION SOURCE

I. G. BROWN*, M. D. RUBIN*, K. M. YU*, R. MUTIKAINEN** and N. W. CHEUNG***

* Lawrence Berkeley Laboratory, University of California, Berkeley, CA 94720
** Technical Research Center of Finland, Espoo, Finland
***Electrical Engineering & Computer Sciences Department, University of California, Berkeley, CA 94720

ABSTRACT

We have used high-dose metal ion implantation to 'fine tune' the composition of Y-Ba-Cu-O thin films. The films were prepared by either of two rf sputtering systems. One system uses three modified Varian S-guns capable of sputtering various metal powder targets; the other uses reactive rf magnetron sputtering from a single mixed-oxide stoichiometric solid target. Film thickness was typically in the range 2000 - 5000 A. Substrates of magnesium oxide, zirconia-buffered silicon, and strontium titanate have been used. Ion implantation was carried out using a metal vapor vacuum arc (MEVVA) high current metal ion source. Beam energy was 100 - 200 keV, average beam current about 1 mA, and dose up to about 10^{17} ions/cm^2. Samples were annealed at 800 - 900°C in wet oxygen. Film composition was determined using Rutherford Backscattering Spectrometry (RBS), and the resistivity versus temperature curves were obtained using a four-point probe method. We find that the zero-resistance temperature can be greatly increased after implantation and reannealing, and that the ion beam modification technique described here provides a powerful means for optimizing the thin film superconducting properties.

INTRODUCTION

In this paper we describe the use of ion implantation to tailor the composition of sputtered thin films of Y-Ba-Cu-O. Attaining the stoichiometric 1:2:3 ratio of Y:Ba:Cu is an important step in forming films with good superconducting properties. Sputter deposition, however, does not always transfer material congruently from target to film.

High temperature superconducting Y-Ba-Cu-O films have been produced by several variations of the sputtering process. Cosputtering - simultaneous sputtering from three sources [1-4] - offers the advantage of independent control over the flux of each metal element. Sputtering from a single target, however, is better suited to large-area uniform coating. Both stoichiometric targets [5-12] and targets compensated for film elemental deficiencies [13-18] have been used.

A number of techniques have been used to achieve more congruent transfer of elements from a stoichiometric or slightly compensated target, including the following: (1) sputtering at high pressures [5, 7-9]; (2) sputtering in pure argon [6,10,12]; (3) sputtering at low substrate temperature [8, 10-13]; (4) off-axis placement of the substrate [6-8]; (5) close source-to-substrate distance [10]; (6) use of a spherical target [11]; (7) very long pre-sputter times [12]. Methods 1, 2 and 4 reduce preferential resputtering from the film caused by negative oxygen ions, electrons, and sputtered neutrals; methods 1 and 3 reduce preferential reemission; methods 1, 5 and 6 produce a more uniform spatial distribution among the sputtered elements; and method 7 establishes steady-state conditions at the target surface. These methods, however, do not give optimum deposition rates, material usage, or film properties. Another option is to compensate the film by ion implantation during or after deposition.

In the work described here, we prepared Y-Ba-Cu-O thin films by either of two rf sputtering systems. One system was a modified Varian S-gun system using three separate 5-inch targets, capable of sputtering various metals or powders. The other system was a Leybold Heraeus Z400 using reactive rf magnetron sputtering from a single mixed-oxide stoichiometric solid target. Substrates of magnesium oxide, zirconia-buffered silicon, and strontium titanate were used in this study. Film thickness was in the range 2000 - 5000 Å. Ion implantation was carried out using a metal vapor vacuum arc (MEVVA) high current metal ion source with a beam energy typically 100 - 200 keV and a mean beam current of approximately 1 mA.

ION IMPLANTATION

The MEVVA ion source is a new source whose uniqueness lies in the high current beams of metal ions that can be produced. Several different embodiments of the source have been made [19-23], including a multi-cathode version in which one can rapidly switch between any of 16 different metallic species, a miniaturized version which could be of value in research situations, and a broad-beam, very high current version. A schematic of the MEVVA II source used for the present work is shown in Figure 1.

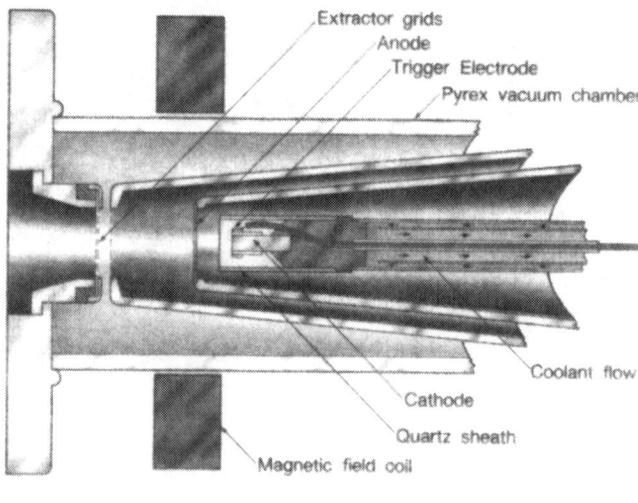

Fig. 1 Schematic of the MEVVA II ion source.

In this source, we use the intense plume of highly ionized metal plasma that is created at the cathode spots of a metal vapor vacuum arc discharge to provide the "plasma feedstock" from which the ion beam is extracted. A dense metal plasma plumes away from the cathode toward the anode and persists for the duration of the arc current drive. The anode of the discharge is located on-axis with respect to the cylindrical cathode and has a central hole through which a part of the plasma plume streams. The plasma drifts through the post-anode region to a set of grids that comprise the extractor - a three grid, accel-decel, multiaperture design.

At present the source is operated in a pulsed mode, with pulse length typically 0.25 msec and a repetition rate of up to 100 pulses per second. The source neither requires nor produces an ambient gas for its operation, and is usually run in a vacuum of around 1×10^{-6} Torr. The source runs cold - there is no oven, and the cathode remains solid. Beams of a wide range of ion species have been produced, including most metallic elements of the Periodic Table [22].

Implantation is done in a broad beam mode with no mass analysis. The implantation target (the Y-Ba-Cu-O sample) was positioned to face the source directly. At the target location, the ion beam is several cm wide. The MEVVA ion beam is relatively pure - ie, it contains ion species of only the cathode material - to better then 99%. The beam ions are in general multiply charged, and the charge state distribution has been studied in detail [22]. The measured charge state spectrum for a Cu ion beam is shown in Figure 2. The presence of more than a single charge state is an advantage in the present study, as it contributes to beam energy spread and consequent flatter implantation depth profile.

RESULTS

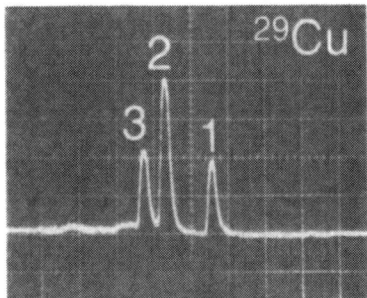

Fig. 2 Charge state spectrum of copper ion beam.

We were able to produce improved superconducting films using the stoichiometric single solid target configuration. Films were prepared on (100) SrTiO$_3$ substrates to a thickness of about 5000 A in pure argon and with a substrate temperature between 60 and 600°C. After deposition the films were annealed in flowing oxygen at 850°C for 1 hour.

The as-deposited film composition was determined by RBS and was found to be deficient in both Ba and Cu for an argon pressue of 1 4 Pa. The Ba to Y ratio was about 1.8 and insensitive to substrate temperature. The Cu to Y ratio, however, decreased rapidly as the temperature was raised from 60 to 600°C. To demonstrate the implantation technique, we investigated two films: film A was deposited at 125°C and had composition Ba:Y = 1.8 and Cu:Y = 2.0; film B was deposited at 360°C and had composition Ba:Y = 1.8 and Cu:Y = 2.5. The films were implanted with Cu with a mean beam energy of 100 keV to a dose of approximately 1 x 10^{17} ions/cm^2. The new Cu concentrations are estimated to be Cu:Y = 2.2 for film A and Cu:Y = 2.8 for film B. After implantation the films were reannealed.

The film resistance vs. temperature characteristic was measured using the four-point probe method. Figures 3 and 4 show the resistivity curves for films A and B (respectively) before and after implantation. For both films the resistance increases with decreasing temperature until 75°K, indicating the presence of a semiconducting phase in the normal state. After implantation and reannealing, both films exhibit a metallic behavior in the normal state. For film A, there is a feature at about 75°K that suggests the presence of a superconducting phase but without sufficient connectivity for the film to exhibit complete superconductivity; after implantation and reannealing, the film remains predominantly semiconducting at high temperatures but at much lower resistance, and there is a broad transition to superconductivity with an onset at 96°K and zero resistance at about 40°K. Film B shows better pre-implantation resistance-temperature properties because the initial Cu content was higher; after implantation and reannealing there is a change to more metallic behavior, and a superconducting transition onset at 95°K with zero resistance at 60°K.

Unfortunately, remnants of the silver paint used to attach the four point probe interfered with the post-implantation set of RBS measurements, and so the before and after RBS data are not available for this preliminary experiment. RBS data will be obtained in future work.

To eliminate the possibility that the effects we have observed could have been caused simply by the second annealing step, we annealed similarly formed films twice, but without the intermediate ion implantation. In these cases the zero resistance temperature of the films decreased by a few degrees only. Thus the improved film properties are due to the change in composition brought about by ion implantation.

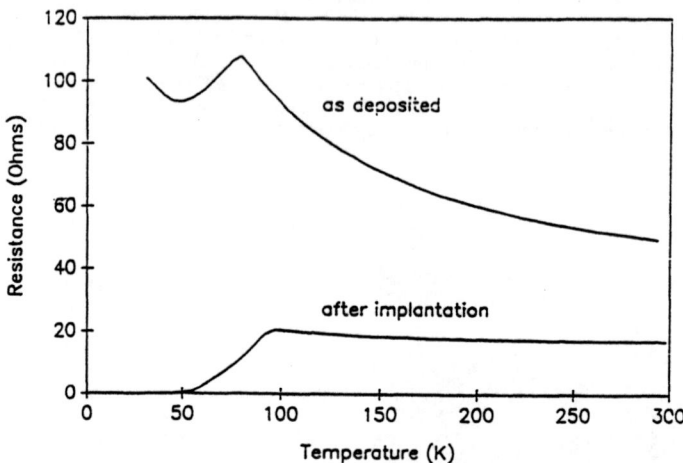

Fig. 3 Resistance vs. temperature of Y-Ba-Cu-O film before and after ion implantation of Cu. Y:Ba:Cu = 1:1.8:2.0 as deposited; the Cu fraction was increased to Cu:Y = 2.2 after implantation.

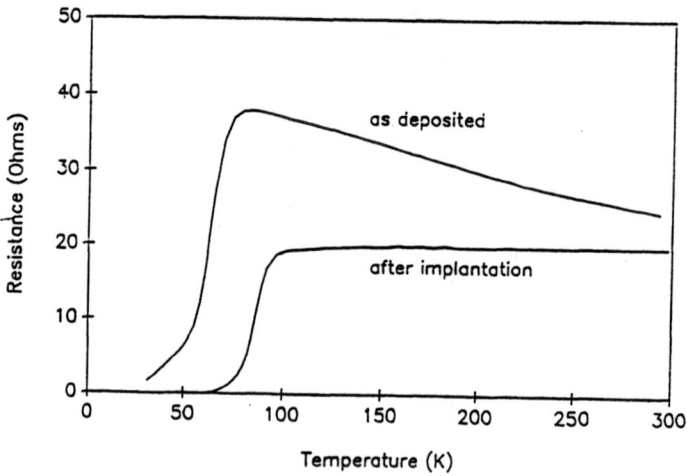

Fig. 4 Resistance vs. temperature of Y-Ba-Cu-O film before and after ion implantation of Cu. Y:Ba:Cu = 1:1.8:2.5 as deposited; the Cu fraction was increased to Cu:Y = 2.8 after implantation.

ACKNOWLEDGMENT

This work was supported by the U.S. Department of Energy under Contract No. DE-AC03-76SF00098.

REFERENCES

1. K. Char, A. D. Kent, A. Kapitulnik, M. R. Beasley and T. H. Geballe, Appl. Phys. Lett. 51, 1370 (1987).
2. R. M. Silver, J. Talvacchio and A. L. de Lozanne, Appl. Phys. Lett. 51, 2149 (1987).
3. M. Gurvitch and A. T. Fiory, Appl. Phys. Lett. 51, 1027 (1987).
4. T. Aida, T. Fukazawa, K. Takagi and K. Miyauchi, Jpn. J. Appl. Phys. 26, L1489 (1987).
5. W. Y. Lee, J. Salem, V. Lee, T. Huang, R. Savoy, V. Deline and J. Duran, Appl. Phys. Lett. 52, 2263 (1988).
6. A. Stamper, D. W. Greve, D. Wong and T. E. Schlesinger, Appl. Phys. Lett. 52, 1746 (1988).
7. N. Terada, H. Ihara, M. Jo, M. Hirabayashi, Y. Kimura, K. Matsutani, K. Hirata, E. Ohno, R. Sugise and F. Kawashima, Jpn. J. Appl. Phys. 27, L639 (1988).
8. S. I. Shah and P. F. Carcia, Appl. Phys. Lett. 51, 2146 (1987).
9. R. L. Sandstrom, W. L. Gallagher, T. R. Dinger, R. H. Koch, R. B. Laibowitz, A. W. Kleinasser, R. J. Gambino,. B. Bumble and M. F. Chisolm, Appl. Phys. Lett. 53, 444 (1988).
10. S. J. Lee, E. D. Rippert, B. Y. Jin, S. N. Song, S. J. Hwu, J. Poeppelmeier and J. B. Ketterson, Appl. Phys. Lett. 51, 1194 (1987)
11. G. K. Wehner, Y. H. Kim, D. H. Kim and A. M. Goldman, Appl. Phys. Lett. 52, 1187 (1988).
12. T. I. Selinder, G. Larsson, U. Helmersson, P. Olsson, J.-E. Sundgren and S. Rudner, Appl. Phys. Lett. 52, 1907 (1988).
13. O. MichR-5%1H. Asano, Y. Kato, S. Kubo and K. Tanabe, Jpn. J. Appl. Phys. 26, L1199 (1987).
14. T. Akune and N. Sakamoto, Jpn. J. Appl. Phys. 27, L2078 (1988).
15. S. Miura, Y. Yoshitake, S. Matsubara, Y. Miyasaka, N. Shonata and T. Satoh, Appl. Phys. Lett. 53, 1967 (1988).
16. H. Ohkuma, T. Mochiku, Y. Kanke, Z. Wen, S. Yokoyama, H. Asano, I. Iguchi and E. Yamaka, Jpn, J. Appl. Phys. 26, L1484 (1987).
17. H. Asano, K. Tanabe, Y. Katoh, S. Kubo and O. Michikami, Jpn. J. Appl. Phys. 26, L1221 (1987).
18. S. H. Liou, M. Hong, J. Kwo, B. A. Davidson, H. S. Chen, S. Nakahara, T. Boone and R. J. Felder, Appl. Phys. Lett. 52, 1735 (1988).
19. I. G. Brown, J. E. Galvin and R. A. MacGill, Appl. Phys. Lett. 47, 358 (1985).
20. I. G. Brown, J. E. Galvin, B. F. Gavin and R. A. MacGill, Rev. Sci. Instrum. 57, 1069 (1986).
21. I. G. Brown, J. E. Galvin, R. A. MacGill and R. T. Wright, 1987 Particle Accelerator Conference, Washington, D.C., March 1987.
22. I. G. Brown, B. Feinberg and J. E. Galvin, J. Appl. Phys. 63, 4889 (1988).
23. I. G. Brown, J. E. Galvin and R. A. MacGill, 1989 Particle Accelerator Conference, Chicago, IL, March 1989.

Author Index

Subject Index